*science
of the earth*

Harper's Geoscience Series
Carey Croneis, Editor

science of the earth

A. J. Eardley
*Professor Emeritus
Department of Geological and
Geophysical Sciences
University of Utah*

*Harper & Row, Publishers
New York, Evanston, San Francisco, London*

Photo acknowledgments: pages 257, 258, 281, 309, American Airlines; pages 403, 404, Michel Cosson; pages 1, 2, 21, 61, 95, 115, 137, 154, 189, 222, 337, 338, 351, 373, Allen Hagood and National Park Service.

SCIENCE OF THE EARTH

Copyright © 1972 by A. J. Eardley

Printed in the United States of America. All rights reserved. No part of this book may be used or reproduced in any manner whatsoever without written permission except in the case of brief quotations embodied in critical articles and reviews. For information address Harper & Row, Publishers, Inc., 49 East 33rd Street, New York, N.Y. 10016.

Standard Book Number: 06–041841–9

LIBRARY OF CONGRESS CATALOG CARD NUMBER: 70–168355

contents

Preface	xi

Part One: The Solid Earth — 1

1. The Earth in the Solar System — 2
   The Solar System — 2
   Planets and Asteroids; Comets
   Origin of the Solar System — 9
   Clues to Origin; Beginning Stages; Later Stages
   Evolution of the Earth-Moon System — 12
   Rotation of Earth and Lunar Tides; Calculating Backward and Consequences
   The Moon's Surface and Materials — 14
   Major Physiographic Divisions
   Moon Materials and Composition — 17

2. Minerals and Igneous Rocks — 21
   Solids—Minerals and Rocks — 21
   Atomic Theory; Isotopes; Combining Properties of Atoms; Nature of Crystals; Crystal Form; Optical Properties; X Rays and Internal Structure
   Mineral Groups and Abundances — 30
   Classification of Minerals; The Silica Tetrahedron and the Silicates
   Igneous Rocks — 32
   Variable Texture; Variable Mineral Composition; Classification; Kinds of Igneous Rock Bodies; Origin of Igneous Rocks
   Volcanoes — 47
   Types of Volcanic Mountains; Types of Vent Eruptions; Causes of Eruptions; Lava Flows from Flanks of Cones; Some Characteristics of Lava Flows; Ash Falls; Productive Soils; Necks and Radiate Dikes; Cauldron Subsidence; Eruptions of Submarine Volcanoes; Prediction of Eruptions

3. Sedimentary Rocks and Their Structures — 61
   Clastic Sedimentary Rocks — 62
   Classification by Particle Size; Origin of Coarse Clastics; Sandstone; Siltstone; Shale

Chemical Sedimentary Rocks　69
　　　　Limestone and Dolomite; Salt and Gypsum
　　　Sedimentary Rocks of Organic Origin　71
　　　　Limestone Reefs; Oil Shale; Coal; Oil and Gas; Chert
　　　Interlayering of Various Sedimentary Rocks　73
　　　　Sedimentary Rocks Interlayered with Volcanic Rocks
　　　Structures of Sedimentary Rocks　74
　　　Common Primary Structures　74
　　　　Cross-Bedding; Ripple Marks; Other Markings; Deformation of Soft Sediments
　　　Joints—Both Primary and Secondary　77
　　　Common Secondary Structures　79
　　　　Anticlines and Synclines; Thrusts and Thrusting; Normal Faults; Faults with Horizontal Displacement; Fault Scarps; Unconformities
　　　Metamorphic Rocks　88
　　　　Definition; Evidence of Change; Common Kinds of Metamorphic Rocks; Causes of Metamorphism; Meaning of Metamorphic Rock at the Surface

4. Earthquakes and Seismology　95
　　　Definitions　96
　　　Examples of Earthquakes　96
　　　　Hebgen Lake Earthquake; Earthquakes in Densely Populated Regions
　　　Method of Recording Earthquakes　98
　　　Reading the Seismograms　98
　　　Locating the Earthquake　100
　　　　Method of Intersecting Arcs; Depth of Focus
　　　Scales of Intensity　102
　　　The Earth's Interior　102
　　　　The Core and Mantle; Low-Velocity Zone in Upper Mantle; Deep-Seated Earthquakes; Moho Discontinuity and Layering of the Crust
　　　Use of Seismology in Exploration　109
　　　The Earthquake Environment　110
　　　　Circum-Pacific Belt; Earthquakes in North America
　　　Predicting Earthquakes　111

5. Magnetic and Gravity Fields of the Earth　115
　　　Earth's Magnetic Field　115
　　　　Earth as a Bar Magnet; Declination; Inclination; Changes in the Magnetic Field; Magnetic Reversals; Source of Earth Magnetism; Magnetism Used in Exploration
　　　Earth's Gravitational Field　125
　　　　What Is Gravity?; Measurement of Gravity; Gravity Increases Toward the Poles; Unit of Gravity Used on Gravity Maps; Reasons for Global Variations in Gravity; Bouguer Anomalies and Correction; Isostasy; Bouguer Gravity Map of the United States; Gravity Anomalies in Exploration

6. Geologic Time and Earth History　137
　　　Dating by Classical Geologic Methods　138
　　　　Superposition and Unconformities

CONTENTS vii

    Fossils as Time Markers   142
        Do Fossils Occur Systematically? Discoveries of William (Strata) Smith; Fossils of the Paris Basin and Their Significance; Guide Fossils and Correlation
    Geologic Maps and the Record   145
        Importance; Methods of Survey; Interpretation
    Methods of Radioactive Isotope Dating   148
        Emitted Rays; Decomposition of the Nucleus; Rates of Decay; Lead Methods; Zircon Lead-Alpha Method; Potassium-Argon Method; Rubidium-Strontium Method; Carbon-14 Method; Fission-Track Dating

7. Weathering, Mass Movements, and River Erosion   154
    Weathering Processes   154
        Frost Action; Solution; Hydration; Soil Formation by Weathering
    Gravity-Induced Mass Movements   159
        Prevalence of Landslides; Factors to Be Considered; Rock Falls, Avalanches, Talus, and Rock Glaciers; Creep of Various Kinds; Slides of Rapid Movement; Landslide Environment
    Rivers and Man   165
        Drainage Patterns; Stream Flow Velocity; Stream Loads (Transportation); Stream Erosion; Relation of Velocity to Erosion, Transportation, and Deposition; Analysis of Stream Activity; Rates of Denudation and Uplift

8. Glaciers and Ground Water   189
    Types of Glaciers   189
    The Glacial Regimen   190
        Nourishment of Valley Glaciers; Rate of Flow; Melting; Features of Continental Glaciers; Glacial Erosion; Features Formed by Glacial Erosion; Glacial Transportation; Glacial Deposits; The Pleistocene and Its Glacial Climates; The Continental Glaciers
    Ground Water   206
        The Hydrologic Cycle; Water in the Pores of Rocks; Ground Water in Rocks Without Barriers to Movement; Ground Water in Stratified Rocks; Advantages of Ground Water over Surface Water; Ground Water in Carbonate Rocks; Springs and Seeps; Examples of Ground-Water Occurrence; Legal Aspects of Ground-Water Use; Geysers in Volcanic Regions

9. Winds and Waves   222
    Wind as a Geologic Agent   222
        Wind Erosion and Transportation; Wind Deposition; Dunes Along Shores; Loess; Transport of Ocean Salt to Continental Areas
    Waves and Shoreline Activity   235
        Waves in the Open Water; Wave Changes in Shallow Water; Waves as a Geologic Agent; Tides; Profiles of Beaches; Deposits Along Shores; Shores of Emergence; Shores of Submergence; The Shoreline Environment

Part Two: The Oceans   257

10. The Chemistry, Dynamics, and Origin of the Oceans   258
    Uniqueness of the Oceans   258

Chemical Nature of Sea Water ... 259
  Ionic Constituents in Sea Water; Salinity; Dissolved Gases
Temperatures of the Oceans ... 263
Ocean Currents ... 265
Deep Circulation and Water Masses in the Atlantic Ocean ... 270
Tides ... 272
Heat Budget ... 274
  Solar Radiation; Elements of the Heat Budget
Origin of the Oceans ... 276
  Problems and Aspects; Constancy of Salinity and Composition; Water and Air from Within the Earth

11. Topography and Sediments of the Ocean Floor ... 281
  Major Geomorphic Divisions ... 282
    Continental Shelves; Submarine Canyons and Fans; Abyssal Plains; Seamounts; Rises; Midoceanic Ridges; Fracture Zones; Island Arcs and Trenches
  Deposits on the Ocean Floors ... 291
    Sampling Methods; Classification of Bottom Sediments; Pelagic Sediments; Terrigenous Sediments
  Mineral Resources of the Marine Environment ... 305
    Petroleum (Oil and Gas); Manganese Oxide Nodules; Other Minerals

12. Evolution of the Continents and the Ocean Basins ... 309
  Constitution of the North American Continental Crust ... 309
    Tectonic Units; Geosynclines and Fold Belts; Convection Currents in the Upper Mantle
  Constitution of the Oceanic Basins ... 319
    Atlantic Ocean Basin; Pacific Ocean Basin
  Major Plates of the Crust Involved in Drifting ... 331

Part Three: The Atmosphere ... 337

13. Characteristics of the Atmosphere ... 338
  Nature of Gases ... 338
  Composition of the Atmosphere ... 339
    Permanent Gases; Water Vapor; Carbon Dioxide; Dust
  General Characteristics of the Atmosphere ... 342
    Mobility; The Gas Laws; Pressures in the Atmosphere
  Stratification of the Atmosphere ... 345
    Divisions; Probing the Upper Atmosphere; Compositional Variations; Temperature Stratification; Other High-Altitude Phenomena

14. The Earth's Atmospheric Environment ... 351
  Solar Radiation and the Atmospheric Engine ... 351
    Heat Budget; Factors Determining Absorption and Reflection; Variation and Effects of Solar Radiation; Conduction and Convection
  The Winds ... 357
    Simple Convective System; Sea and Land Breezes; Valley and Mountain Breezes; Laws Governing Regional Atmospheric Circulations; Surface Winds

CONTENTS

    Condensation of Moisture   361
        Water-Air Contact; Measurement of Water Vapor; Adiabatic Temperature Changes; The Adiabatic Chart; Lapse Rates; Fog; Smog; Clouds; Convection and Condensation in Cumulus; Measuring Precipitation

15. Air Masses and Weather Prediction   373
    General Circulation Pattern   373
        Basic Model; General Global Barometric and Surface Winds
    Air Masses and Fronts   377
        Polar Front Theory; Sources of Air Masses; Classification of Air Masses; Fronts
    Weather from Satellite Photographs   382
        The Photographs and Clouds; Where Clouds Form; Development of the Cyclone; Levels of Air Flow; Daily Weather Maps; Meaning and Use of the Upper Air Maps; Example of Daily Weather Forecasting; Hurricanes and Tornadoes; Jet Streams in the Upper Air; Computerization of Weather Data; Array of Modern Forecast Maps; Final Look at the Tiros Worldwide Photograph

Part Four: Environmental Science   403

16. Conservation, Reclamation, and Management of Natural Resources   404
    Historical Development   404
        Scope of Present Treatment
    U.S. Government Agencies Involved in Environmental Management   406
        Department of the Interior; Department of Agriculture; Department of Health, Education, and Welfare; Department of Housing and Urban Development; National Oceanic and Atmospheric Administration
    Water Management   407
        Water Conservation; Water Reclamation; Disposal of Liquid Waste by Underground Injection
    Land Use and Conservation   418
        Land Ownership and Use in the United States; Soil Conservation; Managing the Runoff; National Forest and Wilderness Areas; National Parks, National Monuments, Riverways, National Shorelines, and Recreation Areas; Highways and Esthetics Policy; Urban Explosion; Conservation of Minerals
    Pesticides in the Environment   434
    Air Pollution   435
        Nature of Air Pollutants; Sources of the Pollutants; Some Bright Spots in the Pollution Picture

Index   443

preface

Science of the Earth is intended for general education courses in geology and earth science. After a brief description of the solar system and its origin, it discusses the solid, liquid, and gaseous divisions of the earth by developing and expanding on their constitutions, processes, and histories.

Since the completion of the preliminary manuscript, the book has undergone several critical reviews, several reorganizations, with condensation of some parts and expansion of others. Many exciting discoveries in planetary and lunar studies, in crustal movements of the earth, in oceanography, and in atmospheric science have taken place since I first began to organize the book; these developments had to be incorporated prominently into its composition. Also, in the past three years the people of the world have become acutely aware of their environment. Renewed concern about objectives prompted the publishers and me to treat environmental science as an integral part of *Science of the Earth*. Inasmuch as no authors of geology or earth science texts, to my knowledge, had organized and presented environmental science in its various aspects, this proved a pioneering venture. The treatment here may thus elicit more comment, both good and bad, than other more conventional parts of the book.

Many teachers of earth science integrate environmental science into each major division of their course as they proceed. I have followed this procedure as far as was feasible but find that a good deal of what is popularly considered environmental science is better considered a summation. At least that part of the subject that I have outlined in Chapter 16 is logically a summary consideration of the foundations laid in previous chapters. The environment consists of both the physical and biological complexes, and inasmuch as this text deals mainly with the physical, it seems best to mention the current popular biological aspects in this last chapter. Also, the controversial, economic, and political aspects of environmental science seem for the most part better left to this concluding section. I have tried to treat objectively the commendable things that have been accomplished to protect the environment as well as the destructive and injurious activities in which man has engaged.

Although the book has been prepared for a one-term course, it should prove suitable for use in a two-term sequence in which the second term intends to emphasize environmental science, especially if laboratory work is interwoven with the text material. The second term could consist of studies of environmental science as they relate to the oceans and the atmosphere.

In the course of defining objectives and subject material, I have consulted many scientists throughout the country and wish to acknowledge their advice and help. I am most appreciative of discussions with my colleagues at the University of Utah, including a former one, Daniel J. Jones, now of the California State College, Bakersfield, California. I also wish to thank George Telecki of Harper & Row for his able editorial direction.

A. J. Eardley

part one
the solid earth

1

the earth in the solar system

The earth is unique in its family of planets because it supports life. Even its closest relatives, Mars and Venus, though similar in size and proximity to the sun, offer at best untenable living conditions. Why the differences? The space effort now involving scientists of all nations centers on Mars, Venus, and the earth's moon. Already, by means of manned spaceships, we are learning much about the moon, and by means of space probes something about Mars and Venus—things we could not have learned by earth-bound telescopic, spectrographic, and radar sensing techniques. The technical achievements that made space travel possible are inspiring in themselves but the new knowledge that will be gained from space research, including new insight into the existence and nature of life, is of very special interest. The results of these investigations make us more conscious of earth problems and, certainly, more discerning in our studies of them.

The solar system

Planets and asteroids

The solar system is made up of a central star (the sun), 9 planets, 31 moons, 30,000 asteroids, and about 100 billion comets, plus innumerable dust particles, gas molecules, and dissociated atoms. The sun contains 99.86 percent of the total mass; of the remaining 0.14 percent, the earth and its moon account for only 1 percent. The sun is the nucleus of the system and by its gravitational field controls the movements of the other bodies. The relative size of the sun and the planets and their order away from the sun are shown in Figure 1–1. Their relative distances from the sun and their orbits are illustrated in Figure 1–2.

The nine planets can be divided into two groups according to their size and density (see Table 1–1). The four small, innermost ones, the terrestrial planets—Mercury, Venus, earth, and Mars—are heavy and solid. The following four, the outer planets, are giants composed mainly of lighter elements. The outermost planet, tiny Pluto, hardly fits either category, and its orbit does not follow the same pattern as that of the other planets.

All planets circle the sun in the same direction and in about the same plane. The earth dips and rises through slightly more than 14° of solar latitude as it circles around the sun, and the other planets, except for Pluto, also revolve about the sun within this limit of variation from the sun's equatorial plane (the disk of revolution). The orbits are elliptical, with a nearest approach to the sun and a most distant position. The innermost planet, Mercury, comes within 28 million miles of the sun at one end of its elliptical orbit and swings out to 43 million miles at the other.

It will be noted in Table 1–1 that the farther each planet is from the sun, the slower it travels and the longer it takes to complete the orbit. Mercury speeds through space at 110,000 miles per hour and takes only 88 days to circle the sun. The earth moves at 67,000 miles per hour and takes about 365 days. The outermost planet, Pluto, down to a walk of 10,000 miles per hour, needs nearly 248 years. The planets are locked in these definite orbits by the gravitational pull of the sun, without which they would fly off in a straight line.

There is a great gap between the inner small planets and the outer giant planets, and the asteroids are found here in a broad belt. They circle the sun the same way as the planets. They are small rough masses of rock and metal and all together have a total mass that is 5 percent less than that of the moon. Ceres is the largest asteroid, with a diameter of 480 miles. Pallas is 300 miles in diameter, Vesta 240 miles, and Juno 120 miles. Today about 30,000 sizable asteroids are thought to exist, ranging from large ones like Ceres to small ones like Icarus, just 1 mile in diameter. Then there are billions of still smaller ones the size of boulders, pebbles, and sand grains.

Jupiter exercises great power over these asteroids and a number have been brought into the same orbit as the giant planet. These are called the Trojans. Nine orbit a respectful and constant distance in front and five the same distance in back, all in a west to east direction. This distance is just one-sixth of the orbital distance and has been shown mathematically to mark a point of gravitational equilibrium.

Jupiter's tyrannical pull on the asteroids sometimes swings them on voyages toward the sun or away toward the outer planets. Icarus's

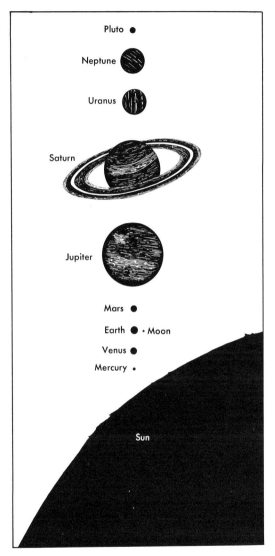

Figure 1–1 Relative sizes of the nine planets of the solar system. Note that Uranus spins with an axis inclined almost vertically to the plane of the orbits of the nine planets. Also, the rings of Saturn, on occasion, when we see the planet directly at its equator, appear as a thin line. Note further the color bands of Jupiter. These bands are now regarded as Van Allen type belts. None of the moons is shown except that of the earth. [*After* The universe, *Life Nature Library, New York, Time, Inc., 1962.*]

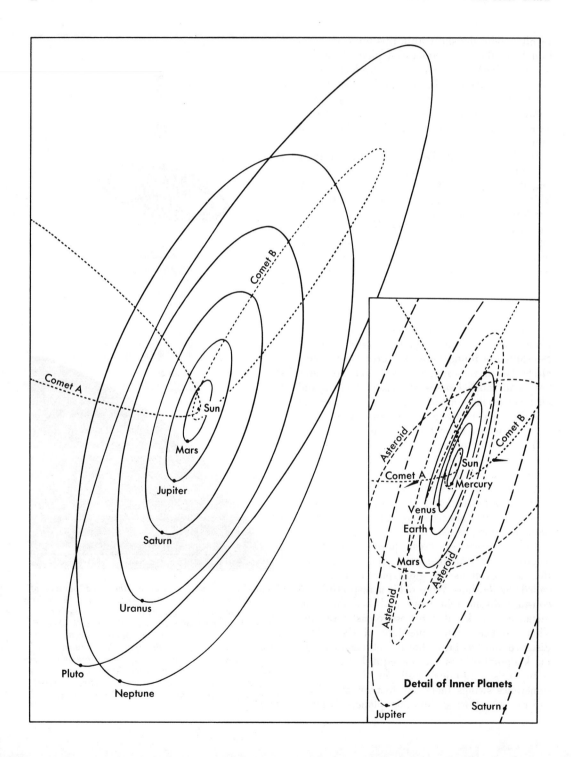

TABLE 1-1 *Members of the Solar System*

NAME	MEAN DISTANCE FROM THE SUN, ASTRONOMICAL UNITS	SIDEREAL PERIOD*	EQUATORIAL DIAMETER, MILES	MASS (EARTH = 1)	AVERAGE SPECIFIC GRAVITY	ROTATION	NUMBER OF SATEL-LITES
Sun	—	—	864,000	332,000	1.42	25–35 days	
Mercury	0.38	88 days	3,100	0.054	4.8	88 days	0
Venus	0.72	225 days	7,700	0.81	4.9	243 days	0
Earth	1.00	1 year	7,927	1.00	5.51	23 hr 56 min	1
Mars	1.52	1.88	4,215	0.11	3.95	24 hr 37 min	2
Jupiter	5.20	11.86	88,700	317.00	1.34	9 hr 50 min	12
Saturn	9.54	29.46	74,000	95.00	0.69	10 hr 2 min	9
Uranus	19.19	84.01	32,000	15.00	1.36	10.7 hr	5
Neptune	30.07	164.8	27,700	17.00	1.30	15.8 hr	2
Pluto†	39.46	247.70	3,600 (?)	0.1 (?)	—	—	0

* The sidereal period is the period of the planet's revolution around the sun measured with respect to the stars.

† Pluto may not be a planet in the usual sense but rather an escaped satellite of Neptune.

orbit is one of these, it appears, and it passes twice as close to the sun as Mercury. Hildalgo whirls out as far as Saturn. Once in an errant way, they may get yanked still farther astray or drawn back into Jupiter's fold. Eros, a cigar shaped rock 15 miles long and 5 miles wide, tumbles end over end and last came within 14 million miles of the earth. Five others have come even closer, with Hermos, in 1937, passing like a jet, only 500,000 miles away.

Asteroids of boulder size collide with the earth an estimated 1500 times each year. These are called meteorites. Geological investigations have pointed out a considerable number of wounds or craters on the earth due to the explosive impact of these stray space bodies. The moon is pitted with many craters, as if it were struck much more often than the earth, but this is probably not so, as will be shown later. When asteroids strike the earth or moon, the impact is like that of a rock thrown into a pool of soft mud—the asteroid is swallowed explosively by the surface, and considerable fragmental material is thrown upward and outward.

Jupiter, Saturn, Uranus, and Neptune are a different species than the terrestrial planets, both in scale and in kind of materials. Jupiter fills a volume 1300 times greater than the earth, its mass is more than twice as large as all the other planets put together, and it has a highly turbulent atmosphere hundreds of miles deep. The atmosphere is present because of Jupiter's high gravity, which is $2\frac{1}{2}$ times that of the earth's. It is believed to be composed of hydrogen, ammonia, and methane and to be very cold. For all Jupiter's monstrous size its density is only a quarter of the earth's. Possibly its innermost core consists of rock or metal, but it may be hydrogen compressed into a heavy metallic state. Jupiter has a retinue of 12 satel-

Figure 1-2 Orbits of the planets and representative asteroids and comets. Note the irregular orbit of Pluto. Note also that the axis of the ellipse of Comet A stands nearly vertical to the planar ecliptic and that the axis of the ellipse of Comet B is approximately in the ecliptic, but that the ellipse itself is about vertical. The asteroids are mostly in a zone between Mars and Jupiter, but some have unusual orbits, as shown. Of the inner planets, only Mars is shown on large diagram. [*After* The universe, 1962.]

lites, 2 of them larger than the earth's moon.

A model for Jupiter, recently proposed,[1] is composed of a core of liquid hydrogen-helium and a mantle of a lower solid hydrogen layer and an upper solid molecular hydrogen-helium layer; above the solid layer is a thick atmosphere made up chiefly of hydrogen, with some helium in the lower strata. The atmosphere has a supercritical temperature. Jupiter displays a red spot on its surface which has been recognized for 300 years, yet a good explanation for it has not yet been realized. The great gravitational attraction of Jupiter is attributed to the liquid hydrogen-helium core, which serves the same way as the liquid iron-nickel core of the earth.

Saturn is an immense half-consolidated world, 95 times as massive as the earth but only seven-tenths as dense as water. Like Jupiter, it has a turbulent atmosphere. It also has three flat rings around its equator, about 10 miles thick and extending 6,000 to 48,000 miles beyond its surface. The rings are believed to be composed of ice-coated grit particles. Saturn has nine conventional moons. The largest, Titan, is as big as Mercury, as orange as Mars, and is the only moon with an atmosphere, which is cold and consists mostly of methane. The outermost satellite, Phoebe, revolves in a direction opposite to that of rotation of its mother planet. Two of the satellites appear to be smooth spheres of pure ice.

Uranus and Neptune are not visible to the naked eye. Uranus has an atmosphere composed mostly of methane, with a temperature 270° F below zero. Its axis of rotation lies nearly parallel with the plane of its orbit, and so also its five moons revolve about Uranus nearly perpendicular to the ecliptic (plane of the earth's orbit). If you were on Uranus you would see all five moons racing along the horizon with startling rapidity, because the planet rotates once in 10 hr and 49 min.

After the discovery of Uranus, astronomers found intolerable irregularities in its orbit and concluded that a planet beyond must be causing the disturbance. Pointing the telescope where the mathematicians predicted, Neptune was discovered in 1846. Similarly Pluto was discovered in 1930.

At the beginning of the space age we are particularly interested in the compositions, atmospheres, temperatures, and escape velocities of the nearby planets Mercury, Venus, and Mars, and of the earth's moon. As for Mercury, it rotates approximately once in 88 days and orbits the sun in the same period. If it were not for the eccentricity of Mercury's orbit, the planet would always keep the same face toward the sun. Because of the orbit's high eccentricity, only about 30 percent of the planet's surface is in total permanent darkness. Infrared light measurements set the surface temperatures on the sunlit side at about 780° F and on the dark side at 70° F. Thus Mercury is considered to have a thin transient atmosphere made up of gases that have only recently emerged from the interior and are too heavy to escape immediately into space. The lighter gases have probably escaped because the gravitational pull is only three-eighths that of the earth. An astronaut would need to attain a vertical speed of 2.6 miles per second to escape from Mercury, as opposed to 7 miles per second needed to escape from the earth. Astronomers think that if a space probe should pass by Mercury and return with photographs, they would show its surface to be pockmarked with craters like those of the moon or Mars.

Venus has an atmosphere and an unbroken cloud cover. Little is known, therefore, of the planet's surface. The dazzling veil of yellowish white clouds conceals not only the surface but also the planet's rotation and the inclination of its axis. By 1956 radio-telescopic measurements showed temperatures on the sunlit surface of up to 1300° F and averaging 800° F. Radar measurements indicate the thickness of the clouds to be 57 km. U.S. Mariner 5 in 1962 made two temperature soundings in the upper 5 percent of the atmosphere, and the USSR's Venera 4 in 1964 may have penetrated the upper 20 percent.

---

[1] R. Smoluchowski, Jupiter's convection and its red spot, *Science*, June 12, 1970, p. 1340.

The latter indicated a temperature of 544° K at a pressure of 19 atmospheres. Radar measurements from the earth combined with Mariner 5 data indicate that the Venusian atmosphere extends 25 km below the Venera 4 data point, and it is calculated that the surface temperature and pressure are 700° K and 100 atmospheres, respectively. The later space probes, Venera 5 and Venera 6, indicate a surface temperature of 740° K. The Russians reported in late 1970 that still another space probe penetrated the hot and thick atmosphere, landed on the Venusian surface, and radioed back information, the nature of which was not reported. A recent earth-bound microwave study suggests a thermal layer at the bottom of the atmosphere with a temperature of 670° K.

The O_2 measurement of the atmosphere obtained by direct chemical analysis by Venera 4 differs by a large factor from that of the land-based spectroscopic method. Venera 5 did not detect the presence of O_2 but did confirm the presence of water vapor. It is agreed that the chief constituent of the atmosphere 25 to 50 km above the surface is CO_2, and that there is little likelihood that the amount of CO_2 here is less than 80 percent. The ionosphere of Venus, which extends outward at least 1400 km from the visible cloud layer, is probably made of He^4.

No magnetic field was recorded by Mariner 5. This is interpreted to mean that Venus's field must be less than 5 to 10 percent of the earth's field. Venus has a retrograde rotation on its axis with a period of 243.16 days. This has been determined by radar reflections which, because they are repeatable, must come from the hard surface. The reflections show the surface to be somewhat smoother than that of the moon, with two circular dark areas of about the size and shape of lunar maria. Since the chemistry and undoubtedly the dynamics of the Venusian atmosphere are so unusual, future space probes to Venus are considered to have top priority. Clearly the major questions that must be answered concern the clouds. What is their composition, and are they condensable? What is their vertical distribution?

Mars is only one-tenth as massive as the earth (the escape velocity is 3 miles per sec, less than half that on the earth) and has an extremely thin atmosphere consisting mostly of CO_2. Mariner 6 and 7 photographs show that Mars is not just an oversized moon, as we were led to believe from the photographs of Mariner 4. Mars is desolate, like the moon, and it has a heavily cratered terrain. But there are also large regions devoid of craters, and a third type of terrain called chaotic, where craters have been erased amidst jumbled ridges unlike anything on the moon or on earth.[2]

Mars has no detectable nitrogen, a basic component of every living cell and the dominant ingredient of the earth's atmosphere. The space probes, Mariner 6 and 7 of early August 1969, revealed that its south polar white cap is ragged at the edges and has a dark splotch at the center. Spectral studies suggest that the white consists of crystalline CO_2, and the surface patterns indicate that it occurs in drifts, thus suggesting the existence of winds in the thin Martian atmosphere. The photographs also show bright puffy spots in the polar snow cap which may be clouds. Surface temperatures range from 75° F down to −100° F in low-lying regions.

A radar scanning technique showed that elevations along latitude 21° N range through 12 km, and a spectroscopic traverse of the CO_2 content of the thin atmosphere along the Martian equator indicated the same range of elevations. It is reasoned that the depressions have the thickest concentration of CO_2. The photographs reveal large light and dark areas which, from the spectroscopic measurements of CO_2, do not correlate with elevation. The irregular but linear markings that the Italian astronomer Giovanni Schiaparelli called "canals" appear to be segments of rubbled rims of craters up to 300 miles across.

The Martian atmosphere, although mostly

[2] Robert B. Leighton, The surface of Mars, *Scientific American*, May 1970.

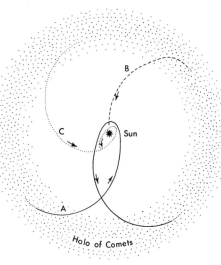

Figure 1-3 *Errant ways of the comets. Comet A sweeps in toward and around the sun, then out to resume approximately its former course. Comet B's course is too close to the sun and is obliterated. Comet C, on circling the sun, is wrenched into a new small orbit, like Halley's comet. [After* The universe, *1962.]*

CO_2, contains appreciable water vapor, which is measurable, and in a vertical column amounts to approximately 0.0035 g/cm². It is thus presumed that if Mars produced its water by "outgassing" from the interior, it has lost most of it to space.

Ultraviolet spectroscopy in the Mariner 6 flight showed that the upper atmosphere of Mars contained ionized carbon dioxide and carbon monoxide, atomic hydrogen, and atomic oxygen.

It may be concluded that the earth is the only hospitable place in the solar system for life, such as we know it. It is hard to conceive that the basic life materials could be other than those composed of hydrogen, oxygen, carbon, and nitrogen, and unless the temperature range is suitable for the existence of water (32° F to 212° F), the necessary chemical reactions can probably not take place. At the high temperatures that exist on some of the planet surfaces, the bonds between carbon and hydrogen would

be ruptured. Mars is the only planet with a remote chance of conditions that would be favorable to life, and this only in the lower plant forms.

The earth's moon is devoid of an atmosphere. Its surface is believed to reach a temperature of 214° F in full sunlight at its equator, −58° F at sunrise, and −243° F on the dark side. The rocks a few feet below the surface are probably perpetually below 32° F.

Comets

The comets are members of the solar system, but odd ones. They are composed of grit and frozen gas particles somewhat like the rings of Saturn and are no more than a few miles in diameter, with a density less than water. Some comets move in the plane of the planetary orbits, but others ignore it entirely and travel in ellipses at all angles to it. We know those particularly which swing in close around the sun on their elliptical paths, because the solar energy vaporizes the outer layers of the comets to form great swollen heads. These are generally plainly visible to the unaided eye. Also some of the vaporized material points outward from the sun in an incandescent tail. At this time the comet may appear as large as the sun; however, it is so attenuated that it is made up of "bucketfuls of nothing." At still closer range the sun's energy creates explosive pockets of gas within the comet, and the tail stretches out even more. The great comet of 1843 had a tail that stretched 500 million miles.

The 100 billion-odd comets presumed to exist roam the icy edges of the solar system as a halo some 12 billion miles out from the sun. This is far beyond the reaches of Neptune or Pluto. Occasionally one is believed to be "kicked" out of its track by the gravitational attraction of a passing star. If the direction is toward the solar nucleus, it has three possible destinies (Figure 1-3). It can make one big loop around the sun and return to its regular path in the halo. It can be dragged so deep into the solar system and so close to the sun that it disintegrates. As a third fate, it can be locked into a new smaller orbit around the sun. Halley's comet whizzes through the disk of revolution of the planets every 75 or 76 years and since 240 B.C. has been noted by the Chinese and Japanese at every passage but one.

An example of a comet that skirted the sun and earth too closely is instructive. The story is known as the "short and wild career of Biela." Biela's comet was first observed approaching the sun in 1772. After its first flirtation with the sun it began reappearing at regular 6½-year intervals. As it approached in 1846 it became two comets moving side by side, and then on its next appearance in 1852, still in its split form, it vanished or, better said, was lost by the astronomers. Twenty years later the whole of Europe was treated to a meteorite shower such as has never before or since been described. As the storm reached England people could see 100 blazing shooting stars a minute. Mathematical calculations have since shown that this was the last of Biela's comet. It had probably crossed close to the earth in past journeys but this time, in playing tag, got drawn into the earth's gravitational net and thus really vanished in a most surprising pyrotechnic extravaganza.

Another remarkable instance of collision of a comet with the earth is now established after many years of mystery. A tremendous explosion shook an area of northern Siberia in 1900 along the Tunguslea River. Trees were felled like dominoes out to more than 30 miles from the blast center. Window glass was broken at a distance of over 100 miles. Earthquakes were felt for 400 miles. Barometers were affected as far away as England. Much fine material that had been shot up into the upper atmosphere produced week-long unusual and beautiful sunsets. Strangely, no impact crater was formed; only small fused pellets, apparently of terrestrial origin, were found in the eye of the explosion, driven into the ground like buckshot. The Soviet Academy of Science, after considerable study, concluded in 1960 that a mighty explosion had definitely occurred and that it was in the head of a comet just before or as it collided with the earth's surface. It was only a small comet compared with other marauders of the solar system; it is estimated to have had a diameter of several miles and to have weighed about 1 million tons. The apparent internal explosion as it reached the earth's surface may be accounted for by the fact that the comet met the earth head-on at a relative speed of 25 miles per sec, instead of overtaking the planet from behind.

Origin of the solar system

During the past 200 years several hypotheses have been proposed for the origin of the solar system. Each has risen to popularity and then fallen, owing to patent impossibilities that later students have exposed. Observed relations considered in the light of known physical laws point to an origin of the sun and planets from a cloud of gas and dust, such as is represented by the Lagoon Nebula of Sagittarius in Figure 1–4.

What are the physical relations and laws that hold and should be noted and explained in any theory of origin? They are listed below.

Clues to origin

Any theory of origin should account for the following observed astronomic conditions.

1. The spacing of the orbits of the planets is nearly regular and follows a geometric progression.

2. The orbits of the planets lie nearly in one plane.

3. The orbits are nearly circular and all planets travel in the same direction.

4. Over 99 percent of the mass of the solar system is contained in the sun, yet the sun has only 2 percent of the total angular momentum of the solar system. Angular momentum cannot be created or destroyed, and hence it must have been present from an early beginning. Its distribution must have resulted from some circumstances related to the system.

5. The farther each planet is from the sun, the slower it travels.

6. *The outer planets are light; the inner ones are heavy.*

7. *The satellites of the inner planets orbit in the same direction as the planets.*

8. *Most of the moons of the outer planets, especially those of Jupiter and Saturn, also orbit in "regular" directions, but some outer ones travel in the reverse direction and have orbits appreciably inclined to the ecliptic. These are certainly exceptional.*

9. *The sun's rotation is in the same direction as the orbital motions of the planets and its axis is only slightly inclined to the average plane of the orbits. Thus we see a system in regular motion—rotations and orbits are generally in the same direction.*

10. *Orbits of comets are highly eccentric and inclined at all angles to the ecliptic.*

The earth has an iron-nickel core, and it is believed that Venus has such a structure. Mercury, Mars, and the moon have lower densities and lower internal pressures. Although an appreciable amount of the iron-nickel phase exists in Mercury and Mars, it is not believed that these planets have a core of iron and nickel. The moon consists almost entirely of the silicate phase.

The infalling meteorites are of two kinds, iron-nickel and silicate; if this is the basic material from which the planets were formed, then the fractions have been partially separated, and one has been lost in variable amounts. An acceptable model of the solar system must take this into account.

That Mars has a high amount of nickel and iron that is not separated into a core suggests that the planet was never molten. Also, it appears that the earth was not molten during its late stages of formation—certain elements that are volatile at the melting temperatures of the silicates are not concentrated on the earth's surface. These are reasons proposed by Urey[3] in favoring the theory that there were low average temperatures during the formation of the inner planets.

Beginning stages

The formulation of modern concepts of the origin of the solar system began with von Weizsäcker of Germany. These were modified by others, particularly G. P. Kuiper of the United States, and the postulate is commonly referred to as the von Weizsäcker-Kuiper hypothesis. It starts with a "globule" or dense region in our galaxy, the Milky Way, such as the Lagoon Nebula in Sagittarius (Figure 1–4). This globule presumably was composed of gas and dust, the so-called cosmic material. It was about 1 light-year in diameter and is assumed to have had a very slow rotation.

The cosmic globule began to shrink under its own gravitation, and as it shrank, its rotational speed increased. About 90 percent of the material eventually condensed to become the early sun, and the other 10 percent remained outside, where rotation caused it to flatten and spread out into a disk. In simple orbital motion the inner part of this disk of gas and dust particles would move faster than its outer part; hence, it is conceived that great turbulent eddies resulted. The moving of one part past another was a source of friction that somewhat slowed down the inner rotating mass. The eddies were the sites of large condensations of smaller clouds. Within them, substantial bodies of asteroidal and perhaps even of lunar dimensions together with their gaseous envelopes developed. These were drawn together through mutual gravitational attraction to form the "protoplanets." The protoplanets were many times larger and more massive than the present planets. The central sun had not yet begun to shine.

The large protoplanets were subject to the tidal action of the sun and, as in the moon-earth

Figure 1–4 Lagoon Nebula in Sagittarius is a nebula of gas and dust something like one from which the solar system is presumed to have evolved. [Courtesy of Mount Wilson and Palomar Observatories.]

---

[3] Harold C. Urey, The origin of the earth, *Scientific American*, Oct. 1952.

relation, kept a constant face toward the sun. Gradually, the heavier materials of each protoplanet settled into a central denser nucleus, leaving a disk of dust and gas around it. In this disk the satellites evolved. As the shrinkage occurred, rotation speeded up, and thus the planets acquired variable rotational velocities faster than the original. It can be shown that this process results in planets at specific distances from the contracting core in accord with stability criteria. These criteria are derived from the interplay of turbulent action, the density of matter in the cloud, and the gravitational tidal action of the core. The spacing is given by Bode's law, which states that every star with approximately the same mass and intrinsic luminosity as our sun will have a belt in which one, two, or three planets will evolve in which the physical and chemical conditions are similar to those of the earth, and where life may evolve. This is called the life belt of the star, and thus it is concluded that every star similar to the sun in the Milky Way galaxy (which has some 300 million similar stars) has at least one planet, and possibly three, that may support life.

Later stages

The compression of the sun under its own gravitation eventually resulted in a heat build-up until it began to shine. Radiation pressure and streams of particles ejected from the sun swept away much of the light gasses surrounding the inner protoplanets. The massive outer protoplanets were able to hold large amounts of hydrogen, helium, methane, and ammonia, while the inner planets lost practically all but the heavy elements. Hence the inner planets are rocky and metallic, whereas the outer planets are largely gaseous. (The earth is judged to contain only one-thousandth of its original protoplanet material, and Jupiter about one-tenth.) Urey concludes that at this stage several factors contributed to bringing about high temperatures for a while: adiabatic compression of the gases, a temporary high temperature of the sun, and collisions of the solid objects. Consequently much material was volatized and selectively lost.

As the material of the earth became more compact under its own gravitation, its interior heated up, owing to compression and supposedly a greater abundance of potassium-40 and uranium-235 than now exist. These radioactive elements were great sources of heat, and the interior of the earth melted to such an extent that the separation of the metallic core and the silicate mantle occurred.

After the separation of nickel-iron core and silicate mantle the earth cooled, with the mantle solidifying from bottom upward, according to Urey. This stage is believed to have occurred about 4.5 billion years ago. The radioactive elements were concentrated in the uppermost layer as this process took place and perhaps caused the outer layer to be remelted. The condition may have resulted in an early differentiation of lighter granitic material from the heavier silicates and the formations of the original continents. Since the oldest rocks yet found in the crust are about 3 billion years old, we might assume that this was the time of the last molten crust and the beginning of evolution of the continents and ocean basins.

Evolution of the earth-moon system

Rotation of earth and lunar tides

It is well established that the rate of rotation of the earth on its axis is gradually slowing and that this is due to lunar tidal friction. But another effect is significant also. The earth rotates and the moon orbits it in the same direction (see Figure 1–5). The earth's rotation drags the high tides east of the position directly opposite the moon, the sublunar point A, to the position B. The force on the moon due to the nearer high tide is greater than the force due to the more distant low tide, so that the net torque is such as to accelerate the moon. Thus the energy lost by the earth by the decreased rate of rotation is imparted to the moon. An increase in angular momentum of the moon means that its orbit will increase in size (the moon will get farther away from the earth), and that it will take the moon longer to orbit around the earth. The increased circumference of the orbit more than compensates for the higher angular momentum. Some recent approximate figures regarding these changes follow. The data for ancient periods come from fossil evidence. Many shell-growing marine organisms build their shells in small increments which produce small waves or ridges. The waves or ridges represent daily growth and occur in modulations or in nearly periodic rhythms. The modulations in modern specimens occur in two rhythms, one with about 360 ridges which is presumed to represent a year's growth, and one in about 27 to 30 ridges, which is a lunar cycle. Fossils of Devonian age (380 million years ago), however, contain 385 to 410 ridges. The smaller modulations in the Devonian shells contain about 30 ridges, which presumably represents growth during a month.

Earth-rotation and moon-orbit variations have been compiled from observations made in historical times, and these prove to be compatible, although variable, with the fossil evidence.[4] It is thought that the slowing of the earth's rotation and the slowing of the moon's orbiting will continue, and that the two rates eventually will approach each other until both the earth and the moon have a period of 55 present days. This will be about 50 billion years hence. At this time the tides will be stationary, and no tidal friction will exist. The moon-earth relation will thus have reached a stable condition.

Calculating backward and consequences

Now let us calculate backward to see what the lunar orbit and terrestrial day looked like several billion years ago. Some rather exciting

[4] Robert R. Nowton, Secular accelerations of the earth and moon, *Science*, Nov. 14, 1969, p. 825.

YEARS AGO	DAYS PER YEAR	DAYS PER MONTH
380 million years ago	410 ±	30.7 ±
300 million years ago	398 ±	30.2 ±
Present	365.25	27.3

conclusions are reached.[5] Both the earth's rotation and the moon's orbital velocity increase, with the moon ultimately catching up with the earth to again produce a condition of stationary tides. But at this time the moon was only at a distance of a few earth radii from the earth and the resultant tide was very great. In fact the great lunar tide would have resonated with the solar tide, and according to George Darwin, the British astronomer and son of Charles Darwin, the effect was to pluck out a chunk of the earth's outer layer and pull it into orbit with the moon. The scar, supposedly, is the Pacific Ocean. This so-called *resonance theory* is not widely held today, for several mathematical reasons, but there seems little doubt that the moon was once precariously close to the earth.

Recently Dr. Horst Gerstenkorn, a teacher in a girls' school in Hanover, Germany, published a paper in which he computed the history of the moon-earth relationship backward to the near-earth position, but instead of a pulling of a mass of the earth away in a tidal bulge, he postulated that, when the moon reached the closest to the earth, which was 2.89 earth radii, momentum was transferred from the moon to the earth, and the moon began to fall away from the earth in larger orbits. This occurred about 1.4 billion years ago.

Continuing the calculations backward, Gerstenkorn concluded that under the reversed torque the moon not only moved farther away, but the inclination increased until a polar orbit

[5] See W. H. Munk and G. J. F. MacDonald, *The rotation of the earth*, Cambridge University Press, 1960, and Donald U. Wise, Origin of the moon from the earth: Some new mechanisms and comparisons, *Jour. Geophys. Res.*, v. 74, no. 25, 1969, p. 6034.

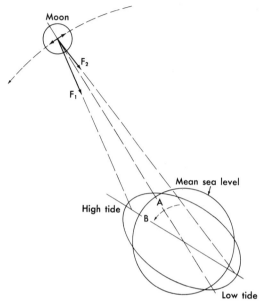

Figure 1–5 *Acceleration of the moon. The earth rotates and the moon orbits in the same direction, west to east. The earth's rotation drags the high tide east of the sun's lunar point A to B. The force F_1, due to the high tide, is larger than the force F_2, due to the low tide, so that the net torque is able to accelerate the moon.*

was achieved. In other words, the moon moved around the earth in a polar orbit instead of the near-equatorial orbit. This happened at a distance of 4.7 earth radii. Further calculations involving this eccentricity carried the moon into a near-parabolic orbit, and thus he concluded that the moon was *captured* by the earth.

The prominent Swedish astronomer H. Alfrén took up Gerstenkorn's calculations and concluded that at the moon's closest approach it was the moon that was disrupted, not the earth, and that much of it rained down upon the earth to form the outer mantle and crust. If the moon were twice as big before it was torn asunder, then an outer shell of the earth would have been built up 50 km thick.

Another point of the Gerstenkorn theory is that the orbit of the moon was retrograde before the polar orbit was achieved. This corresponds to the outer moons of Jupiter and Saturn and thus suggests that they also are captured moons but have not yet been drawn in as far as the earth's moon.

The moon's surface and materials

Major physiographic divisions

The visible lunar surface has been divided into two major divisions, the highlands and the lowlands (see Figure 1–6). The highlands constitute two-thirds of the visible surface of the moon, and their most conspicuous forms are craters. About 30,000 craters are readily distinguishable, and perhaps a million other poorly defined depressions may be craters. The largest crater has a diameter of 146 miles (Clavius), and the others range on down to a few centimeters. Extensive but less conspicuous are rugged relief forms that look more like the peaks and ridges of ancient mountain systems. These generally ring the lowlands and they present steep and imposing fronts toward the lowlands.

About one-third of the moon's visible face is lowlands. These are the dark regions and make the imaginary eyes and nose of the "man in the moon." In early times, the lowlands were thought to be oceans and thus came to be called maria (seas). They are generally smooth plains, but in places they are pockmarked by numerous craters and are wrinkled and cracked.

Craters

Most of the many craters that you see on the photographs of the moon reproduced in this chapter are now generally believed to be impact craters. The shape of the craters is that of explosive impact as testified by ballistic experts, and the distribution is mostly a random scattering, unlike the pattern of volcanic cones on earth.

In Figures 1–6 to 1–9 and 1–11, it will be noted that from a few major craters long rays extend out and appear to mask the other features. These rays are particularly strong when

Figure 1–6 *Full moon. Note emphasis of rays from craters.* [*From* Photographic lunar atlas, *1960.*]

photographed in direct sunlight (central part of full moon), but when the same region is photographed in inclined light (quarter moon), the rays are inconspicuous or even not visible. Compare the crater Tycho in Figure 1–6 with that of 1–7. If a stereo pair of photographs is studied, one showing the rays and one not, then a maximum understanding can be gained of the geology, and it has been concluded that the rays are thin layers of highly reflective material derived primarily from the craters from which they radiate.[6] The good reflection comes from a high sun position, and very poor reflection from a low sun. The ray craters are recognized as very young compared with the nonray craters, with the ray-forming material having been thrown out by the mighty impact explo-

[6] Robert J. Hackman, Photointerpretation of the lunar surface, *Photogrammetric Engin.*, June 1961.

Figure 1–7 Comparatively young Tycho Crater in central part of photograph. This is the great ray crater of Figure 1-6 that appears like the north pole of the moon. The very old and large crater in upper left quadrant has been badly battered by subsequent impacts. It and similar old craters have flat floors, as if partially filled, whereas the fresh and young craters have cup-shaped floors. This fill is called the maria material. Tycho Crater is about 50 miles across. [From Photographic lunar atlas, 1960.]

Figure 1–8 Crater Censorinus is just to the right of and beyond the large crater. Both craters are young, but the smaller is the freshest on the moon's near side. It has explosive debris all around, which appears bright in infrared photographs. [Courtesy of National Space Science Data Center, NASA.]

sion fairly recently, geologically speaking. After a while, either the weathering effect of the solar wind dulls the reflective capacity, or the reflecting particles are covered by dust. The thin skin of fragments is probably still there but no longer visible. It has been suggested that we may be dealing with a beaded-screen effect and that drops of glass, created by fusion at impact, were thrown out, and these reflect the sun's rays most profusely directly back at the sun.

Maria

The conspicuous large dark areas of the moon, such as seen in Figure 1-6, are the maria. They appear as vast plains, with scattered ridges rising a few hundred feet above the general surface, and are commonly interpreted as lava fields.

The three astronaut visits, Apollo 11, 12, and 14, testify beyond doubt that the maria are lava plains, and it has been concluded that previous large basins in the moon have been filled with lava. The rock fill of the moon is identified as basalt, the common lava rock of the earth. When molten, however, it has a greater fluidity than earthly basalt. The eruptions on the moon have probably been from long fissures, from which the highly fluid lava spread over the maria basins for many kilometers. Photographs from the orbiting lunar modules show some indistinct lava flows, but certain individual flows with their lobate fronts can be discerned.

Rock samples from three separate maria have been procured, age determinations made, and the ages of the lava fillings determined. Each basin has been filled at a different time as follows: one at about 4 billion years, one at 3.65 billion years, and one at about 2.5 billion years.

The craters date from 4 billion years to the present, but in each maria are craters of two age groups, those that have been filled or partially filled with lava and are thus older than the lava outpourings, and those that have been

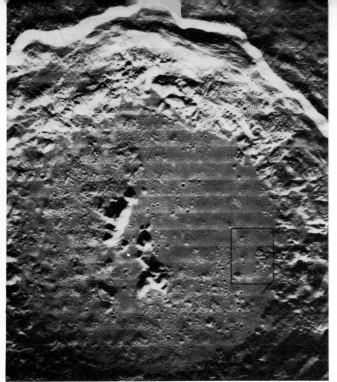
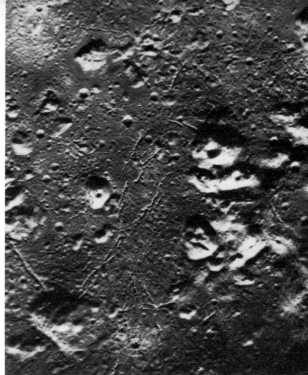

Figure 1–9 Crater Copernicus by low-resolution camera of Lunar Orbiter V. Lumpy blisterlike hills in the crater floor appear to be volcanic phenomena. Those in the small box are shown in Figure 1–10. [Courtesy of National Space Science Data Center, NASA.]

Figure 1–10 High-resolution orbiter view of box in Figure 1–9. Two cratered hills are undoubtedly volcanic, like cinder cones. Note the large blocks on these and other hills. Note also the fractures in the floor material of the big crater Copernicus. [Courtesy of National Space Science Data Center, NASA.]

formed later and are not filled with lava. The second group displays nearly perfect explosive-impact form. These craters exhibit various degrees of "weathering" and consequent softening of form. Weathering on the moon is now regarded as the process of long and continued bombardment of large and small meteorites. The highlands have in places been struck by so many meteorites that the entire surface has been reduced to a lumpy, powdery soil mantle or regalith, now determined to be 3 to 4 meters thick.

Old mountain chains

The oldest relief features are rugged ridges and peaks of apparent mountain chains. The Apennine Mountains of Figure 1–12 are an example. These are partly buried by the maria material and therefore older. They are also older than a long succession of craters. Two large mountain systems nearly encircle the Mare Imbrium and the Mare Serenitatis. In places the mountain system has a linear pattern as it is cut by many faults.

Lava domes

Some blisterlike hills, which have been mapped on volcanic flows and domes, appear in Figures 1–9, 1–10, and 1–11. Some have small depressions in them that may be the vents. The low slopes and low reflectivity of the sun's rays from these domes suggest that they are of basaltic composition.

Moon materials and composition

The examination of the samples from the Sea of Tranquility, brought back by astronauts Armstrong, Aldrin, and Collins on July 24,

Figure 1-11 *Marius Hills region and crater Marius at the upper right. Photograph by Lunar Orbiter II. Note beside the fresh young impact crater, the domelike, somewhat irregular hills which are probably volcanic. Note also the cracks along some of which lava appears to have been emitted. [Courtesy of National Space Science Data Center, NASA.]*

1969, were distributed to a number of teams of scientists from nine nations. Over 500 scientists participated in a most thorough and sophisticated analysis of the 30-pound rock sample.

The material from the maria around Tranquility Base consists of

1. Basaltic igneous rocks
2. Microbreccia, a mechanical mixture of soil and small rock fragments compacted into a coherent rock
3. Lunar soil (Figure 1-13), a diverse mixture of crystalline and glassy fragments in a variety of shapes. Some small fragments of iron meteorites were also found.

Most of the fragments of the lunar soil are similar to the large pieces of (1) and apparently were derived from the same rock forma-

Figure 1-12 *Hadley's Rill, the riverlike structure. Also shown is part of the Apennine Mountain region, which is older than the maria material. The origin of the rill is baffling. A future landing nearby will perhaps solve the mystery. [Courtesy of National Space Science Data Center, NASA.]*

Figure 1–13 *Astronaut Aldrin plants a solar wind collector on the Sea of Tranquility, Apollo XI Mission, Aug. 20, 1969. Note especially the texture of the loose surface material and the astronauts' footprints in it.* [Courtesy of NASA.]

Figure 1–14 *Macroscopic view of an assemblage of lunar fragments chosen to represent most of the rock types in sample from Apollo XI collection.* [From J. A. Wood et al., Special Report 307, Smithsonian Astrophysical Observatory, Plate 1.]

tion, which is thought to be the underlying bedrock, or the rock that constitutes the Sea of Tranquility (Figure 1–14). But some of the small fragments (4 percent) are somewhat different from the basalts; they are called anorthosite. The lunar basalts are rich in titanium whereas the anorthosites are poor in titanium and richer in calcium and aluminum than the basalts. They also have other different properties, and as such represent fragments from an entirely different formation, possibly from the highlands.

The small glassy objects or beads were formed by impact processes, and they contain beautifully preserved microscopic pits as small as 10 microns in diameter, which undoubtedly are the result of impact of tiny high-velocity particles. The impact process is accompanied by local melting, splashing, evaporation, and condensation.

Melting experiments indicate that most of the basalt liquid crystallized at 1210° C to 1060° C. The viscosity of these lunar silicate liquids, called magmas (see Chapter 2), was about an order of magnitude lower than that of terrestrial basalt. This means that the lunar magma would flow more readily, and, because molten basalt already flows rather readily, we can see that the lunar magma would fill a depression smoothly and to a level surface much like water.

The basalts from Tranquility Base, when subjected to high pressure and temperature, have a density far in excess of the average density of the moon. Thus these basalts cannot represent the bulk composition of the moon. Isotopic age determinations of the basalts yielded a uniform age of 3.65 billion years. At this point in time the basalts were melted, probably extruded, and crystallized. A single rock fragment in the Apollo 11 collection gave an age of 4.4 billion years, and one in the Apollo 12 collection yielded an age of 4.6 billion years. The breccia and samples of lunar soil also yielded an age of 4.6 billion years; this is believed to be the time when the original moon crust was formed. The Russians report an age of about 2.5 billion years for the basalts of Oceanus Procellarum; thus, the period of eruption of the maria basalts is known to span about 2 billion years. (The oldest rocks found on the earth are about 3.5 billion years old.) Because the rocks on the moon are over 1 billion years older, the moon may provide important clues about early earth history and possibly even that of the solar system.[7]

The so-called "genesis rock," picked up in the foothills of the Apennine Mountains by astronauts David Scott and James Irwin on the Apollo 15 mission, turns out to have an age of 4.15 billion years, plus or minus 200 million. It is a light-colored, coarsely crystalline aggregate of feldspar crystals (see Chapter 2), and is called anorthosite. Such a rock was viewed as a possible remnant of the moon's original crystallized material, light enough to float to the surface (crust) in the original liquid magma. But we are a little disappointed that it did not turn out to have the projected 4.6-billion-year age.

[7] Some of the above information comes from a speech by Dr. Eugene Shoemaker, chairman of the Division of Geological Sciences, California Institute of Technology.

References

Hess, Wilmot, et al. Exploration of the moon. *Scientific American*, Oct. 1969, pp. 54–75.

Science, January 30, 1970. The moon issue: The entire volume is devoted to the study of Apollo 11 samples. Over 500 scientists report their findings. Reports of the lunar regolith and the seismic experiments are also given. See also *Science*, March 6, 1970, for a report on the samples of Apollo 12.

minerals and igneous rocks

Need we say in these days of lunar spaceships and weather satellites that the earth is round? It comes with an oxygen-enriched atmosphere and extensive oceans. Below the atmosphere and the oceans is the solid earth. The major units of the solid surface of the earth are the continents and the ocean basins. Water has more than filled the basins, and laps over the edges of the continents in many places. The three states of matter, gaseous, liquid, and solid, will thus absorb our attention as we venture into the science of the earth.

The earth is unusual among the family of planets. It has nourished life, something no other planet of the solar system could. Most of us regard it as a very fair place in which to live, and thus we should understand it, and care about its natural beauties and resources. Hopefully, this book will help lay a foundation for this concern.

Solids—minerals and rocks

The solids that make up the earth's crust are rocks. A geologist calls a rock mass a formation if he can trace it for some distance and distinguish it from other rocks. But a rock is commonly an aggregate of minerals or mineral grains, and thus the basic units of rocks are minerals. So we begin by inquiring about minerals.

Atomic theory

As minerals are the building blocks of rocks, so are atoms the building blocks of minerals, and thus it is necessary to take a brief look at the atom.

John Dalton advanced the idea in 1805 that all matter is composed of tiny individual particles, which he called *atoms*. Many discoveries in the past 150 years have added greatly to our understanding of the atom, and a simplified description of it today would be as follows.

1. Atoms are composed of bits of matter (subatomic particles) with empty space between them.

2. The principal subatomic particles are electrons, protons, and neutrons. Electrons have a unit negative charge, protons a positive charge, and neutrons no charge.

3. The particles of atoms are held together by strong forces.

4. Most of the mass of an atom is concen-

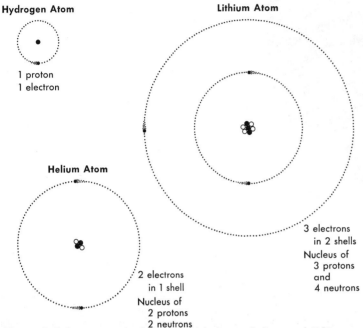

Figure 2–1 *Structure and relative size of hydrogen, helium, and lithium atoms. The electrons of helium and lithium may not orbit in one plane as shown, and the arrangement of the protons and neutrons in the nucleus is not known.*

trated in a minute nucleus, *which has a positive charge.*

5. *The nucleus is made up of tightly packed neutrons and protons.*

6. *There is one electron for each positive charge on the nucleus, and an atom as a whole is neutral.*

7. *The electrons travel around the nucleus at relatively great distances with almost the speed of light.*

8. *The electrons are regarded as circulating within concentric "shells," and each shell has a definite capacity for electrons. For instance, the first shell has a capacity of 2; the second shell 8; the third shell 18; and the fourth shell 32.*

The simplest atom is that of hydrogen, which consists of a nucleus of one proton about which one electron travels. It and the next simplest atoms, helium and lithium, are illustrated in Figure 2–1.

The number of protons, neutrons, and electrons of a few simple atoms is given in Table 2–1.

If the mass of an electron is taken as 1 unit, then the mass of a proton is 1836 units. Likewise the mass of a neutron is 1836 units. The proton has one positive charge, but the neutron possesses no electrical charge. The number of protons in the nucleus is the *atomic number* of the atom, or of the element represented by the atom. An *element* is defined as a substance composed of atoms all of which have the same atomic number. For instance, hydrogen atoms form the element hydrogen, oxygen atoms form the element oxygen, and iron atoms form the element iron.

The infinitesimally small size of the atom is impossible to imagine. The number of atoms in the tip of your pencil is greater than the number of people on earth. In spite of this minuteness, the diameters of atoms have been measured by modern physicists and are expressed in angstrom units (one hundred-millionth of a centimeter, generally written 10^{-8} cm). The comparative sizes of the three simplest atoms are shown in Figure 2–1. Some are larger than those of lithium, such as sodium, potassium, magnesium, and calcium. The size of an atom changes somewhat when combined with other atoms, and the distances between centers of atoms are especially amenable to measurement when they are closely packed and in orderly arrangement, as in crystals.

The weights of atoms in absolute terms are now known, but it is more convenient to use relative weights. The oxygen atom is arbitrarily assigned a weight of 16.000 units (its actual weight is only 2.66×10^{-23} g). Relative to oxygen, hydrogen has a weight of 1.008.

Isotopes

John Dalton postulated that all atoms of the same element have the same weight, but later more precise studies have indicated that atoms of the same element can have different weights. The weight of an atom is dependent almost entirely on the weight of the protons and neutrons in its nucleus, since the weight of an electron is negligibly small. If the number of neutrons in the atoms of a given element varies but the number of protons and electrons remains the same, then the atomic number remains the same but the mass will vary. Atoms of an element that vary in the number of neutrons are called *isotopes*. The isotopes of an element have identical chemical properties and cannot be separated by chemical analysis. Isotopes are recognized and measured by mass spectrographic analysis.

Hydrogen, for example, has three isotopes with the following composition:

NAME	NUCLEUS	ELEC-TRONS	NOTA-TION
Hydrogen	1 proton	1	^1_1H
Deuterium	1 proton 1 neutron	1	^2_1H
Tritium	1 proton 2 neutrons	1	^3_1H

The subscript in front of the symbol is the atomic number (number of protons), and the superscript also in front of the letter is the *mass number* (number of protons and neutrons).

The element oxygen also has three naturally occurring isotopes, which may be expressed as follows: $^{16}_8O$, $^{17}_8O$, and $^{18}_8O$, of which $^{16}_8O$ is the most abundant (Figure 2–2). When hydrogen combines with oxygen to form water (two atoms of hydrogen and one of oxygen), there are thus 16 isotopic combinations possible, but $^1_1H^{16}_8O^1_1H$ is by far the most abundant. If both hydrogen atoms are the ^2_1H *isotope*, then the resulting water weighs 1.1 g/cm³ as compared with 1.0 g for ordinary water. This is called *heavy water*, but it looks and tastes like ordinary water.

Molecules

The union of two atoms of hydrogen and one of oxygen is the smallest unit that possesses the properties of water, and this is called a *molecule* of water. A molecule may be defined as the smallest unit of a compound that displays the properties of the compound. Compounds are combinations of various atoms, and we will discuss them presently.

Combining properties of atoms

Atoms of one kind have long been known to combine with atoms of another kind. When atoms of different elements combine they do so in *definite proportions*, which may differ from element to element. For instance, one atom of hydrogen combines with one atom of chlorine to form a molecule of hydrogen chloride, HCl. An atom of oxygen, however, combines with two atoms of hydrogen to form a molecule of water, H_2O. An atom of nitrogen combines with *three* atoms of hydrogen to form a molecule of ammonia, NH_3. But argon does not combine with hydrogen (or any other element) at all. What aspects of atomic structure explain this behavior?

The combining properties of atoms are related to the number of electrons in the *outer*

TABLE 2–1 *The Structure of a Few Simple Atoms*

ATOM*	SYMBOL	ATOMIC NUMBER	NUCLEUS	ELECTRON CONFIGURATION BY SHELLS		
				1	2	3
Hydrogen	H	1	1 proton	1		
Helium	He	2	2 protons 2 neutrons	2		
Lithium	Li	3	3 protons 4 neutrons	2	1	
Beryllium	Be	4	4 protons 5 neutrons	2	2	
Oxygen	O	8	8 protons 8 neutrons	2	6	
Sodium	Na	11	11 protons 12 neutrons	2	8	1
Chlorine	Cl	17	17 protons 18 neutrons	2	8	7

* Only one isotope of each element is considered.

shell, and these are called *valence* electrons. Valence simply signifies the capacity of an atom to combine with other atoms. Atoms that have the same number of valence electrons have similar combining behavior. Chlorine (Cl), bromine (Br), and iodine (I), for example, have different atomic structures but they all have seven electrons in the outer shell, and each combines with one atom of hydrogen to form HCl, HBr, and HI, respectively.

It is often useful to represent each atom by an electronic dot formula, which consists of the symbol for the element and dots representing the electrons in its valence shell.

ATOM	ELECTRON DOT FORMULA
Hydrogen	H·
Oxygen	:Ö:
Sodium	Na·
Chlorine	:Cl:
Calcium	Ca:

Thus in the above list hydrogen is shown to have one valence electron, oxygen six valence electrons, and chlorine seven valence electrons. In this representation, the letter-symbol is understood to stand for the nucleus and the inner electron shells.

Atoms will combine into pairs or groups if the resulting structure is more stable than the separate atoms. An individual atom of hydrogen, H·, is unstable, but when two hydrogen atoms combine to form a molecule, H:H, they acquire an increased stability. The inert gases have eight valence electrons (helium has only two) and are exceedingly stable. It is recognized that eight is a stable number, and when atoms combine they generally acquire eight electrons in the outermost (or valence) shell. Two electrons in the outer shell constitute a stable condition for the simplest atoms.

The bonding of sodium and chlorine atoms is illustrated in Figure 2–3. Sodium has one valence electron, but its inner shell contains eight electrons. By losing the one outer electron it acquires the stability represented by a completed shell. At the same time, if the chlorine

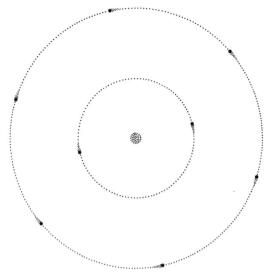

Figure 2-2 Idealized structure of the oxygen atom. The nucleus has eight protons and eight, nine, or ten neutrons. The orbits of the electrons are not in one plane, as represented, and are undoubtedly complex. The oxygen atom is about the size of the helium atom.

atom could gain an electron in its outermost shell, it would increase its stability. Both the sodium and chlorine atoms are neutral, but if the sodium atom should lose an electron (and a negative charge) it would have a unit positive charge. Likewise, if the chlorine atom gained an electron, and a negative charge, the atom would then have a unit negative charge. They would then be written as

$$Na^+ \text{ and } Cl^-$$

Particles that have either gained or lost electrons are called *ions*. Positive ions are called *cations* and negative ions *anions*. The transfer of an electron from a sodium atom to a chlorine atom renders the two atoms of opposite charge, and they attract each other and are bonded together.

Nature of crystals

A crystal is a body in which the fundamental particles are arranged in a regular repeating three-dimensional pattern. The particles are in

Figure 2-3 Structure of sodium and chlorine atoms showing the transfer of an electron from the sodium atom to the chlorine atom. Thus each atom becomes stable but is of opposite charge.

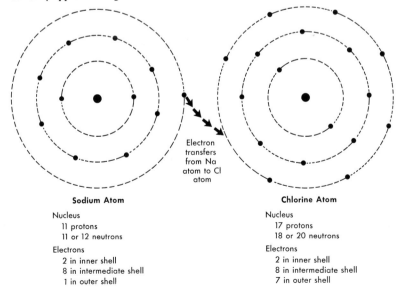

Sodium Atom
Nucleus
  11 protons
  11 or 12 neutrons
Electrons
  2 in inner shell
  8 in intermediate shell
  1 in outer shell

Chlorine Atom
Nucleus
  17 protons
  18 or 20 neutrons
Electrons
  2 in inner shell
  8 in intermediate shell
  7 in outer shell

vibrating motion, but occupy relatively fixed positions. A crystalline solid is made up of crystals. Crystals are bounded by plane surfaces symmetrically arranged, which are the external expression of a definite internal structure. Crystals have the property of growing indefinitely by addition of particles on the outside. The faces often develop unequally, but the angles between them remain constant.

Crystals may be classified as ionic, atomic, or molecular. Most rock-forming crystals are ionic, thus their fundamental particles have positive and negative charges. A good example is the salt crystal (NaCl), which is shown in Figure 2–4. Examples of atomic crystals are copper, silver, and gold. In these there is a geometric pattern that involves the repetition of one basic unit, the atom (upper diagram of Figure 2–5). In molecular crystals, molecules of two or more atoms are arranged in regular repeating patterns (middle diagram of Figure 2–5). Crystals of iodine, sugar, ice, and dry ice (carbon dioxide) are examples.

Since crystals have a regular three-dimensional internal structure of the basic particles they should also have a definite chemical composition. But certain minerals, or crystals, that display similar crystal form and other properties may vary somewhat chemically. Similar sized ions of other elements may substitute in certain amounts for the normal ones (such as Ca ions for Sr ions), and thus the chemical composition of some minerals varies within limits. Each mineral also has its singular physical properties, but in those minerals which have slight variable chemical composition there is also some slight variation in physical properties. True minerals, finally, must occur in nature. A number of compounds have been synthesized that are not known in nature but have all the properties of minerals; yet arbitrarily, they are not recognized as minerals.

It should be appreciated that a certain mineral, whether found in Nevada, Nova Scotia, or New South Wales, has the same crystal, chemical, and physical properties. Quartz from Brazil is the same as quartz from the Black Hills, except perhaps for the size of the crystals and their color. The angles between similar faces are the same, and the internal arrangement of the particles is the same.

Crystal form

Crystals grow by the addition of layers of particles, and in nature the common carrier of the fundamental particles is water. The water becomes saturated with a particular compound, and the compound is precipitated out, first as microscopic crystals. These will be the sites of further precipitation, and the small crystals will thus grow in size. Imagine a salt-water lake in a desert climate. Evaporation will increase the concentration of sodium chloride until saturation is reached. Then, ions of Na^+ and Cl^- will start precipitating out as very small crystals of the mineral *halite* on the bottom. Further evaporation and precipitation results in the growth of the tiny crystals. Some will grow at the expense of others until perhaps large crystals result. Crystals of many varieties grow in restricted places and hence cannot develop characteristic outward crystal form; yet the internal arrangement of particles is constant.

The difference between crystal and mineral may be confusing. As stated, a mineral is composed of a single crystal or an aggregate of similar crystals and hence a crystal is the intrinsic part of a mineral. Minerals have been given names (such as pyrite and quartz); over 2000 have been widely recognized and described (see Figures 2–6, 2–7, and 2–8). Crystals, on the other hand, are known by their natural shapes (hexagonal, octahedral, and so forth). The crystal forms of the minerals have been neatly grouped into six crystal systems, but it will not be possible to go into detail here on this subject. Perhaps some attention can be given to it in the laboratory. Characteristic crystal forms of a few minerals are shown in Figures 2–6 and 2–7.

Every kind of crystal has a characteristic form that is the expression of the internal molecular, atomic, or ionic structure. One of

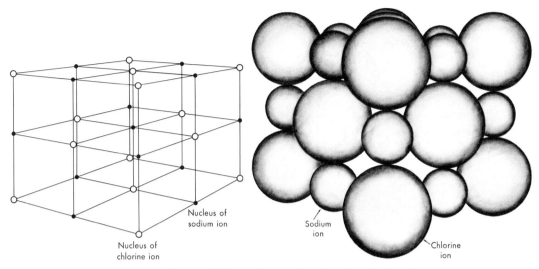

Figure 2–4 Arrangement of sodium and chlorine ions in the mineral halite (rock salt). Left, the positions of the nuclei; right, the outer electron shells.

the earliest discoveries pertaining to crystals was made by Nicolaus Steno (1631–1687), who showed that similar pairs of crystal faces of different specimens of quartz always meet at the same angle regardless of the size or shape of the crystals. There is a constancy of interfacial angles for every crystal variety, and this fact served not only for purposes of identification but also suggested to scientists that crystals had a regular, geometric, internal arrangement of the fundamental particles.

Optical properties

Light rays passing through minerals are affected in several ways, and, especially when thin sections of minerals are made for microscopic examination, most minerals are transparent and amenable to optical analysis. Certain minerals like copper and magnetite are opaque even in thin section, but light can be reflected from their surfaces and studied. It will have penetrated several layers of ions, and the reflected rays may have been altered. A particular ability of certain minerals is to restrict light to vibration in one plane when the light rays are passed through them. Such *polarized* light is then passed through other minerals and singular effects noted. The petrographic microscope, devised to study thin sections of minerals and rocks using polarized light, is a basic instrument of mineral identification and geological rock study.

X rays and internal structure

Electricity was a subject of much fascination to the early physicists a number of whom experimented with the passage of an electric current through gases. Michael Faraday (1791–1867) fitted a glass bulb with electrical terminals, evacuated as much air from it as he could, and then turned on the current. The glass glowed with a fluorescent light that could easily be seen in a dark room. A glass bulb was fitted up with a metal shield, as shown in Figure 2–9, by the German physicist Johann Hittorf in 1869, and a shadow of the shield appeared on the glass at point *A*. It thereafter became apparent that this shadow was caused by rays that travel in straight lines from the negative electrode. These rays were called *cathode rays*. Later William Crookes of England found that cathode rays could be deflected by a magnet,

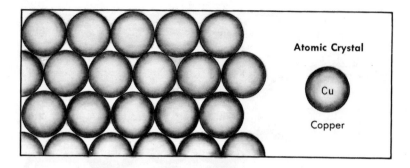

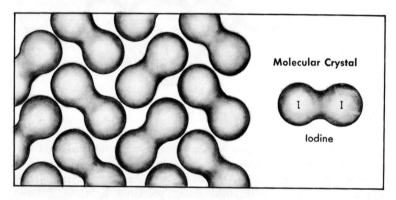

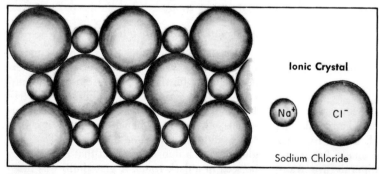

Figure 2–5 *Schematic representation of particle arrangement of atomic, molecular, and ionic crystals. The packing is actually three-dimensional. The ionic arrangement of the sodium chloride crystal is better shown in Figure 2–4. [After Compton, 1958.]*

and the direction of deflection indicated that the rays had a negative charge. In 1897 Joseph Thomson showed that the rays were composed of tiny negatively charged particles traveling at high velocity, which were called electrons.

The German physicist Wilhelm Roentgen discovered in 1895 that other rays come from cathode tubes besides those which cause a visible glow. He was not sure of the nature of the rays and called them *X rays*. They proved to be

Figure 2–6 *A few common crystals in characteristic form, all about natural size.*

a form of energy similar to light rays but having remarkable powers of penetration. It was found that, when X rays were passed through a human body and a photographic film was exposed to them, the bones cast a shadow, whereas the flesh was nearly transparent.

Shortly after 1900, Max von Laue of Germany discovered that a beam of X rays passed through a crystal was scattered, or "diffracted," in a systematic way. This led W. H. Bragg of Great Britain and his son, Sir Laurence Bragg, to the realization that man had been presented with a powerful new microscope that could resolve the structure of matter right down to atomic scale. The growth of X-ray crystallography since 1912 has more than fulfilled the fondest hopes of the Braggs, for it has revealed the way atoms are arranged in many diverse forms of matter and has cast a flood of light on the nature of the forces between the atoms.

The scatter of X rays through a crystal of halite (NaCl) may be photographed; diagrammatically, it will look like Figure 2–10. The photograph is called an *X-ray diffraction* diagram. We have already learned that halite has a cubic crystal symmetry. In an X-ray diffraction diagram of halite, a number of small dots are arranged around a central large dot in a pattern of quadrangle symmetry. It is believed that the small dots are produced by reflections of the rays from the regularly oriented planes of ions within the crystal. The large dot in the center represents the rays that passed through without being bent.

Repetition of the experiment with other minerals gave conclusive proof of the regular packing of the submicroscopic particles of crystals. The X-ray instrument has since been developed into the basic device for determining the internal structure of minerals and for their identification and analysis. Even microscopically small crystals can be analyzed.

Of course, the ultimate particles of minerals cannot be seen by ordinary light microscopes, but the electron microscope comes close to photographing them. Magnifications of nearly one million are possible.

Figure 2–7 Cluster of quartz crystals showing hexagonal prisms and pyramids.
Figure 2–8 Cleavage fragments of halite showing cubic symmetry.

Mineral groups and abundances

Classification of minerals

One common classification of minerals is based on chemical composition. There are the *oxides,* such as magnetite (Fe_3O_4), cassiterite (SnO_2), corundum (Al_2O_3), and ice (H_2O) (see Table 2–2). There are the *sulfide* minerals, such as pyrite (FeS_2), chalcocite (Cu_2S), galena (PbS), and sphalerite (ZnS). Then there are the *sulfate* minerals, which bear the sulfate (SO_4^{--}) ion as a basic part of their constitution. An example is gypsum ($CaSO_4 \cdot 2H_2O$). The *carbonate* minerals contain the ion CO_3^{--}, and its most common combining cation is Ca^{++} to form the mineral calcite ($CaCO_3$). The *chloride* minerals, such as sodium chloride (halite) and potassium chloride (sylvite) are also common. Finally the *silicates* must be mentioned because they are overwhelmingly the most abundant minerals of the earth's crust.

Figure 2–10 Diagram of diffraction of X rays passing through halite.

Figure 2–9 An early type of cathode tube shows the shadow A from the metal cross B. The rays move in straight lines from their source C.

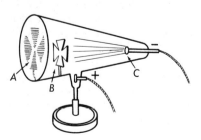
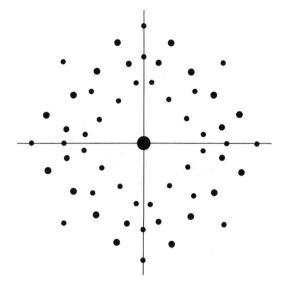

TABLE 2–2 Common Rock-forming Minerals

CHEMICAL GROUP	MINERAL	CHEMICAL COMPOSITION
Oxides	Magnetite	Fe_3O_4
	Hematite	Fe_2O_3
	Limonite	$FeO(OH) \cdot nH_2O$
Sulfides	Pyrite	FeS_2
Sulfates	Gypsum	$CaSO_4 \cdot 2H_2O$
	Anhydrite	$CaSO_4$
Carbonates	Calcite	$CaCO_3$
	Dolomite	$CaMg(CO_3)_2$
Haloids	Halite	$NaCl$
Silicates	Feldspar:	
	Orthoclase	$K(AlSi_3O_8)$
	Plagioclase	$Na(AlSi_3O_8)$ and $Ca(Al_2Si_2O_8)$
	Iron-magnesian minerals:	
	Olivine	$(Mg, Fe)_2SiO_4$
	Augite	$Ca(Mg, Fe, Al)(Al, Si)_2O_6$
	Hornblende	$Ca_2(Mg, Fe)_4Al(OH)_2AlSi_2O_7$
	Biotite mica	$K(Mg, Fe)_3(OHF)_2AlSi_3O_{10}$
	Muscovite mica	$KAl_2(OH, F)_2AlSi_3O_{10}$
	Quartz	SiO_2

They contain the silicate ion (SiO_4^{----}).

The silicate minerals compose more than 95 percent by volume of the earth's crust, which is taken as 10 miles thick in making the estimate. If a thicker shell is considered, the percentage of silicate minerals increases. There are many silicate mineral varieties, but comparatively few make up the great bulk of them. The feldspars, hornblende, augite, the micas, quartz, and olivine are about all that we need to mention or discuss here and in later chapters of the book. These minerals are composed of nine elements, and thus this small number of elements makes up most of the earth's crust. In fact, it is computed that they compose 99 percent of it.

It should be noted that one of the most abundant minerals is quartz, whose chemical composition is SiO_2. It would thus seem to be an oxide, but its internal structure indicates that its basic structure is the silica tetrahedron (see below), and hence it is listed above with the silicates.

The silica tetrahedron and the silicates

By X-ray analysis it is now known that all silicates have one basic unit in common. This is the so-called *silica tetrahedron*, which is composed of four oxygen ions and one silicon ion. The oxygen ions are at the four corners of the tetrahedron, and the silicon ion is in the center as shown in Figure 2–11. It is estimated that this basic mineral unit constitutes 90 percent of all crustal matter. The silicon atom has four electrons in its outer shell. Oxygen atoms have six electrons in their outer shell and thus need two more to complete this shell. The four oxygen atoms bond firmly with the one silicon atom by sharing the four electrons in its outer shell, all thus becoming ions, and produce a larger ion, the silica tetrahedron, with four negative charges. These must be satisfied when the silica tetrahedra join together with one another or with such cations as Ca^{++} and Mg^{++}. That is, since the SiO_2^4 ion has four negative

TABLE 2–3 *Abundant Elements in the Earth's Crust*

ATOMIC NUMBER	ELEMENT	ION	IONIC RADIUS, ANGSTROMS 1/100 MILLIONTH CM	ABUNDANCE IN EARTH'S CRUST WEIGHT, PERCENT	ABUNDANCE IN EARTH'S CRUST VOLUME, PERCENT
8	Oxygen (O)	O^{--}	1.40	46.60	91.97
14	Silicon (Si)	Si^{++++}	0.42	27.72	0.80
13	Aluminum (Al)	Al^{+++}	0.51	8.13	0.77
26	Iron (Fe)	Fe^{++}	0.74	5.00	0.68
		Fe^{+++}	0.64		
20	Calcium (Ca)	Ca^{++}	0.99	3.63	1.48
11	Sodium (Na)	Na^{+}	0.97	2.83	1.60
19	Potassium (K)	K^{+}	1.35	2.59	2.14
12	Magnesium (Mg)	Mg^{++}	0.66	2.09	0.56
22	Titanium (Ti)	Ti^{+++}	0.76	0.44	0.03
		Ti^{++++}	0.68		

SOURCE: Compiled from Brian Mason, *Principles of Geochemistry*, New York, Wiley, 1952, and Jack Green, Geochemical Table of the Elements for 1953, *Bull. Geol. Soc. Amer.*, v. 64, 1953, pp. 1001–1002.

charges, it must join with two Ca^{++} or two Mg^{++} ions.

Note that the oxygen ions are much larger than the silicon; from this it can be seen that, in volume, oxygen must compose most of the crust. In fact, Table 2–3 shows that oxygen by weight is 46.60 percent of the crust, but by volume it is 91.97 percent. These figures include all minerals, although the silicates are the chief ones.

The combination of the silica tetrahedra with themselves and with the cations of iron, aluminum, calcium, magnesium, sodium, and potassium are complex and varied, but the precision with which the structures are now documented is a testament to the great strides that have been made in the understanding of the structure of solids.

The tetrahedra arrange themselves in chains and sheets (Figures 2–12 and 2–13), which in turn are bonded together by the positive metallic ions. The chains of tetrahedra impart cleavage directions to the silicate minerals, such as those of feldspar, hornblende, and augite. The sheets of tetrahedra are weakly bonded together, permitting easy splitting or peeling of flakes or sheets from mica crystals. Note these characteristics on your laboratory specimens.

Later paragraphs concern rocks that have formed by cooling and crystallizing from a molten state. Such rocks are abundant on the earth's surface and even more abundant at depth. As the change from a liquid to a solid state occurred, the silica tetrahedra took form—the liquids were preponderantly composed of silicon and oxygen—and then combined with one another and also with the positive ions of the metallic elements to form the silicate minerals. Granite is such a rock and is composed commonly of quartz (SiO_2), orthoclase feldspar ($KAlSi_3O_8$), hornblende ($Ca_2(Mg,Fe)_{4-}Al(OH)_2AlSi_3O_{10}$), and biotite mica ($K(Mg,Fe)_3(OH,F)_2AlSi_3O_{10}$).

Igneous rocks

An igneous rock is one that was once molten but, as a result of cooling, has solidified. Molten rock has solidified both below the surface and above. The molten rock material is referred to as *magma* if it is below the surface in some kind of chamber or reservoir but as *lava* if it pours out on the surface. After solidification the lava is referred to as lava rock.

MINERALS AND IGNEOUS ROCKS

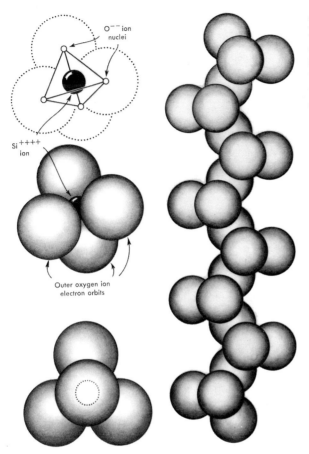

Figure 2–11 *The silica tetrahedron: top, the view emphasizing the structure; bottom, the tetrahedron as used in Figures 1–11 and 1–12.*

Figure 2–12 *Chain of silica tetrahedra. A silicon ion is in the center of each tetrahedron but is not shown, and each tetrahedron shares two oxygen ions with adjacent tetrahedra.*

Variable texture

There are many kinds of igneous rock. This is due to variable mineral composition on the one hand and to variable textures on the other.

By texture is meant the size, shape, and arrangement of the mineral particles that make up the rock. Texture also refers to the condition of a rock, whether it is filled with cavities or is fragmented or badly crushed. The rock may not even be made up of grains but may be glassy. Except for the glassy varieties, all igneous rocks are crystalline, and texture refers mostly to the fineness or coarseness of the mineral grains. If the particles range in size from grains of wheat to kernels of corn, the rock is said to be *coarsely crystalline* (granitoid, granular, or coarse-grained). Granite is a typical example of a coarsely crystalline rock (Figure 2–14). If the crystals are so small that only tiny light flashes from the mineral cleavage faces betray the crystal nature, or if the rock must be studied under the microscope to discern the crystals, then it is said to be *finely crystalline* (felsitic).

Some igneous rocks are composed of large crystals surrounded by a groundmass of small crystals. Such a texture is said to be porphyritic, and the rock is called a *porphyry* (Figure 2–15).

The size of the crystals is determined by the rate of cooling. If the magma cools very slowly, then the ions have the opportunity to migrate to small centers of crystallization (seed crystals) and cause them to grow large, as in a granite. If the magma contains considerable water the crystal growth is even better facilitated, and crystals 1 in. or larger develop. Under very favorable conditions of cooling and in the presence of much water single crystals of certain minerals, like feldspar and mica, grow to a length of several feet. Unusually coarse textures develop within the body of the magma and also in dikes fed by aqueous magma. The very coarse-textured rock is called *pegmatite* (Figure 2–16).

Variable mineral composition

Because of its chemical complexity, magma grows mineral grains of several kinds during the course of cooling. Study Table 2–2. Rock specimens will be shown in class, and you should get acquainted with their appearance and characteristics.

The silicates make up the bulk of crystalline igneous rocks. Two main groups should be noted, the feldspars and the iron-magnesian

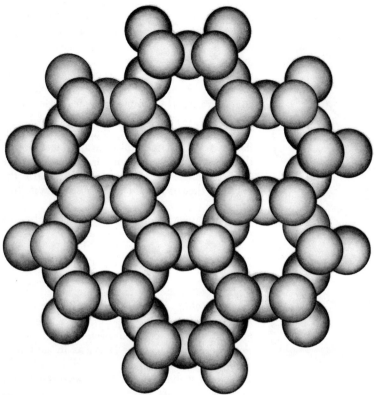

Figure 2–13 Layer of silica tetrahedra. The silicon ions at the center of each tetrahedron of oxygen ions are concealed and not shown. Within the layer three of the oxygen ions of each tetrahedron are shared by adjacent tetrahedra; only one, the upper one in this drawing, is unshared. The unshared oxygen ion therefore has one unsatisfied valence, or needs to share another electron with some adjacent ion in order to be stable. Thus, a bond is provided with another similar layer of silica tetrahedra. The basic structure of mica is a double layer or plate.

minerals. In a freshly broken surface of granite, for instance, the glasslike mineral grains are quartz, the fairly opaque milky white or pink grains are feldspar, and the dark minerals are the iron-magnesian minerals (also called ferromagnesian minerals). Mica grains may also be recognized, the colorless ones being muscovite and the black ones biotite. We should know also that feldspar has two main varieties: the potassium-bearing orthoclase (usually pinkish) and the sodium- and calcium-bearing plagioclase (usually white).

Classification

On the basis of variable textures and variable mineral composition, we can now present a simple classification of igneous rocks. See Table 2–4, where the specific rock names are printed in italic type. We see that granite, for instance, is coarsely crystalline, light-colored, and is composed predominantly of quartz and orthoclase. Dark iron-magnesian minerals are generally scattered throughout it. Gabbro is dark gray and consists of plagioclase and iron-magnesian minerals (Figure 2–14).

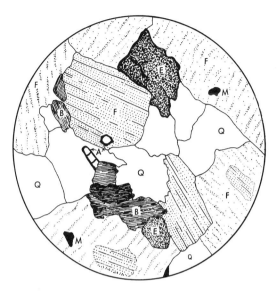

Figure 2–14 Interlocking mineral grains of feldspar (F), quartz (Q), biotite (B), epidote (E), magnetite (M), and apatite (A), such as would form a typical granite. Magnified about 20 times.

The finely crystalline equivalent of a granite is called rhyolite (felsite) and may be light gray to dark gray to reddish. Basalt is a finely crystalline black rock.

A magma of granitic composition, if cooled very rapidly and turned to glass, say on the margin of a lava flow, is called an obsidian (Figure 2–15). A glass of basaltic composition is referred to as a basalt glass. Obsidian is somewhat transparent on thin edges, while basaltic glass is fairly opaque.

At the bottom of Table 2–4 are listed the principal variations in chemical composition. The granite-rhyolite group is rich in silica, contains appreciable potassium, and is low in iron and magnesium. The gabbro basalt group is low in silica, has appreciable sodium and calcium, and is rich in iron and magnesium. All contain considerable aluminum. The diorite-andesite group is intermediate in composition.

The specific gravity of the granite-rhyolite group is about 2.7; that of the gabbro-basalt group is 3.0. A rock not listed but which should be mentioned is *peridotite*, which is made up

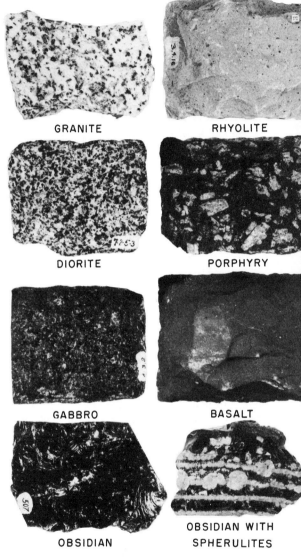

Figure 2–15 Representative igneous rocks.

almost entirely of the iron-magnesian minerals, olivine and augite, and has a density of 3.3. In major perspective, the earth has three outer layers, corresponding to the above densities. The top is a layer of sedimentary and granite-like rocks with a density of 2.3 to 2.7; under this is the gabbro-basalt layer with a density of

TABLE 2-4 *Classification of Igneous Rocks*

OCCURRENCE	TEXTURE	COLOR AND MINERAL COMPOSITION		
		LIGHT	INTERMEDIATE	DARK
		QUARTZ AND ORTHOCLASE	APPROXIMATELY EQUAL AMOUNTS OF LIGHT AND DARK	PLAGIOCLASE AND IRON-MAGNESIAN SILICATE MINERALS
Batholiths, stocks,* laccoliths*	Coarsely crystalline	*Granite*†	*Diorite*	*Gabbro* *Diabase*
Dikes, sills, surface flows	Finely crystalline	*Felsite*		*Basalt*
		Rhyolite	*Andesite*	
Surface flows, crusts of flows	Glassy, vesicular	*Obsidian* *pumice*	*Andesitic glass, pumice*	*Basaltic glass, scoria*
Beds and layers	Fragmental	*Fragmental pumice*	*Tuff* (fine) *Breccia* (coarse)	*Fragmental scoria*
	Chemical composition →	High silica and potassium; low iron and magnesium		Low silica, sodium, and calcium; High iron and magnesium

* Commonly porphyritic, with finely crystalline groundmass.
† The words in italic type are the rock names.

3.0; and under this is the peridotite layer with a density in its upper part of 3.3. The peridotite layer is very thick and is called the mantle. The layers above the mantle constitute the crust. More will be said about these layers in the chapter on earthquakes.

Some magmas contain much water. Under high pressures, such as exist several miles below the surface, the water is as much part of the magma as is silica, alumina, or iron oxide. When this water-rich magma rises in a conduit that feeds a volcano, the confining pressures decrease, and finally the water begins to be released as gas. It fills the magma with bubbles, and as the magma reaches the surface it may expand to a froth. If the magma cools and solidifies in this form, and if it is of felsitic composition, it will result in *pumice*. If it is of basaltic composition, *scoria* will form (Figure 2-17). The thin partitions between the myriad of bubble holes are mostly volcanic glass.

Any lava rock containing bubble holes is said to be *vesicular*. Pumice is an extremely vesicular lava rock, and scoria is generally moderately vesicular.

In some eruptions there is so much highly heated water that the froth is blown into tiny particles high in the air admist a tremendous emission of steam (Figure 2-18). The particles are generally not larger than dust and settle on the countryside in blankets several inches to several feet thick to form a deposit called *tuff* (Figure 2-19). The volcanic dust particles consist largely of glass shards and small pieces of pumice. Fragments of several different minerals are also identified in some tuffs.

Fragments of pumice or scoria the size of buckshot to marbles are called *lapilli* or *cinders* (Figure 2-20). Some cones are built up almost entirely of such fragments, and are called

Figure 2–16 *Pegmatite dike and veins in older layered crystalline rock. Note the very coarse texture of the pegmatite.*

cinder cones (Figure 2–21). The explosions are generally mild but frequent, with each small eruption producing considerable lapilli and perhaps large fragments and building a symmetrical but generally small cone.

Volcanic bombs are common around cinder cones. They are blobs of molten magma which were hurled from the crater into the air, and as they rotated became spindle-shaped, very vesicular in the center, and with crust or skin commonly checked like bread crust (Figure 2–22).

Volcanic breccia is a rock mass composed of large and small particles without sorting or layering (Figures 2–23 and 2–24). Mudflows or slides down the volcanic cone, or possibly an explosion of variable-sized particles, may result in a volcanic breccia. The particles may be so hot that they weld together.

Kinds of igneous rock bodies

We have been concerned with the kinds of igneous rocks in preceding paragraphs. Now,

Figure 2–17 *Scoria. This is a moderately vesicular basalt.*

Figure 2-18 *A new volcano bursts from the Atlantic like the mushroom cloud of an atomic blast. Dwarfing the 115-ft Capelinhos lighthouse and Capelinhos rocks in the foreground, the black sulfurous smoke spurts to height of 1 mi. This picture, taken in October 1957, shows the crater shortly before it disappeared. It was reborn a few days later.*

Figure 2-19 *Layer of ash 20 ft thick recently ejected on the island of Iwo Jima, Japan. The cliff is part of a new explosion crater. [Courtesy of Helen Foster.]*

the modes of occurrence, the shapes and sizes of the rock masses, and the relation of these bodies to adjacent rocks must be considered. There are two main categories of igneous rock bodies. The *intrusive* rocks are those that cooled and solidified below the surface, the magma having made its way upward from some deep-seated source into the upper part of the crust. *Extrusive* rocks are those whose magmas have issued out on the surface.

Intrusive bodies

A *batholith* is a very large mass of coarsely crystalline intrusive rock, generally granitic in composition. The bulk of the Sierra Nevada of California, a range of 70 miles across and 400 miles long, consists of granite or a near relative and is an example of a batholith. The depth to which the granite extends is uncertain, but probably several miles, so that thousands of cubic miles of rock are involved. Batholiths constititute the largest rock units of the crust. They are generally not formed of a single intrusion of magma but of several separate and successive intrusions. They occur in the belts of major crustal deformation or mountain building, generally along the continental margins.

Figure 2-25 is a diagrammatic illustration of a batholith. Where it cuts through or across the sedimentary strata it is said to form a *discordant* contact, at other places it has pushed aside or wedged between the strata and does not cut across them. Such a contact is said to be *concordant*. The sedimentary rocks have generally been deformed by folding and thrusting before the great intrusions made their way up into and through them.

The magma emplaces itself into the rock

Figure 2–20 Pumice fragments of nut size, Japan. [Courtesy of Helen Foster.]

Figure 2–22 Volcanic bombs lying on cinders of cinder cone, Craters of the Moon National Monument, Idaho.

Figure 2–21 Cinder cone of Paricutín Volcano, Mexico. The fall of volcanic fragments stripped trees of foliage. [Photograph by F. O. Jones, U.S. Geological Survey.]

Figure 2–23 Hill of volcanic breccia near Marysvale, Utah.

Figure 2–24 Close-up view of volcanic breccia, Japan.

above by thrusting it aside, by melting or assimilating it, or by a process of blocks dropping off the roof and sinking into the magma. It is difficult to see how some batholiths have emplaced themselves into the vast space that they occupy.

A *stock* is a discordant intrusive but much smaller than a batholith. Most stocks are 1 to 10 miles across as exposed at the surface (Figure 2–26). Most of the silver, gold, lead, zinc, and copper ore deposits of the West are clustered about stocks, and the stocks are regarded as having nurtured the ore deposits by feeding *mineralizing solutions* upward into the adjacent or *country rock*.

Stocks are commonly formed of granite-like rocks, and many are porphyritic, with a finely crystalline groundmass.

A *dike* is a sheetlike mass of intrusive rock, generally in a near-vertical position and cutting through other rocks. It forms when the crustal rocks fracture and pull apart to form an open *fissure*, with the fissure penetrating downward into a magma chamber. The magma under pressure then surges up the fissure, cools rapidly, and crystallizes into a fine-grained rock. Dikes are shown diagrammatically in Figure 2–27. Basalt dikes are shown in a road cut in Figure 2–28.

The magmas that surge up the fissures often wedge between the strata in thin sheets to form *sills* (refer to Figure 2–27 again, and also to Figure 2–29). When lava flows out on accumulating sediments and is later buried by more sediments, it may appear like a sill. A real sill will have "baked" or bleached the sediments both above and below, whereas a lava flow will only have affected the sediments below. A lava flow may also have a frothy or vesicular upper layer, owing to the escaping gas, but not a sill. Since a sill penetrates between the strata and is parallel with them, it is said to be a *concordant* intrusion. Intrusions in certain near-horizontal sedimentary sequences are thickened domed-shaped sills, like those illustrated in Figure 2–30. These are called *laccoliths* and are about 1 mile or so across. They are usually made up of felsite porphyry. Laccoliths were originally thought to have been fed by stubby dikes or circular conduits from below, and perhaps some are, but a new study in the classic locality of the laccolith, the Henry Mountains of Utah, shows that a central pipelike stock through the beds has fed thickened sills out between the sedimentary layers and that each lateral intrusion, therefore, is not a mushroom-shaped body but a partial dome. In either type of intrusion the beds above are domed upward. After a long episode of erosion the domed beds are removed, and the igneous rock of the laccolith is exposed.

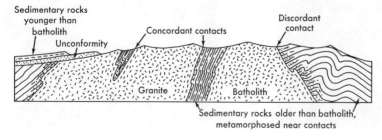

Figure 2–25 Various relations of sedimentary rocks to intrusive rock.

Figure 2-26 Granite (light-colored) stock intrusive into quartzite series (dark, stratified). The contact is covered by talus. At head of Bredefjord, eastern Greenland. [Courtesy of John Haller.]

Since laccoliths are generally porphyritic, it might be well to explain here how this texture may have arisen. The two distinct sizes of mineral grains in a porphyry result from two episodes of cooling, and it is thus believed that in a deep-seated magma chamber the large crystals grew while the magma slowly cooled. Then this mixture of crystals and melt was suddenly injected as a sill or laccolith between relatively cold sedimentary layers. The melt froze quickly, allowing only very little crystal growth, and thus a matrix of fine crystals surrounds scattered large crystals. Stocks, sills, dikes, and flows all may have porphyritic texture. This indicates that there was an early stage of cooling and partial crystallization and then a later stage of quick cooling and complete solidification.

Extrusive bodies

Magma is known to issue at the surface either through *fissure* eruptions or through *vent* eruptions. Cracks or fissures open up in the crust and provide a passage for the magma from some deep reservoir to the surface. Basaltic magma wells up the fissure and then pours out in the form of broad thin sheets on the land surface on one or both sides of the fissure. The Columbia River Plateau is North America's most notable example of fissure eruptions of basalt. There, many flows are piled on top of one another to build a layered accumulation 2000 to 3000 ft thick in places. The basalt flows spread over a large part of the states of Oregon and Washington, and it has been estimated that about 40,000 cubic miles of lava has been erupted.

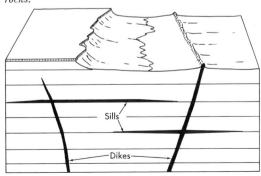

Figure 2-27 Relation of dikes and sills to sedimentary rocks.

Figure 2–28 Basalt dikes cutting metamorphic rocks 10 miles west of Revelstocke, British Columbia. [Courtesy of Ed Schiller and Geological Survey of Canada.]

Magmas of felsitic composition have issued through fissures in tremendous amounts in Nevada and western Utah. This is the Great Basin province. Sufficient water was present to expand and distend the magma into dust-sized particles (principally glass shards) as it issued from the fissures. The emissions were not explosive, as sometimes they are when much hot water vapor is present, but instead, the water vapor was released slowly and was just enough to hold the particles apart, so that the combined water vapor and volcanic dust moved as a heavy incandescent cloud and poured or flowed outward over the land surface from the fissure. This postulated phenomenon occurred when Mount Pelée on the Caribbean island of Martinique erupted in 1903, and a searing-hot dust cloud rolled down the mountainside, consuming all vegetation and animal life in its wake. It overran the village of St. Pierre, and all but one person perished—he, a miserable prisoner in an underground dungeon.

The fissure eruptions in the Great Basin were almost unbelievably large. One emission left a blanket of tuff over 100 ft thick in places across a plane 50 miles wide or more. Such an avalanche probably happened in the course of a few minutes. The front of the cloud may have advanced at the rate of 60 miles per hour or more. The water vapor separated out and was lost by the tuff particles, which often consolidated into a light gray, banded glassy rock called a *vitric*, or *welded*, *tuff*.

Structures of flows

Basalt flows commonly develop a columnar structure, as shown in Figures 2–31 to 2–33. The perfection of the hexagonal columns is amazing in this famous locality, the Giants Causeway in North Ireland. Many basalt flows develop polygonal columns, such as are seen in Figure 2–34, but not so regularly nor so perfectly in six-sided forms as at the Giants Causeway. After the lava has crystallized and solidified, it cools to ground temperature and, in so doing, shrinks. The resulting joints define the four-, five-, or six-sided columns. Why not seven- or eight-sided columns?

Basalt flows also erupt beneath marine waters on ocean floors at times, and there are many examples of ancient flows of such origin. When they do, however, large vesicular rolls or pillows develop, such as shown in Figure 2–35. In cross section the pillows look like those of Figure 2–36.

Figure 2-29 A dark (basaltic?) sill in stratified Precambrian rocks, Banks Islands, Arctic Archipelago. [Courtesy of R. L. Christie and Geological Survey of Canada.]

Figure 2-30 Laccoliths, one fed laterally from a stock and one fed by a conduit from below. Erosion has stripped the sedimentary rocks from the top of the laccolith on the left.

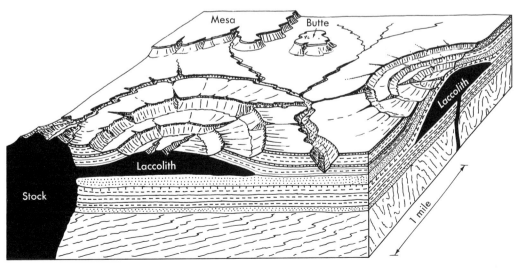

Figure 2–31 Hexagonally jointed basalt of Giants Causeway, Northern Ireland. The slope in the rear is a layer of ancient lateritic soil, 40 ft thick, that rests on the jointed basalt flow; on top of the soil, forming the upland cap, is another jointed basalt flow. [Courtesy of A. M. Gooding.]

Figure 2–32 Close-up of the hexagonal columns of the Giants Causeway. [Courtesy of A. M. Gooding.]

Figure 2–33 Segmented character of certain polygonal columns of basalt. These are the so-called Organ Pipes of the Giants Causeway. [Courtesy of A. M. Gooding.]

Origin of igneous rocks

Magma variations

Because igneous rocks vary in two ways, composition and texture, it is evident that a magma of one composition should lead to one suite of igneous rocks and one of another composition to another suite. And thus we seek to define the common types of magma that have invaded the earth's crust. A study of the most voluminous kinds of igneous rock should tell us something about the common types of magma. These are the extrusive, dark-colored basalts, the associated basalts, and andesites, the extrusive, light-colored felsites, and the batholithic granites and diorites. An example of the vast amounts of fairly uniform basalt, already mentioned, is the Columbia River Plateau lava field.

Figure 2–34 *Columnar jointing of basalt flow cut by Yukon River at Whitehorse, Yukon Territory.*

An example of basalt and associated andesite in great volume is the Coast Ranges of Oregon and Washington, where probably a greater volume of lava rock exists than in the Columbia River Plateau. An example of the light colored felsite is located in the Great Basin where, from central Utah to western Nevada, a layer of these rocks, estimated to have been 2000 to 3000 feet on the average, originally existed. Again this is a mass tens of thousands of cubic miles in volume. An example of granite and diorite rocks is the Sierra Nevada batholith containing 50,000 to 100,000 cubic miles of coarse grained intrusive rocks. As a temporary measure to explain the origin of these magma types before we study crustal structure and the interior of the earth (Chapters 3, 4, and 5) it should be said that the crust consists of two principal layers, the upper granitelike and the lower basaltlike. Beneath the crust is the very thick mantle which consists of a material like peridotite, mostly crystalline and solid but in places partly molten or potentially so. Now, the basaltic magmas are believed to come from the upper mantle or lower part of the crust from reservoirs possibly 15 to 75 km deep. The magmas from which the associated basalts and andesites have crystallized may have come from very great depths, up to several hundred kilometers, where convection currents in the upper mantle plunge downward under the continental margin. The granites and diorites of batholith volume probably originated by the melting of the lower part of the granitelike layer of the crust where it had locally been thickened and had bulged downward into the basaltic layer. This downward bulge or root had been brought into temperatures sufficient to melt the granitelike root. The light-colored rhyolites or felsites probably come from magmas created by the melting of the lower part of the granitelike crustal layer, but not because it developed a downbulge, but because basaltic magmas rose from the mantle to the base of the granitelike layer bringing sufficient heat to partially meet the granitelike layer. All these concepts need support, which will come in later chapters.

Magmatic differentiation

The above great groups of igneous rocks owe their differences directly to the different magmas from which the rocks came. The differences in the magmas were chemical. Such magmas have been called *primary*. It has been well demonstrated that several rock types can originate from one primary magma, even rock types with varying chemical composition as well as texture. The process by which this takes place is called *magmatic differentiation*. In other words, nature can create an andesite and a rhyolite from a basaltic magma.

The change from silicate melts to silicate minerals takes place throughout a long temperature range (from about 1400° C to 800° C), and during this temperature drop a succession of minerals crystallize out, each during a certain temperature range. The succession or order of minerals is shown in Figure 2–37. The first minerals to crystallize out, olivine and calcium-rich feldspar, have more iron, magnesium, and calcium than the still liquid part, and thus if olivine and calcium-rich feldspar are separated

Figure 2–35 Vesicular pillow lava, Newfoundland. This is an occurrence of igneous rock of Cambrian age. [Courtesy of W. D. McCartney and Ed Schiller.]

out, they form a rock more mafic (Fe and Mg) than the original liquid, leaving the remaining liquid enriched in silica, potassium, and sodium. The separation of calcium-rich feldspar and olivine takes place under certain conditions, either by crystal settling (olivine and Ca-rich feldspar are heavier than the liquid and settle out), or the chamber that contains the mush of crystals and liquid is squeezed by crustal deformation, and the liquid is pressed

Figure 2–36 The Purcell lava of Waterton Lakes National Park, Alberta, showing cross sections of pillows. [Courtesy of R. A. Price and Geological Survey of Canada.]

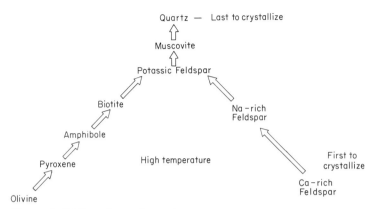

Figure 2–37 *Bowen's reaction series.*

out (a filter press arrangement). Unless the early-formed crystals are removed from the melt by some such process as crystal settling or filter pressing, the olivine reacts with the surrounding liquid as it cools to produce pyroxene, and the calcium feldspar reacts to produce a feldspar richer in sodium. If no crystals are removed from the melt, thus allowing these reactions to go on unimpeded, the whole mass solidifies to a mixture of feldspar and pyroxene with a small amount of olivine—the typical minerals of basalt or gabbro (see Figure 2–38). If the olivine or the calcium feldspar, or both are separated from the melt, perhaps by settling in large quantities to the floor of the magma chamber, reactions between crystals and liquid cannot take place. Olivine contains a higher proportion of magnesium and a lower proportion of silica than the melt as a whole; calcium-rich feldspar contains a higher proportion of calcium and a lower proportion of both sodium and silica than the melt. The liquid left behind will thus be relatively impoverished in magnesium and calcium and relatively enriched in silica, the alkalies, and iron—it is thus no longer of gabbro composition but nearer diorite. If no further crystals are separated, the liquid crystallizes to a rock of andesitic composition, but if the next crop of crystals (now pyroxene and calcium-sodium feldspar) are removed from the liquid as fast as they form, so

that they are not available to react with the liquid, the leftover magma continues to change in composition, ultimately becoming a siliceous, alkali-rich granite. Ten percent or less of the original basalt magma can be transformed to granite; the rest has already accumulated as peridotites, olivine-rich gabbros, diorites, and similar rocks on the floor of the magma chamber.

This is the famous *Bowen's reaction series,* developed by N. L. Bowen in the U. S. Geophysical Laboratory in Washington, D.C. from about 1920 to 1940. But it is now evident that the voluminous granite batholiths could not have derived from a gabbro magma, because if they had, somewhere a more basic rock of much greater volume than the granite would have been created, and this seems to be wanting. However, Bowen's reaction series does give us a much better understanding of the origin of igneous rocks, particularly the unusual silicic varieties, as well as of the physical and chemical properties of the deeper parts of the crust and underlying mantle.

Volcanoes

Volcanic eruptions have captivated, dismayed, and terrified mankind from primitive times to the present. The fiery cataclysms, when seen from a safe distance, are nature's most

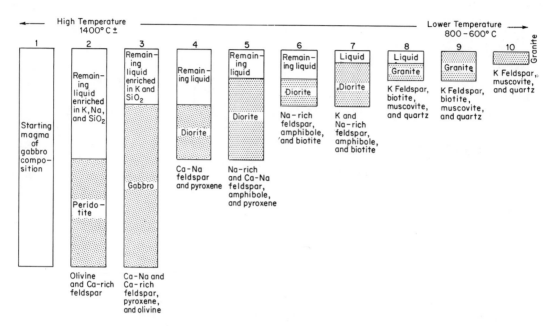

Figure 2–38 Evolution of granite and intermediate rock types from a gabbro (basalt) magma. 1—Original magma of gabbro composition in the lower crust or upper mantle. 2—Same magma but now about half crystallized, resulting in the rock, peridotite. 3—Same magma but now about 75 percent crystallized. 4—The liquid of sample 2 separated from the crystallized portion and then half crystallized. 6—Remaining liquid of sample 4 separated from the crystallized portion and then half crystallized. 7—Same as 5 but 75 percent crystallized. 8—Remaining liquid of sample 6 but separated from the crystallized portion and then about half crystallized. The remaining liquid is about 10 percent of the original magma of sample 1, but much enriched in SiO_2 and K. 9—Same as 8 but all crystallized. 10—The remaining liquid of 8 but now separated from the crystallized portion and then itself crystallized.

spectacular extravaganzas; at the same time, lava flows overrun fertile farmlands, and ash falls bury villages and cities and often extract a fearful toll of lives. Through the centuries man has stood helpless to prevent the eruptions, and as his only recourse has supplicated the deity for help or mercy. Volcanic mountains in many places have come to have a supernatural significance. For instance, the common name for the highest and most exquisite volcanic cone in Japan is Fujiyama, but in a religious sense it is referred to as Fujisan[1] (Figure 2–39). To

[1] In Japanese, *san* means sir, a sign of respect.

propitiate the angry spirit of El Misti, the Peruvian volcano, the Incas built a temple as close to the elemental forces as possible, right in the crater, and often offered up human sacrifices.

In Italy and especially in Japan, the eruptions of certain volcanoes have been chronicled for over 2000 years, and some of the records are very interesting indeed. Vesuvius has had 18 major eruptions since A.D. 79, the last in 1944. Seventy-one eruptions have been recorded at Aso-San since A.D. 864. Historical records in other volcanic regions such as the Aleutians take us back only 200 years, but in

Figure 2-39 *Fujisan, the sacred mountain, Japan.*

Figure 2-40 *Shield volcano. The upper sketch is from a painting in* Volcanoes declare war, *by Naggar, 1945, and is the crater of Mauna Loa, Hawaii. The lower profile is the complex of shield volcanoes of the Island of Hawaii shown in relation to the ocean floor.* [*After Stearns and MacDonald, Hawaii Division of Hydrography, 1946.*]

Figure 2–41 Mount Shishaldin, a composite cone, Unimak Island in the Aleutians, Alaska.

most places, the geologic record is replete with the evidence of many and varied eruptions of over 300 volcanoes around the world.

Volcanoes are commonly classed as active, dormant, or dead. According to the Japanese, whose studies are widely accepted, active volcanoes are arbitrarily defined as having erupted in the past fifty years, whereas dormant volcanoes are those which have erupted in historical times but not in the last 50 years. Presumably dead volcanoes are those which have been inactive historically and display badly gullied and dissected cones, with the form commonly so much modified that a crater no longer exists. The geological evidence points to the conclusion that volcanic activity has ceased for good.

Types of volcanic mountains

Eruptions of the vent type are those in which the magma rises through a more or less circular conduit and issues from a localized vent at the surface. A pile of extrusive rock, which we call the volcanic cone, is soon built up around the vent. The apex of the cone is usually marked by a depression. This is called the *crater*. Most craters where explosive eruptions have occurred are less than 2000 ft across. Those which have emitted chiefly basalt flows may have larger craters, however.

There are three types of volcanic mountains, namely, the *shield* type, the *composite* type, and *lava domes*. Smaller accumulations are *cinder cones* and *spatter cones*.

Shield-shaped domes

When the eruption is chiefly basaltic lava and the lava issues quietly from the central crater to pour down the slope in thin interlacing streams, the sides of the cone have low declivity, and the cone takes the shape of a broad, flattish dome. Numerous small fissure eruptions generally mark the gentle slopes and contribute to the building of a shield-shaped accumulation. The Island of Hawaii is a complex of several shield volcanoes, with Mauna Loa the central and

Figure 2–42 Mount Showa-Shinzan, Hokkaido, Japan. Lava dome formed between 1943 and 1945. [*From* Volcanoes of Japanese archipelago, *Geological Survey of Japan, 1960.*]

highest one. It rises nearly 14,000 ft above sea level and has a crater 2 miles wide and 1000 ft deep. A pit of molten lava in the floor of the crater occasionally wells up and overflows the crater rim, spilling a thin flow down the mountainside. Kilauea is a cone 20 miles away with a crater 9000 ft lower than Mauna Loa. It, too, has an active, and at times seething, lava pit in the floor of the large crater (refer to Figure 2–40).

Geologists are now certain that the Island of Hawaii is entirely a massive and broad pile of volcanic rock built up from the ocean floor, which is 15,000 ft below sea level. The complex of shield volcanoes, therefore, represents a bulky, sprawling pile, 28,000 ft high and several hundred miles across the base. In fact, the several islands of the Hawaiian group are all volcanic and build an immense volcanic "rise" from the central Pacific floor (cross section of Figure 2–40).

Composite cones

Many volcanoes have exhibited intermittent explosive action and passive emission of flows. The cones are thus built of layers of fragmental debris and flows and are called *composite*. If none of the eruptions has been destructively violent, then fairly symmetrical, steep-sided cones are built. Examples are the beautiful snow-capped Fujiyama in Japan (Figure 2–39), Mount Hood in Oregon, Cotopaxi in Ecuador, Kilimanjaro in Tanzania, and Mount Shishaldin in Alaska (Figure 2–41). Such structures are also referred to as *strato-volcanoes*. Composite cones usually consist of rocks of felsitic composition, because the lava they are composed of is stiff (viscous), whereas basalt lava is rather liquid and runs rapidly in thin streams. If water vapor is present, it is released easily from basalt lava.

Lava domes

The term *"lava dome"* is used to denote a mountain that has been built by a very stiff magma. Either the stiff magma domes up surficial layers of rock, or the very viscous magma breaks through and piles up in an irregular

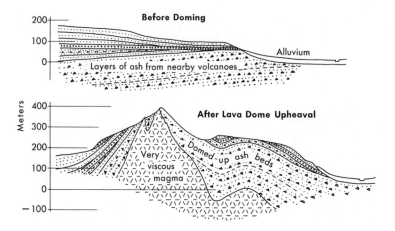

Figure 2–43 Lava dome of Mount Showa-Shinzan. [*From* Volcanoes of Japanese archipelago, *Geological Survey of Japan, 1960.*]

dome much like shaving cream would if ejected from below through a tube to the surface.

Mount Showa-Shinzan, in Japan, is a lava dome that rose in stages from a peaceful cornfield during the period from 1943 to 1945 (Figure 2–42). The area before eruption is shown in Figure 2–43, as well as the dome as it now stands. Lava, or rock at a temperature over 800° C, reached the surface at one or two small points. The rock at these places still records 400° C. Much steam or water vapor was released, as well as volcanic dust at times. Around the steam-releasing cracks or vents some native sulfur collects. Volcanologists are much interested in a number of other elements in small amounts that the water vapor contains. Large river-rounded boulders that made up the rock layers immediately under the cornfield may be seen on the slopes of the dome and even on the very top.

Cinder cones

The habit or conduct of volcanism in some areas is to build numerous small cinder cones (see Figure 2–44). A column of lava seems to bore through to the surface and then cinders are coughed up such that after a few days or weeks a small but delicately fashioned cone is built. This is generally followed by a basalt lava flow, and when this ceases, the activity is over, the volcano is dead, and a new outlet breaks through in another place, perhaps nearby. The Snake River lava plains of southern Idaho contain about 500 such cones. It must be pointed

Figure 2–44 View of a cinder cone in the Snake River volcanic plains, Idaho.

out that some cones here are shieldlike and are composed mostly of lava.

Cinder cones are common and numerous as small satellitic structures on the flanks of large strato-volcanoes. Sunset Crater is a well-known cinder cone in the San Francisco Peaks volcanic field of northern Arizona. It is considerably higher, however, than the small cinder cones of the Snake River field.

Spatter cones

Instead of spitting out cinders, some small vents in volcanic fields throw out blobs of lava and hard chunks of rock. These build rough and irregular piles usually called *spatter cones* (Figure 2–45). There are all manner of variations in cinder and spatter cones, both large and small.

Types of vent eruptions

Besides a classification of cones, such as has just been given, the manner of eruption of different volcanoes has led to another classification.

Hawaiian type

Abundant outpouring of basaltic lava in a passive manner is the rule. Gases are liberated more or less quietly, although lava fountains sometimes erupt in the craters to heights of 1000 ft. The passive eruptions of basaltic lava build cones of the shield type.

Strombolian type

The volcano Stromboli, in the Aeolian Islands near Sicily, erupts in mildly explosive form every half hour or so. Incandescent lava is hurled into the air with each explosion, and a white vapor cloud hangs almost continuously over the crater. The cloud, particularly at night, reflects light from the glowing incandescent lava, and may be seen for many miles around from ships at sea. The volcano has been called the Lighthouse of the Mediterranean.

Vulcanian type

The Island of Vulcano, also in the Aeolian group, is representative of volcanoes that erupt violently. Often the cone is split from top to bottom and lava pours out along the entire fissure; parts of the cone may be blown away. The circumstances necessary to produce such violent explosions are, first, a viscous felsitic magma, and second, the freezing or solidification of a thick crust across the crater floor. Thereupon, the escaping water vapor is trapped below the crust. The orifice is thought to become plugged, and eventually a violent explosion occurs that cleans out the obstruction, and commonly at the same time part of the cone is blasted away.

Mount Vesuvius near Naples had been inactive so long that it was not known to be a volcano when in A.D. 79 an eruption of extreme violence blew at least half of the cone away. Since then a new cone has been built (see Figure 2–46).

Another example of a violent, cone-rending eruption in historic time (1912) is that of Mount Katmai in Alaska. There had been no sign of activity in historic time, and the volcano was considered possibly dead or, at best, dormant. Its cone was deeply gullied by erosion and it looked old. After the tremendous explosion the cone stood shattered, irregular, and containing a large basin. Pumice fragments and glass shards fell on Kodiak Island 60 miles away to a depth of 10 ft or more. The plant cover was buried for miles around, and broad sterile slopes and plains resulted. It was not many years, however, before the plants reestablished themselves, and the previously indigenous animals moved in again.

Peléan type

The Peléan type is like the Vulcanian, but with the additional characteristic that glowing clouds called *nuée ardentes* are emitted. Such eruptions have already been described in connection with Mount Pelée on the Island of Martinique in the Caribbees.

The Strombolian, Vulcanian, and Peléan types of eruption all build cones of the composite variety.

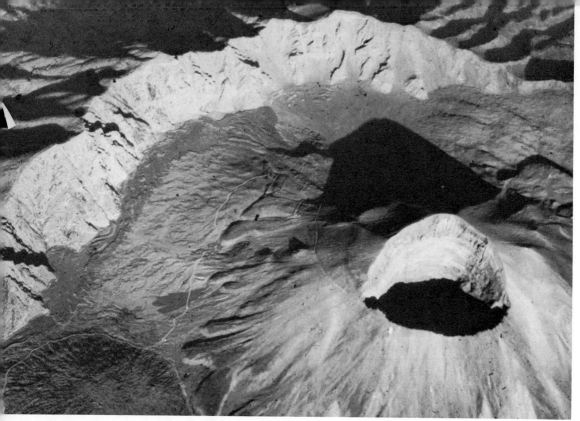

Figure 2–46 *Inner cone of Mount Vesuvius, Italy. Remnant of a very large old cone is called Mount Somma.* [Photograph by Cecil B. Jacobson, U.S. Bureau of Reclamation.]

Cause of eruptions

Molten rock is lighter (has a smaller density) than solid rock, and hence when solid rock melts its volume increases. If melting occurs underground in an enclosed space, the liquid will be under pressure since there is now a greater volume trying to occupy the same space. Thus if there are passageways upward the magma will rise. In the quiet or passive eruptions, perhaps this is all that is necessary to explain the phenomenon, but even in such eruptions there is always a release of much gas in the form of water vapor. The fountains in the Kilauea lava pit and the scoriaceous lavas that pour over the rim attest to the presence of much gas as part of the rising magmas.

The explosive eruptions are entirely due to the release of water vapor. The violence of the eruption is due to several factors, namely, the amount of gas, the degree of plugging of the orifice, and the viscosity of the magma containing the gas. Many variations in the nature of the eruptions and the fragmental ejecta result from an interplay of these factors.

Figure 2–45 *Spatter cone in Craters of the Moon National Monument, Idaho.*

Lava flows from flanks of cones

In both shield and composite volcanic cones a common occurrence is the breaking out of a lava flow on the side or near the base. The drawing of the Three Sisters (Figure 2–47) illustrates such a flow. Even the near-perfect symmetry of Mount Fuji (Figure 2–39) and Mount Shishaldin (Figure 2–41) is marred by the outbursts of lava from the flanks. The general conduct of Japanese volcanoes in a single eruptive cycle is an explosion of ash, or pumice fragments, and then a lava flow. The crater may have been blown out on one side by the explosion, and the flow may have issued from the gap, but more commonly the flow breaks out somewhere down the slope.

Some characteristics of lava flows

On the Island of Hawaii the flow surfaces are classed as *pahoehoe*, or ropy, and *aa*, or blocky. There are many variations of each. Figure 2–48 gives some idea of the interesting types of pahoehoe, and Figure 2–49 shows an aa or blocky flow.

Blisterlike swellings develop in the crust of basalt flows in places. Figure 2–50 shows one in which the crust has cracked apart over the blister. In fact, it looks as if the crust of the blister were about to submerge and be engulfed in lava.

The sides and top of a basalt flow may freeze solid while the central part still flows. In fact the interior may run on out and leave a tube or tunnel that can be walked through for thousands of feet.

Ash falls

Mapping an ash fall

When a volcano ejects fragmented material high into the air, the small and medium-sized particles, especially if they are lightweight pumice, may be carried a considerable distance and deposited over the countryside. Note the dust and steam cloud of Mount Asama (Figure 2–51). After an eruption the limits of the ash fall can be charted on maps, and also its thickness from place to place can be shown by thickness contours. The direction of the wind at the time of eruption determines the position of the ash fall in relation to the vent. Not only may witnessed ash falls be mapped, but older ash falls, perhaps laid down before the region was inhabited, may be distinguished and charted, and much may be learned in this way about the history of the volcano. Figure 2–52 is a remarkable map of several ash falls in the Sapporo-Tomakomai lowland of Hokkaido, Japan. The amount of material ejected differed from time to time, and also, the wind direction was not always the same. The pumice particles ranged in size from $\frac{1}{2}$ to 4 in. in diameter; yet they were carried as far as 10 miles. Some fragmental pumice layers apparently poured down the volcanic cone and spread out on the skirting areas as flows.

Productive soils

Soils that form on volcanic rocks generally support a lush vegetation. This is because of

Figure 2–47 The Three Sisters, in the Cascades of Oregon. A fairly recent outburst of lava partly down the closest cone is evident, as is a new cone with crater. The middle and distant cones are badly eroded and presumably dead.

Figure 2-49 Blocky flow, Craters of the Moon National Monument, Idaho.

Figure 2-48 Two varieties of pahoehoe lava in Craters of the Moon National Monument, Idaho. [Courtesy of Walter Sadlick.]

the mineral nutrients, particularly potassium, in the igneous rocks. When the volcanic rocks are loose ashes, soils are quick to form and easily tillable. Therefore the hills and plains that have been covered by old ash falls are particularly good agricultural areas.

Necks and radiate dikes

Long after the eruptive activity has ceased and the erosional processes carried much of the cone away, the solid resistant *neck* and *radiate dikes* stand etched out in relief. Shiprock of northwestern New Mexico is a notable example. In south Africa some old cones have been removed by erosion entirely, but the solidified neck rock is mined below the surface for its diamonds. These circular, downward-extending structures are called *pipes*.

Cauldron subsidence

After majestic composite cones are built, their fate frequently is partial destruction by collapse. Crater Lake, Oregon, is an example of a volcanic cone that had been built to a great height, like Mount Rainier or Mount Hood, when finally it erupted violently with the emission of incredible amounts of water vapor, ash, and lava and then collapsed. By collapse is meant that the top part of the cone simply subsided into the depths of the crust to form a pit several miles across and perhaps 3000 ft deep (see Figure 2-53). The great pit is called a *caldera*.

The loss of water vapor, ash, and lava from the magma reservoir below is thought to have removed support of the overlying cone, with resultant collapse. In the case of Crater Lake, the caldera was partially filled with lava and pyroclastics (fragmental volcanic rocks) in lingering eruptive activity afterward. A small cinder cone was built inside the caldera, which became Wizard Island when the lake waters collected.

Mount Vesuvius is a notable example of a remnant of a once great volcano and the growth of a new one. Its history is depicted in Figure 2-46. Mount Somma is the highest peak on the remaining rim of the old volcano, but it is the remains of a cone blasted away by a terrific eruption rather than the rim of a collapse caldera. Now, however, as collapse calderas all over the world are recognized, the surviving rim of the early cone is called the *somma*, meaning skin.

Aso-san in Japan, the largest collapse caldera yet found, is a circular area between the somma walls about 35 miles in diameter. Various small cones have been built in the great caldera, where 71 eruptions have been recorded since A.D. 864.

Eruptions of submarine volcanoes

Eruptions from a submarine volcano are rare spectacles, and only very few have been witnessed by man. Some of these eruptions are in every respect the equivalent of an H-bomb explosion. Figure 2–54 shows the eruption of Myojin-sho on September 23, 1952. A ship of Japanese scientists, dispatched to study the phenomenon, evidently approached too close and vanished.

For accounts of other great submarine eruptions see suggested readings at end of this chapter.

Prediction of eruptions

The lower flanks and surrounding plains of volcanic cones are generally areas of considerable population because of the fertility of the soil. In the volcanic archipelagos an entire island is commonly that part of a volcanic pile built up from the ocean floor that projects above sea level. Its inhabitants are crowded into limited shoreline areas with little chance to escape in the event of a major eruption. Needless to say such people are most anxious to know, if possible, when the volcano around which they live is going to erupt.

Fortunately, geologists have been able to do a good job of predicting eruptions. Their predictions may only be a few weeks or days in advance, but this gives sufficient time for evacuation. Long-range (several years) predictions, however, are still impossible or insecure in their reliability.

As magma charged with steam rises in the vent that feeds the volcano and as the confining pressures become less, rumblings from the pent-up gas occur and become more frequent as the eruption stage approaches. Major rumblings may be felt as earthquakes, but these generally

Figure 2–50 Squeeze-up on the floor of Kilauea Caldera. [Courtesy of Gordon McDonald and Helen Foster.]

come too late, because the eruptive activity follows soon or immediately in the wake of strong earthquakes. But by establishing a network of seismological listening stations around a vol-

Figure 2–51 Mount Asama in eruption, 1954. The ash fall of 1783 eruption ruined the harvest of all Kwanto Plain, covering 24,000 sq. km. [Photograph courtesy of M. Minakami, Earthquake Research Institute, University of Tokyo.]

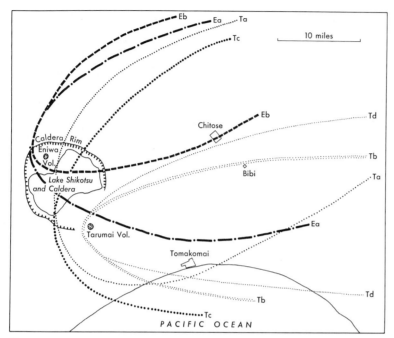

Figure 2–52 Pumice falls from Tarumai Volcano and Eniwa Volcano near Chitose, Hokkaido, Japan. The Shikotsu pumice which underlies all the other beds is not shown. [From Uragami, Doi, and Katsui, Supplement to Guidebook, 1962, International Symposium on Volcanology, Japan.]

canic cone and by watching the buildup of seismic activity, the attendant geologist or seismologist can learn to predict eruptions. He gets to know the habits or conduct of the monster he has been assigned to observe. The listening stations are equipped with geophones, and the earth vibrations that they detect are wired to a central observatory where small seismographs record the information day and night. The progress of rise of the restless magma in the depths of the volcanic cone can actually be followed in some instances.

Preceding an eruption the surface of the cone is sometimes bulged and deformed, but in a manner too slight and gentle to be detected by the unaided eye. For the purpose of detecting deformation of the volcanic cone, instruments known as tiltmeters have been perfected, and they have brought to light the fact that tilting is much more frequent and significant than previously thought.

On the basis of data collected by seismographs and tiltmeters around Kilauea, the internal structure of the volcano and its manner of eruption can be constructed. Recurrent swarms of earthquakes originating about 60 km (40 miles) beneath the crater are taken to mark the forming of magma at this place and to indicate also that the magma is moving upward through perennial conduits to collect in a shallow magma reservoir only 2 or 3 miles below the summit. The growth or inflation of this shallow reservoir causes the summit area to swell, creating sporadic sequences of small earthquakes in the rocks surrounding the reservoir. Swarms of tiny shallow earthquakes mark the progress of penetration of magma into fissures around the expanding reservoir. When

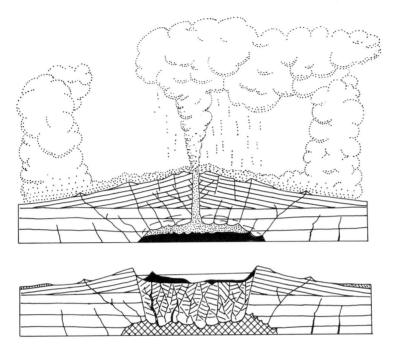

Figure 2–53 Climactic eruptions, above, and cauldron subsidence, below, representing the manner of formation of Crater Lake, Oregon. [After Howell Williams.]

a magma-filled fissure finally splits to the surface, a parasitic eruption occurs.

A major fracture or rift zone exists across the shield volcano of Kilauea, and sometimes the magma leaks through fissures in this rift zone to erupt at the surface. Rapid escape through the rift zone of magma from the expanded reservoir leads to the sudden subsidence of the summit of the volcano, with attendant earthquake activity (thousands of small to moderate local earthquakes) and deepening of the lava level in the great crater.

In this chapter we have studied the constitution and origin of igneous rocks, and the nature of their occurrence. Now, in the environment of the earth's envelope of air and water the igneous rocks weather, and the products of weathering are sediments. Sediments and their hardened products, the sedimentary rocks, blanket a good part of the earth, and are the subject of the next chapter.

Figure 2–54 The late explosive eruption of Myojin-sho, a submarine volcano south of Tokyo, on September 23, 1952, rose about 10,000 ft. The photograph was taken from a distance of 5 nautical mi. [Courtesy of the Tokyo University of Fisheries and Helen Foster.]

References

Bragg, Sir Lawrence. X-ray crystallography. *Scientific American*, July 1968, pp. 58–70.
Bullard, Fred M. *Volcanoes in history, in theory, and in eruption.* Austin: University of Texas Press, 1962.
Holden, Alan, and Singer, Phylis. *Crystals and crystal growing.* Garden City, N.Y.: Doubleday, Anchor, 1960.
Pearl, R. M. *How to know the minerals and rocks.* New York: McGraw-Hill, 1955.
Pough, F. H. *Field guide to rocks and minerals.* Boston: Houghton Mifflin, 1953.
Rittman, A. *Volcanoes and their activity.* New York: Wiley, 1962.
Williams, Howell. *The ancient volcanoes of Oregon.* Eugene, Oreg.: Oregon State System of Higher Education, 1948.
Williams, Howell. Volcanoes. *Scientific American*, Nov. 1951. Reprints may be purchased from W. H. Freeman and Co., San Francisco.

Motion Pictures

Crystals. Educational Services, Inc., black and white, 16 mm, 25 min.
Eruption of Kilauea: 1959 and 1960. U.S. Department of Agriculture, color, 16 mm, 33 min, $8.50.
Volcanoes in action. Encyclopaedia Britannica, black and white, 16 mm, 11 min, $2.25.

3
sedimentary rocks and their structures

Sands of an ocean shoreline, muds of a river delta, debris scattered by cloud-burst floods on the western deserts, or light gray limey ooze of the shallow Bahama Island seas are all *sediments* to the geologist. Although of different types they are loose, easily deformed aggregates. But all are subject to hardening or consolidation over geologic time, and as a result they become *sedimentary rocks*. Certain unconsolidated sedimentary deposits have been traced for scores of miles and have been denoted by formational names, and these, although soft materials, are generally regarded as sedimentary rocks by geologists because of their large size. So when a sediment becomes a rock is probably a subjective affair. A hand specimen is one matter and the size of the deposit is another.

Layers of sedimentary rock blanket three-fourths of the surface of the continents. A multitude of land forms has been sculptured from them by the erosional processes, and the soils that support our agricultural economy developed on them.

Evidences of past life are found profusely preserved in sedimentary rocks, and a record of evolving plants and animals that extends over the last half-billion years has been established. From the fossil record the elaborate geologic time scale has been worked out. We search for oil and gas in sedimentary rocks, because the oily and gaseous hydrocarbons are found only in them. Sedimentary rocks are the stores of common salt, potassium and magnesium salts, rock phosphate, limestone and cement-making materials, and the ceramic clays, all immensely valuable to modern economy. And by the study of accumulating sediments geologists have learned to understand the ancient sedimentary rocks and to analyze the geography of the region landward and oceanward of the sedimentary rock deposit and to recognize the climate of the time.

Sedimentary rocks are classified into three main groups: *clastic, chemical,* and *organic.* Clastic sedimentary rocks are composed of discrete mineral grains or broken-down particles of previously existing rocks. Chemical sedimentary rocks are composed of various precipitates from solution such as common salt (NaCl), gypsum ($CaSO_4 \cdot 2H_2O$), and limestone ($CaCO_3$). Organic sedimentary rocks are such deposits as coal and calcareous reefs. Calcareous reefs as a rock type are limestones, and thus some limestones are of inorganic and some of organic origin.

Clastic sedimentary rocks

Classification by particle size

Clastic sediments are divided into groups, depending on the size of the particles, as listed in Table 3–1.

TABLE 3–1 *Classification of Clastic Rocks*

PARTICLE	SIZE	SEDIMENTARY ROCK
Boulder	Larger than 256 mm	Boulder conglomerate
Cobble	64–256 mm	Cobble conglomerate
Pebble	4–64 mm	Pebble conglomerate
Granule	2–4 mm	Granule conglomerate
Sand	1/16–2 mm	Sandstone
Silt	1/256–1/16 mm	Siltstone
Clay	Less than 1/256 mm	Shale

Accumulations of boulders, cobbles, and pebbles are characterized as coarse. When cemented together to form a rock, they are called conglomerates. Sand grains become cemented to form sandstone, whereas silt and clay particles compact into a close arrangement to form siltstone and shale, respectively.

Origin of coarse clastics

The coarse clastics are commonly formed during torrential floods where swollen, fast-flowing rivers spread the large particles on a gently sloping surface at the foot of a precipitous terrain. Sometimes waves leave a layer or lens of boulders, cobbles, and pebbles along a beach during a storm. In a desert, the wind may carry the finer particles away, leaving a residue of cobbles and pebbles. Such accumulations of coarse particles are then often buried by other layers of sediments and become cemented together to form conglomerates (Figures 3–1 and 3–2). Cementation is discussed below.

The coarse rock fragments have come from preexisting rock of one kind or another that has been broken up by weathering processes and made available in angular fragments to the streams or waves. As the rivers transport the angular rock particles downstream or the waves pound them back and forth on the beaches, the fragments become rounded to various degrees. The boulders, cobbles, and pebbles that we see in conglomerates are commonly fairly well rounded.

Sandstone

Size of particles

Sand grains as specified in Table 3–1 range in size from 1/16 to 2 mm. They are large enough to be distinguishable by the unaided eye, but to see them clearly and distinguish their physical characteristics, a low-powered hand lens, say with a 10-power magnification, is needed. They settle through water readily and are rolled along by moderate winds but are picked up only by strong winds.

Minerals

Most sand grains are rounded or subrounded particles of the mineral quartz. Quartz is generally colorless and clear, and the crystals grow when possible as hexagonal prisms and pyramids. Quartz is rather hard and will scratch glass. It is resistant to solution in most natural waters, and to mechanical abrasion, and hence outlasts other common and abundant minerals in weathering and transportation processes. For this reason immense volumes of quartz grains accumulate to the exclusion of almost all others, and many sandstone formations are composed almost entirely of quartz grains.

Degree of roundness

The quartz grains that make up sandstone derive from rocks like granite, which have been broken up by weathering. Granite is composed of an interlocking mixture of crystal grains of several minerals. Quartz commonly fills irregular spaces between the other minerals, and therefore the quartz grains seldom start out as nicely formed hexagonal crystals but rather as irregular angular, sharp edged little grains. In the weathering of granite the other minerals decompose, freeing the resistant quartz grains to the rivers, waves, and wind. Transportation of quartz grains by these agents causes wear and rounding of the grains, and they are thus found in nature in various stages of rounding (see Figures 3–3 and 3–4).

Cement and color

Some sandstones are purple-red, some brick red, some rusty, some tan, some gray, and some nearly white on freshly broken surfaces. Colors are mostly imparted by the cement. The spaces between grains render sand porous, and water is able to move into the pores and through the rock. Ground water carries certain compounds in solution, such as calcium carbonate, iron oxide, and silica, and as the water passes through the sand, one or more of these materials may be precipitated as a film on the sand grains, and if the spaces between grains are filled, the precipitate becomes a cement that bonds the individual grains together and gives color to the rock (see Figure 3–5). Water picks up considerable CO_2 in percolating through a soil and thus becomes weakly acidic. Then, if this acidic water passes through a limestone formation, it dissolves considerable amounts of $CaCO_3$. Later, if it passes through a porous sandstone where the environment is slightly basic, the $CaCO_3$ is precipitated. If the cement is pure calcium carbonate, which precipitates out as the mineral calcite, then the color of the sand grains will not be changed, and the rock may be light gray. A drop of weak hydrochloric acid on the rock will cause effervescence, which indicates that the cement is calcium carbonate.

If iron oxide is the coating and cementing material, then various shades from tan to rust to deep red will result. The vivid red colors of the Colorado Plateau are due in part to the red hematite cement of the several sandstone formations that crop out in the walls of the deep gorges.

The cement of sandstone may be silica, which is held in alkaline or basic solutions and then precipitated out when the basic condition is neutralized by mixing with weak acid solutions.

Figure 3–1 Thick conglomerate exposed on Weber River, north-central Utah.

Figure 3–2 Close view of conglomerate exposed on Hoback River, western Wyoming.

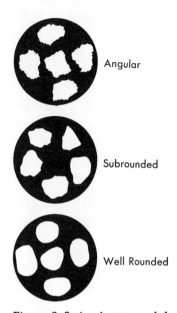

Figure 3-3 *Angular to rounded sand grains.*

Figure 3-4 *Rounded and angular sand grains. Enlarged about 55 times.*

Figure 3-5 *Sand grains cemented together.*

Sandstones whose cement is silica are very hard generally and break with glasslike surfaces. Such sandstones are called *quartzites* ("orthoquartzite," in specific geologic terminology).

Some sandstones whose cement is calcium carbonate or iron oxide are only weakly cemented, and one may rub off some of the sand grains with the thumb. Such sandstones are said to be *friable*. A friable sandstone is not good for construction purposes.

Occurrence

Sandstone occurs in layers, which are also called *beds* or *strata* (see Figures 3-6 and 3-7). The layers may be from a fraction of an inch to many feet thick. In some places, they extend for many miles as sheets of fairly uniform thickness, in others they are lens-shaped, and in still others they represent the sand fill of old river channels and hence are stubby lenses in cross section but winding, narrow bodies in ground plan.

When sand is deposited by rivers, waves

along beaches, or wind, it commonly shows a layering effect called *cross-* or *false bedding.* By this is meant that a sheet of sand is composed of small layers or *laminations* lying obliquely across the main bed or sheet (see Figure 3–7). Cross-bedding forms where a shallow current capable of transporting sand and silt particles enters deeper water, as shown in Figure 3–8. Its velocity is suddenly checked, and the grains are dropped down the slope. The continued growth of this slope by further accretion produces a layer with cross-bedded structure.

The surfaces that separate beds are called *bedding planes,* and these surfaces are commonly marked by *ripples* (Figure 3–9), tracks of various shelled invertebrates and crustaceans that inhabited the sea or lake floor, and less commonly by tracks of birds, reptiles, and mammals.

How sand layers were deposited

As already indicated, sand grains are concentrated and cleaned of silt and clay particles principally by running water, by waves and currents playing along shore lines, and by wind. Rivers, especially in flood time, build bars of various kinds in their channels, levees along their banks, and sheets across their flood plains. Some transport the sand to their mouths where they build large deposits called *deltas.*

In thinking of wave activity along shore lines, we must consider the situation of a sea advancing over the land or a sea withdrawing from the land. These are slow processes but they are happening today and have occurred many times in many places in the past. In the case of an advancing sea, either the land is subsiding or the water level is rising; in the case of a retreating sea, either the land is rising or the water level is falling. If the water level and land are relatively stationary, a belt of sand will form along a fairly stabilized shore. If the sea advances, the belt of sand is extended as a sheet across all the land invaded by the sea. If the sea retreats, the shore line retreats also, and a sheet of sand may be spread across the ex-

Figure 3–6 Beds of various thickness, here severely bent. Near Biarritz, France.

posed bottom. Sheets of sand of the most uniform thickness and widest distribution were formed by advancing or retreating seas. If the position of the shore line holds constant for a while, then various kinds of cross-bedded bars, spits, and cuspate beaches will form, and the sand accumulations will be far from regular. The various types of beach deposits will be discussed in Chap. 8.

Sand dunes built by the wind are made up of large, wedge-shaped units, each with its own cross-bedding (see Figures 3–7 and 3–10). These are usually so distinctive that the deposit can be identified as of wind origin, but the cross bedding of river and shoreline accumulations is usually not distinctive enough to identify the medium of deposition without other characteristics or clues.

Figure 3–7 *Massive cross-bedded sandstone beds of the Colorado Plateau near Glen Canyon Dam. This vertical cliff is about 1000 ft high.*

Siltstone

Size of particles

The size of silt particles is approximately that of dust particles. They are gritty to the touch or between the teeth and are barely visible as individual grains. They show up well under the microscope, however, and may be easily studied by this means. They settle slowly in water and are picked up by the wind and

Figure 3–8 *Origin of cross-bedding and example of cross-bedded sandstone, Grand Canyon of the Colorado River.*

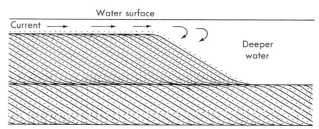

carried hundreds of miles before falling to the surface again. A good rain cleans the air of silt particles.

Unlike sand grains, which commonly consist of quartz, silt grains are made up of several minerals.

Minerals

These are quartz, mica, feldspar, calcite, hornblende, and augite. A number of other minerals occur in minor amounts.

Pure siltstone is rarely found. It is usually mixed with sand or clay, and geologists working in the field usually call a rock that is mostly made up of silt-sized particles a shale.

Occurrence

Silt layers accumulate seaward of the sandy beach zone, and in places they are deposited by heavy mud-laden bottom currents that flow many miles down-slope into deep water. As these turbid bottom currents spread out and slow down, the largest particles that they carry settle out first, the next smaller later, and the smallest last. This produces a layer, composed at the bottom perhaps, of fine sand grading upward into a layer of silt, and then into clay. This is called *graded bedding* (see Figure 3–11).

Shale

Size of particles

An accumulation of clay-sized particles hardens into a rock called shale. Clay-sized grains are so small that they feel smooth between the fingers and, when mixed with water, remain suspended for hours before reaching the bottom in the process of settling. The smooth pastelike layer that forms on the bottom contains much water, and it is principally a matter of losing the water that turns the soft plastic clay into a shale.

Beds of clay or shale are rarely formed entirely of the very small clay-sized particles; silt particles are present in various proportions mixed with those of clay size. Also, more or less

Figure 3–9 Various ripple marks and possible raindrop impressions on bedding surfaces of a sandstone from the Moenkopi Formation, Colorado Plateau. This is a masonry wall.

calcium carbonate may be present. At any rate, when the fine sediment has dehydrated sufficiently to have hardened somewhat, it generally has the fissility to part in thin layers. This parting is generally considered characteristic of shale. If the sediment becomes very hard and does not split readily into thin plates parallel to the bedding surface, then it is called *argillite*.

Minerals

Clay particles are so small that their mineral composition is difficult to determine. Analyses are mostly made with X rays, and we recognize chiefly quartz, sericite, chlorite, hydroxides of iron, and the clay minerals. By clay minerals is meant a group of complex hydrous aluminum silicates. One of these, for example, is *kaolinite*, from which the white porcelains are made. The clay deposits composed mostly of kaolinite are

Figure 3–10 *Cross-bedded Navajo Sandstone in Zion National Park, Utah. [Courtesy National Park Service and Allen Hagood.]*

valuable for the making of many ceramic products, and clay deposits of this kind are much sought after.

Occurrence

Clays are deposited in the central areas of certain lakes at sufficient distance away from shore so as not to be contaminated with coarser particles. They form in arms or embayments of the sea and in waters adjacent to low-lying land. In the central part of fresh-water lakes and in protected basins of the sea, black or gray clay often accumulates. This contains up to 20 percent organic matter and, when dehydrated, forms the so-called oil shales. The organic matter in oil shale is really a solid hydrocarbon, but when heated several hundred degrees it turns into a black, pitchy liquid. The black shales are also the source rocks of much of our natural oil and gas.

Shales commonly occur in thin alternating light and dark layers called *laminations*. These laminations are caused by alternations of layers of coarse and fine particles, by variations in the amount of organic material, or by alternations of calcium carbonate with clay and silt particles. Almost all laminae are produced by seasonal climatic changes during the course of deposition. The temperature and salinity of the water may be affected by the cyclical climatic changes, the silt and clay content brought to the basins of deposition may change with the seasons, and the amount of life and consequent productivity of organic material may also vary.

Color

Shales are of many colors, and in the desert climates their colors may be vivid. The Chinle and Morrison shale formations, both widely displayed in the Colorado Plateau, are particularly noted for their intense greens, yellows, purples, reds, chocolates, and grays. The tones become most intense when the shale is moist after a rain. Some shale formations are made up of beds, each with its own color; such formations are said to be variegated. The colors are due mostly to iron in various degrees of hydration and oxidation.

Vividly colored shales usually contain appreciable volcanic tuff (volcanic dust) that has changed into the clay we call *bentonite*. Benton-

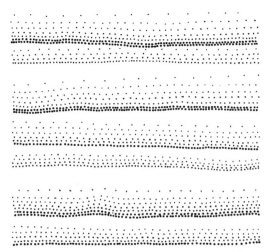

Figure 3–11 Graded bedding. Each layer is made up of coarsest particles at the bottom and finest particles at the top.

ite is highly absorbent of water and swells conspicuously when wetted. The exposed surface becomes slippery and sticky, and ranch roads over these clays when wet are generally impassable by car. Upon drying, the clay surface is left puffy and powdery. The bentonite clay slopes are usually rather barren, even of desert vegetation.

Chemical sedimentary rocks

Limestone and dolomite

Composition

Limestone and dolomite are the carbonate sedimentary rocks. Pure limestone is made up of crystals of calcite ($CaCO_3$) and pure dolomite of the mineral dolomite $CaMg(CO_3)_2$. If the crystals are microscopic in size, the rock appears dense, but if they are visible to the eye, then the rock is said to be crystalline. Commonly limestones contain some dolomite and dolomites some calcite. There are all gradations between the two. Also limestones and dolomites commonly grade into shales on the one hand and sandstones on the other, and we therefore speak of shaley limestones and sandy limestones.

Origin

Most calcium and magnesium carbonate is probably extracted from ocean and lake waters by organisms. The shells of clams and snails and the skeletons of corals are composed of it. Extensive carbonate reefs are forming today, such as the Great Barrier Reef off eastern Australia, and similar reefs were built in many seas and at many times in the past.

Bacteria are another cause of the precipitation of carbonates, leading to the formation of a lime mud or ooze, which collects on the bottom of oceans or lakes. Calcium carbonate is held in solution chiefly in the form of $Ca(HCO_3)_2$, and when CO_2 is removed from the solution $CaCO_3$ is precipitated. Certain bacteria require CO_2 for the life processes. Evaporation of the water may also concentrate the carbonates and cause their precipitation, again in the form of lime ooze.

Some limestone beds are composed of small spherical calcareous concretions the size of sand grains, called *oolites* (see Figure 3–12). In thin section under the microscope they appear as shown in Figure 3–13. Here we can see the small amount of cement necessary to hold the oolites together. The oolites form under beach conditions, where the waves play strongly and where calcium carbonate is being precipitated copiously (Figure 3–14).

Occurrence

We find limestone and dolomite beds or groups of beds inches to hundreds of feet thick widely distributed across North America. They are numerous and widespread in the Appalachian Mountains, in the Ohio, Mississippi, and Missouri Valley regions, and in the great western Cordillera. They have formed in most past geologic ages. With their abundant marine fossils, they indicate that shallow seas spread here and there widely across the continent at numerous times in the past. In the Appalachian Mountains and in many ranges of the western

mountain systems, the beds have been much deformed, elevated, and dissected by erosion. We can thus find marine fossils in the broken and weathered edges and bedding surfaces of the strata on mountain slopes at high elevations.

Salt and gypsum

Interstratified in shale and limestone beds are layers of salt ($NaCl$) and gypsum ($CaSO_4 \cdot 2H_2O$). Geologists have determined that where these deposits occur arms of the invading seas over the continents have been cut off, or nearly so, and these isolated bodies of salty ocean water have evaporated until the sodium chloride and calcium sulfate precipitated out (Figure 3–15) to collect in layers on the bottom. In the process of evaporation of sea water the carbonates precipitate out first, then the calcium sulfate, then sodium chloride, and finally the potassium and magnesium salts. An evaporation cycle would be represented by a layer of limestone on the bottom, over this a layer of gypsum, then of salt, and finally on top a thin layer of potassium and magnesium salt. This sequence may be repeated several times, or an

Figure 3–12 Oolites from Great Salt Lake, Utah, magnified about 40 times.

Figure 3–13 Thin section of oolites from Great Salt Lake.

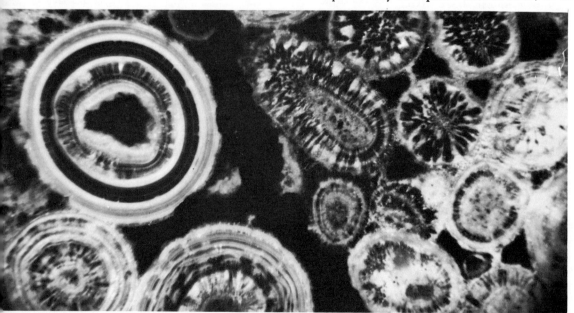

evaporation cycle may become interrupted and only proceed part way.

Interior basins and arms of the sea prone to salt precipitation have occurred in several places in the United States in the past. Salt beds hundreds of feet thick occur beneath the surface in western New York and Michigan. The same is true under a wide region in Kansas, Oklahoma, Texas, and southeastern New Mexico. There are salt layers several thousand feet thick under parts of the Colorado Plateau of Utah, Arizona, Colorado, and New Mexico.

So-called fresh-water lakes have salt in them, if only in terms of a few parts per million, but if one of these lakes falls below its outlet and proceeds to diminish, eventually to dry up, its salts will be deposited as a thin crust, a few inches to a few feet in thickness, on the lowest bottom area. Such a salt bed is shown in Figure 3–16.

Sedimentary rocks of organic origin

Limestone reefs

The Great Barrier Reef off the east coast of Australia is an example of an extensive deposit of calcium carbonate that has been extracted from sea water chiefly by the organisms that live on the reef. Algae and bacteria, among the plants, are abundant; among the animals are corals, bryozoa, sponges, clams, snails, foraminifera, and ostracods.

The algae, through the process of photosynthesis, extract CO_2 from the water and cause the precipitation of crusts of $CaCO_3$, called algal heads (Figure 3–17). The skeletal forms of the corals and bryozoa contribute to the $CaCO_3$ of the reef, and the shells of other forms add still more $CaCO_3$. The waves break up some of these deposits and build them into clastic accumulations, but it is probably only a short time before more precipitating $CaCO_3$ cements the fragments together into a firm rock.

Needless to say, a reef is generally a very porous mass, and long afterward, when buried

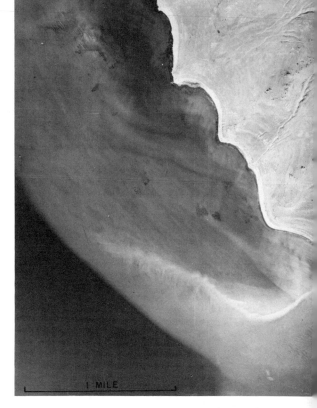

Figure 3–14 Vertical aerial photograph of a sandy shore on the west side of Antelope Island, Great Salt Lake. Sand is made up of oolites.

by other sediments, it is a good container of oil. Much oil and gas have been found in old buried reefs (see Figure 3–18).

Oil shale

Shale in certain places contains so much organic matter that it will burn hesitatingly with a yellow, smoky flame. In an airtight retort, it may be heated, and oil and gas driven off. The so-called Green River shale formation of Wyoming, Colorado, and Utah contains beds from which over 40 gal of oil per ton of rock can be distilled. This deposit formed in a large inland lake of ancient time (Eocene epoch) where algae contributed a great deal to the organic content of the shale.

Coal

Of all sedimentary rocks, coal is obviously of organic origin. Only the ash content can be said to be of inorganic origin. The ash is the clay and sand washed into the peat swamp during

Figure 3–15 Salt precipitating out of saturated water in Great Salt Lake, 1934.

the accumulation of peat from the plants growing there. Upon burial the peat changes to lignite and then to bituminous coal. The change is brought about chiefly by loss of water (see Figure 3–19).

Figure 3–17 Algal heads of Great Salt Lake. The lake level has recently fallen, exposing the deposits.

Figure 3–16 Bonneville salt crust, Great Salt Lake Desert, Utah. Note the high shore lines on the distant mountains, made by a former fresh-water lake.

Oil and gas

Oil and gas occur below the surface in the pores of rocks and are believed to come from organic matter that accumulated along with the bottom clays and lime oozes. The transformation of the organic matter to oil and gas may be caused by the action of bacteria living in the bottom clays and muds. The oil and gas are forced out of the fine-grained sediments into the pores of adjacent porous and generally coarser-grained rocks. The cause of further movement and collection into major accumulations, and the search for such, is the subject of petroleum geology.

Chert

Chert is composed of silica, very finely crystalline, or in part amorphous or hydrous. The term "flint" with which we are all familiar is a synonym or variety. Chert occurs in all colors, the red and yellow varieties, which contain iron oxide, being called *jasper*. In some places, extensive beds of chert form parts of the sedimentary sequence. In other places, we note that nodules of chert one to several inches across are abundant in limestone and dolomite.

It has been presumed that a good deal of the chert, especially the massive deposits, is of inorganic chemical origin; that is, it was precipitated from the ocean waters as a silica gel and later lost its water to form the very hard chert. But now it is realized that the ocean waters are far from saturated in silica, and probably no natural inorganic process would

Figure 3–18 Reefs in limestone strata exposed in walls of deep canyons of Canadian Rockies, Flathead Range, Alberta. [Courtesy of Geological Survey of Canada, through Ed Schiller.]

cause the supposed precipitation. Thus, we are left with the process of *organic* extraction of the silica, such as results in diatom and radiolarian shells.

Interlayering of various sedimentary rocks

Sedimentary rocks generally occur in alternating layers, building sequences many thousands of feet thick in places. Figure 3–20 shows a series of alternating limestone and shale layers through which the San Juan River has cut a gorge. The limestone layers form the vertical cliffs and the shale layers the slopes and notches, and thus the conclusion may be drawn that in an arid climate limestone is more resistant to weatherings than shale.

Some sedimentary sequences are composed dominantly of limestone and shale beds, some of sandstone and shale (Figure 3–21), and some of limestone and sandstone. Sediments have accumulated in some basins to thicknesses of thousands of feet, and since the sediments show signs of shallow-water origin, we can only postulate that the floor of the sea sank as the

Figure 3–19 Bituminous coal bed, about 25 ft thick, near Kemmerer, Wyoming. Shale and sandstone beds overlie the coal bed.

Figure 3–20 *Limestone and shale beds exposed on the steep walls of the San Juan River, Utah.*

sediments accumulated, thus maintaining shallow-water conditions for a long time.

Sedimentary rocks interlayered with volcanic rocks

Some of the thickest known sequences of stratified rocks are those which contain numerous lava flows and beds of pyroclastic material. The sandstones in these sequences are contaminated with volcanic fragments and the shales with ash.

The very thick volcanic-sedimentary successions of strata have accumulated in zones along the Atlantic and Pacific margins of the continent in times past. A zone possibly 200 miles wide along the Atlantic from Alabama to Newfoundland consists of such rocks, now deformed and more or less altered by heat and pressure. A zone over 300 miles wide in California and western Nevada contains vast thicknesses of interlayered volcanic and sedimentary rocks (see Figure 3–22). They are large and distinct geologic provinces.

Structures of sedimentary rocks

Sedimentary layers may be cross-bedded, marked by ripples, dried and cracked, and may show the tracks of animals. Such layers are called *primary structures*, with the meaning that they were formed while the sediments were being deposited, or shortly afterward while the sediments were still soft. The sediments may have settled on a gently inclined surface and flowed a little downhill, or a dinosaur may have wandered over a fresh deposit and left its footprints. Another class of sedimentary rock structures is called *secondary*, meaning that the structures were formed after the sediments had hardened. Examples of these are joints, faults, and folds.

Common primary structures

Cross-bedding

There are several kinds of cross-bedding, which are generally classified according to the agent that formed them. Waves and currents along a shore will drift sand and build bars, spits, and other beach deposits, all of which are made up of cross-bedded layers or lenses. Figure 3–8 is a cross-bedded shoreline deposit of an ancient sea. Rivers fill their channels, and in the process build a type of cross-bedding illustrated in Figures 3–23, 3–24, and 3–25. Still another type of cross-bedding is formed by the wind (Figure 3–10).

Ripple marks

Newly deposited sediments become rippled in many environments. We see ripples in the sand in stream beds, along lake and ocean shores, and on desert dunes. Underwater cameras reveal ripple marks in sediments to great depths. Most such flutings are due to currents, and when thus formed are asymmetrical (see Figure 3–26). Although the ripples pictured in these photographs have formed in unconsolidated sands during the last storm, replicas can be observed in old sandstones (see Figure 3–9). Current-caused ripple marks occur in many

Figure 3–21 Slopes are eroded from shale, siltstone, and the thin-bedded sandstone; cliff-making beds are standstone. Zion National Park, Utah. [Courtesy of National Park Service and Allen Hagood.]

variations, two of which are shown in Figure 3–26.

It has commonly been said that there are symmetrical ripple marks as opposed to asymmetrical ripple marks. Figure 3–27 is an example. Further, it is thought that the symmetrical ripples are due to the water particles rocking back and forth without appreciable forward motion, and thus sweeping the sand or silt particles into symmetrical ridges. Such motion is said to be oscillatory, thus symmetrical ripples have also been called oscillatory ripples. True symmetrical ripple marks are hard to find and thus probably rare, according to Dane Picard, of the University of Utah, who made a special study of ripple marks in general.

Other markings

There are many other kinds of markings on bedding surfaces, and perhaps you will see some on field trips. Watch for mud cracks (Figure 3–28), various rill marks and flutings caused by breaker swash, and tracks (Figure 3–29), trails or borings of the numerous invertebrates that lived in and on the bottom sediments.

Figure 3–22 Idealized cross section of the western half of the continent of North America in ancient times (Paleozoic era), showing the place and nature of the thick accumulation of interlayered volcanic and sedimentary rocks. The basement is ancient foundation rock of igneous and metamorphic nature.

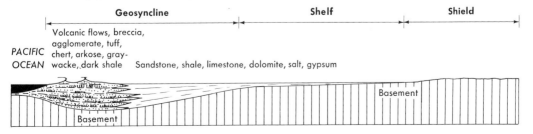

Figure 3–24 *Artificial cuts that reveal the nature of channel-fill cross-bedding.* [Courtesy of W. L. Stokes.]

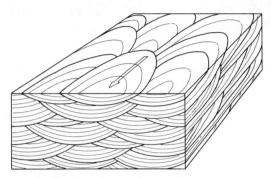

Figure 3–23 *Type of cross-bedding produced by rivers that fill up their channels repeatedly. The arrow indicates direction of flow.* [After Stokes, 1953.]

Deformation of soft sediments

When sediments, especially silt and clay layers, are deposited on sloping bottoms, the force of gravity is often sufficient to cause the weak or almost soupy sediment layer to slump or flow. Its laminations then become folded in complicated patterns. Then more sediments are deposited on top of the deformed mass. These may be sufficiently strong to resist slumping or sliding down the slope, but the next layer may

Figure 3–25 *Channel-fill cross-bedding after the sand has been cemented and eroded to a desert surface.* [Courtesy of W. L. Stokes.]

Figure 3–26 *Two variations of asymmetrical ripple marks. Photographs were taken after a recent storm along the shore of Great Salt Lake. Material is oolitic sand. Currents moved from left to right.*

Figure 3-27 Symmetrical ripple marks in the oolitic sand of Great Salt Lake.

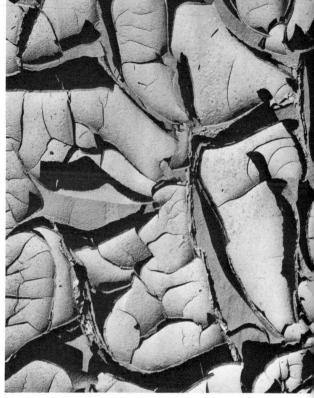

Figure 3-28 Mud cracks on Virgin River floodplain of Zion National Park. Polygons are about 4 in. across. [Courtesy of National Park Service and Allen Hagood.]

yield again. This is the story revealed in Figure 3-30. Of course, long after the total accumulation has lithified or become cemented, we may wonder how the hard, brittle rocks could have been so intricately folded, but we realize that the folding occurred before lithification. This is called a primary structure.

Figure 3-31 is another case of primary deformation in which we can see that the layers yielded more like stiff clay than soup, and some of them broke in blocks or cakes.

Joints—both primary and secondary

Any exposure of hard rock will exhibit cracks. These are called joints. They may be as much as 20 ft apart but generally they are

Figure 3-29 Bipedal dinosaur tracks in shale layer of Kayenta Formation, Zion National Park, Utah. [Courtesy of National Park Service and Allen Hagood.]

Figure 3–30 Contorted bedding due to flow soon after deposition. Soft glacial clays near Gunnar Mine, Beaverlodge area, Saskatchewan. [Courtesy of C. K. Bell and Geological Survey of Canada.]

about 1 to 5 ft apart (see Figure 3–32). In quarrying building and monumental stone, joints always have to be contended with, and it is impossible in some quarries to obtain blocks more than 2 or 3 ft on a side because of the numerous joints. Large monumental blocks are found only in certain places.

Some joints develop soon after the sediments are deposited. Drying and compaction of the fine-grained sediments are partly responsible, but warping and tilting of the unconsolidated sediments may also play a role. Long after the sands have been cemented to sandstones, the lime muds have hardened to limestones, and the clays compacted to shale, the beds may be deformed and new sets of joints superposed on the old. Thus we may see parallel or semi-parallel sets of joints in two or three directions (examine Figure 3–33).

Joints provide the weathering agents access to the rocks, and commonly decay and disintegration along the joints result in fissures and the intricate dissection of the rock surface.

The massive sandstone formations of the Colorado Plateau are jointed to form impres-

Figure 3–31 Primary deformation of plastic sediments, now hard rock in Canadian Rockies. [Courtesy of Canadian Geological Survey and E. Mountjoy.]

Figure 3–32 Weathering along joints in granite, Wasatch Mountains, Utah.

Figure 3–34 Southeast edge of Deertrap Mountain is outlined by joints in cross-bedded sandstone. [Courtesy of Zion National Park Service and Allen Hagood.]

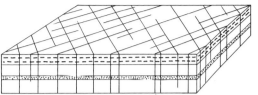
Two Sets of Joints Cutting Horizontal Beds

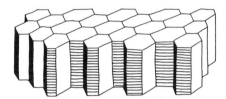
Columnar Joints in Basalt Flow

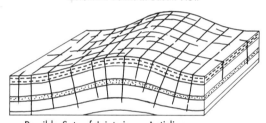
Possible Sets of Joints in an Anticline

Figure 3–33 Various relations and settings for intersecting sets of joints. An anticline is an upfold of the strata.

sive columns, such as are seen in Figure 3–34. When a person is 2 miles above the surface and looking straight down, the pattern formed by erosion along joints is sometimes striking. How many directions of jointing can you discern in the aerial photograph, Figure 3–35?

Joints in volcanic rocks have already been described (see also Figure 3–36).

Common secondary structures

As soon as early geologists learned to identify sedimentary rocks, they also recognized that the layers had been tilted to various degrees, and from this they perceived that the beds had been cast into folds and broken by faults of various kinds and magnitudes. The Appalachian Mountains of Pennsylvania and the Virginias display an imposing series of

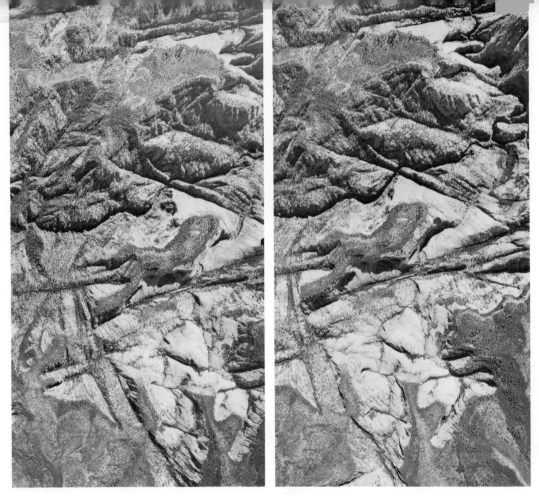

Figure 3–35 *Stereoscopic photographs of an area of the Virgin River drainage, Zion National Park, showing the striking way that erosion has etched out the joints in the massive Navajo sandstone.*

folds, and this was the birthplace of North American geology. Attention was soon given to the geometry of folds and faults, and to their outcrop patterns. These structures, as already defined, are secondary, having been formed generally after hardening or cementation.

Anticlines and synclines

Progressive development

Up-arched strata are called *anticlines*, and down-bent strata are called *synclines*. Anticlines and synclines occur in all sizes from almost microscopic to great folds scores of miles across.

Folds may be gentle and symmetrical, or sharp and asymmetrical. Depending on the intensity of the deforming forces the folds may be gentle, sharp, overturned, or overthrust (see Figures 3–37 and 3–38). In these examples the deforming force acted in a horizontal direction.

Anticlines as described so far are long structures, but others may be circular or dome-shaped. Here the deforming force acted vertically and upward. The oval or circular uplifts are generally referred to as domes (Figures 3–39 and 3–40).

As the strata are folded, the erosional processes begin to cut or dissect the uplifted parts. Soon, just the stumps of the upturned beds remain at the surface, and various outcrop patterns betray various kinds of folds. For in-

Figure 3–36 Multiple flows of basalt and columns that developed in the individual flows. North Creek, Zion National Park. [Courtesy of National Park Service and Allen Hagood.]

Figure 3–37 Progressive development of folds. The deforming force is probably a component of gravity.

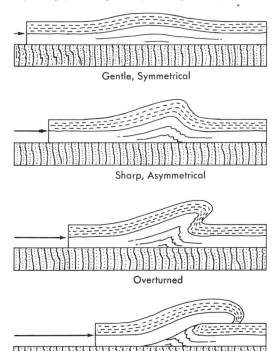

stance, circular outcrops indicate a dome (Figure 3–41), oval outcrops indicate a dome longer than wide (Figure 3–42), and parallel outcrops a long anticline (Figure 3–39). If crests of anticlines and troughs of adjacent synclines are not level, but plunge, then a zigzag outcrop pattern results (see Figures 3–40 and 3–43).

Competent and incompetent strata

The folding of strata is accomplished by sliding along bedding planes, by fracturing with minor adjustments along the breaks, and by plastic deformation. In a series of sedimentary beds some strata deform like brittle substances and break, and some deform plastically by flowing. Both examples are shown in Figure 3–44. If a series of beds is folded and each layer preserves its original thicknesses, the folding is called *competent* (see Figure 3–45a). If the beds thin at the limbs of the folds and thicken at the crests and in the troughs, the folding is said to be *incompetent* (Figure 3–45b). Variations between competent and incompetent folding are common (Figure 3–45c). Thickening and thinning of beds without fracturing and amidst small intricate curlings is plainly a manifestation of plastic deformation. Fractur-

81

Figure 3–38 Overturned fold in the Rockies of the Yukon Territory. [Courtesy of Geological Survey of Canada and L. H. Green.]

ing of competent beds incident to folding may look like Figure 3–46a or b. The incompetent plastic deformation of the shale bed in Figure 3–46b to form slate produces rock cleavage, which is a form of incipient fracturing. It is disposed parallel to the axis of the fold, as shown in Figure 3–46b. Small-scale fracturing of a sandstone bed takes an outward symmetrical pattern. The axial plane is the direction of lengthening of the fold; the other direction, at right-angles, is one of shortening or compression.

Plastic deformation, sometimes called rock flowage, occurs readily in gypsum, salt, and coal, and is commonly conspicuous in shale and limestone. When the deforming forces act on shale or limestone under the confining pressure of several thousands of feet of overlying strata, slow plastic deformation takes place, as in the incompetent type of folding. Also, when temperatures are somewhat elevated and water is present, even rocks like granite or rhyolite seem to become plastic and deform as if they were viscous liquids. Such conditions obtain near the contacts of batholiths and fairly deep in the siliceous crust. Note the photographs of gneiss in Figures 3–64 through 3–67. The fact that such rocks are now exposed and can be observed is the result of uplift and much erosion.

Décollement folding

The cross section of Figure 3–47 shows a rather intensely folded upper series of beds resting on basement rock that is little affected. The basement rock has been deformed, but at a much earlier time than the folding of the upper series. The basement rock, long after its deformation, was eroded and bevelled to form an extensive erosion surface. On this surface the upper series of sediments then accumulated. Finally, a force, presumably the downhill component of gravity, caused the upper series to slide downhill, gliding on the old erosion surface, and in the process it wrinkled or folded. This is called *décollement folding*. The surface on which the gliding occurred is called the décollement surface or plane.

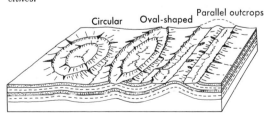

Figure 3–39 Outcrop patterns of various kinds of anticlines.

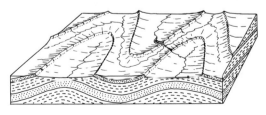

Figure 3–40 Outcrop pattern made by plunging anticlines and synclines.

Figure 3–41 *Erosional pattern of nearly circular dome. Here the strata have been domed up by a cylindrical plug of gypsum that rose from below. The gypsum plug penetrated the beds as it domed them up and extruded at the surface. The central rough area is gypsum, and the circular belts are upturned and eroded sedimentary rocks. Ellef Ringnes Island, Canadian Arctic. [Courtesy of Geological Survey of Canada and W. W. Heywood.]*

Figure 3–43 *Anticline plunging to north, southwestern Montana.*

Figure 3–42 *Outcrop pattern around an elliptical uplift, Canyon Range, Yukon Territory. [Courtesy of Geological Survey of Canada and Ed Schiller.]*

Diapiric folding and salt domes

A thick bed of gypsum or salt flows with the least provocation (directional pressure). These materials have a lower specific gravity than shale, sandstone, or limestone, and hence, when buried, are unstable. Salt buried by shale and sandstone is like a layer of oil covered with a layer of water. If you could arrange such a situation, the oil would funnel up through the water, spread out on top, and thus reverse the layering. Salt with a specific gravity of 2.14 will dome up and actually pierce the overlying beds of sand and shale, which have specific gravities of 2.4–2.6, and a variety of piercement-type flowage structures may form (see Figures 3–41 and 3–48).

Shale, especially shale with thin interbeds of gypsum or salt, is unstable under certain conditions, and various domes of faulted piercement-type structures result. These are called *diapirs*.

Thrusts and thrusting

In the progressive development of folds, the stresses and strains on the folding rocks finally become so great that fractures develop. Further movement of the sheets of rocks results in an *overthrust* fault, as shown in Figure 3–37. Faults are fractures in the rocks along which the blocks on either side have moved relative to each other. The southern Appalachian Mountains and the Rocky Mountains of Alberta, British Columbia, Montana, Wyoming, Colorado, and Utah exhibit impressive examples of sheets of rocks that have been thrust along fractures or shear surfaces. The evidence is usually clear that older rocks have been thrust up and over younger ones. Thrust masses of fairly flat *sole* have been recognized and

Figure 3–44 Highly deformed coal in the core of an anticline. The coal has deformed as if plastic, the overlying beds of sandstone and limestone as if brittle substances. Canadian Rockies of British Columbia. Coal, like gypsum, deforms easily. [Courtesy of Geological Survey of Canada.]

mapped. A great thrust sheet of note is one along the east side of Glacier National Park; it is called the Lewis overthrust or simply the thrust. It may be traced from the east mountain front up Many Glacier Valley at least five miles before it dives downward to the west. The structure is represented schematically in Figure 3–49. The scenic Chief Mountain is an erosional remnant or *klippe* of the thrust sheet. Similar remnants of other thrust sheets are shown in Figures 3–50 and 3–51.

The décollement shown in Figure 3–47 is also a flat or gently inclined thrust surface; it is an example of structures that have given rise to a difference of opinion about the cause of thrusting. Were these thrust sheets pushed ahead by a horizontal force from the rear, or did they arise by downhill gliding as detached masses? Many thrust sheets are now thought to be large-scale landslides, and geologists are aligned on both sides of the controversy. The question is a basic one, as we shall see when we discuss the origin of mountains.

Normal faults

Normal faults are those in which the relative movement or displacement of the blocks is up and down, and the down-dropped block is on the side toward which the fault plane dips (see Figures 3–52 and 3–53). On a large scale there are the rift valleys of Africa and the ranges and basins of the Great Basin of Utah and Nevada. In these places the earth's crust seems to have been stretched with the development of the normal faults (see Figures 3–52 again, and 3–54 and 3–55). Part of the Rhine Valley through the Black Forest is a rift valley, more generally called a *graben*, meaning trench. A valley formed by faults is called a graben.

Faults with horizontal displacement

By horizontal displacement is meant that the blocks on either side of a fault shifted horizontally in respect to each other, as shown in Figure 3–56. A number of great faults are of this kind. The San Andreas Fault of California is an outstanding example. It starts in the Gulf of California and extends up the Coast Ranges to Cape Mendocino north of San Francisco, a distance of 800 miles or more. Some geologists postulate that the block or slice of crust on the ocean side has moved 350 miles to the northwest. An aerial photograph of a small part of the San Andreas fault is shown in Figure 3–57.

Figure 3-45 Types of folding: A, competent; B, incompetent; and C, competent except for thinning on the overturned limb. [After Badgley, 1959.]

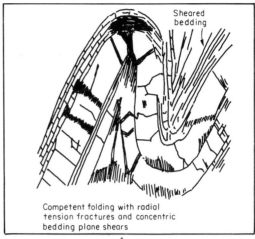

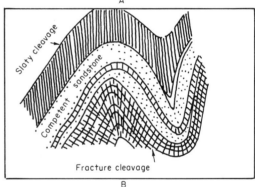

Figure 3-46 Types of shearing and fracturing to accommodate folding of sedimentary rocks. [Adapted with modifications from Badgley, 1959.]

Such faults are straight over long distances and the fault plane is nearly vertical.

Fault scarps

When a displacement along a fracture breaks through at the surface, a fault scarp is produced. Movements are generally sudden and range from 1 in. to 50 ft. The jolting and destructive earthquake at Hebgen Lake near the west entrance of Yellowstone National Park was the result of a vertical movement of 10 to 25 ft along the Hebgen Lake Fault. The scarp, which formed instantly, is 16 ft high on the

Figure 3–47 Cross section showing folds of the Jura Mountains, Germany. The strata above the basement rocks have been strongly folded without the basement's being affected. This type of folding is called décollement.

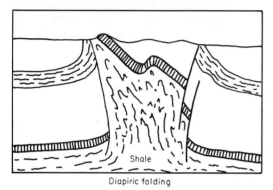

Figure 3–48 Flow of rocks with excessive plasticity. [In part after Badgley, 1959.]

Madison River (Figure 3–58). Many of the imposing ranges of Nevada, eastern California, and western Utah owe much of their height to displacement along normal faults, but it is certain that the total displacement of several thousand feet, which we can measure, is a summation of many small displacements. Like the Hebgen Lake faulting, each movement was only a few feet. The individual small movements occur at intervals of hundreds and even thousands of years, so the growth in height of a range is a slow process.

It must be evident that if two great masses of rock rub along each other, much crushing, shattering, grinding, and even polishing may occur. The badly broken rock is called *fault breccia,* and the striated and polished surfaces are called *slickensides* (see Figure 3–59).

Unconformities

When a sequence of stratified rocks is folded or faulted and much erosion occurs, the beds are commonly trimmed off so that we see the beveled edges of inclined layers in the outcrop. If more sedimentary layers are deposited across these eroded stumps of beds, then the structure produced is called an unconformity. There are many kinds of unconformities, but all are in-

Figure 3–49 Thrust sheet after considerable erosion.

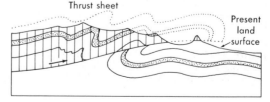

Figure 3–50 *Old strata (Devonian and Mississippian ages) are thrust over young strata (late Cretaceous age). See Figure 6–1 for geologic time scale.* [Courtesy of Geological Survey of Canada.]

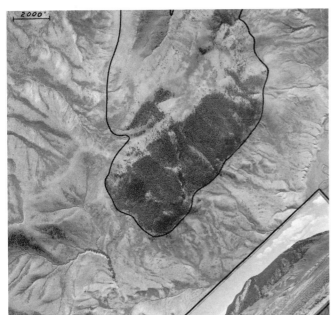

Figure 3–51 *Klippe in west-central Colorado. Precambrian rock has been thrust over Cretaceous strata.*

cluded under a simple definition: Unconformities are buried erosion surfaces. Any kind of rock may be undergoing erosion, and if the rock is buried by new sediments, or even volcanic rocks, an unconformity is created.

An unconformity is illustrated in Figure 3–60, where the beds below the old erosion sur-

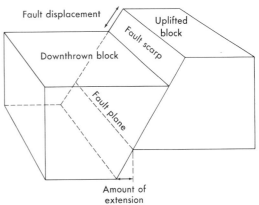

Figure 3–52 *Parts of a normal fault.*

Figure 3–53 Faulting of volcanic ash beds in Japan.

face are inclined and the beds above are nearly horizontal. Another unconformity is shown in Figure 3–61, where the beds above and below are both horizontal, yet with an old erosion surface and soil separating them.

Metamorphic rocks

Definition

Metamorphic rocks are changed or altered sedimentary or igneous rocks. Since there are numerous kinds of sedimentary and igneous rocks and also since the degree of metamorphism may be variable, there are many kinds of metamorphic rocks. The rock may be so intensely altered that its original identity is obscure. In fact, several kinds of sedimentary rocks and igneous rocks, after intense alteration, may end up as approximately the same metamorphic rock.

In the process of metamorphism, the overall chemical composition of some rocks has remained the same—no elements or compounds have been added or removed—in others, considerable chemical change has taken place.

Evidence of change

Probably the most immediate evidence of change is the fact that there are many rocks that cannot be classed among the common sedi-

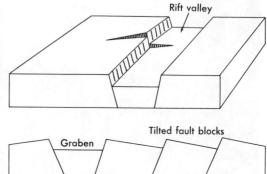

Figure 3–54 Normal faults resulting in rift valleys (graben) and tilted fault blocks.

Figure 3–55 West front of Wasatch Mountains near Nephi, Utah. This is in large measure a dissected fault scarp. [Photograph by H. J. Bissell.]

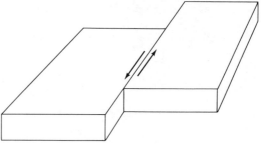

Figure 3–56 Fault with horizontal displacement.

Figure 3–58 A 16-ft-high scarp on Madison River, southwestern Montana, along Hebgen Lake Fault.

Figure 3–59 Slickenside surface of uplifted block of Wasatch fault. The surface had been protected by sand and gravel deposits, which, when quarried, left the striated and polished scarp in good shape. The weathering agents are now eating into it.

Figure 3–57 San Andreas Fault, southern California. [Vertical aerial photograph by Fairchild Aerial Surveys, Inc.]

mentary or igneous rocks. We therefore must assume they are metamorphic rocks. Direct evidence of change may be seen in a marble, for instance, which is made up of visible mineral grains of calcite, all tightly interlocked. We have never seen a lime sediment being deposited in this manner or a limestone develop from lime ooze in this coarsely crystalline nature, so that we presume that a limestone has been affected in some way to cause the microscopically small calcite grains to consolidate into larger crystals. The process is called recrystallization. We have little doubt that this is the origin of some marbles when we find them associated with other metamorphic rocks, or can trace a bed of limestone into marble.

Again, if we examine a series of sharply folded sedimentary strata, we see that some beds particularly have thickened in the crests and troughs of folds and have thinned on the flanks (see Figure 3–62). This indicates plastic flow of the beds. Plastic deformation is prominent in limestones and is helpful in inducing recrystallization to marble. It is also com-

Figure 3–60 Unconformity seen in canyon wall, Yakataga District, Alaska. [Photograph by D. J. Miller, U.S. Geological Survey.]

Figure 3–61 Unconformity between Dakota Sandstone and Morrison Shale north of Boulder, Colorado. Immediately beneath the sandstone is a reddish-brown soil about 2 ft thick that formed on the shale before burial by the sand. [Courtesy of Walter Sadlick.]

mon in shale and creates cleavage at an angle to the bedding (refer to the middle and lower drawing of Figure 3–62). The beds become somewhat welded together and fail to split apart; however, a new cleavage direction is set up in the rock that is along planes approximately parallel to the axial plane of the fold. This is called slaty cleavage, and the rock thus formed from the shale is called slate (Figure 3–63). Actually, there has been a transformation on a microscopic scale of some of the clay minerals that make up the shale to a mica

Figure 3–62 Relation of slaty cleavage to folds: top, plastic flow of shale beds upon sharp folding; middle, position of slaty cleavage; bottom, relation of cleavage to bedding.

Figure 3–63 Slaty cleavage across fine-grained beds, Moe River, Quebec. [Courtesy of Geological Survey of Canada and C. K. Bell.]

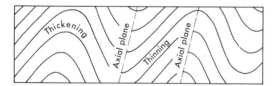

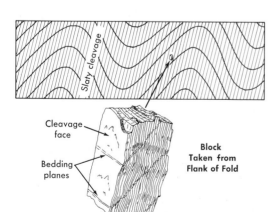

TABLE 3–2 *Kinds and Characteristics of Common Metamorphic Rocks*

NAME	TEXTURE	CHIEF MINERALS	DERIVATION
Marble	Granoblastic*	Calcite	Limestone
Quartzite	Granoblastic	Quartz	Sandstone
Slate	Slaty	Sericite, quartz	Shale
Metafelsite	Slaty, phyllitic	Sericite, quartz, feldspar	Felsite
Metabasalt	Slaty to schistose	Sericite, feldspar, amphibole	Basalt
Metagraywacke	Granoblastic to phyllitic	Feldspar, amphibole	Graywacke
Phyllite	Slaty, phyllitic	Sericite, quartz	Shale, tuff
Schist	Schistose	Muscovite, quartz, biotite, plagioclase	Shale, tuff, felsite
Gneiss	Gneissose	Feldspar, quartz, mica, amphibole, garnet	Granite, felsite, shale

* Granoblastic means granular texture caused by recrystallization.

mineral called sericite, with the tiny sericite plates all oriented in the direction of flow. This establishes cleavage, or the property that permits the slate to be split into thin sheets. The overall chemical composition has not been altered, however.

Another clear evidence of alteration of a previously existing rock can be seen in some metamorphic conglomerates. In these, pebbles have been drawn out into spindle-shaped forms, all having their long axes parallel. This attests to plastic deformation of the rock.

In the study of certain sandstones under the microscope, it can be seen that the sand grains have been lengthened into spindle form and often have been thoroughly broken in the process. The rock then becomes a *quartzite*.

Common kinds of metamorphic rocks

Table 3–2 lists a few common metamorphic rocks, with their texture, chief mineral constituents, and the rocks from which they are derived.

Examples of these rocks will have to be handled and observed in class or laboratory in order to gain a proper appreciation of them. Comments on a few varieties follow.

Quartzite

As described above, when the sand grains are somewhat crushed, drawn out, and rehealed, the rock exhibits a glassy or vitreous surface and is a true metamorphic rock. However, certain sandstones are composed of quartz grains cemented by silica. Although not metamorphosed in any way, a freshly broken surface is glassy because the sand grains break across instead of separating where cemented and the rock may not be distinguishable with the naked eye from a true quartzite. Technically, such a sandstone is called an orthoquartzite. Its constitution can easily be discerned under the microscope.

Slate

Slate develops from shale that has been deformed and caused to flow plastically. As previously noted, shale beds, when deformed under considerable pressure, thicken in the crests of anticlines and in the troughs of synclines, and the deformed rock develops the property of cleavage across the old bedding surfaces. Some slates have cleavage so well developed that slabs $\frac{1}{4}$ in. thick and 3 ft across can be split

Figure 3–64 *Banded gneiss occurring in northern Manitoba. [Courtesy of Geological Survey of Canada and Ed Schiller.]*

off. These make excellent shingles. Slate comes in gray, green, and various shades of brown and red, and the colored varieties are generally desired. A slate roof is expensive but lasts a lifetime or more.

The cleavage is due to the development of microscopic sericite grains (a form of mica) all oriented in one plane—the cleavage plane.

Phyllite

Phyllites were once shales that developed so much sericite that a micaceous sheen was imparted to the cleavage surfaces by the abundant tiny mica flakes. Cleavage may take a direction across the bedding or parallel with it.

Schist

A schist is an intensely metamorphosed fine-grained sedimentary or igneous rock. Mica is dominant and appears as flakes large enough to be clearly visible. The flakes are packed together in a parallel and possibly wavy arrangement, wrapping around other minerals that may be present, such as quartz, feldspar, amphibole, and garnet. In *mica schists*, muscovite mica is dominant, but in other schists biotite mica and chlorite are prominent. Chlorite imparts a green color to the rock.

Gneiss

A gneiss is a coarsely crystalline rock that looks like granite except that the light and dark minerals are segregated into thin layers or lenses (see Figures 3–64 to 3–67). The layers are commonly folded or crenulated. The minerals are much the same as in granite, with the

Figure 3–65 *A gneiss composed of stretched pebbles. It was formerly a conglomerate. Cross Lake, Manitoba.* [*Courtesy of Geological Survey of Canada and Ed Schiller.*]

Figure 3–66 *Crenulate gneiss of Precambrian age.* [*Courtesy of Geological Survey of Canada and Ed Schiller.*]

feldspars especially abundant. Quartz, amphibole, mica, and garnet make up a good part of the rock. Gneisses are commonly the most voluminous types in metamorphic rock terranes and may have been derived from many different rocks, such as granite, shale, slate, and schist. Their texture is said to be *gneissose*.

Causes of metamorphism

Pressure and heat are the common causes of rock transformation. Pressure in one direction, if great enough, will either shear the rock or cause it to flow plastically. Besides directional pressure, rocks come under the influence of confining (all-sided or hydrostatic) pressure when buried below the surface. At a depth of several miles the confining pressure is great, and when subjected to directional pressure in addition, most rocks yield plastically. If the temperature is elevated, the plastic deformation is great. These are the conditions that effect recrystallization, say of limestone to marble, or the transformation of shale to slate or phyllite.

The temperature increases approximately at the rate of $1°$ C for every 100 ft of depth and confining pressure at the rate of about 120 lb per square in. for every 100 ft of depth, so that we can approximately compute the temperature and pressure conditions that prevail at any depth. It is easy to see that at depths of 20,000 to 100,000 ft below the surface the rocks are subjected to conditions that might easily cause their alteration.

In addition to elevated temperatures and pressures water, if present, is influential in furthering recrystallization and mineral reconstitution. Water may also carry various compounds in solution and bring new materials to the rock that is undergoing change, thus resulting in minerals that would not form otherwise.

Geochemists have constructed presses in which carefully sized pieces of rock are placed and subjected to confining pressures up to those that prevail 50 miles below the surface. Also, the temperatures may be elevated several hundred degrees. In other presses the influence of water can be determined. It has thus been found that certain minerals form and exist (are stable) at certain temperatures and pressures, other minerals at other temperatures and pressures. The *stability ranges* are said to be determined. Now, when a metamorphic rock is found to be made up of a certain group or association of minerals, it may be known at what pressure (depth) and at what temperature the minerals were formed. At least, this is the experimental goal, and much progress has been made.

We observe zones of metamorphism next to the contact of intrusive igneous rocks. Dikes and sills bake red shales gray and turn soft coals to anthracite or coke near the contact. Magmas react with limestones below the surface, and sometimes the more unusual associations of minerals are formed, such as those of the iron and copper ore deposits.

Meaning of metamorphic rock at the surface

Extensive terranes of gneiss and schist now occur at the surface, and we are sure that the metamorphism that caused these rock types occurred several miles below the surface. It may only be concluded, therefore, that several miles of rock have been eroded away from such areas. Where the granites of batholiths are exposed at the surface a great deal of erosion has also occurred. In constructing the geologic history of a region, this is an important item. It is also meaningful to those who explore for various kinds of ore deposits.

Further, these metamorphic rocks have given us a better appreciation of conditions deep in the earth's crust; with this understanding, we shall see in later chapters how the crust is deformed, mountains are built, and magmas are formed.

Figure 3–67 Granite masses (light) intruding gneiss (dark and stratified), Rushmore Memorial, Black Hills, South Dakota.

References

Badgley, Peter C. *Structural methods for the exploration geologist.* New York: Harper & Row, 1959.
Billings, Marland P. *Structural geology.* 2d ed. Englewood Cliffs, N.J.: Prentice-Hall, 1954.
Heller, Robert L., ed. *Geology and earth sciences sourcebook.* New York: Holt, Rinehart and Winston, 1962.
Pearl, R. M. *How to know the minerals and rocks.* New York: McGraw-Hill, 1955.
Pettijohn, F. J. *Sedimentary rocks.* 2d ed. New York: Harper & Row, 1957.
Pough, F. H. *Field guide to rocks and minerals.* Boston: Houghton Mifflin, 1953.
Schrock, R. R. *Sequence in layered rocks.* New York: McGraw-Hill, 1948.

Motion Picture

The Bahamas: Where limestone grows today. Houston: Humble Oil Refining Co., color, 16 mm, 40 min.

earthquakes and seismology

From the layman's point of view, an earthquake is the rumbling, shaking, rolling, and jolting of the earth under and around him. Earthquakes are very real to many people because about 220 truly great shocks and 1200 strong shocks occur around the world per century, and thousands of small earthquakes occur annually. From the scientist's point of view, an earthquake is the vibration of the earth due to various kinds of waves that travel through and around it. Thus two major objectives in the study of earthquakes are clear: first, we should define the belts where destructive earthquakes have occurred and attempt to predict future occurrences; second, we should learn as much as possible of the interior of the earth by means of the earthquake waves. Much progress has been made in defining the internal constitution of the earth. Also, the belts of earthquake activity have been mapped, so we know approximately where future earthquakes will occur, but we have not yet quite arrived at the capability of forecasting the specific time and place. Much attention is now being given to this problem because of its importance to life itself.

Scientists have turned with vigor to the study of the interior of the earth in order to increase their understanding of the crust and the surface. Mountain building, or the deformation of the earth's crust, has always been a fascinating subject to geoscientists, and they have learned much about the architecture of the mountains. But it was not until the nature and dynamics of the earth beneath the crust were recognized that they began to fathom the cause of mountain building.

An analysis of earthquake records has been the chief source of information of the earth's interior. Geophysicists no longer have to wait for natural earthquakes to occur, however, but simulate artificial earthquakes by means of explosives. The wave records of the powerful

Figure 4–2 *Faulting in Red Canyon, tributary to the Madison River, as seen from the air, incident to the Hebgen Lake earthquake. [Courtesy of U.S. Geological Survey.]*

Figure 4–1 *Fault scarp formed at the time of the Hebgen Lake earthquake, near West Yellowstone, Montana.*

nuclear blasts have proved particularly valuable in discerning the earth's interior.

Definitions

The science of earthquakes is called *seismology*, the person who specializes in seismology is a *seismologist*, the instruments designed to record the vibrations are *seismographs*, and the records they write are *seismograms*.

Many large earthquakes have been marked at the surface by a fresh escarpment a few inches to 40 ft high or by open fissures. At the site of others there has been horizontal slipping of the rock masses on either side of a master fracture. Such evidence led geologists many years ago to believe that earthquakes are caused by sudden movements or displacements along faults (examine Figures 4–1 and 4–2). However, many small and even some major earthquakes have occurred without the development of a scarp at the surface, and for these it is somewhat difficult to prove that displacement along a fault caused the earthquake. There can be no doubt that the source of the waves is very localized, because the seismograms indicate the source as a fairly definite point. If no fresh fault scarp forms, we can generally relate the earthquake focus logically to an underground fault by assuming that the motion was considerably below the surface (see Figure 4–3 for such a situation). Here the rocks are sufficiently elastic or plastic to absorb the motion above the focal point without displacement at the surface. In the same way that movement may be localized along a fault in the horizontal direction, so it may be limited in the vertical direction also.

The localized place of slipping on the fault and the source of the earth waves is called the *focus*, and the vertical projection of the focus to the surface is called the *epicenter* (Figure 4–3).

Examples of earthquakes

Hebgen Lake earthquake

At 11:37 P.M. on the night of August 17, 1959, a displacement occurred along two faults in the Hebgen Lake and Madison Canyon area of southwestern Montana, just north and northwest of the west entrance of Yellowstone National Park. There are numerous summer cabins and dude ranches around Hebgen Lake as well as several U.S. Forest Service camp sites in the canyon. Although the height of the tourist season was at hand, it was fortunate that the area was sparsely inhabited because the earthquake was of frightful proportions. The first intense shaking lasted for several minutes, and

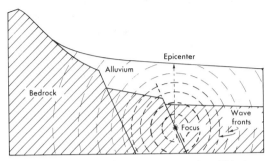

Figure 4–3 Relation of focus to epicenter. This is an example of a fault displacement not reaching to surface.

Figure 4–4 A 16-ft-high scarp through the Cabin Creek camp site on the Madison River, which formed suddenly on the night of August 17, 1959.

then, at approximately one-minute intervals, the ground continued to heave and writhe throughout the rest of the night. Aftershocks that were easily recognizable occurred for two weeks.

The earthquake was recorded on seismographs as far away as New Zealand. Fresh scarps up to 20 ft high were formed. Giant waves were set up on 7-mile-long Hebgen Lake, which rolled from one end of the lake to the other and back several times, like water in a bathtub would do if one end were lifted a few inches and then set down again. Each time the wave, or *seich*, hit the concrete dam the waters sloshed over the top like a giant breaker, 4 ft high and 721 ft wide, the length of the dam. This caused floods down the canyon, but the dam miraculously held, although it was badly cracked, and the spillway shattered. Some bench marks near the dam settled between 18 and 19 ft. After the quakes subsided, it was evident that the water on the north side of the lake had encroached about 8 ft on boat piers and cabins, whereas on the south side, docks and boats were 8 ft out of water.

Many new springs burst forth, old springs increased in flow, and all streams in the area discharged more water than formerly for several weeks. This was due to the compaction incident to the shaking of the water-bearing sand and gravel beds that fill the basin. The compaction excluded some of the water.

Several landslides carried parts of the black-topped highway into the lake. A number of tourists were thus trapped between the slides. Huge rock fragments were dislodged from the steep canyon walls and came crashing down to the highways and canyon bottom. Possibly a hundred tourists were bedded down for the night at the Cabin Creek camp site when a 16-ft-high fault scarp formed directly through the place (Figure 4–4), and some giant boulders crushed two of the campers.

The vibrations touched off a tremendous landslide at the lower end of Madison Canyon (see Chapter 7).

Earthquakes in densely populated regions

Earthquakes in densely populated regions often take a great toll of lives, primarily because buildings collapse on the inhabitants and secondarily because of fires and giant ocean waves. The earthquake in Iran near Tehran in September, 1962, claimed over 4000 lives, mostly in village houses made of mud and straw bricks. The Chilean earthquake of May, 1960, claimed many lives through falling roofs and walls and through landslides down steep mountain slopes that overran parts of several villages. In addition, the sea withdrew from one of the embayments and then rushed back in a wall of water up to 35 ft high, drowning hundreds of people. In one village alone, 500 people were drowned.

The terrible Lisbon, Portugal, earthquake of

1755 was much the same. Many inhabitants were in church on All Saint's Day when the ground began to shake and writhe with jarring shocks that lasted six minutes. When it was over, most of the buildings in the city had crumbled and crashed in rubble, killing a large segment of the population. Here, as in Chile, the water withdrew from the harbor, then rushed back in a wall 50 ft high and drowned thousands more. Fires broke out to complete the destruction. By nightfall 60,000 people had perished.

The San Francisco earthquake of April, 1906, was noted for its cause and the resultant fires. A slip of about 20 ft on the well-known San Andreas Fault in a horizontal direction was the cause of the earth tremors. Fires started in numerous places. The earth waves broke the water mains, and firemen and citizens alike were helpless to fight the many blazes, which went uncontrolled. Property losses exceeded $400 million which would be several billion dollars in terms of the present dollar value. This earthquake and many others since in California have made the public there very conscious of proper building regulations to safeguard against destruction by earthquakes and of precautionary engineering measures in the construction of dams and bridges.

Methods of recording earthquakes

Pictures and chandeliers swing and trees sway during an earthquake. Utilizing this phenomenon, which is the pendulum mechanism, an instrument can be devised that records the earth waves. If a pendulum is hung from a frame that is anchored to the rocks of the earth's crust, then, as the waves pass by, the frame moves back and forth but the pendulum tends to stand still (Figure 4–5, top). It appears, however, as if the pendulum were swinging and the frame standing still. The "swinging" of the pendulum in relation to the frame can be recorded on a rotating drum either mechanically by a series of amplifying levers or, better, by a light beam. It is evident that a free-swinging pendulum would be difficult to manage. It must be controlled to swing in one direction so its oscillations can be recorded (Figure 4–5, middle). But since the earth tremors may come from any direction, we must set up two seismographs, one with the pendulum swinging in an east-west direction and one in a north-south direction (Figure 4–5, bottom). With the two pendulums in operation, the approximate directional source of the earth waves can be determined (study Figure 4–6).

The mechanical principle of the seismographs just described pertains to the recording of the horizontal component of motion of the wave trains that radiate from earthquake foci. Special instruments, called vertical seismographs, must be constructed to record the vertical component of motion (Figure 4–7), and modern seismograph stations are equipped with both horizontal and vertical seismographs (see Figure 4–8).

Actually, there are several kinds of earth waves with several frequencies. There are long-period and short-period waves, and different instruments must be constructed to record each. As in the case of radio waves, there is more or less background noise that at some stations obscures the incoming waves of earthquakes. Automobiles, trucks, trains, and heavy machinery contribute to this background noise, so that seismographs constructed to record the very small waves cannot be used. With a very low background noise, amplification of 400,000 can be obtained. This means that the oscillations written on the seismogram are 400,000 times as large as the actual earth waves. At such amplification a nearby earthquake, although very small, might throw the oscillations off the record sheet.

Reading the seismograms

In order to read a seismogram, it is necessary to understand the kinds of waves and how

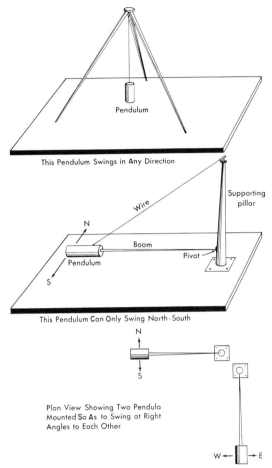

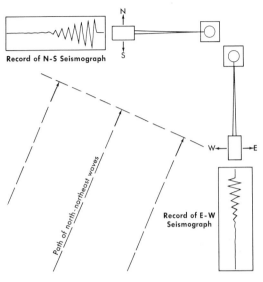

Figure 4-5 *The mechanical principle of the horizontal seismograph.*

Figure 4-6 *Simplified version of seismograms recorded by two horizontal seismographs for a certain wave train. The N-S seismograph records greater amplitudes than the E-W seismograph for the waves traveling to the north-northeast.*

they are propagated. There are two basic kinds of waves recorded by the seismograph that are propagated through the earth: the pressure type (P) and the shear type (S). By pressure type is implied the transfer of energy by compression and expansion, as exemplified by striking a long rod of steel on one end with a sledge hammer and noting the bounce that is imparted to an object in contact with the other end. A compression wave travels down the bar. Compression waves travel with a velocity of about 5.5 km per sec through granite, 6.5 km per sec through basalt, and about half as fast through unconsolidated sediments.

In shear waves the rock particles vibrate transverse to the direction of travel. If a rope is tied to a post at one end and you hold it taut at the other, then give it a sharp shake, a wave proceeds from your hand down the rope to the post. It is reflected back up the rope to your hand, then back and forth until it dies out, owing to the internal friction of the rope.

L waves are a third type of wave that follows the surface. Their motion is complex. These waves are recorded with the greatest amplitude on seismograms. Thus, for our purposes, we have to deal with P and S waves traveling along the surface.

Now, if an earthquake occurs at a point as shown in Figure 4-9, waves will radiate in all directions. Both pressure and shear waves will travel into the earth and will pursue a curved

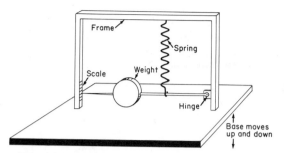

Figure 4–7 Principle of the vertical seismograph. A weight is suspended on a spring, and the weight tends to remain stationary as the vertical waves effect the base and frame. The movements of the weight relative to the frame are amplified and recorded either optically or electronically (indicated by scale).

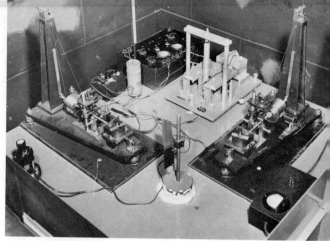

Figure 4–8 Seismographs for detecting earthquakes in the seismograph laboratory, Department of Geological and Geophysical Sciences, University of Utah.

path to the seismograph station because of the increase of wave velocity with depth.

The P waves travel 1.73 times as fast as the S waves and reach the seismograph station first. The S waves arrive next, and the L waves last. A seismogram for a somewhat distant earthquake will then look like the one shown in Figure 4–10.

The difference in time between the arrival of the P waves and the S waves is a measure of the distance between the earthquake and the seismograph station. Tables are available with which the distance of the quake can be calculated. The basic travel-time curves of the three waves are given in Figure 4–11.

Figure 4–9 Nature of wave propagation from an earthquake focus in the earth.

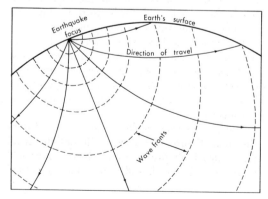

In the case of the Kermadec earthquake, whose seismograms, as recorded in Salt Lake City, are shown in Figure 4–10, we note that in addition to the arrivals of the P and S waves, PP and PPP, and SS and SSS are designated. These letters represent reflected P and S waves. In Figure 4–12 it will be seen that from the Kermadec focus one path leads directly although in a curved line to the seismograph station. The first arrivals of these waves are indicated by P and S on the record. Another path reaches the seismograph after having been reflected once from the surface. Waves along this path arrive after those of the direct path and are labeled PP and SS. Still another path is one of two reflections, and waves pursuing it reach the station still later. They are labeled PPP and SSS. These several waves give you some idea of the complexity of wave trains that are recorded, and in reality it is still more complex. Seismologists are not just very skillful in deciphering the complex seismograms but have developed the theory of elastic waves through rocks and through the earth to a high degree of understanding.

Locating the earthquake

Method of intersecting arcs

With two horizontal seismographs placed at right angles to each other, we can approximately determine the line along which the earthquake is located. That is, we can tell, for

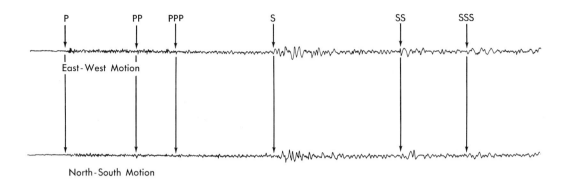

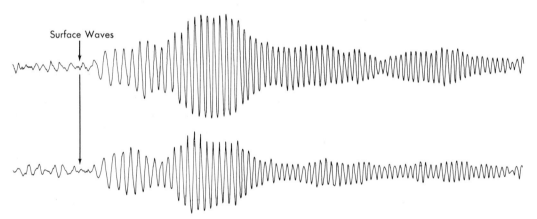

Figure 4–10 Kermadec Islands earthquake of February 27, 1955, as recorded on horizontal seismographs at the University of Utah. The lower two records are a continuation on the right of the upper two. [Courtesy of S. T. Algermissen, with wave arrivals determined by him.]

example, that a particular earthquake lies along a line running from northeast to southwest. However, we cannot tell the actual direction, northeast or southwest, from which the waves originate. If, in addition, a vertical seismograph is available, then in theory at least, we can also pinpoint the direction.

It takes at least three stations to make an accurate fix on the epicenter. One way to do this is to take a map and draw an arc from each station the distance away the earthquake was determined to be. The arcs should intersect in a point on the surface that is the only place that satisfies the data of all three stations. This is the epicenter (see Figure 4–13).

Depth of focus

The foci of some earthquakes are far below the surface, but the determination of the depth is generally an intricate job.

The principal method is to compare the times of arrival of the reflected waves, PP, and those

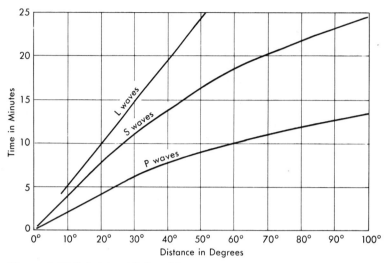

Figure 4–11 *Velocities of P, S, and L waves.*

of the direct waves P. If the focus is near the surface, the difference in time of arrival will be less than if the focus is deep-seated. Tables have been prepared for use in such cases. Earthquake foci are classed as shallow, intermediate, and deep-seated and will be discussed later.

Scales of intensity

The intensity of an earthquake may be measured roughly by the degree of damage or the extent to which it was felt at the epicenter and at varying distances away from the center. Twelve degrees of intensity were set up by the Italian seismologist Mercalli and in somewhat modified form these are now used as a simple guide. Lines drawn through the points of equal intensity, known as *isoseismal* lines, generally surround the epicenter and, in fact, locate the epicenter. Figure 4–14 is an example. When an earthquake rocks a certain city, you go to the local newspaper office and note all communications regarding the earthquake. Plot these on a map according to Mercalli's scale, or a more simplified one of your own design, and soon you will perceive where the center of greatest destruction or disturbance is. It is this simple.

However, the Mercalli method is subjective and approximate. It would be better to measure the *energy* released, if possible and compare earthquakes on this basis. Professor C. F. Richter, of the California Institute of Technology, has proposed a scale of this sort. Several seismograph stations must participate and be equipped with a specified seismograph, so that records are uniform. Then the largest horizontal trace (amplitude) is measured at each station. From these it can be computed what the maximum trace would have been for a station located 100 km (about 60 miles) from the station. This is the standard of comparison. A scale of these measures is used (the Richter scale) in which the magnitude is based on the logarithm of the amplitude. This means that an earthquake of magnitude 4 records with an amplitude of about ten times greater than one of magnitude 3, and one of 5 about ten times more than one of 4.

The earth's interior

The core and mantle

Of all geophysical methods of determining the nature of the earth below the observable

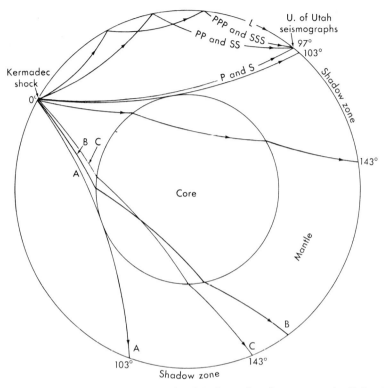

Figure 4-12 Paths of the Kermadec Islands earthquake waves to the University of Utah seismograph station, and the effect of the core of the earth on the wave paths, producing the shadow zone for P and S waves. The Kermadec epicenter is about 97 degrees away from the University of Utah station, and the shadow zone extends from 103° to 143° away from an epicenter. The P and S wave paths to the University of Utah station, therefore, just missed the core.

skin or crust, seismology is the most useful. Of course, all avenues of research must be integrated for maximum successful results, but the study of waves that have penetrated deeply into the earth is the best and most informative guide.

Perhaps the most positive information that earth waves have given us is that the earth has a core that is decidedly different in physical properties from the shell around it. Waves that pass through the core are slowed down or damped out. From Figure 4-15 it can be seen that the P waves increase in velocity to a depth of about 2900 km and then suddenly fall off to about their near-surface velocity. The S waves are absorbed or damped out completely at this depth. They have reached the outer boundary of the core. The paths of travel of the waves through the earth are made complicated by the presence of the core, as can be seen in Figure 4-12, which is a cross section of the earth. It will be seen that path A, just tangent to the core, is not deflected but that paths B and C penetrate the core and, in doing so, are *refracted* like light rays passing from the air into water. The interesting aspect of the refractions is that no waves can pass through the earth and reach the surface between A and C. This zone

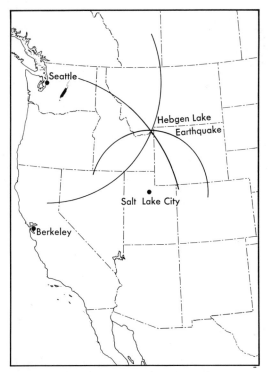

Figure 4–13 Intersection method of locating an earthquake. Arcs were described from Berkeley, Seattle, and Salt Lake City seismograph stations with radii equal to the distance the shock occurred from each.

around the earth begins at 103° from the focus and extends to 143° from it. It is known as the *shadow zone* and in itself is further evidence of the core at 2900 km below the surface.

The damping out of the S waves and the refraction of the P waves by the core indicate that it has the properties of a liquid. The core, furthermore, is now thought to have an inner solid division, again because seismograms indicate an increase in velocity of the P waves that have passed through this inner core.

As to the composition of the core, three main lines of evidence converge on a single conclusion, namely, that it is composed chiefly of nickel and iron. The first comes from a calculation of Von Jolly of Munich in 1878, in which he showed that the earth has a specific gravity of 5.7 (now reckoned to be 5.5). Since the density of rocks at the surface averages about 2.7, the core must be unusually dense to compensate. The second line of evidence comes from the infalling meteorites, which are of two general kinds. The metallic ones, principally of nickel and iron, are believed to indicate the composition of the core, and the stony ones, of iron-magnesian silicates, suggest the composition of the mantle. The third line of evidence comes from consideration of the rotational inertia of the earth. Since seismology fixes the limits between the crust and mantle and the mantle and core, the shells of the earth must have certain densities in order to yield a rotational inertia equal to that determined astronomically. The core is thus believed to be composed of iron and nickel with surface densities of 7.9 and 8.6, respectively. Owing to the great pressure in the core, these densities are raised to about 9.5. The outer and greater part of the core is probably molten and the inner and smaller part is solid.

Walter Elsasser, of the United States, and E. C. Bullard and H. Gellman, of Great Britain, have proposed that the earth's magnetic field is maintained by circulation of the earth's fluid, electrically conducting core. It is similar to a self-excited dynamo in which the electric current generated by induction in the rotor provides the necessary magnetic field. It is called a magnetohydrodynamic dynamo.

Low-velocity zone in upper mantle

The great thick shell from the base of the crust (5 to 70 km below the surface) to the core is called the *mantle*. By volume it makes up the greater part of the earth.

Near the top of the mantle a low-velocity zone has recently been detected. If the wave velocity in the mantle increased uniformly with depth, then the wave paths from an earthquake focus would be as shown in the upper diagram of Figure 4–16. Beno Gutenberg of the California Institute of Technology thought he recognized a low-velocity zone at about 100 to 200

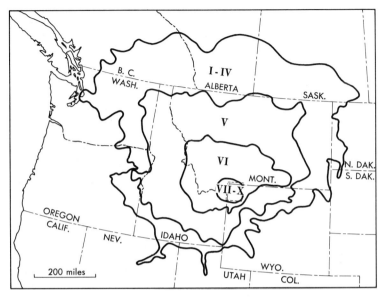

Figure 4–14 *Isoseismals of Hebgen Lake earthquake of August 17, 1959. This is according to the modified Mercalli scale.* [From U.S. Earthquakes, 1959, U.S. Coast and Geodetic Survey.]

km below the surface because the waves did not follow the theoretical paths. In a study of the seismograms of underground nuclear blasts in Nevada and New Mexico, Don L. Anderson, also of the California Institute of Technology, has clearly defined the low-velocity zone, as indicated by the travel paths in the lower diagram of Figure 4–16. The shadow zone is one in which the amplitude of P waves reaching the surface is decreased. The shadow zone extends horizontally from about 100 km from the focus to a distance of about 1000 km away. Beyond 1000 km the amplitude picks up again.

The low-velocity zone is also verified by other types of analysis of seismological data and is of world-wide occurrence, both under the oceans and under the continents. It is probably of utmost significance to crustal deformation and volcanic activity because it represents a layer where the rocks are about to change from the crystal to the liquid state if, indeed, they have not already done so in some places. The melting curve of rocks from the surface down,

as deduced by Anderson, is reproduced in Figure 4–17, where it will be seen that in the low-velocity zone the rocks are at the incipient melting point. Here, then, is a source of the primary magmas. Here also is a very weak layer where not only isostatic adjustments can take place readily, such as in response to the melting of ice caps or the building of great deltas, but possibly along which the entire outer 100-km shell can slide on the inner earth. The much discussed postulates of continental drift and polar migration might thereby be reasonably explained (see Chapters 4 and 11).

Deep-seated earthquakes

Until thirty-five years ago the mantle below the crust was supposed to be too weak to break and cause earthquakes, but since then many earthquake foci have been fixed at considerable depth in the mantle. They are known best in the Kermadec Island arc, the Japanese and Kurile Island arcs, and in the Andes of Peru. In each of these places a deep trench on the ocean floor

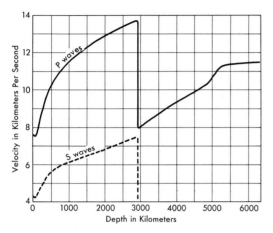

Figure 4–15 *Velocities of waves within the earth. The sudden drop in velocity of the P waves and the damping out entirely of the S waves mark the outer boundary of the core. The leveling out of the P-wave velocity curve near the center of the earth is due to the liquid core.*

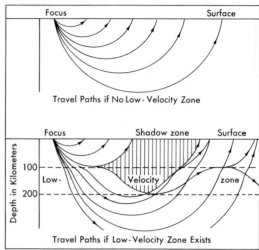

Figure 4–16 *The shadow zone, or zone of decreased amplitude of L waves, is due to the low-velocity zone 100 to 200 km below the surface. [After Don L. Anderson, Scientific American, July 1962.]*

lies in front of the island arc or, as in the case of the Andes, in front of the continental margin. The deepest yet recorded occurred at 690 km below the surface. The number per year down to 500 km is shown in Figure 4–17 (far-right graph). The earthquakes, especially those below 70 km, lie fairly close to an inclined plane which projects upward to the trench at the surface and dips under the island arc or continental margin (see Figure 4–18).

Figure 4–17 *Physical properties of the outer 500 km of the earth, showing the zone of low velocities and low strength. Temperature and strength curves are largely theoretical. [After Anderson, 1962.]*

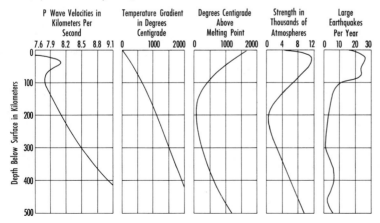

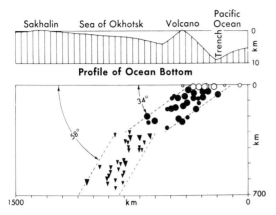

Figure 4–18 *Shallow, intermediate, and deep-seated earthquakes in the region of the Kurile volcanic arc. [After Benioff, 1955.]*

The foci are classed as shallow-surface down to about 70 km, intermediate from 70 to 250 or 300 km, and deep-seated from 300 to 690 km.

Since the foci define a large surface that dips to 45° to 60° under the island arc and since the seismograms indicate upward movement of the overriding block, Hugo Benioff, of the California Institute of Technology, has postulated that the great plane is in effect a reverse fault along which the horizontal strain in the upper mantle is being relieved. It is now believed that the thin oceanic crust is being thrust under the continental crustal margin due possibly to the sweep of convection currents in the upper mantle. (Refer to sea-floor spreading in Chapter 11.)

Moho discontinuity and layering of the crust

Any abrupt change in the physical properties of the rocks below the surface causes the earth waves to be reflected and refracted. Such a boundary is called a *discontinuity*. Other than the mantle-core discontinuity the most discussed one is the Moho. The word Moho is a short version of Mohorovičić, the name of the Yugoslav seismologist who discovered the discontinuity.

The Moho discontinuity lies at an average depth of about 35 km under the continents and 5 to 10 km under the ocean floors and is arbitrarily regarded as the boundary between the crust and the mantle (see Figure 4–19). By various studies of earth waves, both refracted and reflected, a sharp increase in velocity is noted below the discontinuity. Immediately above, the P waves have velocities of 6.5 to 7.0 km per sec but just below they travel at 8.0 to 8.2 km per sec. These velocities correspond to those through basalt and peridotite or eclogite when measured in the laboratories, and this with other evidence has led geologists and geophysicists to postulate that the layer of rocks above is made up of basalt or its more coarsely equivalent gabbro, and the upper part of the mantle of peridotite or eclogite. Eclogite has the composition of basalt but consists of high-density minerals. Basalt has a density of 3 and peridotite and eclogite of 3.3.

In fact, numerous seismological analyses have led to the recognition of several layers in the crust. Basically there are three layers, the upper consisting of unconsolidated and consolidated sedimentary rocks (low-velocity layer), an intermediate layer of igneous and metamorphic rocks, generally referred to as the granitic or silicic layer (intermediate-velocity layer), and the lower basalt or mafic layer (higher-velocity layer). More recently a still higher-velocity layer (7.6 km/sec) has been discovered by several investigators under parts of the intermountain region of western United States. The layers with their respective velocities are compared in Table 4–1.

The Moho discontinuity has attracted much attention because answers to important questions about igneous activity and mountain building in the broad sense can only be found when we know more about the nature of the rocks above and below it. Perhaps the recognition of the low-velocity zone in the upper mantle has taken some of the luster from the Moho, but earth scientists still believe it is very important to know for sure what conditions prevail at the discontinuity. The value of such information is appraised sufficiently highly by

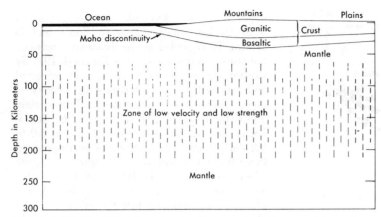

Figure 4–19 The Moho discontinuity, the crust, and the zone of low velocity in the upper mantle. Note difference in crust of continents and oceans. [Modified from Anderson, 1962.]

the National Science Foundation to convince it to finance preliminary efforts in the design of drilling equipment. It will be necessary to drill in 10,000 to 12,000 ft of water through about 5 miles of oceanic crust to reach the Moho, and the project, known as the Mohole, has attracted much attention. An experimental effort was successful in drilling through about 2000 ft of

TABLE 4–1 *Velocity Layers of the Earth's Crust*

DEPTH, KM	OCEANIC CRUST	NORMAL CONTINENTAL CRUST	CONTINENTAL CRUST WITH 7.5–7.7 LAYER
	6.5–7.0 km/sec Basaltic	1.8–5.5 km/sec Sedimentary rocks	5.73 km/sec Stratified rocks?
10			
	8.0–8.2 km/sec Mantle	5.8–6.1 km/sec "Granitic" layer	6.33 km/sec Crystalline rocks?
20			
30		6.5–7.0 km/sec Basaltic layer	
40			7.59 km/sec Transition layer, Basalt to peridotite
50			
60		8.0–8.2 km/sec Mantle	
70			
			7.97 km/sec Mantle

SOURCE: J. W. Berg et al., Seismic investigations of crustal structure in the eastern part of the Great Basin, *Bull. Seis. Soc. Amer.*, v. 50, 1960, pp. 511–536.

sediment on the ocean floor into hard basalt and recovering a basalt core. But the Mohole project became plagued with very high costs, politics, and personalities, and was set aside. In its place has come the Glomar Challenger project (see Chapter 9).

A most significant discovery in recent years has been the so-called 7.6 km/sec velocity layer under the western mountain systems. As shown in Figure 4–20, this layer reaches a depth of 75 km and extends from Denver to the Sierra Nevada where the upper granitic layer is thicker than usual and the basaltic layer is thinner. A wave velocity of 7.6 km/sec is intermediate between that in the basaltic crust and in the upper mantle, and thus may represent a transition in either physical or chemical properties. If the upper part of the mantle consists of silicates high in iron and magnesium (peridotite), it may be melting and producing a basaltic magma. The basaltic liquid would be mixed or interspersed with the crystal grains of the peridotite to produce the intermediate layer. If, on the other hand, the mantle is composed of a high density form of basalt (eclogite), it may represent the conversion in a solid state to a lighter density aggregate of silicate minerals. In either case expansion would have occurred, and the crust would have been lifted. The more geologists study and analyze the structures of the granitic crust the more they are becoming convinced that the many primary deforming forces have acted vertically, and thus a plausible mechanism for uplift, such as the 7.6 km/sec velocity, is very attractive. Most of the individual ranges of the Rocky Mountains, such as the Front Range of Colorado, the Uinta Range of Utah, and the Big Horn Range of Wyoming, appear to be giant blisters, as if liquid basalt had risen from the mantle and intruded the lower part of the granitic layer.

Use of seismology in exploration

Seismologic techniques have been basic in the discovery of a vast amount of oil and gas in the past 30 years. It is not that we have to wait for an earthquake to use seismology, but rather, we stimulate earth waves by explosions. Further, it is not oil and gas that we detect but geological traps that might contain the hydrocarbons.

Since oil and gas occur in sedimentary rocks we have to deal with strata; some are good reflectors of earth waves and some poor. If a good reflector layer in the sedimentary sequence is found, then folds and faults might be searched out underground by seismic reflection techniques. Methods of seismic exploration have been perfected for both land and water surveys. Where water exists it is the strata beneath the water that are the objects of attention. An example of the reflections from strata beneath the ocean is given in Figure 11–20. This is a continuous seismic profile obtained by a ship cruising in places over fairly deep water. Small bursts of energy either by an electrical sparker or by a gas gun in the water send trains of waves downward through the water and into the bottom sediments or sedimentary rocks. Reflections come from several layers. One or more sparks or shots are set off each second as the ship travels and thus a record is produced which appears almost like a continuous profile of the bottom layers. The method has been so well perfected that the results are amazing, as can be seen in Figures 11–20, 11–25, and 11–29.

On land, more difficulty is encountered in getting the energy into the ground. A small drilling rig bores holes 100 to 200 ft deep, an array of receiving instruments (jugs) is strung out along the ground, and then dynamite is detonated in the bottom of the hole. The reflected waves to each jug are recorded in such a way that the depth of the reflecting stratum can be judged, and also approximately the dip at that place. Needless to say, the modern seismograph truck carries expensive and sophisticated equipment and is costly to operate, but the results are usually well worth the cost. It is also apparent that the data can only be interpreted by a geophysicist or geologist trained in petro-

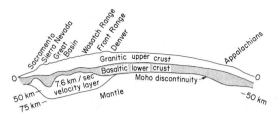

Figure 4–20 *Crustal structure across the United States showing the 7.6 km/sec layer especially. The sketch shows the curvature of the earth. The sedimentary veneer on the granite crust is not shown.* [From Pakiser and Zietz, 1965.]

leum geology, which is a major profession in itself.

The earthquake environment

Circum-Pacific belt

The island arcs and continental margins of the great Pacific Ocean make up the earth's dominant earthquake belt. Tens of thousands of respectable earthquakes have been recorded in this long zone. Figure 4–21 shows only the largest of the shallow earthquakes that occurred between 1904 and 1943. The large dots represent earthquakes of 8½ to 7¾ intensity on the Richter scale, and the small dots those of 7.7 to 7.0 intensity. These great earthquakes define the zone very well, but if all the smaller shocks were recorded, it would be fairly solid with dots.

Coincident with the circum-Pacific earthquake belt is a volcanic belt. Most of the world's volcanic fields of late geologic time, including the volcanoes that are now active, are in this belt. It is a necessary corollary of major earthquake and volcanic activity that this is a belt of most active crustal deformation. We are confident that mountains are being built in parts of this belt as energetically and rapidly today as at any time and place in the geologic past. If we would understand the growth of ancient mountain systems, this is the place to study the phenomenon. Here the earth's crust is continually creaking and groaning in the throes of deformation, while other major continental and oceanic regions seem fairly or quite stable.

Earthquakes in North America

Baja California and the Imperial Valley and Coast Ranges of California are part of the circum-Pacific belt of earthquake activity and display the greatest seismicity in North America. Movements along the San Andreas Fault or its several branches and related subparallel faults can generally be identified as the cause of the shocks (see Figure 4–22). According to Richter,[1] the San Francisco Bay area has displayed the greatest seismic activity in historical times, with the Coast Ranges of northwestern-most California second, and the Los Angeles and Imperial Valley areas third.

Another active zone of earthquakes runs through central Utah, southeastern Idaho, western Wyoming, and western Montana into the Rocky Mountain Trench of British Columbia (see Figure 4–23). The Hebgen Lake earthquake of 1959 is in this zone. Whereas in California the slippings on the faults are mostly horizontal, in the Rocky Mountains the motion is vertical, and a dozen or more very fresh scarps of 10- to 90-ft displacement, such as seen in Figure 4–24, can be found in the Great Basin of Utah and Nevada. Some of these scarps have formed in historic times, but some predate the arrival of white man yet they are still so fresh, that they must be only a few hundred years old.

The western part of the Great Basin just east of the Sierra Nevada is seismically very active, just as is the belt through western Montana, Wyoming, and Utah.

Seismicity is rather inconsequential in the Great Plains, the Mississippi Valley, the Appalachian Mountains, the Atlantic seaboard, and in all of Canada east of the Rockies. However, three major earthquakes have beset parts of this vast continental region and may suggest the possibility of more major shocks near the same

[1] Charles F. Richter, *Elementary seismology*, San Francisco, W. H. Freeman and Co., 1958.

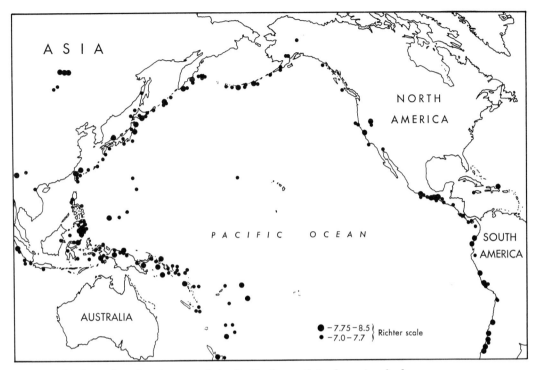

Figure 4–21 Zone of earthquakes around the Pacific Ocean. Only the major shocks that occurred between 1904 and 1943 are plotted. [From Gutenberg and Richter, 1945.]

centers. Short notes about an intense earthquake in Quebec in 1663 were written by the Jesuits. The area of the St. Maurice River, a tributary of the St. Lawrence, was the center of the activity. Landslides were particularly described.

The New Madrid earthquakes of southern Missouri in 1811 and 1812 shook the region for a year. In Louisville, Kentucky, 200 miles away, 8 shocks were severe and 1874 were felt over the year's time. An area in the river lowlands 140 miles long and 35 miles wide sank 3 to 10 ft, water rushed in, and new lakes, swamps, and bayous were formed.[2]

Charleston, South Carolina, was the site of another bad earthquake in 1886. A few minor vibrations preceded the major shock, which lasted for 70 sec. Then aftershocks were felt for two years. The size of the area affected was very large, and a large amount of energy was released. However, the damage, although considerable, was not so great as might have been expected.[3]

Predicting earthquakes

The U.S. Coast and Geodetic Survey publishes bulletins that itemize almost all earthquakes year by year. When one peruses these accounts, he wonders if there might be a certain frequency or periodicity in place and time to the shocks, and actually a number of efforts have been made to find the formula or law that defines the rhythm and would lead to the prediction of the next quake. Certainly, this is a

[2] Perry Byerly, *Seismology*, Englewood Cliffs, N.J., Prentice-Hall, Inc., 1942, pp. 89–90.

[3] Ibid., pp. 90–91.

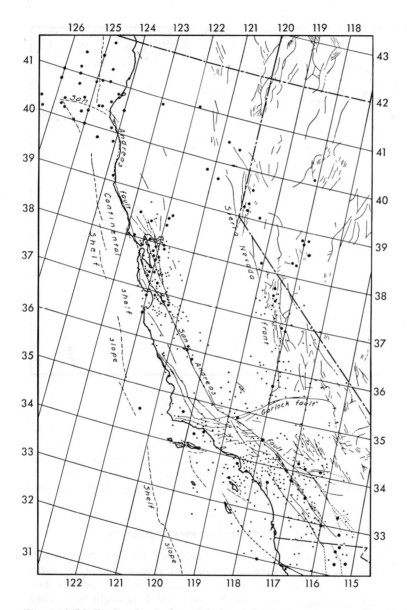

Figure 4–22 *Earthquake shocks and faults of California and western Nevada. The faults are those generally considered to have suffered late Pleistocene or Recent activity (about the last hundred thousand years). Earthquakes above the magnitude of 5 are shown by large dots, those below by small. [Earthquakes compiled from Byerly, 1940; Gutenberg, 1941; Byerly and Wilson, 1936, 1937; and tables supplied by C. F. Richter.]*

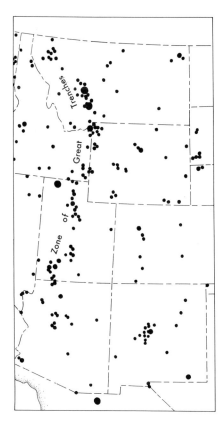

Figure 4–23 *Earthquake epicenters of the Rocky Mountain region showing coincidence of zone of concentrated seismic activity and the belt of trenches. [Taken from a map compiled by G. P. Woolard from U.S. Coast and Geodetic Survey reports.]*

worthy endeavor, but unfortunately little system has been noted in nature's conduct in regard to earthquakes. Davidson's *Studies on the Periodicities of Earthquakes of 1938*, shows that they occur in greatest proportion in the summer, particularly in July. The tendency, however, is not strong enough to warrant its use for prediction, according to Byerly.

The present approach to earthquake prediction lies in the belief that the rocks around the epicenter begin to deform prior to the earthquake. This then, would be a forewarning of the earthquake if the deformation can be detected. It would be manifested in (1) tilts and strains in the epicentral region, (2) a general increase in the number of small seismic events, and (3) changes in the physical properties of rocks near the fault as they are strained. In order to detect such changes a new generation of highly sensitive instruments has been developed, and various plans have been conceived for deploying the instruments in the earthquake-prone region.

A noted example of detection of anomalous tilt and strain preceding an earthquake was that of the Niigata earthquake of 1964 in Japan. It had a magnitude of $7\frac{1}{2}$; not large, but very destructive. The sensing instruments were vertical strainmeters installed in shallow holes over a zone 10 km long. A vertical expansion of the ground of 0.3 to 0.4 millimeters was detected by 15 of the 20 instruments. The instruments were approximately 70 km from the epicenter and 20 km from one end of the aftershock zone.

We have been impressed by the exploits of Apollo XI and the use of laser beams to measure precisely the distance from earth to moon. It has been explained that these measurements from day to day might possibly prove to be a forewarning of a large earthquake somewhere in the earth. It has been observed that times of pronounced secular polar shift (wobble of the earth) coincide with large earthquakes, and there is the hypothesis that large earthquakes excite the earth's natural wobble. Thus, there might be a forewarning in the form of an increase of wobble before the earthquake, and perhaps we can detect this build-up by the precise laser-beam measurements. Laser-beam measurements will also be used in detecting the shift of the ground in premonitory strains of the crust adjacent to active faults.

Finally, it should be said that the electrical conductivity changes as the rocks are deformed. This property might also be exploited.[4]

---

[4] Refer to Frank Press and W. F. Brace, Earthquake prediction, *Science*, June 17, 1966.

Figure 4–24 Fresh fault scarp at base of Mount Nebo, central Utah. Two stages of slipping are evident, the combined displacement being about 90 ft.

In general, homes of the masonry type are least suited and frame houses are best suited for withstanding earthquakes. The main requirement is that the roof be securely fastened to the walls. Traditionally, American homes of brick construction simply have the roof plates sitting on the walls without being securely anchored. In the event of an earthquake, the walls fall away, leaving the roof to cave in.

Most public buildings, at least in the past decade, have been built according to earthquake code specifications and should be reasonably safe. But there are some unexpected exceptions to this conclusion. For instance, the Venezuelan city of Caracas was hit by a recent earthquake and most of the city escaped with little damage. One small section of high-rise structures, however, was almost completely demolished. The tall buildings became pancakes of rubble. An engineering and geologic study revealed that the earthquake vibrations seemed to focus in this one spot, we must say because of unknown subsurface geologic conditions, and were especially intense. More than this, the natural frequency of vibration of the buildings that collapsed was the same as the frequency of the earthquake vibrations (the vibrations were in resonance), and this intensified the shaking of the buildings until they collapsed. Engineers can now measure the natural frequency of vibration of a building. It is also possible to calculate this frequency before the building is constructed and to take measures to change it, if necessary.

References

Earthquake prediction. *Tectonophysics*, v. 6, 1968. Special issue.
Richter, Charles F. *Elementary seismology*. San Francisco: W. H. Freeman and Co., 1958.

Motion Picture

What's inside the earth? Planet Earth Series, WSC-161, color, 16 mm, 13 min.

5
magnetic and gravity fields of the earth

Earth's magnetic field

Earth as a bar magnet

If we place a short bar magnet on a piece of white cardboard, sprinkle dark iron filings around it, and then tap the cardboard or the table on which the cardboard rests, the filings will arrange themselves in a pattern that betrays the invisible magnetic field. Figure 5–1 is a photograph of a large chunk of magnetite with iron filings on and around it, and you see that the field consist of an axis with a pole at each end, and looping lines, called *lines of force,* between. In three dimensions the field looks like a thick doughnut with the magnetic axis passing through the hole. One pole is positive, the other negative. If you hold two bar magnets in your hand, it becomes immediately evident that similar poles repel each other and opposite poles attract each other.

It has been known since 1600 that the shape of the earth's magnetic field is approximately like that of a bar magnet or the piece of magnetite shown in Figure 5–1, but the compass was known long before the earth's magnetic field was discovered and mapped. The early compasses were made by floating a small blade of lodestone (magnetite) on a sliver of wood in water. The sliver would swing around to a given position, quiver a bit, and then hold this orientation. If needles are rubbed on lodestone, they become magnetized, and if hung and balanced by a delicate thread, they all seek the same orientation. The positive end of the compass needle always points toward the negative pole of the earth's magnetic field, and the needle also aligns itself with the lines of magnetic force, which in the polar areas approach a vertical direction.

The chunk of magnetite shown in Figure 5–1 is somewhat longer than thick and also irregular, yet its magnetic field appears to be regular. If the same piece of magnetite were turned or ground to a sphere, resembling the earth, it would still display the same bipolar, or dipole field. Figure 5–2 is the same as 5–1, except that a sphere (circle of paper) masks the irregular chunk of magnetite, and now you can see the magnetic lines of force in relation to the earth's surface.

Declination

From Figure 5–2 it is apparent that the compass needle at most places on the earth's surface

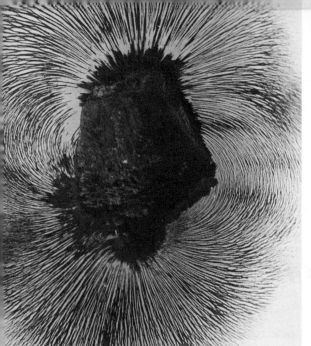

Figure 5–1 *Lodestone (magnetite) and its magnetic field shown by sprinkling iron filings on a table around it.*

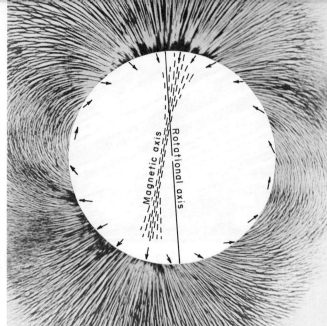

Figure 5–2 *A close replica of the earth's magnetic field. This is the same as Figure 5–1 but has the earth sketched in. The north geographic and magnetic poles are at the top. The arrows show the inclination of the compass needle.*

seeks an inclined position parallel with the lines of magnetic force as they exist at that place. This inclination of the needle will thus have two components—the horizontal direction and the angle of dip. If the north magnetic pole were located precisely at the north pole of rotation, the compass needle would always have a true north direction, but unfortunately this is not the case. The north magnetic pole lies at about Lat. 75½° N and Long. 100° W, and the south magnetic pole at about Lat. 66° S and Long. 148° E (1965), and thus the lines of magnetic force depart from the meridians. They do this progressively more toward the magnetic poles. The angular divergence of the compass needle from the true north is called *declination*, or by the mariners *variation*.

The lines of equal declination of Figure 5–3 were taken from the U.S. Naval Oceanographic Office Map of Magnetic Variation, 1965 (map 1706). Rather than lines of force directed toward the magnetic poles, lines are drawn connecting points of equal declination. These are called *isogonics*. In order to use a compass effectively one must know the declination from place to place and hence an isogonic map is needed. In the polar projection of Figure 5–4, for instance, the compass needle in southern Greenland points about 36° W of the pole. The declination is said to be *west so many degrees*. At Point Barrow, Alaska, the declination is 25° E. The line drawn southerly from the north magnetic pole through Canada and the United States is labeled *no variation;* east of it the compass has a west declination, and west of it the needle has an east declination. In Maine the declination is about 20° W, in Washington it is about 22° E, and in Alaska it ranges from 20° to 30° E.

A good compass can be set for any declination. For instance, if you are using a compass in an area that has a declination of 20° E, the north line of the instrument must be set such that the magnetic north is 20° E of the due north line (see Figure 5–5).

Inclination

The inclination of a compass needle at any place is the angle between the horizontal and the needle measured in a vertical plane containing the local line of magnetic force (local magnetic meridian) and the needle. This will be better understood by referring to Figure 5–6. The vertical plane through the line of force is labeled *abcd*. The angle of inclination is labeled *I* and is called the dip. The horizontal component of the magnetic force is *ab* and the vertical component is *ad*. It is evident that at the magnetic equator the horizontal component is the same as the total force and is at its maximum, and as the magnetic poles are approached the horizontal component decreases to zero. For the vertical component the opposite is true. Thus maps can be produced that show the horizontal and vertical intensities of the earth's magnetic force. In all inclination maps the north-seeking point of the compass needle points downward north of the magnetic equator, and upward south of the magnetic equator (study Figure 5–2 again).

Examples of magnetic inclination or dip maps are given in Figures 5–7 and 5–8. If a compass needle is balanced and supported by an axle, as shown in Figure 5–9, and held vertically parallel to the line of force, it will record the dip. This instrument is held in the hand and used to find and delineate magnetic ore bodies. Near such bodies, the dip and vertical component of force are conspicuously different within short distances and can be used to great advantage in exploration.

The smoothly curved and regularly spaced contours of the various U.S. Naval Oceanographic Office and U.S. Coast and Geodetic Survey maps do not show the local irregularities that are known to exist, but represent the smoothed and regular magnetic field determined by mathematical computations. Local detailed surveys reveal many variations from these idealized contours, but the idealized contours are necessary to fill in the large areas of sparse data. They assist also in the studies of the constant changes that are occurring in the earth's magnetic field.

Changes in the magnetic field

Observations at fixed points have been carried out for 300 years, and these with many others over shorter periods have led to the realization that the magnetic field changes continually in both intensity and position. Some changes are slow and occur over long periods; these are *secular* changes. Others are more abrupt and irregular, and are called *short-period* changes. The short-period changes are of minor significance in the magnetic force field; they occur in the upper atmosphere and are due to external causes, such as solar radiation cycles with resulting ionic perturbations. What concerns us most in this chapter is the source of magnetism inside the earth where 96 percent of the magnetic field originates. Figure 5–10 shows the secular changes recorded in London (1576–1955), Paris (1617–1940), and Boston (1723–1955). The London and Paris changes suggest that the magnetic pole is very slowly precessing about the geographic pole, but the Boston change is far less definitive. In fact, on a worldwide basis it is not yet possible to recognize the nature of the change. It seems to be concentrated in centers of subcontinent size, but the change is in one direction here and in another there. In still another place the change is nil. It is thought that these secular changes, poorly as they are understood, are related to internal changes taking place within the earth in definite centers and at great depth.

In addition to the secular changes, the entire pattern of secular variations seems to be shifting from east to west. This is occurring at a rate of about $0.18°$ per year, which means that the pattern completes a circuit of the earth every 2000 years.

The magnetic field of previous geological ages has been frozen in igneous rocks by the orientation of any crystals with magnetic qualities while floating in the still liquid magma and conforming to the existing field of force. Upon complete crystallization the magnetic field is thus locked in place in the rock, and so it remains although the earth's field may shift later on. The magnetic field may also be frozen in place in accumulating sediments by the orientation of magnetite grains in conformity with the magnetic lines of force while the sediment particles were in partial water suspension and somewhat free to adjust. Thus in these rocks the

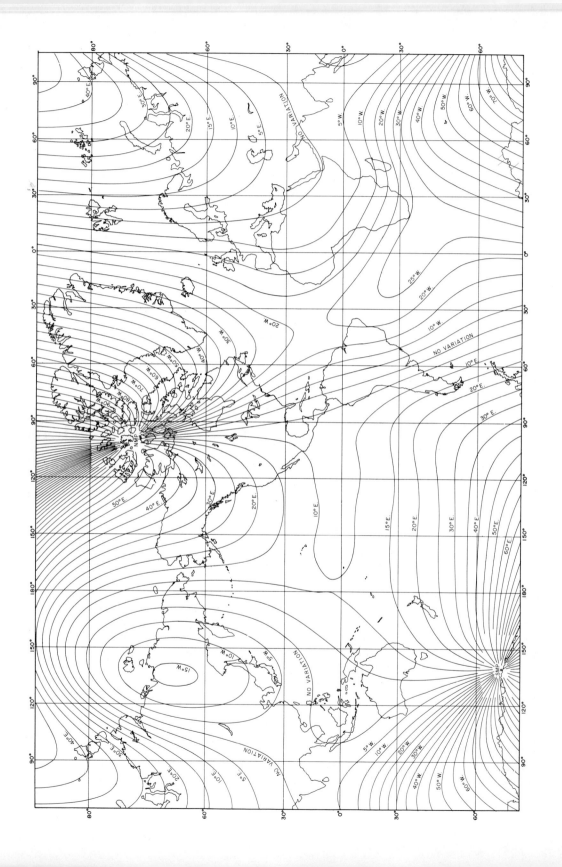

MAGNETIC AND GRAVITY FIELDS OF THE EARTH 119

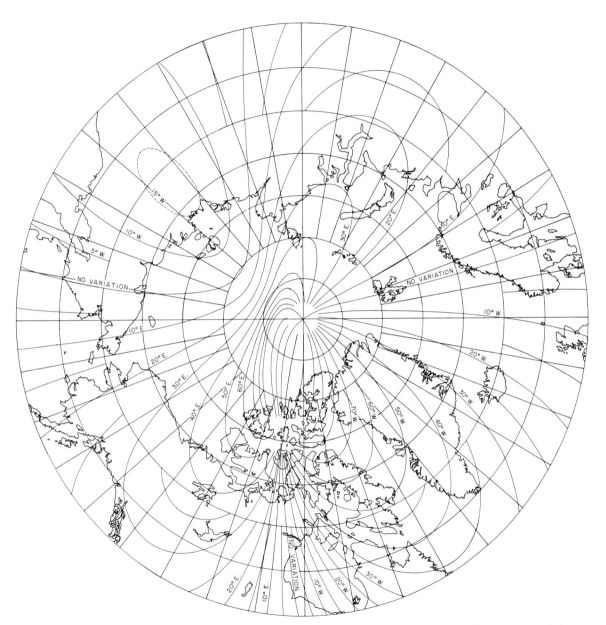

Figure 5–4 North polar projection of the world showing the magnetic variation, or declination, as of 1965. [After U.S. Naval Office Map 1706N.]

Figure 5–3 Mercator projection of the world showing the magnetic variation, or declination, as of 1965. [After U.S. Naval Office Map 1706N.]

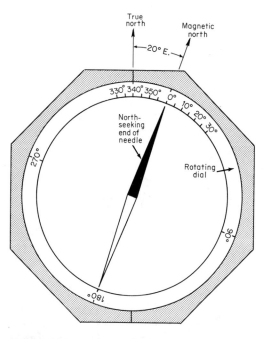

Figure 5–5 *Illustrating the setting of the rotating dial of a compass for a declination of N 20° E. Now, when the north-seeking end of the needle points to 0°, the north line of the compass points true north.*

magnetic orientation remains as a remanent or legacy of an ancient magnetic field. Methods have been devised to detect and measure these ancient fields separate from the present day magnetic field, and from them the position of the north pole has been calculated or projected for various times during the past billion years of earth history. This has lead to the concept that the poles have wandered appreciably, and although the theory has not been embraced by all geologists and geophysicists, it stands as an attractive concept. It is one of the chief bases for the theory of continental drift, according to which the continents were once connected and then split and drifted apart. At the same time the entire crust of the earth was shifting over the interior, and thus, the crustal position of present magnetic poles and the rotational poles are not now where they used to be. The shifting of the magnetic north pole over geologic time is reconstructed in Figure 5–11. The secular changes of historical times are small in comparison with this migration in geologic time. During all this time the magnetic and geographic poles were probably not far apart, perhaps rarely as far apart as they are now.

Figure 5–6 *Declination (D) and inclination (I); A, east declination and rather steep inclination; B, west declination and gentle inclination.*

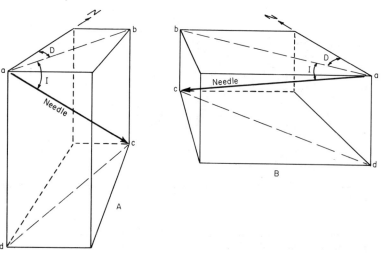

Magnetic reversals

One of the surprising discoveries about the earth's magnetic field was that the dipole had reversed itself a number of times. The record of these reversals has been left in the rocks, and the geologic times of major switches have been fairly well documented. One of the ways in which the discovery was made was as follows. Magnetometers were towed by ships at sea, and very large areas of the ocean floor were thus accurately surveyed. One of these early surveys is reproduced in Figure 5-13. The first rewarding result was the recognition of faults with great horizontal displacements. Here we see a fault extending from point A to A'. The reason for these surprising ribbons of anomalies was not perceived for a few years, when it finally became apparent that they represented successive fissure eruptions of basalt on the ocean floor, which were magnetized first toward the north magnetic pole and then toward the south. This will be related more fully in Chapter 11, but suffice it to say at this point that the magnetic reversals helped chronicle the progressive widening of the Atlantic Ocean basin and were important in the development of the theory of sea-floor spreading and continental drift.

Source of earth magnetism

The source of most of the earth's magnetic field lies within the earth. Before considering the source of the dipole field, it should be pointed out that the north and south magnetic poles are not exactly antipodal, and the magnetic axis does not pass through the center of the earth. The south magnetic pole is stronger than the north magnetic pole, indicating that the effective center of the dipole is nearer the south (refer again to Figure 5-2).

The following are some of the possible sources of the earth's magnetic field.

1. It might be due to a permanent magnetization; physicists believe, however, that this is not possible because the temperature of the interior is above the Curie point, the temperature above which a magnetized substance ceases to be magnetic.
2. It may be an inherent property of massive rotating bodies.
3. It is due to electrical currents.

A discussion of these possibilities is beyond the scope of this book, and it must suffice to say that considerable evidence points to electrical currents and to the top layer of the *core* of the earth, which is liquid and metallic, as the source. The major part of the magnetic field is thought to originate in the metallic, fluid part of the earth's core. The secular changes seem to originate in the upper 50 to 100 km of the fluid core. There, it seems, two eddies or more rotate at a constant angular velocity about a fixed axis, and set up a self-exciting dynamo, thus yielding the dipole magnetic field of the earth. This is the so-called *dynamo* or *hydrodynamo* theory, proposed independently by E. C. Bullard of Great Britain and W. M. Elsasser of the United States (see Figure 5-12).

According to hydrodynamo theory of the earth's magnetic field, three conditions are necessary to produce and maintain a magnetic field:

1. Large dimensions. The large size of the earth's metallic fluid core satisfies this condition.
2. Convection currents. Convection currents are assumed to exist within the earth's fluid core.
3. Rotation. The earth's rotation lines up the convection currents and the three conditions together give us the dipole field.

It is interesting to note that attempts to measure the magnetic field of the planet Venus showed a weak field, which substantiates the belief that Venus rotates slowly, once in 243 days. The requirement of large dimensions ensures that the magnetic field will not decay too rapidly, so that it can be rejuvenated and perpetuated by convection currents and rotation.

The secular variations, like those of the main

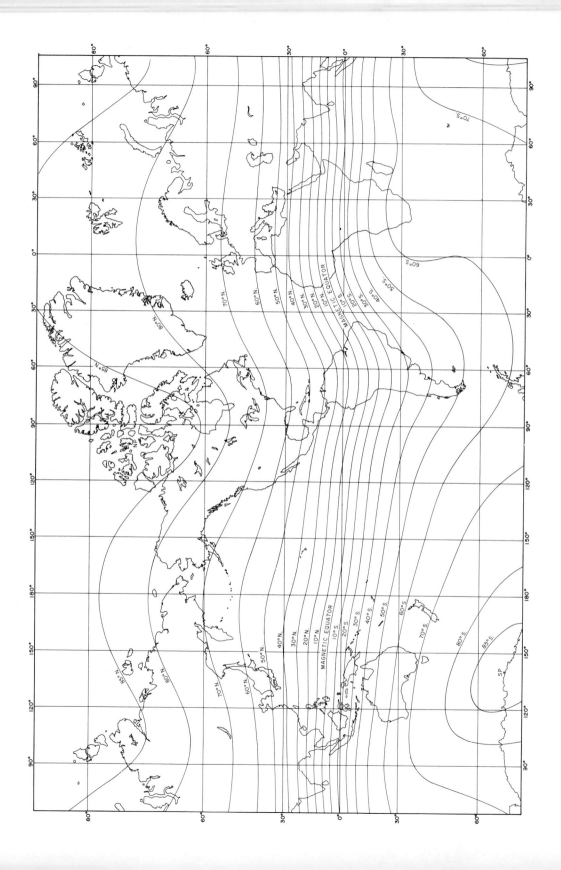

field, seem explained in the hydrodynamo theory by supposing that local eddies occur near the outer surface of the core. Likewise, the slow westward drift of the entire secular variation pattern, previously described, appears to be due to electromagnetic coupling between the mantle and the core. The drift, since 1900 at least, closely parallels the curve of irregularities of the earth's rotation as defined by the length of the day. This observation indicates that the mantle rotates faster than the top layers of the core, and the westerly drift can be explained by electromagnetic torques adjusting the balance of angular momentum between the core and mantle.[1]

Magnetism used in exploration

Our discussion of the magnetic field thus far has centered on its origin and what it could tell us about the internal constitution and processes of the earth. The magnetic field may also be used effectively in the search for mineral deposits and in the analysis of faults and various rock masses of the crust.

A number of portable instruments have been designed for magnetic surveys, either for the measurement of the horizontal or vertical components, or of the total field intensity. One of these instruments is the dip needle. Its first commercially successful application was in the exploration for iron ore deposits that escaped visual recognition because of burial under swamps, heavy soils, glacial deposits, or alluvium. The dip needle was simple to understand and operate, and although it was not very sensitive, it was sufficiently so to detect magnetic material associated with iron ore bodies. The dip needle was also used successfully in the Keweenawan Peninsula of Lake Superior to detect the presence of concealed basalt ledges that indicated the presence of copper deposits. It was a tough job to chart and map parallel lines of traverse through swamps, windfalls, and bush, to fight mosquitoes and blackflies, to take dip needle readings at regular intervals and record them in a notebook. Then came more sensitive instruments for detailed overland surveys. Finally, magnetometers were developed that could be towed on cables by airplanes or by ships throughout the oceans. These instruments have electronic devices for continuous recording. In the case of the airborne magnetometers, their position at every second was determined by aerial photographs of the land surface. The position of ship-towed magnetometers was accurately determined on maps by new radar methods (see map, Figure 5–13).[2] These maps revealed surprising and significant information about the ocean floors.

The areas mapped by airplane and ship are immensely larger than any that man on foot could have mapped. In such large surveys the conspicuous variations of intensity of the magnetic field, called *anomalies*, are the particular points of concern. What do they mean? It must be understood that anomalies are the result of several influences: (1) the principal dipole field of the earth, which in itself is quite complicated; (2) secular variations; (3) short-term periodic variations; (4) various electrical currents that may be passing through the rocks; (5) the local magnetic qualities of the rocks themselves. The local rock masses may concentrate or repel the principal field or they may have a remanent field different in direction and intensity from the existing principal field. The

---

[1] Jacobs, Russell, and Wilson, J. Tuzo, *Physics and geology*, New York, McGraw-Hill, 1959, pp. 425; Howell, B. F., Jr., *Introduction to geophysics*, New York, McGraw-Hill, 1959, pp. 399.

[2] The property of the geomagnetic field contoured in Figure 5–13 is neither declination nor inclination but rather *total intensity*. This is proportional to the torque or twisting force that turns a compass needle toward the magnetic lines of force. The numerical values are in units of gammas and are relative to an arbitrary datum. (The actual values of the total intensity are about 50,000 gammas.)

Figure 5–7 Mercator projection of the world showing the magnetic inclination, as of 1965. [*After U.S. Naval Office Map 1700.*]

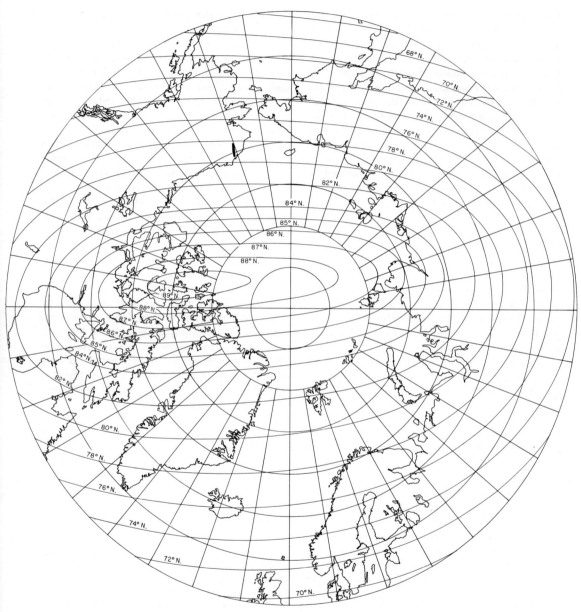

Figure 5–8 *North polar projection of the world showing the magnetic inclination, as of 1965. [After U.S. Naval Office Map 1700N.]*

size of a specially magnetized rock mass and its distance below the surface can be calculated.

An example of an airborne magnetometer survey over land is given in Figure 5–14. The prominent anomaly in the northeast corner of the map is associated with a high-angle fault along the southwest margin of an uplift called the Uncompahgre. Also, apparently, the crystalline rocks below the layers of sedimentary rocks are particularly magnetic, and these, together with the fault, produce the strong local field. It is well known that the geologic features of this area trend in a northwest direction. The fault just referred to and a prominent graben structure in the southwest corner of the map both trend to the northwest, and so do the anomalies of the magnetic field trend in a general way in this direction. Thus, we can say the magnetic anomalies suggest the trend of the structures, but still we cannot identify the kind or nature of the structure from the magnetic anomalies alone. If, however, we compare the magnetic anomalies with the gravity anomalies discussed below (see Figure 5–23) and interpret the two kinds of anomalies together, much better definition of the geologic structures is possible.

Earth's gravitational field

What is gravity?

About 300 years ago Isaac Newton propounded his system of mechanics in which he explained and described the laws of motion. To account for the fact that objects fall to the earth he postulated that there exists a force of attraction between all bodies in the universe, which is called gravity. The celebrated earth-apple relation has been recounted many times—an apple falls from a tree to the earth because of the mutual attraction of the two for each other. Since the earth is so much greater in mass than the apple, the apple appears to do all the moving, but actually the earth is pulled an infinitesimal amount toward the apple.

The earth has a gravitational field, in which the gravitational forces vary from place to place. The force of gravity can be measured accurately and the earth's gravitational field has been fairly well charted. This field, however, has no relation to the earth's magnetic field. What physical phenomenon gravity is in terms of light, radiation, electricity, or what not, has not yet been discovered.

Newton's law of gravitation states that two bodies attract each other with a force directly proportional to the product of their masses and inversely proportional to the square of the distance between them. The moon and the earth attract each other with a force proportional to the product of their masses and inversely proportional to the square of the distance that separates them. Why then, do they not pull each other together? We will come to this question later.

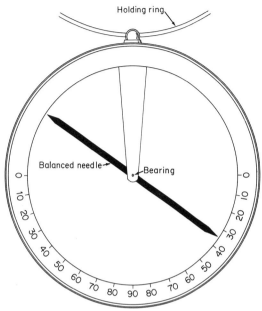

Figure 5–9 Dip needle compass such as used in searching for magnetic ore bodies.

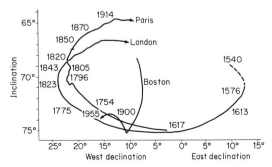

Figure 5–10 Secular changes in declination and inclination at London, Paris, and Boston.

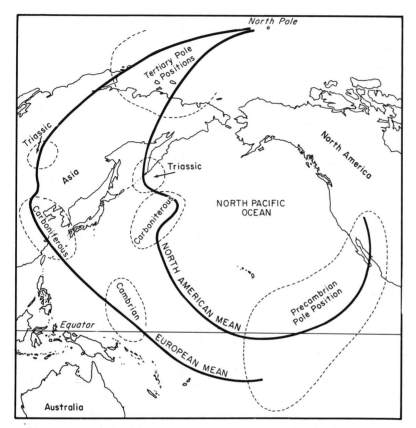

Figure 5–11 Heavy lines are mean polar wandering routes as determined from European and North American remnant magnetic observations. The areas indicated by Precambrian, and so on, are those in which projections of polar positions fall from the various places where remnant magnetic readings have been taken. [From Runcorn, 1962.]

Measurement of gravity

If an object is dropped in a vacuum chamber near sea level where there is no air resistance to the fall, it will accelerate at a rate of approximately 32 ft per sec per sec (about 980 cm per sec per sec), and this acceleration is called 1 g. If an astronaut is resting on the ground the force of gravity is pulling him downward toward the center of the earth with an acceleration of about 1 g. If he is propelled outward with an acceleration of 32 ft per sec per sec he will have an additional g pulling on his body.

To return to the question why the moon and the earth are not drawn together, the answer lies in the fact that the moon is in circular motion around the earth. The orbital motion of one body such as the moon, around a central body, such as the earth, and the role of gravity in holding the moon to a constant circular course was the problem that Newton attacked, and in solving it mathematically he felt that he demonstrated that gravity was universal. This is now considered a law. If the moon were not in circular motion it would move toward the earth (the earth would also move toward the moon) with an acceleration of 32 ft per sec per sec, but the velocity vector, tangent to the circular path of the moon, results in a movement toward the earth at the rate of separation the moon would have from the earth if it were free to streak off in a straight-line course. The demonstration of this principle is somewhat beyond the present book's objectives, but the same principle holds for spaceships in orbit, either around the earth or around the moon. In such a condition g is nil, and objects float around freely in the spaceship. The astronaut in orbit would seem to be in free fall, and his body would be weightless.

Gravity increases toward the poles

We have stated above that the acceleration of gravity at the surface of the earth is *about* 32 ft per sec per sec. This is necessary because the acceleration is greater at the poles than at the equator, and thus we must stipulate some particular place where a standard of gravity can be set up and used for world-wide measurements. Potsdam, Germany, is the place so recognized, and the latest exact computations there yield a figure of 981.2663 cm/sec^2. It was discovered that gravity increases toward the poles when a pendulum clock, precisely set to mark time at Paris, France, lost 2½ minutes per day when taken to Chayenne, French Guiana. This was the famous experiment of Jean Richer in 1671. Chayenne is at lat 5° N whereas Paris is at lat 49° N. The time that a pendulum takes to swing back and forth is dependent upon the acceleration of gravity. In fact, a most precise method of measuring gravity is by means of a pendulum mechanism—the only trouble is that you must count the pendulum beats for possibly 24 hours to make *one* measurement. But with a coiled spring instrument (spring gravimeter) fairly accurate relative measurements can be made in a few seconds, and this is important when measurements must be recorded in a dense network of stations for purposes of mineral or oil exploration. Figure 5–15 is a schematic drawing of a so-called stable type of spring gravimeter, and Figure 5–16 of an un-

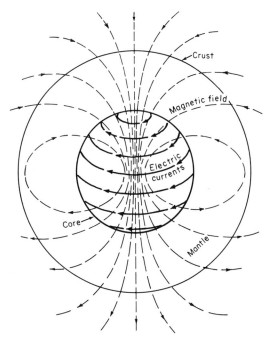

Figure 5–12 Possible electric currents in outer layer of core that produce the earth's dipole magnetic field. [*After Runcorn,* Scientific American, *1965.*]

stable type. The latter is a more sensitive instrument because a third force is called into play which intensifies any change in gravity. In Figure 5–16, when the balancing weight is equal to the pull of the mainspring the beam will be horizontal, and the auxiliary weight attached to a bar from the beam will be vertical and hence will have no effect on the balance, but when the force of gravity tilts the beam, then the auxiliary weight is shifted and exerts a moment of rotation to the beam. In other words, it reinforces the change in gravity.

After Richer had published his results Newton pronounced that the earth was shorter from pole to pole than across at the equator. What he was saying was that because the acceleration of gravity is greater at the poles the distance to the center of the earth must be shorter. The latest measurements give the polar axis as 7,899.98

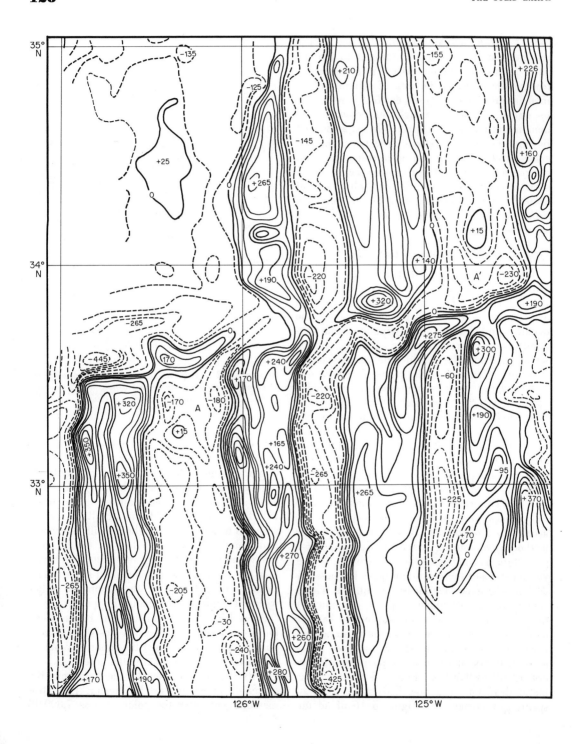

miles (12,713.82 km) and the equatorial axis as 7,926.68 miles (12,756.78 km). The polar axis is thus 26.70 miles shorter than the equatorial, and the earth is called an oblate ellipsoid (*oblate* meaning the poles are flattened as opposed to *prolate*, meaning sharpened). It is by means of gravity that we can measure the departure of the earth from a true sphere and determine the flattening of the poles.

A level line is determined by the level bubble in a surveyor's instrument, and the position of the bubble is normal to the force of gravity. Since the force of gravity is toward the center of the earth, successive level readings from the equator to the pole would describe the departure of the earth from a true sphere. A level line at any latitude must be tied by astronomic observation to the earth's axis of rotation or, expressed in another way, the latitudes are determined by astronomic observation, and if the earth is flatter at the poles than at the equator, the measured distance between latitudes should become greater as we approach the poles. This is explained in Figure 5–17. That the distance between latitudes becomes greater the farther north we go was settled by expeditions to Lapland and Peru sponsored by the Royal Academy of Sciences of Paris in 1736 and 1737.

Unit of gravity used on gravity maps

The unit of gravity for standard measurements is an acceleration of 1 cm/sec^2 and is called the *gal*, after Galileo. But the acceleration of gravity over the United States ranges through about ½ gal, and since gravimeters are easily capable of measuring better than one thousandth part of a gal, the *milligal* is the common unit used. Thus the gravity contours of the gravity map of the United States, Figure 5–18, are in milligals. The so-called zero gravity contour corresponds approximately to the gravity at Washington, D.C., which is 980.080 cm/sec^2.

Reasons for global variations in gravity

Let us suppose that the earth were perfectly spherical and composed of concentric layers of homogeneous material arranged with increasing density inward. Under these conditions gravity would be the same in all places at the surface, providing only that one correction be made. This is for the variation of acceleration due to the rotation of the earth. Such an effect is greatest at the equator and decreases northward and southward. But the earth is an oblate ellipsoid, and thus a further correction to account for different distances to the center of the earth must be made in computing gravity at any place. The normal sea-level values of gravity at 15° intervals northward or southward of the equator, taking into account the rotation of the earth and the polar flattening, are as follows:

DISTANCE FROM EQUATOR	CM/SEC^2
0°	978.049
15°	978.394
30°	979.388
45°	980.629
60°	981.924
75°	982.873
90°	983.221

Figure 5–13 Total magnetic intensity of an area off the California coast, recorded by a magnetometer towed by a vessel at sea. Contour interval is 50 gammas, solid contours being positive and dashed contours negative. The east-west discontinuity marks the Murray ocean floor escarpment, which is regarded as a fault zone of horizontal displacement. This magnetic pattern suggests that the horizontal displacement has been from A to A', a distance of 84 nautical mi. and leads to the concept of faulting on the ocean floor and the theory of ocean floor spreading. See Chapter 11.

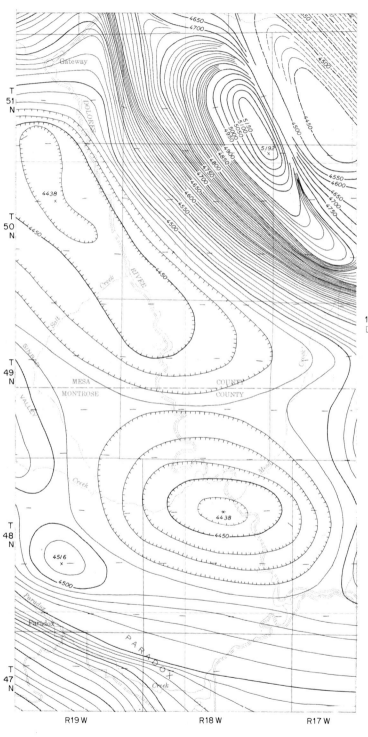

Explanation

Magnetic contours and flight transverse

Contours dashed where data are incomplete; contours show total intensity relative to an arbitrary datum

Magnetic contour enclosing area of lower magnetic intensity

Measured maximum or minimum intensity within closed high or closed low

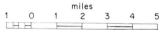

Contour interval 1 milligal

It must be evident from Newton's law of gravity that the higher we go above sea level the smaller will be the pull of gravity, and thus in order to compare gravity at sea level at one place with gravity in a high-level mountainous place a further correction for elevation must be made. In order to do this, the following data, based on sea-level measurements at the equator, are applicable.

FT ABOVE SEA LEVEL	CM/SEC²
20,000	976.169
15,000	976.639
10,000	977.108
5,000	977.579
0	978.049

Bouguer anomalies and correction

Pierre Bouguer had concluded that the mass of the Andes above sea level would exert its own gravitational pull, and this he proved in an eighteenth-century expedition to Peru by observing the deflection of the plumb line. In Figure 5–19 the relation of ocean basin, sea level, and mountains (the Andes) are shown, but the plumb line (normal to the level line) does not point to the center of the earth. It has been attracted a little laterally by the gravitational field of the Andean mass. If we could dig a sea-level canal through the Andes, the water would stand a few tens of feet above sea level through the Andes, and this water level, if projected through the continents and over the ocean basins would establish a new surface slightly different from the oblate ellipsoid. It would have low bumps under the high plateau and mountainous regions and would sag a little over the ocean basins. This bumpy surface has been called the *geoid*.

And thus, in order to reconcile the calculated (theoretical) value of gravity with the value we actually measure in any one place we must

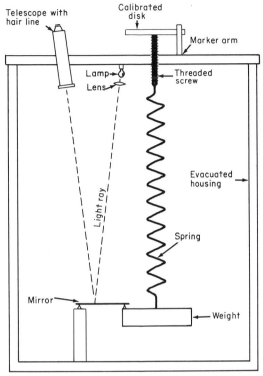

Figure 5–15 Schematic drawing of a spring gravimeter (stable type). [After B. F. Howells, Jr., 1959.]

make a further correction in the computed value to conform with the geoid surface. This is the Bouguer correction. The map of Figure 5–18 shows the variations of gravity over the United States with the four corrections thus far discussed: (1) the correction for rotation of the earth; (2) the correction for the flattening of the poles; (3) the correction for altitude; (4) the Bouguer correction for the geoid surface.

In connection with the geoid surface, it has now been discovered in tracking artificial earth satellites—whose orbits are determined by gravity, speed, and original projectory, as well as by the equatorial bulge and polar flattening —that the earth does not have quite the general

Figure 5–14 Aeromagnetic map of the Uravan area, Colorado. [Reproduced from U.S. Geol. Surv. Prof. Paper 316, by Joesting and Byerly, 1958.]

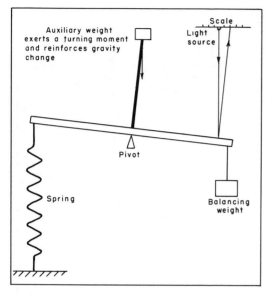

Figure 5–16 *Schematic drawing of an unstable type of spring gravimeter.*

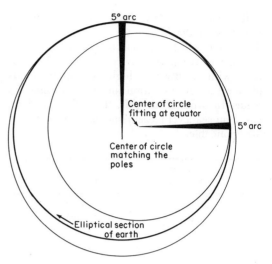

Figure 5–17 *The curvature of the earth at the poles requires a larger circle than that of the equator. Hence, the distance between latitudes is greater at the poles than at the equator.* [*After Strahler, 1963.*]

oblate ellipsoid shape that geodesists had previously assigned it. The North Pole is about 50 feet higher than it was formerly thought to be (compare this with the polar radius of 13 miles less than the equatorial), the South Pole is about 50 feet lower, and the northern middle latitudes of the southern Hemisphere about 25 feet higher. Thus the newspapers have proclaimed that the earth is pear-shaped, but really, not so you would notice it.

Isostasy

An observation similar to that of Bouguer in the Andes was one in India by the British geodesist Everest. He was precisely measuring the lengths of degrees of latitude in the Indo-Ganges Plain, and discovered that the plumb line was not drawn as much toward the Himalaya Ranges as he had calculated (see Figure 5–20). What does this mean? According to Archdeacon Airy, an associate of Everest, it meant that the density of the rock mass under the Himalaya Ranges was less dense than that under the Indo-Ganges Plain, as illustrated in the upper sketch of Figure 5–21. Airy viewed the crustal make-up under the Indo-Ganges Plain and Himalayas as shown in the lower drawing of Figure 5–21.

Geologically, this can only mean that as the Himalayas are eroded and the waste is deposited in the Indo-Ganges Plain, the root of the mountain mass rises and the crust under the plain sinks. In other words, as mass is removed from one part of the crust it rises, and as mass is added to another, it sinks. As a corrolary, the subcrustal material must be weak, potentially fluid, and must flow horizontally from under the plain to under the mountain. (See Figure 5–22.) That the crust is sufficiently weak to be depressed or uplifted under these influences, and moreover, that the subcrustal material is weaker still, in fact will flow under long-continued stress, has come to be considered almost as a geological law. The equilibrium state of the crust, which essentially is floating and at rest on the subcrust, has been called *isostasy*. Isostasy is not a mountain-building force nor a process, as some have considered it, but a state of equilibrium. By means of this state and concept

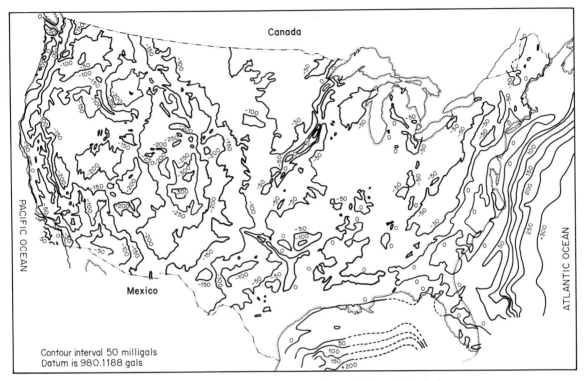

Figure 5–18 Bouguer gravity anomaly map of the United States. [After G. P. Woolard and H. R. Joesting, 1964.]

crustal structures and processes can be intelligently undertsood.

Bouguer gravity map of the United States

The map of Figure 5–18 is the result of many years of field work and compilation, particularly of Professor George P. Wollard. It serves to indicate a number of crustal features as well as a base for local gravity surveys. The original map with all its details may be procured from the U.S. Geological Survey. The obvious features are the large positive anomalies of the Atlantic Ocean and the Gulf of Mexico, the modest negative anomalies of the Appalachian Mountains and associated plateaus, and the high negative anomalies of the western Cordillera. Then there is a belt of positive anomalies extending southerly from the western end of Lake Superior. From the colored geologic map of the United States that probably hangs on your classroom wall, you will see that this is a region of sedimentary rocks, not evidently different from those on either side. But at Duluth the positive anomalies are due to exposed thick, ancient, gabbro sills—rocks with above normal density—and so it is postulated that the same rocks exist southward under the sedimentary veneer to form the belt of positive gravity anomalies.

The modest negative anomalies under the Appalachians suggest that the mountain roots are shallow, while under the Cordillera they should extend to considerable depth, and the crust should be thick.

Gravity anomalies in exploration

An example of a gravity map made to assist in geologic interpretations is given in Figure

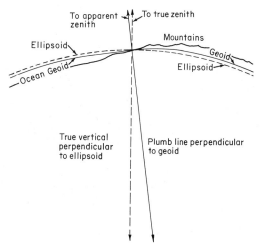

Figure 5–19 *Highly exaggerated cross section of an ocean and continent with a mountainous terrain showing relation of ellipsoid to geoid. [After Heiskanen, 1958, and Strahler, 1963.]*

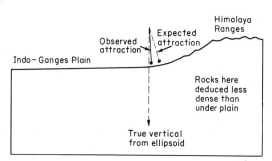

Figure 5–20 *The Himalaya Ranges attract the plumb bob less than expected.*

cates that more is involved than just faulting and confirm the deduction that the rocks underlying the valleys within the grabens are of unusually light density. Assuming that the light-density rock is rock salt, which has been demonstrated by well-drilling, the possible subsurface structures are shown in Figure 5–24. It is confidently postulated that Paradox Valley, especially, is underlain by a piercement type of salt dome with a base at least 10,000 feet deep. The bounding graben faults turn out to have originated when the top part of the salt dome was dissolved by ground waters, and as a result the surficial rocks somewhat subsided.

In this and the previous chapter the three basic physical approaches to the study of the earth have been reviewed, those of earthquakes,

5–23. The high negative anomalies occur in Sinbad and Paradox valleys, and confirm the geologic observations that the valleys are grabens. The magnitude of the anomalies indi-

Figure 5–21 *The Airy hypotheses of isostasy and mountain roots. The crust consists of variable length blocks but all of the same density.*

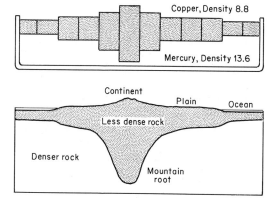

Figure 5–22 *Isostatic responses to erosion and sedimentation.*

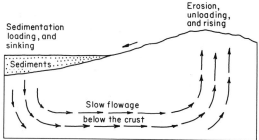

Figure 5–23 *Bouguer gravity anomaly map of part of the Uravan area, Mesa, and Montrose Counties, Colorado. [From H. R. Joesting and P. E. Byerly, 1958.]*

miles

Contour intervals 10 and 50 gammas
Flown 500 feet above surface

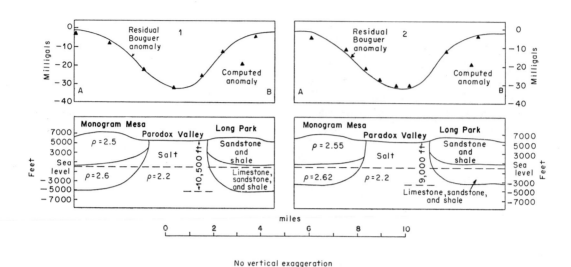

Figure 5–24 Interpretations of the Paradox Valley gravity anomaly. [From Joesting and Byerly, 1958.]

magnetism, and gravity. The study of surface features in the traditional geologic manner would have stagnated had it not been for geophysical help. Geophysical tools and theories have not only let the way in discerning earth constitution and processes but have also been of great help in the search for mineral deposits and oil and gas accumulations.

References

Dobrin, Milton B. *Introduction to geophysical prospecting.* 2d ed. New York: McGraw-Hill, 1960.
Howell, B. F., Jr. *Introduction to geophysics.* New York: McGraw-Hill, 1959.
Jacobs, Russell, and Wilson, J. Tuzo. *Physics and geology.* New York: McGraw-Hill, 1959.
Runcorn, S. K. *Continental drift.* New York: Academic Press, 1962.
Strahler, A. N. *The earth sciences.* New York: Harper & Row, 1963.

geologic time and earth history

The history of the earth has been of basic concern to scientists over the past 100 years. How old is it? When were the Alps, Appalachians, and Rocky Mountains built, and in what order? What has been the nature of climatic changes, and if the climate changed, how many times and when? How long ago did the dinosaurs live, and what was the earth like in those days? Have the oceans and continents always had the same shape and arrangement? Have the continents been invaded by the seas? In fact, can we trace the evolution of the continents and ocean basins from early beginnings? And most absorbing is the procession of life on the earth from remote times to the present. Do we have sufficient fossil evidence to establish the main events in the adaptation of life to changing climates and geographies? Do the changes in plant and animal forms that we recognize in fossils occur systematically through geologic time and support the theory of evolution?

To start the study we will numerate the major time divisions as they have been recognized by geologists and then we will explain how this time scale was chosen.

The main divisions of geologic time are called *eras*. Five eras are recognized: the Archean, Proterozoic, Paleozoic, Mesozoic, and Cenozoic. The eras have been further divided into *periods*, the periods into epochs, and some of the epochs into ages. The geologic time scale in these units is shown in Figure 6–1, which also includes the ages in years and the episodes of mountain building that occurred in North America. It is interesting to note that this scale is internationally used, and there is no controversy about its major divisions. The unsolved problems regard the transition times between periods and epochs, the exact ages in years of the many historical details that characterize each unit, and, of course, the filling in of many more details yet unknown. The details are especially vague or wanting in the first two eras.

The establishment of this time scale has consumed the efforts of thousands of geologists over the past 175 years. It is a scale built upon the relative position of events—such as times of deposition of sedimentary sequences, igneous intrusions and extrusions, mountain building, ocean basin development, world-wide glaciation, and the times of many critical events in organic evolution. The names of the periods had been standardized and a wealth of information about each had been gathered together in scientific journals before much accurate information on their age in years was available. It was not until the past 25 years that geochemists found ways of attaching absolute ages (time in years) to the divisions of the geologic time scale. As these ages in years have been resolved, it is interesting to note that not one event in the time table has been found out of order. Some periods have proved a little longer than expected and some events started a little sooner than had been suggested, but their original order remained.

Dating by classical geologic methods

Superposition and unconformities

The relative time scale was established by classical geologic methods before the development of absolute time determinations by isotope dating. By classical methods we mean the methods of the field geologists and paleontologists who observed and mapped rock formations in an effort to discover their relationship to each other, which then provided clues about their relative ages. These men were aided by the discovery of fossils. Fossils found in a formation in one area could be compared with similar fossils found elsewhere, and strata containing similar fossils were thought to be of similar age. Since fossils are found in sedimentary rocks, it was the places of fossiliferous, well exposed, and generally thick sequences of sedimentary rocks that gave the names to the periods. For instance, Cambrian comes from *Cambria*, the Latin names of Wales, and Jurassic from the Jura Mountains in Germany.

If we consider a series of sedimentary rock layers, we realize that the one at the bottom must have been deposited first, and the one at the top last. The bed on top is therefore younger than the bed underneath. This observation is so obvious that it hardly needs to be mentioned; yet it is the most basic guide in establishing a series of geological events. It is an indisputable law of geology, called the *law of superposition*.

Of great importance in the establishment of the geologic time scale is the principle of *uniformitarianism*. By uniformitarianism is meant that natural processes, generally acting slowly and in the manner and intensity of those that we observe today on the earth's surface, are sufficient to account for past geologic changes. We should interpret the past by the present! Most earth scientists regard this concept as a principle and it has been especially cited by many writers because a contrary concept held sway before the scientific era. This is the doctrine of *catastrophism* which holds that all geologic changes have been sudden, catastrophic in nature, and often terribly destructive. Such a view has religious roots. Nowadays, we must conclude that past changes have occurred in the manner of those we see today, such as the erosional processes treated in this chapter and in Chapters 7 and 8, and mountain building treated in Chapters 2, 3, and 12, but in addition there have been sudden natural spasms, some of catastrophic proportions, which have been superposed on the uniform pace of processes and rates. Catastrophic events that must be recognized are severe earthquakes, large meteorite impacts, hurricanes, tsunamis (tidal waves), and great floods. Climatic changes, which have been several in the geological past, are generally slow in coming and leaving, but some have climaxed suddenly, it would seem, with the extinction of many species of life. These, too, are considered catastrophic by some.

Figure 6–2 is a simple example of superposition. Let us start at the bottom (oldest) layer and list the series of events that can be discerned from the strata.

The oldest deposits here exposed are river terrace gravels containing fossil remains of the mammoth, now extinct. Pictures of mammoths were drawn on the cave walls of France and Spain by men of the Stone Age, and therefore, the great beasts lived contemporaneously with prehistoric man.

On the old terrace gravels was a coarse ash fall (scoria), and on this layer a forest grew and a soil developed. Then came another volcanic ash fall (pumice, this time), which burned and charred the forest in the process of burying it. It is evident, therefore, that the

GEOLOGIC TIME AND EARTH HISTORY

Era	Period	Epoch	Millions of Years Ago	Mountain-Building Episodes (Orogenies)
Cenozoic	Quaternary	Recent	0.001*	Coast Range (W) †
		Pleistocene	1	
	Tertiary	Pliocene	11	
		Miocene	25	
		Oligocene	40	Basin and Range (W)
		Eocene	60	
		Paleocene	70	Laramian (W)
Mesozoic	Cretaceous		135	Nevadian (W)
	Jurassic		180	Palisadian (E)
	Triassic		225	
Paleozoic	Permian		270	Appalachian (E)
	Pennsylvanian			
	Mississippian		350	Antlerian (W)
	Devonian		400	Acadian (E)
	Silurian		440	Taconian (E)
	Ordovician		500	
	Cambrian		600±	
Proterozoic	Late		800±	} Lesser orogenies (W)
			1000±	Grenvillian (E)
	Middle		1300±	Lesser orogeny
	Early		1700±	Hudsonian
			2500±	Kenoran
Archean				

† (W) Indicates the orogeny occurred in western part of continent; (E) In eastern part
*10,000 years

Figure 6–1 Geologic time scale.

mammoths of the river terrace lived some time before this pumice fall. Finally, a soil developed on the pumice layer.

Then a thick pumice flow spread over the area. On this was deposited a layer of sand, gravel, and loam (silty sediments) in which shell heaps and other evidence of early man's existence have been noted. This layer was probably spread by the shallow sea invading the land and then withdrawing. Enough time elapsed after the sea withdrew to form a fourth soil.

Now, from another volcanic center more pumice falls blanketed the area, and from ^{14}C analysis this occurred about 5000 years ago. Time elapsed and still another soil formed. Then, a succession of pumice falls from a third volcanic center occurred. An interim of soil formation occurred between each, and the last soil has been developing since A.D. 1739.

Now let us look at a series of sedimentary strata in the Colorado Plateau. The sequence is portrayed diagrammatically in Figure 6–3.

The formations are analyzed as follows, beginning with the lowest:

1. Dark gray mud, now shale, deposited on a shallow sea floor fairly remote from the shore

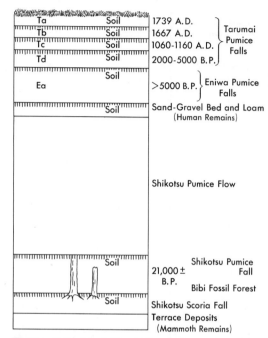

Figure 6-2 *Ash beds of three volcanoes and nine eruptions at Bibi, Hokkaido, Japan. B.P. means years before the present.*

line. Marine fossil types indicate shallow marine water. An arm of the ocean thus spread over this part of western North America at the time the dark gray muds were accumulating.

2. The sea retreated bringing the shore line with its sands and river deposits across the former mud bottom, thus spreading layers of sand over the mud. The coast was low and swampy in places, and peat beds formed, later changing to coal.

3. The sea advanced, and the area again became one of shallow marine deposition of muds. The shore line was fairly distant.

4. The sea again withdrew, and the shore-line sands were spread across the bottom.

5. In still another oscillation the sea advanced, and the area again became one of shallow marine mud deposition. The three advances of the sea and the two withdrawals, thus far described, may be due to elevations and subsidences of the land or to rises and falls of sea level.

6. Advancing and retreating shore line.

7. The sea withdrew, and the low-lying region became one of river flood-plain and swamp deposition. The swamps were extensive, lasted for a long time, and thick beds of peat accumulated in them. Gigantic dinosaur footprints are common in tops of the coal beds, which were once the swamp peats.

8. River deposits of sand and conglomerate. Evidently mountains were being uplifted nearby, and floods of sand and gravel were spread over the swamps and broad flood plains.

9. Extensive river flood plains, on which red, purple, chocolate, green, and gray muds were spread, succeeded the sand and gravel. The colors are due to the erosion of tropical or subtropical soils in which iron oxides of several kinds produce the various colors.

A few dinosaur skeletal remnants occur in the lower part of this formation, but nowhere has anyone found dinosaur fossils in higher and younger beds. This, then, is apparently the time of extinction of the dinosaurs and is taken arbitrarily to mark the termination of a great era of geologic time, the Mesozoic, and the beginning of another, the Cenozoic.

10. Somewhere nearby an uplift occurred, and a body of fresh water was impounded in which lime mud or marl was the chief deposit. Fresh-water snails and clams are found in the lime muds, which are now limestone.

The above examples of interpreted histories by the classical method are but two of very many that have been documented. The first represents a span of time in Japan which we might call almost modern, geologically, and the second takes us back to the Cretaceous Period in western North America. Before discussing how we can relate these two formations, the one in Japan and the other in North America, let us consider unconformities and their interpretation.

The top drawing of Figure 6-4 shows that a lower series of beds, 1, was eroded and then covered by the beds of series 2. If the fossils in series 1 indicate that the beds were deposited in shallow ocean water, then uplift of the beds

GEOLOGIC TIME AND EARTH HISTORY   **141**

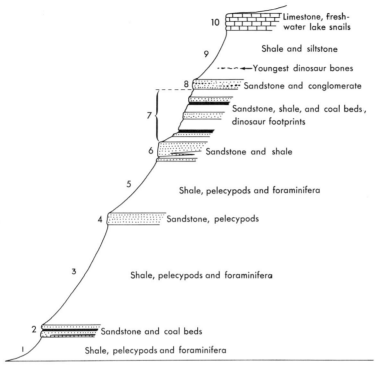

Figure 6–3 *Succession of formations exposed on an escarpment in the Colorado Plateau.*

above sea level must have occurred, whereupon they were dissected by streams. The erosion surface was subsequently buried by series 2. Now if series 2 is the flood-plain deposit of rivers, then we need only postulate a change in the regimen of the streams from one in which they eroded to one in which they deposited. Lowering of the gradient by the tilting of the land, change in climate, or an uplift in the headward reaches of the river system might have caused the streams to change from erosion to deposition. If, however, the upper series of beds contains marine fossils, then we must postulate subsidence below sea level after the uplift. Thus, it can be seen that a simple unconformity, like that shown at the top, contains a significant story.

Referring to the unconformity in the second drawing, we see that, first, series 1 was deposited, then folded, then eroded, and then covered by series 2. In the third drawing the same sequence of events occurred, but in addition and lastly, the beds were faulted. In the fourth drawing we note that the lower series was folded, then faulted, then eroded, before being covered by series 2.

We read the fifth drawing as follows: first, deposition of series 1; second, folding of series 1; third, erosion of the folded beds; fourth, deposition of the upper series 2; fifth, tilting of the upper series and all rocks underneath; and sixth, erosion of the tilted beds.

Magmas that penetrate or intrude are younger than the rocks so intruded. Dikes, sills, laccoliths, stocks, and batholiths are all younger than the rocks in which they are emplaced. Lavas and all pyroclastic rocks extruded on the surface are younger than the rocks that they cover. If the lavas and pyroclastic rocks are cut by dikes, then the dikes are younger than the

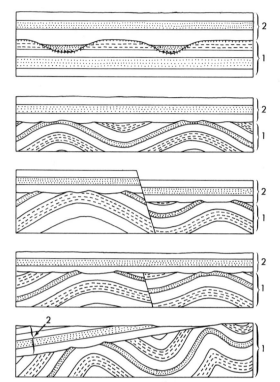

Figure 6-4 *Sequences of events connected with unconformities.*

extrusive rocks. Turn back to the chapter on igneous rocks and volcanoes, and observe some of these relations in the diagrams there.

Fossils as time markers

Do fossils occur systematically?

Fossils have been our major medium of correlation, especially for sedimentary sections long distances apart in which there are no common recognizable beds. All the major plant and animal groups are represented in fossil form. Perhaps you have been on fossil-collecting trips and have discovered that specimens are abundant in some formations, scattered in others, and probably absent in still others. Now, the important question arises, Is there a system, a recognizable order in the occurrence of fossils, or do we find them in haphazard and unpredictable occurrences and associations? The answer is, Yes, there is a meticulous order and system in the occurrence of fossils. If we will only search for all the forms and details, and study and classify the specimens carefully, this order will become clear. The principle was recognized by William Smith of England 150 years ago. It is absorbing to read how he came about the discovery.

Discoveries of William (Strata) Smith

William Smith was born in 1769. Eight years later his father died whereupon his uncle, a crotchety farmer, unmarried and with little interest in anything but manual labor, agreed to raise the lad. William's early collecting of "pundibs," or fossil clams, and poundstones, or fossil sea urchins, as well as his interest in geometry in school, was nonsense to his uncle. At eighteen he became a surveyor's assistant, and from then on his eager interests were allowed expression:

For three years he held the rod and dragged the chain, but he also observed different kinds of soil, noted "agricultural and commercial appropriations" of the land, and studied borings for coal. . . . At twenty-two he began independent work which led to a detailed subterraneous survey of some important coal mines. It showed that certain rocks always lay in a definite order, a fact that the tradition-bound miners stubbornly refused to accept.[2]

At twenty-five William Smith became surveyor and engineer of the Somerset Coal Canal, and for six years, while surveying its course, he examined the strata through which it was being excavated. He noticed the dip of the beds, and by knowing the succession, he was able to predict the formations across which the canal would be extended. He was always fascinated with the fossils that almost everyone dismissed

[2] Carroll L. Fenton, *Giants of geology*, Garden City, N.Y., Doubleday, 1952, p. 71.

as curios. He made the very basic discovery that each formation or lithologic unit had its own peculiar fossils and that he could rely on this occurrence. As far as he traced a certain group of shale beds, for instance, they had the same fossils; no irregular mixtures of fossils of other beds appeared. This was of great help to him, because in a succession of interlayered sandstone and shale formations the several shales were much alike, and the sandstones were also alike, but not the fossils. By the fossils he could tell whether or not he was in the bottom of the sequence or near the top. He identified one formation by its thick poundstones, another by its thin poundstones, another by its pundibs, and still another by its petrified nuts. He was able to use the fossils intelligently even without scientific classification and name.

Years later he became a respected member of the British scientific community, and in connection with the printing of his geologic map of England in 1815, which was a remarkable and pioneering achievement, and the first of its kind, he wrote:

I have devoted the whole period of my life . . . to prove that there is a great degree of regularity in the position and thickness of all these strata; and although considerable dislocations are found in collieries and mines, and some vacancies in the superficial courses of them, yet . . . the general order is preserved; and . . . each stratum is also possessed of properties peculiar to itself, has the same exterior characters and chemical qualities, and the same . . . fossils throughout its course. I have, with immense labour and expense, collected specimens of each stratum, and of the peculiar fossils, organic remains and vegetable impressions, and compared them with others from very distant parts of the island, with reference to the exact habitation of each, and have arranged them in the same order as they lay in the earth; which arrangement must convince every scientific or discerning person, that the earth is formed as well as governed, like the other works of its great Creator, according to regular and immutable laws, and which form a legitimate and most important object of science.[3]

Fossils of the Paris Basin and their significance

"Strata" Smith, as he became affectionately known, pointed up the significance of fossils so clearly that a number of able scientists soon began systematic collecting and made new discoveries. The beds of the Paris Basin in France yielded many fossils, especially the series of beds which had come to be known as the Tertiary. Three main groups of formations were recognized in France and Germany in these early days of the science of geology, the Primary (oldest), the Secondary (intermediate), and the Tertiary (youngest). In 1883 Sir Charles Lyell's book *Principles of Geology* appeared, and one of the notable contributions in it was the discussion of the Tertiary strata of the Paris Basin and the fossils in them. He recognized four divisions.

In the uppermost and youngest division, containing several hundred different species of shells, 90 to 95 percent of them are species still living and found today especially around the Mediterranean, whereas 5 to 10 percent are not found and evidently are extinct. In the next older division of strata (those next lower in the sequence) 35 to 50 percent are represented by living forms, and 50 to 65 percent are extinct. In the next older division only 17 percent are still living, and in the lowest and oldest division only $3\frac{1}{2}$ percent are represented by living forms. Besides the marine mollusks there were vertebrate fossils, and these were recognized as species not now living.

From the geologic facts gained in good measure from the Paris Basin, Lyell published several concepts that are now regarded as truisms:

1. Changes in plants and animals occur gradually. The older theory of sudden and profound changes in the organic world, lacking in phylo-

---

[3] Ibid., p. 71.

genetic order and due to mysterious or cataclysmic causes is "calculated to foster indolence and to blunt the keen edge of curiosity."

2. Certain species have greater longevity than others. Some are quite susceptible to changes in climate, particularly temperature changes, and vanish under the new conditions. Other species survive the changes and live through the time consumed in the deposition of hundreds or even thousands of feet of strata.

3. Some species are able to migrate freely and follow the course of climatic change, becoming extinct in the original area but surviving in another. One species, Melania inquinata, a fossil in the lower beds of the Paris Basin, and extinct in the European locale, still lives in the Philippine Islands.

4. By comparing fossil assemblages with modern assemblages and by noting the environmental conditions under which the present species live, the climates of the past can be deciphered. It is clear that the climate of the Paris Basin was tropical at times during the Tertiary, whereas now it is temperate.

5. Lyell affirmed the conclusion of Strata Smith that fossils could be used for correlation. The "blue clay" of London and the "coarse white limestone" of Paris are marked by a considerable variety of similar, peculiar, extinct species of testacea and by a paucity of living forms. He concluded that the two formations are of the same age, and this has been amply verified since. This realization enabled him and other geologists to relate sedimentary successions across the English Channel, and by the same reasoning fossil-bearing sequences have been related across the oceans.

Guide fossils and correlation

Great collections of fossils have been accumulated from strata the world over since the time of Smith and Lyell, and much is known about the types of plants and animals that lived through long spans of earth history and about those which had a very short life span. The time ranges of thousands of species are now well known, and the procession of life forms, as they appeared on the earth, changed, and died off is fairly well documented. It is an absorbing story but must be left for other courses.

Those fossil species which can be used best for correlation must have two chief characteristics. They must be short-lived and must have had a widespread distribution. It helps little if an animal or plant evolved and became extinct in a certain basin of sedimentation but did not spread to other areas of sedimentation. Likewise, if a certain species spread along shore lines to many places around the world but changed little over long spans of time, it cannot be used for correlation. For practical purposes it helps considerably if the individuals of the species were very abundant and fossilized well. This means that if the fossils are easy to find they are readily useful. It is discouraging to have to search for rare specimens to solve practical problems of correlation.

The most abundant and useful forms for correlation are trilobites, the shelled animals (the brachiopods, cephalopods, clams, and snails), and the microscopic forms, the foraminifera. In fresh-water lake and river flood-plain deposits plant and vertebrate fossils are important. Sedimentary rocks originally deposited in marine waters are very abundant in almost all mountain ranges, in plateaus, where we can examine them in outcrops, and under plains, where they are penetrated by wells drilled for oil and gas. It is by means of guide fossils in such marine strata that we tie the geologic history of one continent with another across the oceans.

How did the shelled forms and the trilobites migrate from one continent to another? Since similar species are confined to shallow water today, we conclude that related forms in the past were also limited to shallow water, and thus shallow waterways along banks and continental shelves must have existed from one continent to another at times in the geologic past. These were the migration routes for brachiopods, clams, snails, corals, bryozoans, and many others. Species that lived in the upper layer of the oceans and composed the plankton were free-swimmers or drifters in the ocean currents and spread from one continent to another without being restricted to shallow-water migration routes. The cephalopods, graptolites, and foraminifera were of this group. Thus, any short-lived species of this group makes an excellent guide fossil. Another Smith, James P. Smith, professor at Stanford University about fifty years ago, had been working on the many species of elaborate cephalopod shells, and among the thousands of described species he recognized that a singular assemblage of species from the Himalaya Mountains was almost precisely the same as an assemblage from the Wasatch Mountains in Utah. We know now that these fossils were deposited at a time in earth history that we might call medieval in the evolution of life, and we conclude that at that time the ocean currents transported the floating cephalopods in a way that distributed the same species in Asia and western North America. Earlier studies had brought out the striking similarities of trilobites in New England and the Maritime Provinces with those of the British Isles. These forms migrated along shallow-water connections, possibly by way of the North Atlantic, or as some believe, the shallow sedimentary basins were continuous at a time when Europe and North America were in close proximity (see Chapter 12).

With increasing understanding of the fossil record, geologists in Europe and North America gained greater insight into earth history. Soon they were proposing names for the major and minor divisions. These names, as we have mentioned, were derived from local areas in which major discoveries were made. Certain intervals were first studied and better known in England, others in Germany, others in France, and somewhat later, still others in North America. There was the need for a standard geologic time scale that geologists all over the world would recognize and could use as a scaffold on which to build and relate their local histories. For purposes of study of world-wide climatic changes, intercontinental plant and animal migrations, and widespread mountain building, we must be sure of the time at which the events occurred.

Geologic maps and the record

Importance

Strata Smith's concern was not only with the systematic occurrence of fossils in the numerous formations of England but with the possibility of charting the outcrops of the formations on a map. He came to realize that almost all the formations he had observed were fairly uniform in character and thickness over considerable distance and that they could be traced by simply learning their characteristics and "walking them out." He had noted futile efforts to farm certain areas, which, when he examined them, always turned out to be on a certain formation. Now, if a geologic map were available, the lands on this formation could be easily avoided:

By a classification of soils according to the substrata good practical farmers may choose such as are best suited to their accustomed mode of management, and they may thus be tempted to transfer useful and well-established practices in husbandry to many parts of the same stratum as are still highly susceptible to improvement.[4]

Smith noted also that "immense sums of money were imprudently expended in searching for coal and other minerals" in formations other than those that contain these substances. He noted that the shipping canals, so vital to transportation of bulk materials, could be built easier and the water supply and water surplus problems could be more readily solved if the formations were considered. By means of his geologic map not only the formation at the surface at any one place could be told, but successive formations underlying the surface could also be confidently predicted. This should help in the drilling of water wells and in the search for limestone, brick and pottery clay, fertilizer rock, as well as coal, oil, and gas. Nothing has been more basic than the geologic map to the mineral and industrial economy of the prosperous nations of the world. As he wrote:

[4] Ibid., p. 3.

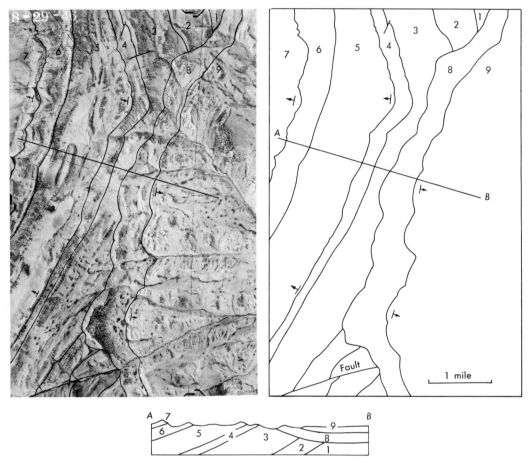

Figure 6–5 Left, vertical aerial photograph with contacts inked in; right, corresponding geological map; bottom cross section A–B. Oldest formation is labeled 1, youngest 9.

The wealth of a country primarily consists in the industry of its inhabitants, and in its vegetable and mineral productions; the application of the latter of which to the purposes of manufacture, within memory, has principally enabled our happy island to attain her present preeminence among the nations of the earth.[5]

The basic purpose of the U.S. Geological Survey is to prepare a detailed geologic map of the country and to have it printed for all to use. Progress has been appreciable but slow, and many states have helped attain the goal by supporting mapping projects of their own. Although the geologic map is a basic tool in exploration and development, it has still another highly important use and purpose, namely, the assistance it gives in working out the geologic history of a region. This is the main point of the present discussion of geologic maps.

[5] Ibid., p. 71.

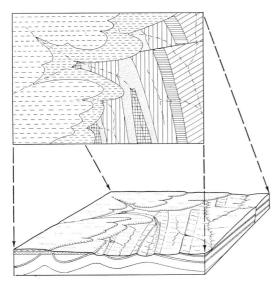

Figure 6-6 Angular unconformity in perspective and in map view. Contacts of the upper formations are continuous, and contacts and fault of the lower formations are discontinuous.

Methods of survey

Geologic maps were originally made by standard surveying practices, and almost all geologists until twenty years ago had to be trained in surveying. They spent almost all their time in the field making maps and precious little time on the rocks themselves. About thirty years ago vertical aerial photography made a dramatic impact on the work of the field geologist, and nowadays, base maps of all kinds are made principally in the laboratory from photographs. This new science is called *photogrammetry*. The geologist now takes the photographs into the field, notes the contacts of the different formations directly on the photographs, and then these are compiled by the photogrammetrist on base maps. The photographs commonly reveal the trend of the strata better than they can be seen in the field and usually show elaborate detail. They save the geologist endless tramping and climbing and lead him to the most critical places for observation.

Figure 6-5 is an example of a vertical aerial photograph (left) with contact lines between the recognizable formations drawn on it. It becomes, then, a geologic map, which would be printed as shown on the right.

Interpretation

The succession of formations can be recognized on a geologic map. Folding and faulting of the beds is evident, and intrusive and extrusive relations can be made out. Often obscure from field observations are unconformities, but these will be detected if a map is made by tracing out the formations. Study Figure 6-5 and its accompanying cross section. Figure 6-6 is a block diagram and geologic map of an angular unconformity. An early series of beds was folded, eroded, and then covered by a later series. During the modern erosion cycle some of the younger beds have been removed, leaving an irregular escarpment. The geologic map pattern may be analyzed by noting that the contacts of the upper series of formations are continuous but that those of the lower series terminate, or are discontinuous, against the basal beds of the upper series. Now this is the relation that will emerge if an angular unconformity exists between two sequences of beds when their contacts are traced out in the field. From casual inspection, without mapping, the relation may not be realized.

The geologic relations of Figure 6-6 are much the same as those in west-central England, where Strata Smith first worked. The older folded and faulted series contained the valuable coal beds, which with the advent of the steam engine were becoming basic to England's rising industrial economy, and Smith knew this. No wonder he was indignant that miners would waste money seeking coal in the upper series. He never quite realized, however, that the folding of the beds of the lower series attested to a major episode of mountain building in England and continental Europe. The mountain building is now known to have occurred in an extensive belt through southern England and

Ireland, France, Germany, and other adjacent countries, and it occurred after the lower series was deposited and before the upper series was laid down.

The working out of stratigraphic sequences, the collecting of fossils, and the recognition of structural and igneous-rock relations goes hand in hand with the geologic mapping of a region. All activities must be integrated carefully to obtain a satisfactory understanding of the history of the region.

Methods of radioactive isotope dating

Although the relative time scale has served amazingly well as an outline of earth history, geologists have always had the urge to find out exactly in years how long ago certain events occurred. In the past twenty-five years this goal has been partially achieved in a major scientific advance. The key was found in unstable isotopes whose half-lives and constant rates of decay are known. It is seen in Figure 6–1 that we have to deal with tens of thousands, millions, and billions of years. The figures stagger the imagination, like astronomic distances. It is not necessary to memorize them, but in round figures they become second nature to a geologist like the names of the periods of the time scale themselves when he works with them a while. The common methods of absolute age determination are as follows.

Emitted rays

With the discovery of cathode rays, which cause fluorescence in vacuum tubes, and the invisible penetrating X rays, it seemed natural that someone would seek to find out what kinds of rays are given off by phosphorescent substances. Many minerals will glow in the dark after exposure to light. In 1896, while working on this problem, the French physicist Antoine Becquerel found out much more. He discovered that a substance containing atoms of uranium continually gives off rays even without exposure to light. Later, other workers found that the spontaneous rays were of three different kinds.

1. Alpha rays—small particles of fairly high velocity with a positive charge
2. Beta rays—small, penetrating particles of high velocity with a negative charge
3. Gamma rays—a form of highly penetrating invisible light rays with no electrical charge

These properties were determined by passing the rays through a magnetic field. The beta rays undergo the greatest deflection, the alpha rays are deflected in the opposite direction and to a smaller extent; and the gamma rays are unaffected.

Additional work by Becquerel, the Curies, Ernest Rutherford, and others established that alpha rays were helium nuclei and beta rays were electrons. Alpha rays can penetrate up to 0.002 cm of aluminum, beta rays up to 0.2 cm of aluminum, and gamma rays up to 100 cm of aluminum.

Decomposition of the nucleus

Rutherford and Frederick Soddy, two British scientists, proposed in 1902 that the radioactivity of uranium involved a decomposition of the atomic nuclei and that the alpha, beta, and gamma rays were emitted from the nucleus in the course of the transformation. The first step in the spontaneous decomposition of uranium involves the conversion of uranium to thorium, with the emission of an alpha particle, and may be represented as follows:

$$^{238}_{92}U \rightarrow {}^{234}_{90}Th + {}^{4}_{2}He$$

The lower left subscripts are the atomic numbers (number of protons in the nucleus), and the upper left superscripts are mass numbers (sum of protons and neutrons). It should be noted that the sum of the atomic numbers for Th and He equals the atomic number of U and also that the sum of the mass numbers of Th and He equals the mass number of U. This kind of a change is called a *nuclear chemical change* or *transmutation*, because the nuclei of the atoms are altered and thus the identity of the atoms is changed. In ordinary chemical reactions only the outer electrons are affected, and the identity of the atom is preserved.

The second step in the transmutation of uranium involves the decomposition of the thorium isotope into a protactinium isotope, with the emission of a beta particle. In the complete process there are 14 successive steps involving intermediate isotopes that are unstable, finally resulting in the stable lead isotope 206. Table 6–1 shows the 14 steps and also the *half-life* of the unstable isotopes of the different elements that result.

Rates of decay

It should be observed that radioactive decomposition takes place at a constant rate that is independent of the temperature, the pressure, and the state of chemical combination. The rate of decomposition depends only on the number of radioactive atoms present. In view of the constant rate of decomposition, it is convenient to express it in terms of the half-life. By this is meant the time necessary for one-half of the original isotope to decompose; for uranium 238 this time is 4.5 billion years. Then one-half of the remaining unchanged half will decompose in the next 4.5 billion years, and so on. This is represented in Figure 6–7.

Because the proportion of the various isotopes contained in a rock sample can be measured and the rates of decay are known, radioactive dating has proved to be immensely valuable in establishing the exact age of rocks and thus in deciphering the past history of the earth.

Lead methods

Table 6–1 shows that ^{238}U disintegrates radioactively in a series of 14 steps to the lead isotope 206 and that half of a given amount of

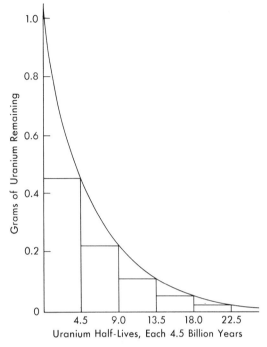

Figure 6–7 Graph of radioactive decay of 1 g of uranium.

^{238}U disintegrates in 4.5 billion years. In other words, the half-life of ^{238}U is 4.5 billion years. The percentage of the amounts of ^{238}U and ^{206}Pb in a sample of a uranium mineral, such as uraninite, may be determined by means of a spectroscope. On the assumption that all ^{206}Pb has been formed by radioactive disintegration since the original uraninite mineral was formed, the ratio ^{206}Pb/^{238}U related to the half-life of ^{238}U is a measure of this time.

However, the uranium ore minerals contain a second uranium isotope, ^{235}U, which is six times more radioactive than ^{238}U and which has a half-life of 890 million years. It produces the lead isotope 207 nearly six times faster than ^{238}U produces ^{206}Pb. Furthermore, thorium isotope 232 is also present in some uranium ores and minerals and disintegrates into the lead isotope 208. It is therefore evident that a

TABLE 6–1 *Uranium-Radium Series of Radioactive Disintegrations (Simplified)*

ATOM	SYMBOL	EMITTED PARTICLE	HALF-LIFE
Uranium	$^{238}_{92}U$		4.5×10^9 years
	↓	α	
Thorium	$^{234}_{90}Th$		24.5 days
	↓	β	
Protactinium	$^{234}_{91}Pa$		1.14 min
	↓	β	
Uranium	$^{234}_{92}U$		2.7×10^5 years
	↓	α	
Thorium	$^{230}_{90}Th$		8.3×10^4 years
	↓	α	
Radium	$^{220}_{88}Ra$		1590 years
	↓	α	
Radon	$^{222}_{86}Rn$		3.82 years
	↓	α	
Polonium	$^{218}_{84}Po$		3.05 min
	↓	α	
Lead	$^{214}_{82}Pb$		26.8 min
	↓	β	
Bismuth	$^{214}_{83}Bi$		19.7 min
	↓	β	
Polonium	$^{214}_{84}Po$		1.4×10^{-4} sec
	↓	α	
Lead	$^{210}_{82}Pb$		22 years
	↓	β	
Bismuth	$^{206}_{83}Bi$		5.0 days
	↓	β	
Polonium	$^{210}_{84}Po$		140 days
	↓	α	
Lead	$^{200}_{82}Pb$		Stable

SOURCE: Charles Compton, *An introduction to chemistry*, New York, Van Nostrand Reinhold, 1958, p. 195.

sample of uranium mineral may yield three lead isotopes from three different sources. All three ratios are separate measures of the time that has elapsed since the mineral was formed.

Therefore the ratio of $^{206}Pb/^{207}Pb$ is also a measure of age and has the advantage of not requiring an analysis of ^{238}U or ^{235}U. The relative abundances of ^{238}U, ^{235}U, ^{232}Th, and the lead isotopes can be determined spectroscopically. If all three ratios yield tolerably close age figures, then the determination is felt to be fairly secure; but if the figures are different, as has happened, then the causes of the discrepancies must be considered, and this is beyond the scope of the present volume. A complication, which is really a refinement, in the lead method of age determination is the discovery of still a fourth lead isotope, ^{204}Pb. This isotope is believed to be original lead built into the mineral at the time it was formed.

We owe much to the pioneering work of Arthur Holmes, of Great Britain, whose work on the age of the earth and in assigning dates to the geologic time scale is reviewed in his paper of 1956, "How Old Is the Earth?"[1] To this question his answer is 4.5 billion years, and it comes from the lead-lead isotope ratio.

The main trouble with the uranium-lead method is the scarcity of appropriate uranium minerals that can be analyzed. They come principally from pegmatite dikes, and these do not serve well in defining the boundaries of the periods of the geologic time scale. We need radioactive minerals that occur in igneous and sedimentary rocks in abundance and on a widespread basis. Such minerals have been recognized, and dating methods to match have been developed. They are described below.

Zircon lead-alpha method

The mineral zircon ($ZrSiO_4$) is used in the lead-alpha method because, it is reasoned, the atoms of zirconium are too small (0.82 Å) to have permitted the much larger lead atoms (1.32 Å) to occupy any of their positions in the crystal lattice during the original cooling of the magma and the crystallization of the zircon grains. Uranium atoms might be present but not lead. Any lead that is detected by spectroscopic analysis has formed by disintegration of ^{238}U, ^{235}U, or ^{232}Th. Alpha-ray counters are used to determine the amount of helium given off per milligram of zircon per hour. The approximate age is then found by the equation $t = CPb/a$, where $C = 2480$, Pb is lead in parts per million, and a is the number of alpha particles (helium atoms) per milligram of zircon per hour. The amounts of each lead isotope are found spectroscopically, and then ages are determined for each isotope. These commonly vary appreciably. However, much useful information has been obtained, even though the age determined may be in error by as much as 25 percent. Zircon grains can be extracted and concentrated from almost all igneous rocks and from many sandstones, arkoses, and graywackes, and hence, the method is applicable to many rocks that can be dated relatively by fossils.

It has been found, for one thing, that the zircon method may yield a greater age than other methods; this probably indicates that the zircon grains, say in a sandstone, came from a much older igneous or metamorphic rock than the sandstone in which they occur. Also this may mean that zircon grains found in a schist are part of the original volcanic material and have not been changed in a later metamorphism, which produced mica grains from which the age of metamorphism is determined by the potassium-argon or rubidium-strontium method.

Potassium-argon method

The isotope ^{40}K is the only radioactive part of common potassium and constitutes only 1/8400 part of it. ^{40}K undergoes a dual transformation: one part to ^{40}Ca and one part to ^{40}A. The ratio of ^{40}Ca to ^{40}A is about 0.1235. The half-life of ^{40}K is 1,310 million years. It is now known that biotite mica retains the ^{40}A that is developed in it by the disintegration of ^{40}K better than the potash feldspars and hence it is used, if possible, in the determination. Biotite mica is present in many rocks, hence the potassium-argon method has proved very important. Specially built mass-spectrometers are necessary to determine the isotope abundances.

Analyses of the entire rock (whole rock analyses) for the potassium-argon ratio have proved successful particularly in regard to

[1] *Trans. Edinburgh Geol. Soc.*, v. 16, pp. 313–333.

lavas that are 1 to 5 million years old. The great value of such analyses has been realized in the work on sea-floor spreading. (See Chapter 12.) Argon begins to form and is trapped as soon as the basaltic magma crystallizes on the ocean floor, and if the crystals are not heated again in some way, the accumulation of argon is a function of the amount of potassium present and the length of time since the decay and entrapment processes began.

Rubidium-strontium method

Rubidium-87 disintegrates to strontium-87 with the emission of beta rays, with a half-life of 4.9 to 5.0×10^{10} years. Rubidium-bearing minerals are the feldspars and micas, and hence many igneous-rock bodies can be dated by determining the proportions of ^{87}Rb and ^{87}Sr that exist in these minerals.

A common mineral in marine sediments, glauconite, forms at the time the sediments accumulate and contains both potassium and rubidium. It is thus amenable to both potassium-argon and rubidium-strontium methods of age analysis.

The dates determined by both the potassium-argon and strontium-rubidium methods are said to be minimum dates. This means that the ages assigned to the minerals are the smallest that they could possibly be; the true ages are probably somewhat greater. The method has undergone refinement and testing and is now widely used, especially for the very old rocks of the crust.

Carbon-14 method

Neutrons produced by cosmic rays convert atmospheric nitrogen to radioactive carbon, ^{14}C, which then combines with oxygen to form carbon dioxide. This radioactive carbon dioxide, along with normal carbon dioxide, is taken up by plants and hence by all living matter. The ^{14}C gives off beta particles and reverts to nitrogen. In 1947 W. F. Libby successfully analyzed carbonaceous items of archeological antiquity by counting the beta radiations and computing the amount of ^{14}C remaining. He thus opened the door to the determination of many important dates in the past 30,000 years.

Most ^{14}C analyses have been made on organic matter of some sort since Libby's work. But it is realized that some CO_2 from the atmosphere becomes part of freshly precipitated limestone, $CaCO_3$, thus the CO_2 in the newly formed limestone is removed from the atmosphere and its ^{14}C atoms proceed to decompose radioactively thereafter, supposedly without further exchange with the atmosphere. The age of certain limestone deposits, if comparatively young, might therefore be determined by the ^{14}C method. Some of the results match organic matter analyses but some are quite divergent and irregular, and dates from $CaCO_3$ must be regarded carefully. The irregular results are possibly due to chemical replacement of new $CaCO_3$ for the old by circulating water, but there is considerable disagreement on the subject of the variability of ^{14}C analyses on $CaCO_3$.

When nitrogen atoms of the atmosphere are struck by cosmic particles they are transformed to an isotope of carbon, $^{14}_{6}C$, which is unstable or radioactive. The nitrogen atom of the atmosphere has seven electrons and a nucleus of seven protons and seven neutrons; when struck by a cosmic particle, it has one of its protons knocked out, leaving six. And this is the number of protons of the carbon atom! The reaction is written as follows:

$$\text{Cosmic particle} + {}^{14}_{7}N \rightarrow {}^{14}_{6}C + {}^{1}_{1}H$$

(See Chapter 1 for designations of chemical symbols.) Thus, the nitrogen atom becomes an isotope of carbon, and an atom of hydrogen is created. The cosmic particle is considered a proton, which then combines with an electron knocked out of the atom of nitrogen to form an atom of hydrogen.

Natural carbon contains about 99 percent of the stable isotope $^{12}_{6}C$, about 1 percent of the stable isotope $^{13}_{6}C$, and a very small percentage of the radioactive isotope $^{14}_{6}C$. There is one $^{14}_{6}C$ atom in about a trillion carbon atoms; nevertheless the radioactivity of natural carbon is measurable. And since $^{14}_{6}C$ isotope is radioactive it disintegrates at a steady rate. It changes back to $^{14}_{7}N$ by acquiring a new proton. The rate of disintegration has been accurately determined, and is such that half of the atoms of $^{14}_{6}C$ will have changed back to $^{14}_{7}N$ in 5568 years. Then half of the remaining half will have changed over in the next 5568 years, and so forth. Rates of radiodecomposition are stated in terms of half-life periods, and thus the half-life of $^{14}_{6}C$ is 5568 years.

The isotope $^{14}_{6}C$ in the atmosphere is soon oxidized into CO_2, but this does not change the identity of the isotope; it is still radioactive. And the CO_2, thus formed, enters the life cycle of plants and animals. All living things have the same proportion of radioactive and stable carbon, and this proportion remains constant as long as the animal or plant is in contact with the atmosphere. When the plant or animal dies, however, the ^{14}C is no longer replaced, and the inherent radioactivity gradually decreases. For instance, the carbon radioactivity of the living wood next to the bark of a tree will show no decline in radioactivity, but as succeeding growth rings, say of the very old redwoods or bristlecone pines, are measured, their radioactivity steadily decreases. Organic matter covered by muds in some natural process of sedimentation, or in refuse piles of early human habitations, is protected from the atmosphere, and its radioactivity gradually declines. Thus the age of the materials may be determined by radiocarbon dating.

You can see that if a piece of carbonaceous material is 5568 years old, half of the radioactivity is gone, and if it is 11,000 years old, only about $3/4$ of the activity is lost. If it is 16,500 years old, only $1/8$ of the original activity is present, and soon the activity becomes so weak that it is difficult to record. In fact, ^{14}C dating is useful only to 30,000 years; beyond that other radioactive isotopes must be utilized.

Fission-track dating

A useful method of dating the low-potassium basalts is the so-called fission-track method. This method is probably more reliable than the potassium-argon method for dating the relatively young basalts near the central trenches of oceanic ridges, and has been used successfully for such purpose. Submarine lavas often form pillows with glassy sheaths, in which an interesting phenomenon may be observed if the glass is properly prepared. Uranium 238 atoms in the glass emit neutrons that travel through the glass and leave damage trails. These trails through the glass may be seen microscopically as thin lines if they are made more conspicuous by leaching. The glass is ground, polished, leached, and then observed by reflected light under a microscope. The number of trails, or the trail density, is a function of the amount of ^{238}U in the glass and the time since the glass first formed.

References

Geological Survey of Canada, Paper 62–17, 1960, through Paper 66–17, 1967. K-Ar age determinations of the Canadian Shield.

Goldick, S. S., et al. Geochronology of the mid-continent region, United States. *Jour. Geophys. Res.*, v. 71, 1966, pp. 5375–5388.

Hurley, Patrick, et al. Test of continental drift by comparison of radiometric ages. *Science*, v. 157, Aug. 1967.

Knopf, Adolf. Measuring geologic time. In *Study of the earth*, ed. J. F. White. Englewood Cliffs, N.J.: Prentice-Hall, 1962.

weathering, mass movements, and river erosion

The paint on an automobile, when subjected to the hot sun and seasonal rains, gradually deteriorates. You may note pits on the fenders and forward parts of the hood due to sand grains and pebbles that have hit the car. We call these processes weathering. Rocks are also subject to weathering, and, in fact, the modern automobile finishes are more durable and resistant to weathering than some rocks. Other rocks, of course are fairly resistant, but none escape serious deterioration over a long time.

The removal of weathered particles by surface processes is called erosion. Mass movements, such as downhill creep of soil and landslides, together with running water, remove and transport tremendous amounts of rock material from one place to another. The weathering processes are static, or take place *in situ*, whereas the erosional processes are dynamic in that they represent motion or transport. The weathering and erosional processes result in most of the land forms whose shape and origin we are about to investigate.

Weathering processes

Frost action

When analyzed, the weathering processes are found to be of chemical and mechanical nature.

We use the word *decomposition* to refer to chemical weathering of rock and *disintegration* to refer to mechanical weathering. In decomposition we deal with solution or chemical reaction, whereas in disintegration we deal with mechanical breakdown, such as cracking, wedging apart, and separation of mineral grains from one another.

All consolidated rocks are jointed, as we have seen in Chapter 3. Some rocks are so badly broken that pieces larger than a foot across are hard to find. This condition makes such rock masses particularly susceptible to frost pry. Water seeps into and fills the cracks, and then if the weather turns cold, the water may freeze. When water turns to ice its volume increases, and the expansive force is great. Automobile radiators and even engine blocks crack open if allowed to freeze, and so also the rock fragments are pried apart as the cracks are widened by the ice.

In climates where a single deep freeze occurs in the fall and lasts until spring, only a little disintegration results, but in some climates alternate freezing and thawing may take place up to fifty times a year, and there rocks are severely disrupted. We find that the angular blocks between joints have been wedged apart, heaved, and thrown into a rubble of

Figure 7-1 The foreground is littered with frost-pry fragments, called a "felsenmer," or rock sea. View eastward across Yellowstone Creek, Uinta Mountains, Utah. [Photograph by U.S. Geological Survey, and courtesy of Max Crittenden.]

jagged fragments, such as seen in Figure 7-1.

The places where alternate freezing and thawing occur most frequently in a year are those near the timber line in high mountain ranges. Here it freezes nearly every night in the fall and spring, and in the daytime it thaws. The result is a prolific generation of loose rock fragments called *talus*. Steep mountain slopes are conducive to downhill sliding and tumbling of these loose fragments. They converge at the base of the cliffs into fan-shaped accumulations called *talus fans* or *talus cones* (Figures 7-2 and 7-3).

In the arid Colorado Plateau we note an absence of or very little talus at the base of many of the high imposing cliffs. This may be due to a paucity of rainfall and only a few alternations of freezing and thawing each year, but it is generally due to the fact that the cliffs are formed of friable or poorly cemented sandstone and that the sandstone grains break apart easily, thus producing fragments so small that the wind and occasional thunder showers remove them about as fast as they accumulate. We commonly see large scattered blocks of sandstone broken loose from a sandstone formation in various stages of sliding down a shale slope below. The farther down the slope, the smaller the blocks become, owing to disintegration into sand grains and the removal of the sand grains by the wind (examine Figure 7-4). Frost pry probably helps dislodge large blocks of sandstone, but the shale underneath the sandstone also weathers so rapidly that it undermines the sandstone and leaves an overhanging ledge, likely to cave off.

Solution

Limestone and its metamorphic equivalent, marble, are fairly susceptible to weathering. In humid climates the rate of weathering of limestone is conspicuously greater than that of shale, sandstone, the igneous rocks, or the common metamorphic rocks other than marble (Figure 7-5). Dolomite is dissolved away in major proportions but not generally as rapidly as limestone.

Pure limestone is made up of mineral particles of calcium carbonate that are not in themselves very soluble in pure water. But

Figure 7–2 Middle and lower slopes almost completely blanketed with talus of volcanic rocks. This is Sewell at the site of Braden, the famous copper mine in the Chilean Andes. Freezing and thawing occur many times a year. [Courtesy of Kennecott Copper Corporation.]

if carbon dioxide is present, it combines with water to form carbonic acid. In such a solution the limestone is particularly soluble.

The formation of carbonic acid in water is described by the following equation:

$$H_2O + CO_2 \rightarrow H_2CO_3$$
$$\text{water} \quad \text{carbon dioxide} \quad \text{carbonic acid}$$

Now the acid reacts with the calcite to form calcium bicarbonate, which is held in solution and is carried away by the moving ground waters:

$$H_2CO_3 + CaCO_3 \rightarrow Ca(HCO_3)_2$$
$$\text{carbonic acid} \quad \text{calcite or calcium carbonate} \quad \text{calcium bicarbonate}$$

In most limestone terranes, particularly in semihumid and humid climates, the limestone layers are so extensively channeled and tunneled by circulating groundwaters that elaborate networks of caves result. The carbon dioxide, so essential to solution, is derived in part from the atmosphere while the waters are at the surface, but more abundantly from decaying vegetation in the soil as the surface waters filter through to become ground water.

In desert regions limestone and dolomite formations are resistant to weathering and form conspicuous ridges and cliffs, whereas in humid climates they dissolve away rapidly and usually result in valleys. This is due to the sparsity of rainfall and soils in the desert and to the abundance of rain and the production of much carbon dioxide in the thick soils of the humid regions, respectively.

Figure 7-3 Talus cone in Glacier National Park [Photograph by H. E. Malde and courtesy of U.S. Geological Survey.]

Hydration

All rocks containing feldspar minerals are easy prey to the chemical process called hydration. Such rocks as the granites, felsites, arkoses, and graywackes contain much feldspar and thus decay badly, particularly in humid climates. The reactions are complex, with various clay minerals forming from the feldspars together with the loss of potassium, sodium, calcium, and magnesium ions that pass off in solution. An example of the alteration of plagioclase feldspar is as follows:

$$CaAl_2Si_2O_8 \cdot 2NaAlSi_3O_8 + 4H_2CO_3$$
$$\text{calcium feldspar} \quad \text{sodium feldspar} \quad \text{carbonic acid}$$
$$+ 2(nH_2O) \rightarrow Ca(HCO_3)_2 + 2NaHCO_3$$
$$\text{water} \quad \text{calcium bicarbonate} \quad \text{sodium carbonate}$$
$$+ 2Al_2(OH)_2Si_4O_{10} \cdot nH_2O$$
$$\text{clay mineral}$$

It can be seen that the clay mineral that forms is a hydrated aluminum silicate but that carbonic acid is necessary to remove the calcium and sodium ions.

In the above conversion of feldspar to a clay mineral, considerable expansion occurs, and this is the chief cause of decay. In the case of a granite composed of feldspar, quartz, and ferromagnesian minerals, the feldspar particles are converted to clay with expansion, the bonds between the particles are loosened, and the rock crumbles (Figure 7-6). The clay is washed

Figure 7-4 Sandstone talus disintegrating with the sand blown away by the wind. Note the general absence of talus except for the cone of broken rock. This probably is a landslide from the cliff above, but the rock fragments are rapidly breaking down into sand grains and being removed. Near Glen Canyon Dam, Arizona.

away, and a residue of quartz grains remains. This marks the beginning of quartz sand. But what happens to the ferromagnesian minerals?

An iron-magnesian mineral like biotite is altered to a hydrated clay mineral plus a hydrated iron oxide called limonite. As in feldspar alteration, the potassium and magnesium ions are carried away in a bicarbonate solution, and also a little silica is lost in solution. The limonite stains the crumbling granite rusty and makes the resultant clay yellowish or rusty too.

Even in arid climates granites decay somewhat, but in humid climates the alteration proceeds from the surface down for 10 to 100 ft or more. Bulldozers and power shovels can generally cut and move the decayed rock without the use of explosives. A hand pick sinks an inch or two into the "rotten" rock. The chief visible result is crumbling. Large boulders of granite display a crumbling surface, and the bases of steep slopes are marked by the granular debris of the decaying rock.

Some rocks shed scales or shell-like peels in the process of weathering. The layers range from $\frac{1}{4}$ in. (Figure 7-7) to several feet in thickness (Figure 7-8). Most commonly they occur in granite, but in places also in sandstone. Scaling of granite is due to feldspar alteration, but in sandstones containing little feldspar another cause must be sought. It may be heating and cooling, but we are not sure. The process of scaling or shedding of peels is called *exfoliation*.

157

Figure 7–5 Solution pitting of a boulder of white marble. The surrounding boulders of gneiss have been rounded by abrasion but not pitted by solution.

Figure 7–6 Weathered granite surface in the Wasatch Mountains, Utah. Joints are spaced about 15 in. apart.

From the foregoing discussion it is clear that the rates of weathering are highly variable, and that the kind of rock is one factor and the climate another that contrive to produce the variable rates. In this connection the geologist would like to know how fast cliffs are receding in the Colorado Plateau or how rapidly the high mountain peaks of the Sierra Nevada are being reduced. This is a difficult research project, but we are certain of two things: first, that the rate is real and appreciable, and second, that our calculations will carry into thousands and perhaps millions of years.

Soil formation by weathering

What is soil?

To the soils scientist any unconsolidated material that will support plant growth is a soil, but to the soils engineer plants are immaterial; he is interested in the ability of the unconsolidated material to support structures of one kind or another, and it is unfortunate that he refers to the unconsolidated material as soil. A true soil to the geologist is a result of weathering and contains products of mechanical and chemical alteration mixed with more or less organic matter near the surface.

To be more specific, any rock (granite, sandstone, shale) or unconsolidated material (floodplain silt, alluvial-fan debris, dune sand, or lakebed clay), if exposed to the atmosphere for a few centuries, will develop a profile of weathering, and the upper part of this profile will come to contain organic material because of the activity of plants and animals growing and living in it. If a granite weathers, and the weathering products accumulate, as they will on a flat or gently inclined surface, we note that the solid granite is broken into fragments or blocks, the blocks decay into clay, quartz grains, and limonite, and considerable organic material becomes added to these products. We see this transition in road cuts, trenches, and pits and call it the *soil profile*. The first results of weathering, the blocky fragments, are referred to as the *rock mantle*. Study the soil profiles of Figures 7–9 and 7–10.

Factors that determine type of soil

Soils are complex and of many kinds. They have been extensively studied and elaborately classified, and they truly deserve all this attention, because they are the most basic and essential material that we have to deal with. Crops may be grown year after year when the soil is maintained, whereas a mineral deposit, once mined, is gone forever. Soil, if not protected and safeguarded, can also disappear, and man will have lost a most valuable source of food, cotton, lumber, and energy. Soils can be remade, but it may take centuries to do so. They may disappear by erosion in a few years by careless farming.

Climate is the chief factor in determining the type of soil that will form. The kind of parent rock and the time that the weathering processes have been active are also important. In the early stages of soil formation the kind of parent rock is strongly expressed in the soil type, but given time under a certain climate, all rocks gradually are transformed into soil of uniform composition. The kinds of plants that grow on the soil are also said to help determine the

Figure 7–7 Exfoliation of thin crust in granite of Yosemite National Park, California. Granite had previously been abraded and striated by glaciers. [Courtesy of U.S. Geological Survey and G. K. Gilbert.]

properties of the soil, but plants too are dependent upon the climate. Also topography is listed as a factor, and by this is meant whether the land surface is flat and swampy, hilly and well drained, or one of steep slopes. All these conditions are, of course, significant.

Soil processes

Basically the same processes that were discussed under weathering are operative in the formation of soil. Frost action, hydration, solution or leaching, and oxidation are the most important. Earthworms, insects, and burrowing rodents mix and turn over the soil and afford percolating water better transit through it. Soils, called *podzols*, develop in forested terranes, *chernozems* develop under grasslands, and *desert soils* in sagebrush country.

In a warm humid climate like Cuba, with the parent rock being serpentine, that is, high in iron, the silica and alumina are mostly leached out for tens of feet below the surface, and the remaining soil is rich enough in iron oxide to be mined as iron ore. Such a soil is called a *laterite*. In certain other tropical regions the silica and iron are largely leached out, leaving a soil rich in alumina (the soil is called *bauxite*), and this is mined as aluminum ore.

Soils as indicators of past climates

Besides the incalculable value of soils in agriculture they help the geologist interpret the climates and history of the past. Commonly soils are buried by accumulating sediments and thus preserved for future study. For instance, in

Figure 7–8 Slabbing-off effect of checkerboard sandstone wall. In geological terms, this is an aeolian sandstone with its conspicuous cross-bedded units. The fresh patch in the upper part of the photograph had rently peeled off (exfoliated). Here the exfoliated peel first cracks or joints at right angles to the cross-bedding. The cause of the shallow jointing is not clear. Zion National Park, Utah. [Courtesy of National Park Service and Allen Hagood.]

Figure 7–10 we see two soils. The lower one seems to be a semitropical lateritic soil developed on a red sandstone. It was buried by lake deposits of gravel, sand, and silt, and then the present modern soil, which reflects a temperate, semiarid climate, developed on the lake deposits.

Gravity-induced mass movements

After rock and mineral particles have been weathered loose they commonly come under the influence of forces that move them. This is particularly true if they occur on slopes. The prevalent force is gravity, and all detached particles or blocks tend to move downhill. The forces that act on a loose particle on a hill slope under the influence of gravity are diagrammed in

159

Figure 7–9 Podzol at 8000 ft in the Uinta Basin, Utah. It has developed on jointed flat-lying beds of sandstone and siltstone.

Figure 7–11. The downhill component of gravity causes downhill movements in many ways, and these are the subject of the next several sections.

Prevalence of landslides

Landslides occur everywhere but particularly in mountainous areas. They are found in many forms both large and small, young and old. The large and recent slides are evident to everyone, but the old ones may escape recognition until the trained geologist points them out. Because of the unstable foundation that landslides commonly afford to highways and buildings, and also because of their prevalence, engineers have become very cognizant of them. They either avoid them or must take precautionary measures to contain them. Some states regulate construction on them and require qualified geologic reports to guide planners and engineers.

Factors to be considered

Three variable factors are important in determining the type of mass transfer of rock material. The first is the size of the rock particles. They may range from huge blocks to sand grains, silt, and clay. The second is the amount of moisture present. Fine particles may be so wet that they flow like soup, but talus may be rather dry, yet move like a stiff liquid. The third variable is the rate of movement. Some landslides may accelerate momentarily to attain free-fall velocity, whereas others creep very slowly. These three factors have been combined in the diagram of Figure 7–12 in an attempt to relate the several kinds of down-slope mass

Figure 7–10 Soil in an excavation in the piedmont of the Wasatch Mountains, Utah. The soil at the top is marked by "horizons" A, B, and C. This is the soil profile and is about 4 ft thick. It has formed in about the last 15,000 years on lake-shore beds since the lake disappeared. (1) represents the lake beds below the upper soil profile, and (2) represents an ancient soil developed on a sandstone labeled (3).

movements. The rapid movements are mudflows, certain landslides, avalanches, and block falls. Those of slower but measurable velocities are rotational slump block slides and glide block slides, rock glaciers, and talus accumulation. Still slower movements are classed as creep phenomena. These are imperceptible to the eye but the results are recognizable. The rates of movement in the diagram are related to moisture conditions and size of particles, and the results have some merit even though they are not entirely satisfactory. For instance, wet conditions are tied to fine particles by the

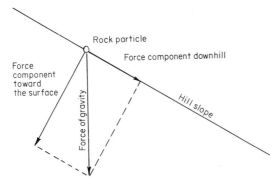

Figure 7–11 Components of gravity acting on a rock particle on a slope.

nature of the diagram, but it would be impossible to show a block fall under moist conditions. This relationship does not occur very often in nature, thus the represented variables are mostly real and commonplace. A fourth variable not shown is most important when the water freezes and we have to deal with the so-called *permafrost* conditions of the high latitudes, and the special kind of downhill movement called *solifluction*.

Rock falls, avalanches, talus and rock glaciers

Rock falls will be defined as blocks of rock subtended by joints that break away from vertical or near-vertical cliffs and fall to the valley floor. Such an occurrence is depicted in Figure 7–13. Vertical cliffs usually retreat by rock falls.

A rock avalanche, as distinguished from a snow avalanche, is a rushing or streaming of rock fragments about the size of talus, down a steep mountain slope or funneled down a steep gully. They generally originate from a steeply eroded mountain mass of badly shattered rock. By badly shattered rock is meant a rock mass that has been broken by many closely spaced joints, inches to a foot apart. Such rock suddenly gives way under the duress of gravity and rushes down-slope amidst much dust and a frightening roar. Moisture may have helped to induce the avalanche, but each occurrence must be investigated to see the extent to which water participated. The only one this author has witnessed seemed to be composed of perfectly dry rock, with the associated crushing and grinding producing clouds of dust.

We have already studied talus, but it must be included in any classification of mass movements because, after frost-pry, it consists of fragments in motion due to the pull of gravity. When talus accumulates in large volumes it sometimes begins to flow downhill much in the manner of a glacier (Chapter 8) (study Figure 7–14). Rock glaciers are common in the high upper basins or amphitheaters of the Rockies. It is possible that interstitial water and ice may help the flowage along; in fact, it may be the key to the process, but a cut has never been made in a rock glacier to investigate the process.

Creep of various kinds

By creep is meant the imperceptible, though measurable, movement of soil and rock material in a surface layer. The layer involved in downhill creep may be a foot thick or possibly as much as 30 feet thick. It involves the soil layer particularly, but in places also the rocks that reach to the surface.

If we consider a particle of soil in the thin surface layer, it may be lifted and settled alternately many times, as illustrated in Figure 7–15. It may be lifted when the soil warms and expands, and settled when it cools and contracts. Plant roots may lift the particle; burrowing animals disturb it. Freezing and thawing accomplish the same cycle. Each time the particle is elevated normal to the surface, it tends to settle either vertically downward due to the pull of gravity or it moves a bit downhill due to the downward component of gravity. This cycle, repeated many times, accounts in one way for creep. The term *colluvium* is used in Figure 7–12 to denote aggregates of hillslope wash material of various mixed sizes. Colluvium may also be subject to creep, like soil. Usually the greater the moisture the faster

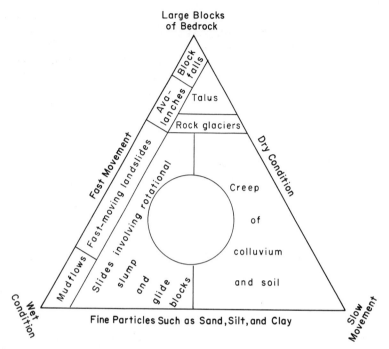

Figure 7-12 *Various kinds of mass movements in relation to three variables—large to small particles, wet to dry conditions, and rapid to slow movements.*

the creep. When water accumulates beyond a certain amount the soil or colluvium suddenly flows to produce one of the forms of landslide, later described, or a mudflow.

Thin beds of shale or siltstone are also subject to creep by downhill bending (see Figure 7-16).

Slides of rapid movement

One of the most newsworthy landslides of recent years in the United States occurred in the lower end of Madison Canyon, northwest of Hebgen Lake and West Yellowstone, Montana. It was tipped off by the vibrations of the Hebgen Lake earthquake of August 17, 1959, and an estimated 80 million tons of rock rushed down the south canyon wall to the bottom with almost free-fall velocity, and up about 300 feet on the opposite wall. In the process the rock mass split and spilled up and down the canyon (see Figures 7-17 and 7-18). Many campers in the area were injured, a few of them died, and 19 more are believed buried beneath the slide. The slide formed a dam behind which a new 4-mile-long lake has formed, covering the highway and several favorite camp sites at the canyon bottom.

It can be seen in the photographs that the rock mass broke away from a fairly well shattered wallrock or was soon completely broken up when it started to move so that it literally flowed as one mass. There are many blocks of rock as large as a table, but they are mixed with much finer material in a topsy-turvy mess. Much of the uprooted timber seemed to float on top. There was much fine material because the dam has proved to this day fairly impervious to water, and not much of the water has leaked through. That the mass accelerated practically to a free-fall velocity is attested to by the great mass that was carried up to 300 feet on the other side. A pressure wave of air in the form of a wind front was reported with a speed of possibly 100 miles per hour, which was carried up and down the canyon from the dam. It has been suggested that the rock fragments caught up much air and were cushioned by it somewhat, and thus the fluidity of the moving mass was greatly increased. As the large volume of fractured rock settled across the bottom and on the other side the air was expelled, hence the sudden gust of wind.

Another example of a great landslide of a few seconds' duration is shown in the stereo pair of photographs in Figure 7-19. A favorable situation is here illustrated for landslide formation: a high plateau formed by a volcanic cap on sedimentary beds. The plateau top is bounded by a precipitous cliff, from which the most recent slide mass has come. The sharp, somewhat parallel ridges below the bounding cliff represent the last slide, but this occurred before the first settlers arrived, because there is no written record of it. The sharp slide ridges are sparingly covered with timber, but away from the fresh slide material are subdued, heavily wooded hills that are interpreted to be the remnants of an older slide, possibly still greater in size.

The fresh slide ridges just below the main cliff are probably rotational slump blocks, as illustrated in Figure 7-20. This is a common way in which landslides occur. However, the lower and outer part of the slide mass responded by flowing, and a tongue poured down a canyon for about three miles.

The Alaska earthquake of March 27, 1964, whose epicenter was 80 miles east-southeast of Anchorage, triggered very destructive landslides, and Anchorage, the largest city of Alaska, bore the brunt of the property damage.[1] The damage was caused partly by direct vibrations as the trains of earth waves passed by. Ground cracks came into existence mostly at the heads of landslides and were capricious in their destruction of buildings, but the most

[1] The material on the Anchorage landslides is from Wallace R. Hansen, *The Alaskan earthquake*, U.S. Geological Survey Professional Paper 542-A, 1966.

Figure 7-13 A huge block of sandstone fell, tumbled, and landed squarely on this truck on November 11, 1947. A second chunk in this block fall hit another truck and broke it in half. Another block went through the roof of the maintenance shed where the trucks were parked and completely devastated a new pickup truck. The men, having a coffee break at the time, were not injured. [Zion National Park Service and Allen Hagood.]

damage was wrought by the landslides themselves. The landslides of Anchorage occurred chiefly along the escarpment or bluffs facing Knik Arm. A good example is Figure 7-21 and the map in Figure 7-22. A depressed zone, called a *graben*, suddenly came into existence, along with numerous subparallel cracks and small escarpments in the higher tableland back of the bluff line, and a pressure ridge or ridges evolved in the lowlands in front of the bluffs and adjacent to the tidal flats. The graben, it is clear, is a pull-away structure at the head of the landslide mass, and the pressure ridge is a small thrust structure at the lower toe of the moving mass. A cross section of the slide, prepared from accurate mapping and geologic field work, is shown in Figure 7-23. The chief culprit in the mass movement, besides gravity, is the clay layer that underlies a sand and gravel surface layer. The clay has been named the "Bootlegger Cove Clay" by geologists, and a schematic representation of the slide is given in Figure 7-24. It is clear that a certain layer within the clay yielded almost like grease, or became liquefied, and that the slide was of a type called "translation gliding." This differs from the mechanism of rotational slumping, previously described.

The depth of the "sensitive" clay layer has been determined by drill holes and by geo-

Figure 7-14 *Rock glacier at Hart Lake, Yukon Territory. So much talus has accumulated here that it is flowing like a glacier. The upper part of the cliff is mainly limestone and the lower part is conglomerate. The front of the rock glacier is about 1000 ft across. [Courtesy of Geological Survey of Canada and P. Vernon.]*

physical log records, but in the absence of drilling and geophysical data, a simple calculation can be made for any translation slide. It is called the *graben rule*, and is illustrated in Figure 7-25. The cross-sectional area of the graben (A) divided by the amount of horizontal displacement (l) equals the depth of failure, or the depth of the slip surface. For instance, one of the grabens is 11 ft deep, 100 ft wide, and thus has an area of 1100 sq ft; as measured by surveys before and after the slide, the horizontal displacement was 17½ feet; the calculated depth would thus be 63 feet below the upper land surface. This turns out to be about the depth that drilling and geophysical surveys indicate.

In the many slides that the Alaskan earthquake of March 17, 1964, set off, there are examples of all sizes and kinds, but the translation gliding type is exemplary. It is a type of rapid motion taking place under very moist conditions.

Landslide environment

Regions of precipitous relief generally have many landslides, both young and fresh, and old and subdued. The landslide environment is most obvious to anyone who becomes acquainted with the signs and looks for them. Although no statistical study has been made, it appears that the slides increase in number and occupy more of the ground surface the farther north one goes from Arizona to Montana. Landslides are abundant in the Rockies of Utah, Wyoming, and Montana. The geologist must seek out the undisturbed bedrock outcrops because these tell him of the underlying structure, such as folds and faults, but if he should read the strike and dip of the beds in a large rotated block of sedimentary rock involved in a landslide, he will err in diagnosing the structure.

Landslides of rapid movement are conspicuous to all, but those of slow movement are not so obvious. A slowly moving landslide may be recognized by its rotational slump blocks, upper pull-away escarpments, small grabens, hummocky topography, and lakes empounded in basins between the rotational blocks and flow masses (refer again to Figure 7-19). These are generally old slides that occurred before anyone was around to record them, or even during the last glacial stage 25,000 to 10,000 years ago. We really do not know whether most of these slides moved rapidly or slowly, but we surmise that some of them moved slowly because they are creeping downhill now. We might not have known about some of these had we not built highways and homes on them. The slow movement keeps the sections of highways that cross them continually in disrepair, while cracks in the foundations and walls of homes keep recurring. Of course, there are portentous signs of slow movement in some slides, such as cracks developing in the ground and occasional rumblings and small local earthquakes, as if a rapid slide were about to be released. A highway or railroad cut through an ancient slide mass may remove support from the slope above, and thus

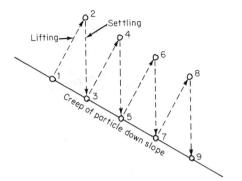

Figure 7–15 Movement of a rock particle downhill during lifting and settling cycles, incident to the pull of gravity.

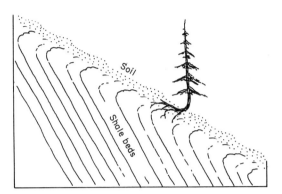

Figure 7–16 Creep of soil and shale beds.

may reactivate the movement. A river may remove the toe of a stable slide and thus cause further movement. Canals and irrigation ditches, even lawn sprinkling systems, add new and unaccustomed moisture to the soil and underlying rock, and may reactivate old slides as well as cause new ones if the gravity and slope conditions are propitious.

Recognition of slides is one thing, but if construction *must* proceed on or across the slide, what to do about it is another. We cannot, for economic reasons, avoid all landslides, and if we must build on them we should know the risk we are taking and try not to aggravate the slide. Each slide presents different problems, and it is therefore essential that a qualified geologist be consulted if there is a potential landslide.

Bad landslides in California have resulted in laws making mandatory professional geologic reports before the start of construction. With the current growth of population, cities *must* expand, but the cities can spread intelligently and on a calculated-risk basis. Unfortunately, the risks involved in most urban spreads are not known, let alone calculated. Foundation stability and subsidence problems in new areas should be studied. The American Institute of Professional Geologists is urging every state and city to demand professional reports on new building sites which would consider such problems as ground stability before allowing development. It also counsels prospective home purchasers to seek a professional report on the geologic hazards of the area.

Rivers and man

Streams, small and large, have consumed man's attention and shaped his destiny in significant measure for many thousands of years. Stone Age man fished the streams and trapped along them. His villages were situated on them, and they were his routes of travel as well as his barriers. He was tied to them for drinking water; yet in flood time they devastated his villages and drove him away. The early Egyptians depended upon the spring floods for arable fields and later learned to divert the water for irrigation. Today man is harnessing the rivers with his greatest of engineering projects. He is building dams for water storage and flood control. Millions of dry acres are being irrigated. Electricity is being generated for new industry, and beautiful lakes are being created for recreation. He is also making the larger waterways more navigable and building many bridges across them. Our purpose will be to understand how streams operate, what they have done to the landscape in the geologic past, and to point out how the geologist and engineer apply this knowledge in controlling and using them.

Figure 7-17 *Landslide at lower end of Madison Canyon, caused by Hebgen Lake earthquake. The rock mass broke away from the crest of the ridge and steep canyon wall.* [Courtesy of U.S. Geological Survey.]

Drainage patterns

First, let us take a broad view of the continent and a map on which the rivers and tributary streams are well shown. We will see that most major streams with their tributaries have a pattern resembling an elm or an oak. The pattern is called *arborescent* or *dendritic* (see Figures 7-26 and 7-27). In the Appalachian Mountains from Pennsylvania to Alabama, however, the tributaries and main streams take courses approximately at right angles to each other or make right angle jogs. This pattern is called *trellis* or *angular* (Figure 7-26). In certain restricted areas we see that the drainage is radially outward, such as from a single high mountain, and this is referred to as radial. In certain restricted settings a number of tributaries are somewhat parallel, and this pattern is called *parallel* (Figure 7-28).

Stream flow velocity

We should note that streams are confined to channels with banks, they flow downhill only, they are often muddy, which means they are

Figure 7-18 Madison Canyon landslide viewed from downstream looking upstream. The slide came from the right side (south) and flowed across the valley and up the left side about 300 ft. Note the buried highway and dry river bed. A 4-mi-long lake formed back of the slide dam. [Courtesy of U.S. Geological Survey.]

carrying particles of clay, silt, and sand, and their flow is turbulent and irregular. Some streams flow slowly, some rapidly; the rate of flow varies along the course of any individual stream; they all flow more rapidly at times of high water than low water.

Velocity of flow is determined principally by four factors: the gradient or slope of the channel, the shape of the channel, the volume of water flowing in the channel, and the roughness of the channel. If there were no retarding influences, the flow velocity would accelerate to incredible speeds in a short distance. The steeper the gradient and the greater the amount of water, the faster the flow; on the other hand, shallow and rough channels hinder the flow and check the acceleration.

Gradient

Over a distance of one mile the lower Mississippi River loses only a few inches of elevation, whereas one of its distant tributaries in the Rocky Mountains falls several hundred feet in 1 mile (Figure 7-29). The gradient of a

 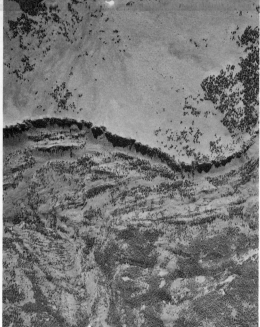

Figure 7–19 Stereo photograph of a landslide skirting the lava plateau of Thousand Lake Mountain, High Plateaus of Utah.

stream is measured in inches or feet per mile in English-speaking countries and in centimeters or meters per kilometer in most other countries. Almost everyone has observed the constant attempt of the water to hurry along in mountain streams of high gradients where it plunges over waterfalls and rapids only to be obstructed by large boulders, fallen tree trunks, channel bends, and deep pools. Low gradients are generally synonymous with sluggish flow.

Roughness of channel

As just mentioned, boulders on a stream bed obstruct the flow, especially at low-water stages. Roughness of the bedrock floor of the channel also hinders flow. A series of beaver dams slows the overall flow considerably. A winding or very crooked channel deters the flow. The factor of roughness is difficult to assess quantitatively, but its significance is most obvious.

Shape of channel

The shape of a canal or flume in cross section which offers the least drag or resistance to flow and in which for a given gradient the water flows fastest is a semicircle (see Figure 7–30). This contains or passes the most water per area of channel surface and hence offers the least resistance to flow. A wide, shallow channel presents more surface area to less water and hence more drag on the water. Such channels are generally characteristic of low gradients. A deep, narrow channel, like the wide, shallow channel, retards the flow more than the semicircular one, but the deep, narrow channels are generally full of water in mountainous areas and have steep gradients; hence the rather high velocities of flow (see Figure 7–31).

Volume of water discharge

The more water in a given channel, the faster the flow. The quantity of water passing a given point on the stream bank per unit of time is called *discharge* and is generally given in cubic feet per second. This measure is significant in comparing the activity of different streams.

The velocity of flow has been emphasized in the foregoing paragraphs because much of a stream's erosional and transportational power depends on velocity. This will be brought out in the following paragraphs.

Stream loads (transportation)

Methods of transport

A stream transports rock and mineral matter

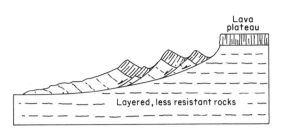

Figure 7-20 Rotational slump blocks in landslide formation.

in three ways: in suspension, along the bottom of the bed or channel, and in solution.

Suspended load

If you take a handful of dirt from the garden and drop it in a bucketful of water, you will immediately perceive that there are different sizes of particles or grains and that the larger ones sink to the bottom rapidly, the intermediate-sized ones sink more slowly, and the small ones settle very slowly. In fact, some particles remain suspended and keep the water roily or turbid for hours.

The particles that settle slowly and tend to remain suspended are the silt-sized and clay-sized. Silt-sized particles will settle in a bucket of quiet water in an hour or so, but clay-sized particles may remain in suspension for hours or days. As has been noted, stream flow is turbulent. There are eddies, vortices, and boils in the water's movement. It falls and tumbles and is deflected to the right and to the left by various obstacles. The upward eddies are forever lifting the silt and clay particles, and thus the particles never get a chance to settle out. Even considerable sand-sized particles are swept up in the eddies and move downstream an appreciable distance before settling out. Samples of the Missouri River water at Kansas City, for instance, were taken from top to bottom, and the particles that make the water muddy were analyzed for size. Figure 7-32 shows the results. Clay-sized particles are in equal abundance from top to bottom; silt-sized particles become slightly more abundant near the bottom; but sand-sized particles increase

Figure 7-21 Air view (looking south) of Native Hospital slide showing graben and pressure ridge. The scar of an older landslide is transected by the slide of March 27, 1964. [Photograph by U.S. Army. From Wallace R. Hanson, U.S. Geological Survey, 1966.]

conspicuously from top to bottom and are being moved in large amounts in the lower 3 ft of a stream some $11\frac{1}{2}$ ft deep. It is evident that clay and silt are transported by suspension, and sand only partly.

Since stream flow is turbulent, the water remains roily or muddy for many miles. If just a small percentage of the weight of the water were suspended load, even a small stream would transport a large amount of material. About 2 million tons of sediment are carried each day by the Mississippi River into the Gulf of Mexico.

Bed load

If windows are constructed in experimental flumes and the movement of the grains observed on or near the bottom, it is seen that only swiftly flowing streams develop enough turbulence to pick up sand grains from the bed of the stream. It is necessary here to distinguish fine, medium, and coarse sand grains, because much of the fine sand follows the silt in suspension, but almost all the medium and coarse sand grains roll along the bottom. As the ve-

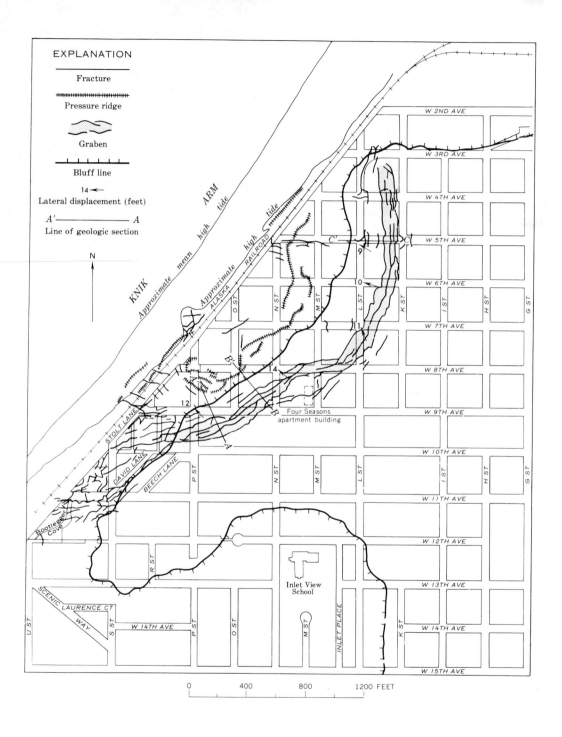

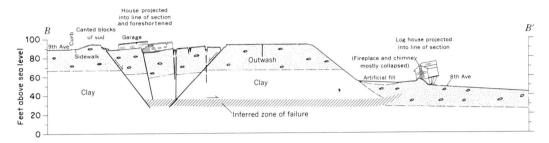

Figure 7-23 Cross section of the L Street slide, Anchorage. Refer to Figure 13-26.

locity of flow is increased, we note a change from rolling and short leaps of a few grains to spasmodic movement of groups of grains to a smooth general streaming of many grains. It will be seen that there develops a transitional zone between bed-load transport and suspension transport of a few inches to a few feet, depending on the velocity of the stream and the nature of the grains affected, in which it is difficult to distinguish the mode of transport. Undoubtedly the medium and coarse sand grains in this zone are proceeding by jumps.

Dissolved load

Water that emerges as seeps and springs contains considerable material in solution, and such water contributes in good measure to almost all streams. The stream water also dissolves some material directly from its channel bottom and from the particles in mechanical transport. The chief compound in solution is calcium carbonate ($CaCO_3$), but ions other than those in $CaCO_3$ are common, such as Na^+, K^+, Mg^{++}, Cl^-, and SO_4^{--}. The major rivers carry only a few hundred parts per million of dissolved matter, but some of the smaller streams in arid regions carry several thousand parts per million and are distinctly saline to the taste. It has been estimated that more than half of all materials carried by river water from continent to ocean is dissolved load.

Velocity of flow, which is so important in the transportation of discrete grains of rocks and minerals, has little to do with solution. Once a material is in solution, except in unusual conditions, it stays there whether the stream flows rapidly or slowly.

Some streams are brown with organic acids in solution derived from the decay of plant material.

Stream erosion

Abrasion

The movement of rock particles along a stream bed accomplishes three things: the particles themselves are worn or abraded, the rock floor of the channel is abraded, and the banks are undercut and cave off into the stream.

The particles, whether sand grains, pebbles, cobbles, or boulders that are moved along, strike one another as well as the bottom and banks, and gradually their corners and sharp edges are chipped off, and they become rounded. The farther the particle has been transported, the rounder it gets.

At the same time the channel bottom is chipped and scratched as the particles tumble and bounce along. In mountain streams where the gradient is steep, large heavy boulders bash the bottom, especially in flood time, and abrasion is rapid. We note the effect in the downcutting of the stream channels (Figure 7-33). Canyons or gorges that have near-vertical walls are clear examples of the stream's power to cut downward, almost as if a giant saw had been

Figure 7-22 L Street slide area, Anchorage, Alaska. [Reproduced from *Wallace R. Hanson*, 1966.]

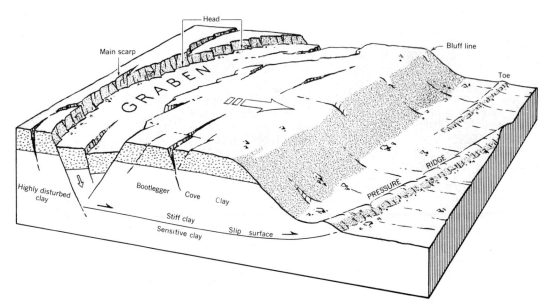

Figure 7–24 *Schematic diagram of a translatory slide.* [*Taken from Wallace R. Hanson, 1966.*]

cutting through the bedrock. From gullies in a cornfield to the Grand Canyon of the Colorado a basic activity is downcutting, and in hard rock it is principally an abrasive phenomenon.

Although the chief activity is downcutting, the banks are undercut in places, and slabs and blocks of rock cave off to become part of the stream's load. Perhaps the blocks are too large to be moved by the stream, but in such event they themselves will be abraded by other passing particles, like the channel bottom, until they are small enough for some flood to move them

Figure 7–25 *Illustration of the graben rule.*

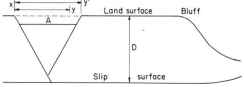

Lateral displacement, $L = xy'(\text{after}) - xy(\text{before})$
Cross sectional area of graben, A
Graben rule for determining depth of slip surface, D
$D = A/L$

on. Caving under waterfalls also helps along the general process of channel abrasion and downcutting (study Figure 7–34).

Solution

Channel erosion is helped by solution especially in limestone terranes. The limestone ($CaCO_3$) is taken into solution especially if the stream contains much carbon dioxide.

Abrasion, in producing small particles from large, helps the process of solution because small particles present more face area per volume than large particles.

Source of a stream's load

Downcutting soon produces such steep valley walls that caving occurs, and this adds to a stream's load. In fact, most arterial streams are busily engaged in transporting the material supplied by the tributaries, and the tributaries are engaged in transporting the material brought to them by rainwash. Only part of the stream's energy can be expended in downcutting; in fact, downcutting is common only

Figure 7–26 *Drainage patterns: left, arborescent, with local radial (R); right, trellis in central area with arborescent in upper and lower reaches.*

along steep gradients by relatively fast-flowing streams.

Relation of velocity to erosion, transportation, and deposition

Factors determining maximum size of particles

If a stream flowing at a certain velocity is just able to move coarse sand grains along the bottom, how much larger grains will it move if its bottom velocity is doubled? The answer is, many times. There are several reasons. With twice the velocity each water particle hits twice as hard, and also twice as much water strikes the face of the sand grain in each second. The water impact therefore increases as the square of the velocity. This means that, if the velocity is doubled, four of these particles placed end on end can be moved. This is shown by the use of cubes in Figure 7–35. If these four cubes, end on end, are built up to one large cube, which also can be moved by the doubled velocity flow, then the size of the particle is sixty-four times as great; or, expressed in another way, the diameter of the particle is increased four times. It can also be shown that if the velocity is tripled, the diameter of the particle that the stream is able to move is increased nine times. This, of course, means that surprisingly large boulders can be moved if immersed in a rapidly flowing stream.

Add to the factor of impact the frictional drag of the water passing over the particle, and a stream with increased velocity attains still more potency. Add, further, the factor of buoyancy, and we begin to appreciate the power of a raging stream. If a pebble of quartz weighs 2.7 g in air, it weighs only 1.7 in water; hence, in the matter of buoyancy alone, running water has a much easier time than wind in moving rock particles.

Experimental results

Figure 7–36 shows the results of experiments with running water and its ability to pick up, retain in transport, or drop (deposit) particles. It must be appreciated that the coordinates of the graph (velocity and grain size) increase logarithmically, and therefore the size of the particles that a stream is able to move increases very much faster than the velocity.

Another point that is a bit surprising is that clay particles, once settled into a compact arrangement on the bottom, are hard to pick up and that the velocity must be increased to do so. Once in suspension they stay in suspension, however, with minimum velocities. The upper

Figure 7–27 Arborescent drainage. Example in southwestern Montana. [*Vertical aerial photograph.*]

Figure 7–28 Parallel drainage. Example in southwestern Montana. [*Vertical aerial photograph.*]

limit of the "transportation zone" marks the velocities necessary to pick up particles and to accomplish erosion. The lower limit represents those velocities which are necessary to just barely keep the particle in motion. If the velocity falls below this minimum, the particle is dropped or stops rolling, and deposition results. All problems of erosion and deposition of rivers and canals must be considered in the framework of this basic graph.

Graded profile and base level

If faulting across a stream should suddenly drop the downstream block and elevate the upstream block, the stream's activity would be upset. Figure 7–37 shows a dislocated stream profile. The stream, in tumbling over the fault scarp, immediately gains velocity and starts to erode or deepen its channel at this point. The down-cutting of the channel proceeds upstream through stages 1, 2, and 3, as indicated. Since the channel on the down-thrown block has had its gradient lessened, the stream there becomes sluggish and incompetent to transport the load it formerly handled. Not only this, but the channel deepening upstream from the fault will have added an additional load, which, under the circumstances, the stream below the fault is not able to carry away, so that sediment is dropped in the channel or valley bottom. This deposition will occur in stages 1, 2, and 3 downstream to match the progress of erosion upstream. In the long run, a smoothly curved longitudinal profile is re-established as it was before faulting. A stream that has established a smoothly curved concave profile is said to be *graded*, but this is an ideal condition toward which streams work. Every stream is attempting to achieve a gradient nicely adjusted to its volume, load, and the bedrock irregularities over which it flows.

In the profiles of Figure 7–37, the stream is shown to empty into the ocean, and the horizontal datum line is at sea level. Any stream that empties into the ocean is controlled by the ocean level because the stream cannot erode its channel more than a few feet deeper than the ocean's surface. In fact, a moment's reflection will show that the profile approaches sea level as a base, and as this base is approached, the gradient of the lower part of the profile becomes gentler, and down-cutting slower. The ocean level is thus referred to as the stream's *base level*. If a stream empties into a lake, the lake is the stream's base level, or if a stream flows out onto an arid flat plain and dries up there, the plain is the base level.

The concept of base level helps explain the hyperbola-like curve of the normal longitudinal profile.

Figure 7–29 *Stream with steep gradient and cascading falls in the Rocky Mountains.*

Figure 7–31 *Narrows of the Snake River below Jackson Hole, Wyoming. The channel here is about 20 ft wide and 60 ft deep with increased flow velocity.*

Headward erosion

Erosion is most vigorous in the steep upper reaches of a stream, and because of this, the longitudinal profile is lengthened. However, a smooth curve is maintained in the manner shown in the several stages of Figure 7–37. An illustration of a stream extending itself by *headward erosion* into a fairly flat upland on a small scale is shown in Figure 7–38, and on a large scale in Figure 7–39.

Rivers in flood stage

Boulders 10 to 15 ft in diameter are rolled out of gullies in cloud-burst floods. Floods resulting from dam breaks have moved huge

Figure 7–30 *Shapes of channels in relation to velocity of flow.*

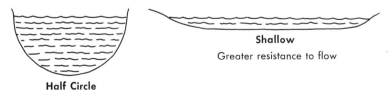

Half Circle
Least resistance to flow

Shallow
Greater resistance to flow

175

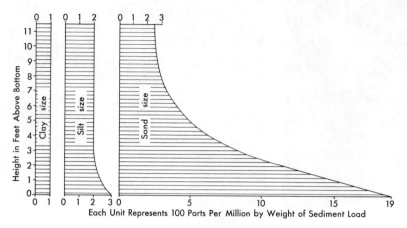

Figure 7–32 *Distribution of load in a stream, from top to bottom. The vertical scale shows the depth of water in the channel, and the horizontal scale indicates the amount of sediment being transported. Of the sand fraction, the grains at the top are nearly all fine sand in size, but at the bottom of the channel they are half fine sand and half medium and coarse sand. This graph shows a sampling of the Missouri River at Kansas City.* [*After Straub, in* Hydrology.]

blocks of the ruptured concrete. An often cited example is that of the St. Francis dam in southern California, which broke in 1928. The great mass of water, suddenly released, carried a block of concrete which measured 63 by 54 by 30 ft a half mile downstream.

The ability to move particles is called *competence*. The greater the velocity, the larger the particles moved, hence the greater the stream's competence. The total load carried, however, is a matter of a stream's *capacity*. Capacity depends on volume as well as velocity.

Analysis of stream activity

Longitudinal profile

If the gradient of a stream is charted from head to mouth, the graph obtained is called a *longitudinal profile*. The gradient is generally steepest at the head and gentlest at the mouth. Streams in the Rocky Mountains fall 5000 to 7000 ft in fifty to one hundred miles, and then, on the Great Plains or other valley floors, they fall only 500 ft in two hundred miles. The entire profile approximates a smooth concave curve which looks like a hyperbola (see upper curve of Figure 7–37). Although every stream shows irregularities and departures from this hyperbolalike profile, almost all streams approach it to varying degrees.

Natural and artificial base levels

The Mississippi River lives a worried existence, attempting to keep adjusted to changing conditions. Undoubtedly, farming of the many fertile acres of its drainage basin has added immensely to the load, and the river has trouble keeping its channels clear. Then there are the flood times and the low-water times, and the river under each condition finds itself out of kilter with the previous situation to which it was adjusting.

Not only this, but sea level, and hence base level, has fluctuated in the past. The ocean's level has risen and fallen as much as 300 ft several times in the past half million years or so (during the climatic fluctuations of the ice age), and the river, frantically it would seem, trenched its valley during lower sea level in adjusting to the lower base level and filled up the trench with sediment again during the return to high sea level. Two cycles of down-cutting followed by filling are shown diagrammatically in Figure 7–40. The surface labeled 1

Figure 7–33 *Pothole in hard rock of stream bed, worn by sand, pebbles, and cobbles being whirled around in eddies of the stream, Yosemite National Park.* [Courtesy of U.S. Geological Survey and F. E. Matthes.]

Figure 7–35 *By doubling the velocity of flow, the impact is four times greater, and theoretically the stream can move a particle having four times the diameter or sixty-four times the volume.*

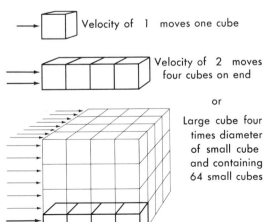

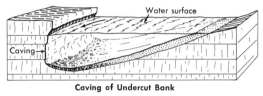

Figure 7–34 *Examples of erosion by caving. The lower diagram is a cut across an asymmetrical stream channel.*

is the valley under graded conditions, presumably when the sea level was high or normal. The sea level dropped, and the river trenched its valley to surface 2. Later, the sea level rose, and the river filled the trench with sediments, labeled A. The sea level fell a second time, and the river trenched its valley again, but not necessarily in exactly the same place as before. The new trench is labeled 3. Then the sea level rose again, and the river followed suit by filling in the second trench with sediments, labeled B. This time, by chance, the trench or valley fill was not built up quite so high as the first fill.

When a dam is built across a river, such as the Boulder Dam on the Colorado near Las Vegas, Nevada, a lake (Lake Mead) is impounded back of the dam, and the lake creates a new base level for the river upstream from the lake. The lake becomes a settling tank for the river's sediments, and the dam thus creates an *artificial base level.*

It is important to measure the volume of sediment that the Colorado River is carrying into Lake Mead, for in time, the sediment will fill the reservoir, and as this occurs, the water-storage capacity of the dam will decrease. Here are some data on the Colorado River and Lake Mead:

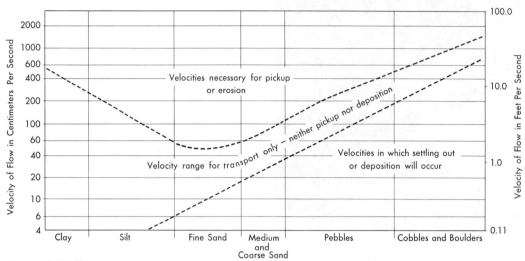

Figure 7–36 *Velocities necessary to erode, transport, or deposit grains of different sizes. Refer to Chapter 2 for size of clay, silt, sand, etc.* [Compiled with simplification from Hjulstrom, Recent marine sediments, *and* Sundborg, Geografiska annaler, *1956.*]

1. The water originally backed far up the Colorado River canyon into the narrow Lower Granite Gorge, but in the first 14 years (1935 to 1948) the river's sediments filled the gorge up to lake level for 43 miles downstream from the original point of entry into quiet water.

2. The heavy turbid currents of the river dive under the clear water and flow along the old canyon bottom on occasion all the way to the dam, a distance of 120 miles. They thus carry the river's sediments along the entire length of the reservoir and particularly have filled the inner deep gorge.

3. The sand particles settle out immediately,

Figure 7–37 *Longitudinal stream profiles.*

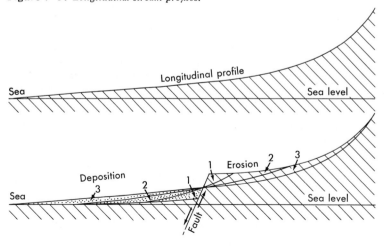

Figure 7–38 *Headward erosion in the arid Colorado Plateau.* [Photograph by Hal Rumel, Salt Lake City.]

but the silt and clay particles are carried by the turbidity currents the entire length of the reservoir (see Figure 7–41).

4. In the first 14-year period of the reservoir's existence some 2 billion tons of sediment was deposited in the lake. About half is deposited immediately at the point of entry of the river into still water, but half is spread out along the entire deep bottom by the turbidity currents. The total water-storage capacity has thus been reduced by about 5 percent.

5. If 5 percent of the storage capacity is lost in 14 years because of sediment fill, then half of the storage capacity will be lost in 140 years, and, by this time, the usefulness of the great dam and reservoir will be seriously impaired.

Grain size, load, and longitudinal profile

Studies of several large rivers have resulted in the generalization that the size of the grains transported gradually decreases downstream. This fact helps to explain the decrease in gradient of the longitudinal profile, because the load, reduced to smaller particles by constant wear, is more easily transported or can be transported at lower velocities.

Meandering streams

Streams seldom pursue straight courses. Even over short stretches they seem to disdain the straight-away channel and develop a sinuous or winding path. The bends are generally picturesquely rounded and curved. Such streams are said to be *meandering* (study Figure 7–42). Many observations and experiments have led to the conclusion that the streams given most to intricate meandering are those whose banks and bed are in unconsolidated alluvium. For instance, during the postglacial rise of sea level the Mississippi River deposited a thick layer of sand and silt across the floor of its broad valley, and now this sand and silt material fosters the tendency of the great river to meander. It is true that some streams pursue meandering courses in hard rock, such as the Green, Colorado, and San Juan Rivers in places in Utah, but such examples usually reflect complicated geologic histories that will be taken up later.

Leonardo da Vinci, about 450 years ago, perceived the activity of meandering streams and pointed out in connection with the River Arno that erosion and deposition were occur-

Figure 7–39 *Vertical aerial photograph of headward erosion into a rolling upland, Book Cliffs, east-central Utah. What is the meaning of the curving subparallel lines in the rolling upland?*

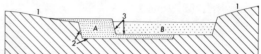

Figure 7–40 *Results of fluctuating base level: 1, original valley surface; 2, entrenched valley due to lowered base level; A, sediments deposited in entrenched valley, with rising base level; 3, second entrenchment due to lowering of base level a second time; B, deposits with rising base level.*

ring and where. Representative of Newton's first law of motion, moving water particles tend to keep in the direction in which they are going and hence, in rounding a bend, the current impinges upon the outer bank (refer to Figure 7–43). Here, undercutting of the bank occurs. Small cuts by small rivers and large cuts by large rivers are numerous and obvious wherever the rivers follow meandering courses.

Now the sediment eroded from the bank is carried downstream by the current and most of it comes to repose in the slack water of the inside of the next lower bend (Figure 7–43). See also the bars deposited by the Yukon on the inside of a great curve (Figure 7–44). The cutting and filling proceeds to such an extent that the channel migrates here and there across the entire river flood plain.

Small rivers have small meanders, and large rivers have large meanders. In fact the radius of curvature of the meander bends is two to three times the width of the river. This ratio is probably determined by the volume of water discharged and by the type of sediment being moved along the channel.[2] If a stream is transporting a high percentage of its total sediment load in the form of sand and gravel (bed load)

[2] S. A. Schumm, Meander wavelength of alluvial rivers, *Science*, Sept. 29, 1967.

it will have wider curves, but if the load is largely fine material being transported in suspension, then the curves will be smaller. Other factors possibly important are gradient and width of channel, but these will seek adjustment to the discharge and the sediment load. The problem is difficult because of variable discharge and because of local factors such as entering tributaries with different characteristics.

As the meanders develop into more complete circles some of them eventually intersect, and cutoffs occur. The stream, encountering a small but sharp drop at the cutoff, follows this more direct course and forsakes the horseshoe-shaped channel, which stands partially full of quiet water. The ends of the cutoff channel become silted up, and thus a lake, called an *oxbow lake*, is born. The cutting across meander bends is often hurried along by floods. The stream overflows its banks, with the overflow taking the direct course to the adjacent bend and eroding out a direct chute or channel.

If the base level of a meandering stream is lowered and the gradient and velocity are thus increased, the stream will erode its bed, and the meanders will become *entrenched* (study Figures 7–45 and 7–46).

Braided streams

Meandering streams are basically eroding streams, but under certain conditions a stream's major activity is turned to deposition, and the channel or valley is filled with sediment. If, for some reason, a stream's tributaries start bringing in a greater load than it can transport, it drops part of its load. The channel fills up, and

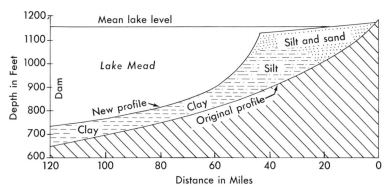

Figure 7–41 *Deposits by the Colorado River in Lake Mead back of Boulder Dam. [After Gould, S.E.P.&M., 1951.]*

the stream overflows its banks. It may find several low spillways over the banks and break into as many *distributaries*. Each small branch channel soon becomes filled, and its flow spills over to form more branches. Such a stream is shown in Figures 7–47 and 7–48. It is said to be *braided*. The easily eroded soil from a newly plowed grassy turf may overload the stream and result in a braided condition. The melt waters from a glacier are generally so heavily laden with debris that the stream is not able to carry the material along the gradient to which it was adjusted before the glacier formed, and a braided condition results.

Natural levees and flood plains

In the lower part of the longitudinal profile of many rivers the valley is generally broad. There spring floods have overflowed the banks of the river and deposited layers of mud on each side. This builds the *flood plain*. Associated with the flood plain are the *natural levees*, which are low ridges that border the river and stand a few feet above the flood plain. Flood stage is reached when the river rises to overflow the levee and to pour out over the flood plain.

The levee is built at times of flood of the coarsest material the river is carrying. At the moment the channel-confined current spreads in a thin sheet over the bank, its velocity drops and the coarsest particles of its load settle on the bottom. The finer particles are carried out over the flood plain, but almost all the material is deposited immediately adjacent to the channel, and hence a levee is built. Levee building and channel fill continues in places to such an extent that the river surface rises above the flood-plain elevation. In this condition if the river in flood spills over the levee, it might cut a channel through it, and much of the flow would then debouch on the flood plain, resulting in serious flooding.

The levees built by the river are natural, in contrast to man-made embankments built to confine rivers in high-flood stage.

Deltas

The deposit of the Colorado River in Lake Mead is called a *delta*. The term refers generally to the part of the deposit at the river mouth and not to the turbidity current deposit far out in deep water.

Great deltas like the Mississippi are fairly complex. The river divides into distributaries, with each distributary building its own smaller delta, and in total a pattern reminiscent of a bird's foot is formed. The water of each distributary jets into the quiet water of the Gulf, and deposits of clay, silt, and sand, such as are shown in Figure 7–49 result.

The coarsest particles, sand at the river mouth, are dropped at the point where the current enters still water. This deposit forms a base over which the current then flows, only to de-

Figure 7–42 *Meandering Hay River, northwestern Alberta. Note the crescentic bars of the shifting meanders, meander cutoffs, and oxbow lakes. [Courtesy of Geological Survey of Canada and Ed Schiller.]*

posit more sand at the new point of entry. A tongue of sand is thus built out by each distributary, and as you might deduce, each distributary is immediately underlain by a tongue or bar of sand. On the sides and underneath is silt, and out still farther in deeper water the finer clay particles settle.

Terraces

Stream terraces are benches along the sides of river valleys. Good examples occur on either side of Madison Valley in southwestern Montana (Figure 7–50). They are common along almost all rivers that have reached fair adjustment but then have been energized or *rejuvenated* by uplift of the land. Entrenchment ensues, then broadening of the entrenched meanders, until a new valley floor is established with terraces on either side. Rising and lowering of base level in several cycles cause rather complex sets of terraces (see Figure 7–40).

Valley widening

When the land is uplifted with respect to the sea or lake into which the stream flows, the stream will cut a sharp, V-shaped valley into the uplifted terrane. Downcutting is the chief process. As the longitudinal profile becomes graded, downcutting becomes slow or minimized, and the processes of valley widening become dominant. The processes of frost action and talus formation, landslides, and rainwash erode away the valley sides. In arid regions the valley walls generally retreat in the manner illustrated in Figures 7–51 and 7–52. The slope remains fairly constant, and a distinct angular break in cross profile between valley wall and valley floor is maintained. This break is called the knick point (see Figure 7–52). In many places of our western arid and semiarid regions, changes in climate in the past 25,000 years (during and following the last ice age) and overgrazing in the past 75 years have resulted in vigorous changes in the streams' regimen. At times filling (alluviation) of valley bottoms and at other times entrenchment of the streams in the previous deposits have resulted in complex terraces and have contrived to obliterate the simplicity of the valley forms, which might otherwise exist.

In humid climates the valley sides become rounded with the upland, and a more rolling and less angular topography results.

Fault scarp features

Erosion attacks the uplifted block of a fault vigorously and soon cuts a number of V-shaped clefts in it, in the manner shown in Figure 7–53. If the fault displacement amounts to several thousand feet, as is the case in many ranges of the Great Basin of Utah, Nevada, and eastern California, then the clefts become canyons, deepest at the front of the uplifted block. As erosion proceeds, the canyons extend headward into the uplifted block, and the old land surface becomes much dissected. Eventually, there are no remnants of the old surface, but a series of nicely aligned triangular faces mark the fault scarp. Such a mountain front is pictured in Figure 7–54.

The deposits at the foot of a fault scarp are cone- or fan-shaped, with the apex of each fan at the mouth of the canyon or gully in the uplifted block. It is clear that the debris has spewed or flooded from the canyon. These de-

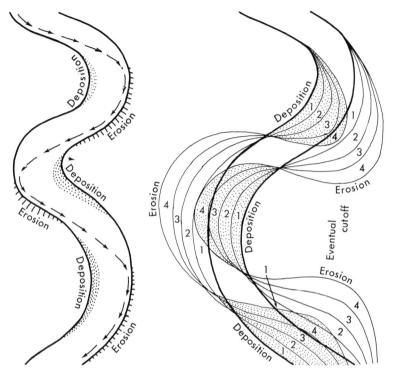

Figure 7-43 The process of meandering by erosion and deposition (left), and successive stages in the development of meanders (right).

posits are called *alluvial fans* (Figures 7-53 and 7-54). They occur in all climates but are especially clear and impressive in arid regions where block faulting has occurred. Like deltas their rock fragments are coarsest at the top or apex and finest at the periphery, and also like deltas this general distribution is altered by long radiating channels of coarse debris encased in fine material. The debris of alluvial fans is not so well sorted as in deltas and contains large boulders. Alluvial fans, however, are good reservoirs for underground water, and considerable attention has been given to their structure and "plumbing system."

Dissected erosion surfaces

Many of the high ranges of the Rocky Mountains display nearly flat, gently rolling, or inclined surfaces at high elevations (see Figure 7-55). These are generally remnants of erosion surfaces fashioned by streams as wide valleys or piedmont slopes before the range was uplifted. With uplift, vigorous headward erosion sets in, which dissects the range and, perhaps,

Figure 7-44 Meander bend on Yukon River, showing bar deposition on inside of curve. Wide rivers have large meander bends.

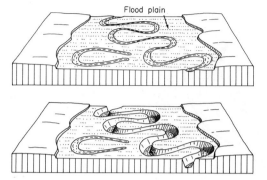

Figure 7–45 *Meandering stream with oxbow lake above, adjusted to base level. Below, base level lowered suddenly with entrenchment of stream in meandering course.*

Figure 7–46 *Entrenched meandering stream, Hoback River, western Wyoming.*

leaves remnants of the old surface here and there, generally at the drainage divides.

The flat-topped ridges of the Appalachian Mountains are classic examples of such an old erosion surface. The history there has been as follows: first, folding; second, extensive erosion across the folded strata to form a surface with only gentle, low divides and wide valley bottoms (a *peneplain*); and third, uplift of the entire region with dissection and entrenchment

Figure 7–47 *Braided stream in Mount McKinley National Park, Alaska.*

of the streams. Parallel ridges and valleys were etched out of the parallel hard and soft upturned or folded strata, but the flat tops of the parallel ridges reflect the old erosion surface (see Figure 7–56).

Rates of denudation and uplift

Rates of denudation

By rate of denudation we mean the rate at which the combined weathering and erosional processes combine to reduce the land surface. Such rates allow us to calculate the time, say, that the Colorado River has taken to erode the Grand Canyon, or the time the Delaware has been at work in the Appalachians. The most immediate method is to measure the rate of sediment accumulation in a reservoir, and spread this sediment, by calculation, over the watershed that drains into the reservoir. Various corrections must be made but results such as the following have been obtained: the drainage region of the Colorado River is being denuded at a rate of 6.5 inches every thousand years; the overall rates of the large drainage

Figure 7–48 *Braided stream and its deposits in a mountainous V-shaped valley. Compare with Figure 7–47.*

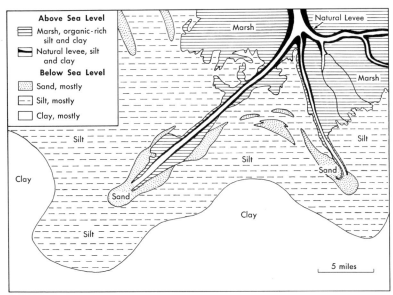

Figure 7–49 Southwest part of Mississippi birdfoot delta, showing distribution of sand, silt, and clay deposits. [Simplified from Fisk et al., 1954.]

Figure 7–50 Terraces along the Madison River near Varney, southwestern Montana.

Figure 7–51 Process of valley widening in arid lands.

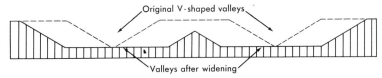

Figure 7–52 Angular break (knick point) between valley wall and valley floor, Badlands of South Dakota. [Courtesy of Kenneth G. Smith.]

Figure 7–54 Aerial photograph of dissected fault scarp and alluvial fans near Dillon, southwestern Montana.

basins of the United States range from 0.1 to 0.3 ft per 1000 years. The maximum rates occur in semiarid climates (10 to 15 in. of rainfall) and increase exponentially with relief. The headward reaches of streams are noted to be denuding much faster than the lower reaches.

An interesting method for measuring the rate of denudation is available in the Cedar Breaks area, Utah, where some bristlecone pines are so old that their roots have been conspicuously exposed by erosion (see Figure 7–57). By coring the trees and counting the growth rings, the age of the trees can be determined. The age of the oldest tree cored was 2480 years. By measuring from the ground up to where the roots leave the trunk, the amount of denudation is determined. By dividing the age of the tree by the amount of denudation the rate of denudation is determined. Forty-one trees were plotted according to age and amount of denudation, and the average rate was found to be 11.5 in. per 1000 years.

Rates of uplift

The Coast Ranges of California are noted for vigorous mountain building going on today. Survey bench marks have been reoccupied several times in the past fifty years, and the progress of uplift was definitely detected. Figures are numerous and substantial and indicate

Figure 7–53 Forms resulting from dissection of a fault scarp. Note remnants of upland surface, V-shaped gorges and triangular faces in uplifted block, and alluvial fans on downfaulted block.

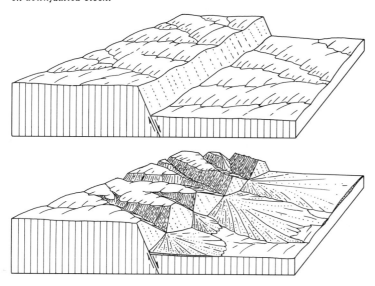

Figure 7–55 Continental Divide in Estes Park, Colorado, showing remnant of high, old erosion surface.

an average rate of uplift of 25 ft per 1000 years. Compare this with an average rate of denudation of 0.25 ft and a maximum rate of 3 ft per 1000 years, and you are immediately struck with the fact that uplift, or mountain building, is a number of times faster than denudation. In certain small areas underlain by easily eroded shale and silt the rate of denuda-

Figure 7–56 Characteristic slope profile, above, and the concept of parallel retreat of slopes (1 to 2 to 3).

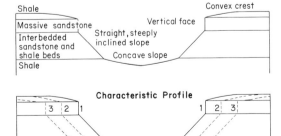

Figure 7–57 Bristlecone pines at Cedar Breaks, Utah, expose their roots because of denudation.

tion is as high as the rate of uplift in parts of the Coast Ranges, but such examples are certainly exceptional.

The average rate of uplift in the central part of Fennoscandia in the past 7000 years has been measured to be 48 ft per 1000 years. This is a region where a thick and extensive ice cap had previously depressed the crust, and the crust is now recovering. The Lake Superior region is rising at the rate of 16 ft per 1000 years for the same reason. Many other examples of rates of uplift are known, and each demands individual consideration, but it is evident that rates of uplift, for a while at least, are too great to be matched by the processes of erosion.

References

Bradley, W. H. *Geomorphology of the north flank of the Uinta Mountains.* U.S. Geol. Survey Professional Paper 185-I, 1936.

Eckel, E. B. *Landslides in engineering practice.* Washington D.C.: National Research Council, Highway Research Board Special Report 29, 1958.

Judson, Sheldon, and Ritter, Dale F. Rates of denudation in the United States. *Jour. Geophys. Res.*, v. 69, n. 16, 1964.

Kellog, C. E. Soils, Scientific American, July 1950.

Leopold, L. B., and Langbein, W. B. A primer on water. U.S. Geol. Survey, Miscellaneous Report, 1960.

Leopold, L. B., and Wolman, M. G. River meanders. *Bull. Geol. Soc. Amer.*, v. 71, pp. 769–794, 1960.

Sharpe, C. F. S. *Landslides and related phenomena.* New York: Columbia University Press, 1938.

Schumm, S. A. Disparity between present rates of denudation and orogeny. U.S. Geol. Survey, Professional Paper 454-H, 1963.

Schumm, S. A. The development and evolution of hillslopes. *Jour. Geol. Educ.*, v. 14, n. 3, 1966.

Schumm, S. A. Meander wavelength of alluvial rivers. *Science*, Sept. 29, 1967.

U.S. Department of Agriculture Yearbook. *Soils.* 1957.

Motion Picture

Work of rivers. Encyclopaedia Britannica, black and white, 16 mm, 16 min.

8
glaciers and ground water

Types of glaciers

Ice is a brittle substance when struck sharply, but in large masses it behaves like a viscous liquid. The ice in many glaciers is over 1000 ft thick, and when such a mass is unconfined, such as on a mountain slope, it starts to flow. Gravity is the sole force acting on it and causes it to flow. The streaming of ice in glaciers is an extremely obvious phenomenon and has been measured in many places. Thus, a glacier may be defined as any mass of ice on land that is flowing.

There are two principal types of glaciers: *continental* glaciers and *valley* glaciers. The continental glaciers are great sheets of ice, called ice caps, that cover parts of continents. The earth has two continental glaciers at present: one covers most of Greenland and the other all of Antarctica save for a small window of rock and the peaks of several ranges (see Figure 8–1). The Greenland ice sheet is over 10,000 ft thick in the central part and covers an area of about 650,000 sq miles. The Antarctic sheet has been sounded, in one place at least, to a depth of 14,000 ft and spreads over an area of 5,500,000 sq miles. This is larger than conterminous United States in the proportion of 5½:3. It is calculated to store 7 million cu miles of ice, which if melted would raise the ocean level 250 ft.

Valley glaciers are ice streams that originate in the high snow fields of mountain ranges and flow down valleys until they melt (Figure 8–2). Some flow all the way to the sea, where they break up into icebergs and eventually melt in the ocean (Figure 8–3). In certain places the valley glaciers flow down the mountain valleys to adjacent plains and there spread out as lobate feet (Figure 8–4). These are called *expanded-foot glaciers*. Generally the sprawling feet of several valley glaciers coalesce to form one major sheet, and this is called a *piedmont glacier*.

Most valley glaciers range in length from a few hundred feet to streams 50 miles long. Some of the valley glaciers of Antarctica, nourished by the vast ice cap, are much longer still. During the past ice age many of the ranges of the western United States supported valley glaciers 10 to 30 miles long, but now there are only shrunken remnants, or the glaciers are gone entirely.

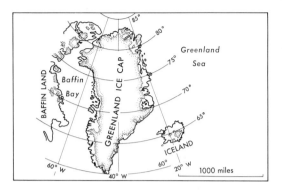

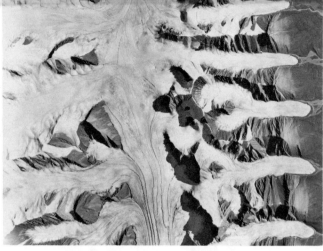

Figure 8–2 *The polar-type glaciers of Ellesmere Island are frozen tight at their bottoms. Hence, meltwater runs as streams on surface. Note that glacial fronts nestle in terminal moraines, indicating that glaciers are at most advanced position. Note also outwash streams.* [*Courtesy of Geological Survey of Canada and Ed Schiller.*]

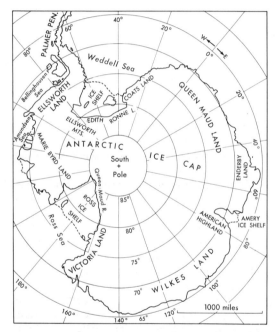

Figure 8–1 *Continental glaciers of Greenland and Antarctica. The ice cap of Greenland extends to the dotted line. Antarctica is all ice except for the peaks and ridges of a number of mountain ranges.*

The glacial regimen

Nourishment of valley glaciers

If more snow falls in the winter in the high parts of a mountain range than melts in the summer, then, year after year, an accumulation builds up. The snow of one year, buried under the snow of the next, turns granular, and after a few years and under the weight of several years' snow falls it turns to ice. This ice, when in sufficient volume, starts to flow downhill under the pull of gravity. It does not take much ice before flowage starts, as may be seen from the small glaciers in Figure 8–5.

Important in the nourishment of a glacier is a good-sized catchment basin. This is situated at the head of a mountain valley or canyon and consists of a large area made up of slopes that drain into the valley head. Here the snow from a large area feeds into a single valley. A number of small valley heads, tributaries of a main valley, may have their own snow fields and small glaciers, and these flow together to form one major ice stream a number of miles long.

It is evident that if the number of inches of snowfall is the same in two catchment basins, the larger basin will nourish the larger and probably longer glacier. In general, snowfall depends on three principal factors: (1) the higher the range, the greater the snowfall; (2) the farther north and south from the equator, the greater the snowfall for the same altitude; (3) proximity to the ocean means notably greater precipitation in certain places. For instance, at an altitude of 12,000 ft along lat 40° N in Colorado and Utah, only small snow-

Figure 8–3 Hubbard Glacier, Alaska. [Courtesy of Bradford Washburn and Duncan Stewart.]

fields last through summer, but at 6000 ft along lat 50° N in the Canadian Rockies, glaciers several miles long occur. Also along the Pacific Coast from the Olympic Mountains in Washington to Mount Saint Elias in Alaska, precipitation is especially heavy, although the temperatures are not particularly low, and this results in great snowfields and extensive glaciers, especially in British Columbia and Alaska.

Rate of flow

Numerous measurements of the rate of flow of glaciers have been made. The procedure is to drive a line of stakes across a glacier and then chart their position every day, month, or year thereafter, depending on the rate of movement. In all cases the center of the flow moves the fastest, and in time the straight row of stakes is bent into a curve, convex downstream. The centers of several glaciers have been observed to flow a few inches to more than 150 ft per day (see Figure 8–6).

It will be seen in later paragraphs that glaciers transport large tonnages of rock fragments from high along their courses to their lower ends. An unusual example of transportation is found in the Bossons Glacier of the Alps. Two climbers plunged to their death in a crevasse, their bodies were frozen in the ice, and, because of the flow of the ice, the bodies were released forty-one years later at the melting end of the glacier several miles below.

Melting

A glacier is nourished by snowfall in the higher elevations, flows to lower and warmer climates, and there melts. Melt-water streams discharge from the glacier from the surface, from passageways within, and from the bottom. The lower mile or two of the large glaciers are generally in a state of dissipation owing to melting, and the front or snout of some glaciers appears tattered and irregular (Figure 8–7).

Figure 8–4 Expanded-foot glaciers of eastern Baffin Island. Glaciers are at maximum stage of advance. Note extensive rolling upland surface that is dissected by stream valleys and etched by cirques. [Courtesy of Geological Survey of Canada and Ed Schiller.]

The discharged water is laden with silt and clay particles, free of organic material, and is light gray. The fine particles are commonly called *rock flour*, and the light, chalky melt-water is called *glacial milk*. One may be driving along a mountain stream, not knowing that it is melt-water from a glacier, but such can be surmised readily by the chalky color of the water. Once seen, the tone will be remembered.

The terminus of the glacier is not just the place of final melting of the ice but also of unloading of much debris that the glacier has carried down from higher altitudes. This will be considered presently.

The front of a glacier remains in the same place if the rate of forward flow is equal to the rate of backward melting. Corollary statements may be made, namely, that if the forward flow exceeds the rate of melting, then the front or toe

Figure 8–5 Glaciers from Mount Marcus Baker literally cascading down steep slopes. [Courtesy of Bradford Washburn and Duncan Stewart.]

of the glacier advances, and if the forward flow is less than the rate of melting, then the front of the glacier recedes. Valley glaciers in almost all mountains have been receding consistently in historical times, and the mean ocean level the world over has gone up a few inches in the last century, undoubtedly owing to the release of water stored in the glaciers. The near worldwide recession of valley glaciers is taken to mean a slight warming up of the earth's annual temperature.

Figure 8–7 Glaciers of Mount St. Elias Range, Canada. The main glacier has stagnated and melted away, leaving only rock debris, once in transport. Note moraines of receding tributary glaciers. [Courtesy of Geological Survey of Canada and Ed Schiller.]

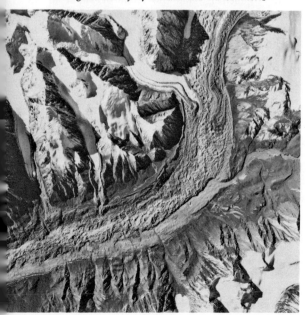

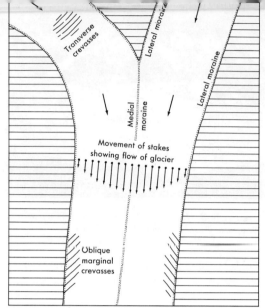

Figure 8–6 Nature of flow, crevasses, and moraines of valley glacier.

Features of continental glaciers

Continental glaciers are fed by snow that forms extensive annual layers. The layers compact into granular and then solid ice within a depth of 50 ft or so. The annual layers are still evident in the solid ice.

Studies during the International Geophysical Year show that in Antarctica the accumulating snowfall in terms of water per year is greatest in the coastal or marginal areas and becomes progressively less toward the high interior plateaus. At the South Pole the accumulation is 6 to 7 cm of water per year (2½ in.), whereas on the Ross Ice Shelf it ranges from 14 to 21 cm per year. The average across Marie Byrd Land and Ellsworth Land is about 18 cm per year, and in the coastal area just south of the Bellinghausen Sea it may be as much as 60 cm per year, or 23½ in.[1] Ice has accumulated on the interior plateaus until great streams pour down through the gaps in the Queen Maud Range into the Ross Ice Shelf. The Leverett, Robert Scott, Amundsen, Liv, Shackleton, and Beardmore Glaciers are some of these streams that have been charted, with the Beardmore

---

[1] M. B. Giovinetto, Mass accumulation in West Antarctica, *Trans. Amer. Geophys. Union*, v. 42, n. 3, 1961, p. 386.

probably the largest. It is 140 miles long and in places 25 miles wide. Some recent maps made from vertical aerial photographs of an area at the base of the Palmer Peninsula in the Ellsworth Mountains show many typical valley glaciers heading in snow fields in the crest of the range. The valley glaciers flow into the vast ice cap fields, from which some major ice streams resolve and flow probably toward the Weddell Sea. The Nimitz Glacier is one of these (Figure 8–8); it is about 10 miles wide and at least 55 miles long. It appears to be nourished partly by ice from the ice cap and partly from the valley glaciers of the Sentinel Range (part of the Ellsworth Mountains).

The Greenland ice cap is hemmed in by marginal ranges, through the passes of which the ice pours in great valleylike glaciers. Their appearance and activity is the same as that of valley glaciers, but they are fed by ice from the interior ice cap. Some reach the sea and discharge angular icebergs which float with jagged peaks above the water.

The nourishment of the continental glacier of Antarctica, as described, with numerous ranges jutting up through the ice, and with many large and small topographic irregularities below the ice, must give rise to a very complex flow pattern. Ultimately, of course, the ice reaches the sea, where it dissipates. The major problem of Antarctic glaciology is that of the ice budget. How much ice is being added each year? Is the great ice cap decreasing or increasing in size? These questions have been only partially answered to date.

A recent report by Péwé and Church[2] summarizes the condition of the glacier margins at several places and concludes that in the period 1911 to 1958 the ice flow was about in equilibrium; the fronts were neither advancing nor retreating. This suggests that the ice budget was in balance during this period and that the ice cap was neither building nor wasting away.

[2] T. L. Péwé and E. Church, *Glacial regimen in Antarctica*, University of Alaska, Department of Geology, Reprint Series No. 15, 1962.

Glacial erosion

Glacial erosion takes place primarily through abrasion. Blocks of rock, commonly large and angular, are frozen in the ice; they are thus firmly wedged in the ice and are ground over the bedrock over which the ice moves. This has been likened to a filing or rasping action. Where glacier snouts have retreated, the gouged, grooved, and polished surfaces of rock over which the ice formerly flowed may be seen. These abrasion effects are a permanent record of past glacial activity. See Figure 8–9, which shows the effects of a glacier that existed about 10,000 years ago.

The blocks of rock that the glacier uses as tools to grind the bedrock away are themselves abraded. They develop grooved and striated faces (see Figure 8–10). Such boulders are common in the terminal deposits of the glaciers. Rivers, the wind, and the waves do not have the particular ability to use rock fragments as rasps or to fashion faceted and striated boulders. Such features are trademarks of glacial action.

As a glacier passes over a bedrock hill or protuberance it grinds away on the side of approach (Figure 8–11), but it is said to "pluck" rock fragments from the leeward side. The ice penetrates the open fractures and in effect freezes around the fragments. Then, as it moves ahead, the fragments are pulled or plucked from their place, and a jagged steep leeward slope results (see Figure 8–12). In the following description of glacial erosional forms the processes of abrasion and plucking will be further elaborated.

Features formed by glacial erosion

The most conspicuous form developed by a valley glacier is the U-shaped valley. Mountain valleys due primarily to stream erosion are V-shaped with zigzag course (see upper diagram of Figure 8–13). After a glacier has flowed down a valley for some time it will have abraded a U-shape in the bottom of the V, as indicated in the bottom diagram. Also the jutting side ridges around which the stream made

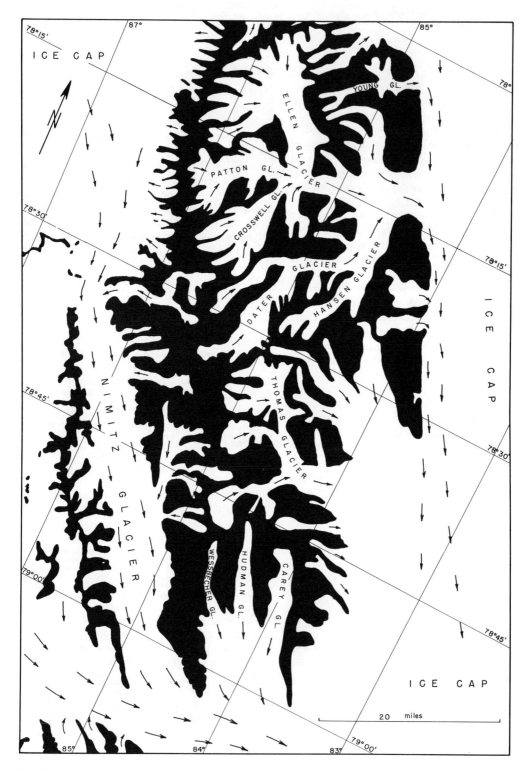

Figure 8–9 Glacial striations (several directions) and chatter marks (cusps, right to left) on bedrock in northern Manitoba. [Courtesy of Geological Survey of Canada and Ed Schiller.]

Figure 8–10 Scratched and faceted boulders used as abrading tools by glaciers. Each about 1½ ft long. [Courtesy of Dr. Duncan Stewart.]

its zigzag course will have been truncated and the canyon straightened so it follows a smoothly curving course (see the bottom diagram again).

Figure 8–14 is an example of a U-shaped valley with a broadly curved course. It acquired this form because a valley glacier flowed down the canyon during the ice age. The valleys of the Alps of Switzerland are classic examples, and there the processes and forms of glaciation were first recognized. Many of our mountain valleys in New England and in the western Rockies are U-shaped, at least in their headward sections, and attest to one or more episodes of glaciation sometime in the past.

In a mountain region that has been glaciated the tributary valleys may enter the main valleys over a cliff face and are called *hanging valleys* (see H in Figure 8–13, and see also Figures 8–15 and 8–16). Although the surface of the tributary glacier and the main glacier are con-

Figure 8–11 Sheep backs, just west of Whitehorse, Yukon Territory. The French term for these glacially abraded hard-rock hills is rôches moutonnées.

Figure 8–12 Plucking effect of glacier. The movement, which was toward the viewer, abraded the top slope smooth but plucked blocks from the leeward slope. The rock is granite in the Wasatch Mountains, Utah.

Figure 8–8 Glaciers of the Sentinel Range, Ellsworth Mountains, Antarctica, showing the relation of valley glaciers to the ice cap. The ice cap contributes to the flow of the Nimitz Glacier. [Taken from the Vinson Massif and Nimitz Glacier maps of the U.S. Geological Survey, 1960.]

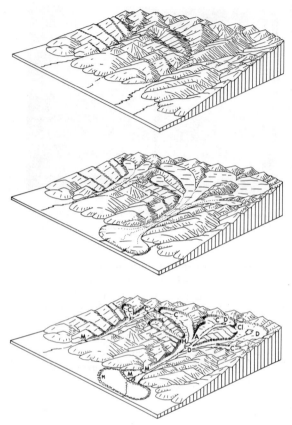

Figure 8–14 Little Cottonwood Canyon (left) in Wasatch Mountains is a smoothly curved U-shaped valley due to glacial erosion. Bell Canyon (right) is also one of glacial erosion, with spoon-shaped terminal moraine at base.

Figure 8–13 Valley glaciers and glacial features. Top, before glaciation; middle, during glaciation; bottom, after glaciation. C, cirques; D, rock basin lakes; H, hanging valley; A, arête ridge; M, terminal moraine, Cl, col.

through these thin bridges that climbers may fall.

The bergschrund separates bedrock and ice in the manner illustrated in Figure 8–18. There is much thaw-water in a summer day dripping down the crevasse, and undoubtedly rock fragments are pried loose by frost action, but it is also believed that there is considerable plucking going on. At any rate, a near-vertical wall is eroded that curves around the upper edge of the catchment basin, and when the glacier melts, we

Figure 8–15 Relation of tributary glacier to main valley glacier, showing particularly the origin of the hanging valley.

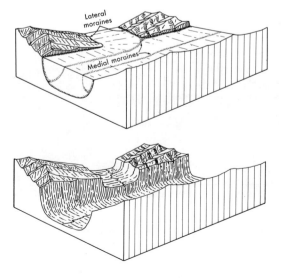

fluent, that is, flow together at a common elevation, the floor of the main canyon is abraded more deeply than the floor of the tributary glacier. Hence, when the glaciers are melted, the tributary U-shaped valley is suspended above the main valley.

In the catchment area of a valley glacier near the upper limit of the snow field a gaping fissure or crevasse commonly forms. This is the *bergschrund* of the Alpine geologist and mountain climber (Figure 8–17). It is a dangerous crevasse because snow cornices commonly develop and even bridge across it, and it is

Figure 8–16 *Hanging valley and waterfall in Yosemite Valley, California. View is up the U-shaped glaciated valley with El Capitan on left and Bridal Veil Falls on right. Half Dome is in the center distance.* [Photograph by F. E. Mathes, U.S. Geological Survey.]

find a large amphitheaterlike basin with very steep walls around three sides at the head of the U-shaped valley. Such a basin is called a *cirque* (see *C* in Figure 8–13 and see Figure 8–19, which shows a cirque wall etched into a gently sloping upland surface).

From the cirque on down the glaciated U-shaped valley, pockets or small basins may have been ground out of the bedrock by the ice, and these are later filled with water when the glacier has melted. These small lakes are called *rock-basin lakes,* and those within the cirque are called *cirque-basin lakes.* Some people use the Scottish word *tarn* for the rock-floored lakes in cirques (see *D* in Figure 8–13).

In some glaciated mountains, a central high peak is surrounded by cirques that feed radiating glaciers. The cirque walls have sharpened the slopes of the peak on all sides until it stands as a commanding buttress or sentinel of the region. Such a peak is called a *horn,* after the Matterhorn, of Switzerland.

Two glaciers or cirques may erode toward each other to the extent that the dividing ridge becomes like a knife-edge and is marked by jagged spikes of rock. Such a ridge is called either a *comb ridge* or an *arête* (see *A,* Figure 8–13).

Where two crescent-shaped cirque walls have been eroded to the point of intersection (*Cl,* Figure 8–13) a pass between the two may have developed. This is called a *col.*

The valley glaciers of certain coastal ranges, in flowing down to the sea, may have eroded the lower parts of their U-shaped valleys below sea level, and with the melting of the glaciers these parts become flooded with seawater. They are characterized by the steep valley walls rising directly from the water and are called *fjords.* Figure 8–20 shows one of the many scenic fjords of Norway.

Glacial transportation

Rock and mineral fragments are transported directly by the ice and also by the melt waters from it. The most evident material in transport is along the margins of valley glaciers, where angular fragments, large and small, pried loose by frost action on the steep slopes above, collect as talus on the ice. This is a *lateral moraine,* shown in Figure 8–21. At the juncture of two valley glaciers, or the tributary with the main ice stream, the two lateral moraines merge to form a train of debris near the center of the glacier. This is called a *medial moraine.* There will be as many medial moraines as there are

Figure 8–17 Bergschrund near the head of a catchment basin. Mount Deception, Alaska. [Courtesy of Bradford Washburn and Duncan Stewart.]

tributaries, and all are being transported to the terminus of the glacier.

Both valley and continental glaciers take up great volumes of loose material from the surface over which they flow and transport it to the glacier terminus. Soils, sand and gravel deposits, previously deposited glacial material, and bedrock fragments are frozen into the ice and carried or dragged to the melting front of the glacier. Some of the Greenland glaciers are heavily charged with rock debris, which they could only have procured from the surface over which they flow. The angular, fresh fragments were obtained, at least in part, by plucking.

The grinding of rock flour and its transportation by melt waters have already been mentioned. The streams that discharge from the front of glaciers also carry much sand and

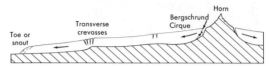

Figure 8–18 Longitudinal section of valley glacier from bergschrund to snout.

gravel and deposit the material in lakes and depressions at the ice front and down the valley beyond the glacier (Figure 8–22). The pebbles and cobbles are commonly rounded, in contrast to the angular fragments transported directly by the ice. The ice may pick up already rounded cobbles and boulders, and in certain deposits these are more abundant than angular fragments.

Glacial deposits

Types of material deposited

Glacial deposits may be classified on the basis of the type of material contained in them, their outward size and form, and their position relative to the ice front. The types of material are of two general kinds. One kind is deposited directly by the ice and is a heterogeneous mixture of silt, sand, and large fragments, either rounded or angular or both. This material is unsorted and unstratified (Figure 8–23) and is called *till*. The other kind of glacial deposit is formed by melt-water streams and is sorted and stratified into layers of silt, sand, and gravel. It is said to be of glacio-fluvial origin (fluvial pertains to streams) and is sometimes referred to as *stratified drift*. All kinds and forms of glacial deposits are called collectively *drift* or *glacial drift*.

Moraines and glacial lakes

The term *moraine* is used to denote the deposits by the ice, particularly till, at the terminus. Glaciofluvial material may be included in the terminal moraine, and the lateral and medial moraines in transport on the ice are also included in the definition of moraine.

Both continental and valley glaciers leave considerable deposits in the form of terminal

Figure 8-19 Cirque wall with much talus that has accumulated since the ice melted. A horn is in the distance. Uinta Mountains, Utah.

moraines. The longer the ice front remains steady, neither advancing nor receding appreciably, the larger the terminal moraine. Thus, the terminal moraine marks the position of a stationary ice front at some time in the past. The hill and lake country of the Great Lakes states is due to the terminal moraines of the continental glaciers that formed in northern Canada and spread in large lobes southward into the Missouri, Mississippi, and Ohio River drainages. It was Louis Agassiz in the period of 1845–1873 who first recognized the deposits as those of glaciers and laid the foundation for the working out of the history of the ice age in North America.

The relief of the deposits in a terminal moraine of a continental glacier is generally about 100 ft high, but some hills may be 300 ft high. There are many undrained depressions in the terminal moraine that harbor small lakes and bogs. Minnesota's "ten thousand lakes" are of glacial origin. The streams in the morainal complex seem to have a confused and aimless pattern, and only after considerable geologic time do they establish a system of trunk streams and tributaries by erosion and deposition that drain the area effectively.

The terminal moraines of valley glaciers are generally crescent-shaped and extend across the valley in hummocky form (Figures 8-13 and 8-23). As the ice front recedes from a terminal moraine, a lake is commonly impounded between the moraine and ice (Figure 8-24). It rises to overflowing, and then the overflow stream cuts a gorge through the moraine (Figure 8-25). As the gorge or outlet is cut down,

Figure 8-20 Geiranger Fjord, Norway. Photograph taken from Eagle Pass. Note hanging valley and waterfall. [Courtesy of Norwegian National Travel Office, 505 Fifth Ave., New York.]

the lake is drained. It is, therefore, evident that these "morainal dam" lakes are ephemeral. Generally those which exist today in our glaciated mountain valleys were once higher and larger but are now in the process of being drained and also being filled with sediment.

The process of withdrawal of glaciers is usually marked by periods during which the front is stationary, when small terminal moraines are built. A succession of these moraines is evident in some places and is referred to as a *recessional moraine*. A scattering of till, as the front withdraws rather rapidly, is called a *ground moraine*, and the volume of such material in the Great Lakes region is undoubtedly very great.

In places the fronts of the great continental ice cap lobes have advanced and ridden over their terminal, recessional, and ground moraines. In doing so, these hummocky deposits have been dragged out and streamlined into elongate hills, with their long axis in the direction of ice movement. These hills are called *drumlins*. They vary considerably in size and have been characterized as tens of feet high,

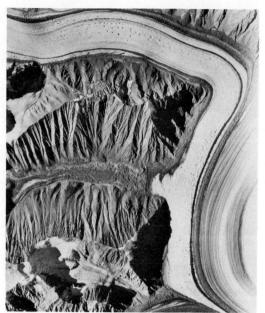 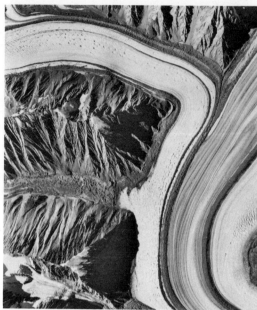

Figure 8–21 *Glaciers of Mount St. Elias Range, Canada and Alaska, showing lateral and medial moraines, crevasses, and small stagnant glacier (center of photographs). Use lens-type stereoscope for viewing.* [Courtesy of Geological Survey of Canada and Ed Schiller.]

hundreds of feet wide, and thousands of feet long. They commonly occur in swarms (Figure 8–26) and are numerous in the Great Lakes region of Canada and the United States. Drumlins consist for the most part of till.

Glaciofluvial deposits

The streams discharging from glaciers breach the terminal moraines and spread stratified silt, sand, and gravel in front of the moraines. Spread out beyond the moraines of continental and piedmont glaciers are deposits similar to broad alluvial fans, called *outwash plains*. Outwash plains are also built inside a previously deposited terminal moraine between the ice and the moraine, but only where the melt water can drain away and is not impounded into a lake. In such a situation, blocks of the retreating ice front become isolated and surrounded or even buried by stratified deposits. These blocks of ice later melt and leave depressions in the plain called *kettles*.

The swollen streams from valley glaciers are almost always overloaded with silt and sand, and although their rate of flow is fairly rapid, they cannot carry all their load and proceed to drop the coarsest fragments. Alluviated valleys result, with picturesque braided streams being the rule (see Figure 8–22).

A surprising deposit of sand and gravel is found in some areas covered by continental glaciers. It is a long, winding ridge, usually about 25 to 50 ft high and 50 to 200 ft wide. Some have such an even crest that they look like an artificial railroad grade winding across the countryside (see Figures 8–26 and 8–27). They are called *eskers*. They occur behind the main terminal moraine in the area of the ground moraine and seem to require a special condition for their formation. If the ice were in active flow, such deposits would have been smeared out at the base of the advancing gla-

Figure 8–22 Outwash gravels from glaciers upstream. Note terminal moraine ridge around snout of glacier just beyond wooded spur. Refer to Figure 8–25.

Figure 8–23 Terminal moraine of present Teton Glacier, northwestern Wyoming.

cier, so that it is postulated that at times sections of an ice lobe become dead or stagnant. During this time aggrading streams that flow in tunnels at the base of the glacier fill or nearly fill their ice-walled passageways with sand and gravel. Then, with the complete melting of the stagnant ice, the winding eskers are left standing above the general ground moraine.

Glacial lakes that are fed directly by melt water often receive a thin dark layer of silt in winter and a thick light gray layer in summer.

Figure 8–24 Vertical aerial photograph of terminal moraines of piedmont glacier in western Montana. Note also the lake impounded back of the second moraine.

They constitute a year's deposit. Such a pair of layers is called a *varve*. Lake deposits containing hundreds and even thousands of varves have been dissected by later stream erosion and, thus exposed, the varves have been counted and used in the same manner as the annual growth-rings of trees in deciphering the history of certain glaciers. By an intricate system of correlation of several varved deposits the time that has elapsed since the glacier made its last stand has been approximated.

The Pleistocene and its glacial climates

Toward the end of the Tertiary the climate of the earth began to cool, and eventually great ice caps developed on several continents and almost

Figure 8–25 V-shaped cut through moraine. Beyond the braided stream behind the moraine is a glacier terminus. Near Columbia ice field on highway between Banff and Jasper, Alberta.

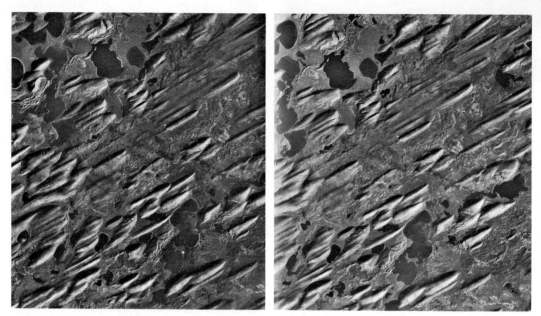

Figure 8–26 *Swarm of drumlins south of Lake Athabaska, Alberta. Ice sheet moved from northeast to southwest. Note esker in upper left corner of photographs. Use lenstype stereoscope for viewing.* [Courtesy of Geological Survey of Canada and Ed Schiller.]

all high mountains became extensively glaciated. This glacial stage, however, vanished and soon, geologically, another glacial stage developed. All told, four major glacial stages have been recognized with intervening warmer episodes. These four glacial stages occurred approximately within the last million years, and the present time would appear to be another interglacial episode. They constitute the Pleistocene Epoch.

As the earth's climate turned cool and warm a number of times, as the glaciers waxed and waned, and as the sea level fluctuated over several hundred feet, the biological world was profoundly affected. The adaptations, migrations, and evolution of new forms during this time constitute to many persons' minds the most exciting period of earth history. The Pleistocene is preeminently the age of man, for during this time he made his major evolutionary strides and spread to many parts of the world. Human cultures and religious systems began to take shape near the end of the Pleistocene, and the great and small political units evolved. Lastly, the stage was set for the age of steel and machinery, utilization of natural energy, and science.

The continental glaciers

Scandinavian ice cap

An ice cap centered over Scandinavia during the Pleistocene and spread southeastward across the Baltic Sea to Moscow and beyond, southward across the North German Plain to Berlin, and westward across the North Sea to and across Great Britain. The limits are marked by the terminal moraines and associated features shown in Figure 8–28. It has taken geologists of the North European countries many years to trace out these moraines in the field. It will be noted that four moraines of different

Figure 8–27 *Esker near Fort Ripley, Minnesota. [Courtesy of William S. Cooper.]*

ages are charted. The oldest moraine, labeled 1, lies farthest east, south, and west, and shows the position of the most extensive spread of the ice lobes of the ice cap. Three younger terminal moraines lie successively behind the outer moraine. The different ages are determined by such criteria as the following.

1. *Degree of soil formation.* The oldest moraine will have the thickest and most mature soil.

2. *Degree to which the drainage system has been developed.* The youngest moraine will be very hilly or hummocky, with many lakes and swamps and a completely aimless stream pattern, if any at all. The oldest moraine will be much subdued and will have very few lakes and a well-integrated drainage system.

3. *The relative position of the four terminal moraines is in itself an indication of relative age.* In places it can be seen that one till rests on another, with such features as a layer of wind-blown silt or a soil between. The stratigraphy of glacial deposits has played an important role in the deciphering of the several stages of glaciation.

If one moraine is deposited on another, with a thick and mature soil between, having been developed on the lower moraine, it must be

Figure 8–28 *Four stages of the Scandinavian ice cap. The hachured lines represent the terminal moraines of each stage; 1 is the earliest glacial stage and 4 is the latest.*

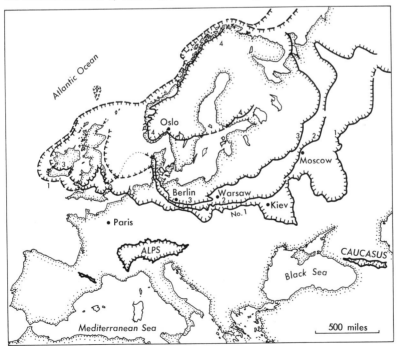

STAGES		Estimated duration in years on basis of soils
WISCONSIN GLACIAL	Mankato Cary Tazewell Iowan	50,000 (C^{14} dates)
Sangamon Interglacial		150,000
ILLINOIAN GLACIAL		100,000
Yarmouth Interglacial		300,000
KANSAN GLACIAL		200,000
Aftonian Interglacial		200,000
NEBRASKAN GLACIAL		200,000
Earlier little-known glacials and interglacials		

Figure 8–29 Pleistocene stages of the Great Lakes region. Ages are approximate and taken from several sources.

concluded that the ice front withdrew a considerable distance to the north or disappeared completely for a long time while the soil formed. In fact, the science of soils is so advanced that soil scientists can say about how long it took to form the soil in question and what kind of climate existed while the weathering was taking place. With this ability and with the evidence of fossils, it is concluded that the climate warmed up appreciably between ice advances and that it stayed warm for periods longer than those during which the glaciers existed. Hence, the interglacial stages are as important as the glacial (see Figure 8–29).

By means of varves (laminations in fine sediments of glacial lakes) it has been determined that the ice sheet that deposited moraine 4 of Figure 8–28 retreated north of Stockholm about 9000 years ago. By means of carbon-14 dating, this advance of moraine 4 is believed to have started about 25,000 years ago and thus had its existence during the period 25,000 to 9000 years before the present. The other three glacial stages are much older and, from the evidence of soil formation, probably lasted several times longer.

North American ice cap

The Canadian Shield of North America was the center of ice accumulation during the Pleistocene, and its lobes overran much of New England and flowed down the depressions of the Great Lakes into western New York, Michigan, Ohio, Indiana, and Illinois to the Ohio River and across Minnesota and the eastern parts of North and South Dakota to the Missouri River (see Figure 8–30). The ice spread westward to the Canadian Rockies and merged with the valley glaciers of the Rockies. The moraines of the continental glacier meet the moraines of the valley glaciers of Glacier National Park on the high plains just east of the park in northwestern Montana. Here the stages of glaciation of the mountains have been related in part to the stages of glaciation of the continental glacier.

There may have been two centers of ice accumulation, one west and one east of Hudson Bay. The ice sheet spread northward to the islands of the Canadian arctic, where several small ice caps remain today as well as many valley glaciers from mountain ranges. The

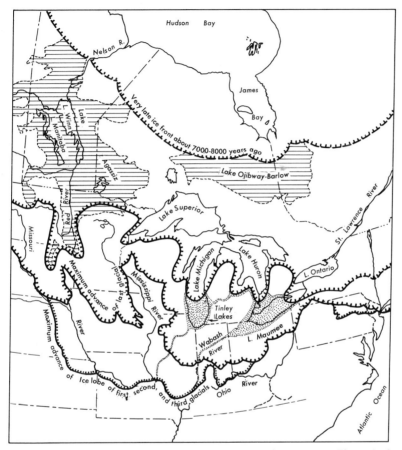

Figure 8–30 Ice sheets and former lakes of the Great Lakes region. The early Great Lakes are stippled, and the later large lakes of Manitoba, Ontario, and Quebec are ruled. [See glacial map of the United States east of the Rocky Mountains, U.S. Geological Survey, 1969.]

Greenland ice cap was probably larger than that of today but did not coalesce with the great Canadian cap. The northern coast of Labrador is one of deep fjords down which valley glaciers discharged into the sea.

The Canadian Shield, except in protected depressions, has been abraded clean of soil and alluvium by the ice cap, and in many places where by chance the surface has been left covered by a little ground moraine, the striations and groves are still intact, although where exposed they have generally crumbled away. The shield's rocks are igneous and metamorphic, such as granites, gabbros, and gneisses. Some rather unusual ones crop out in specific localities. Examples are the iron formation rocks, the Keweenawan native copper-bearing volcanic rocks, and the jasper conglomerate. *Erratics* of these formations are found on the Paleozoic limestones and shales of New York, Ontario, Michigan, and other Great Lakes localities, and by charting the distributions of erratic boulders derived from some singular known locality, the path of ice flow may be determined.

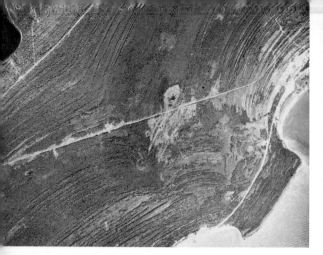

Figure 8–31 Shorelines of higher lake levels, ancestral to Lake Michigan.

Glacial lakes

Synchronous with the glacial stages in a number of places, large fresh-water lakes came into existence. As the glacier fronts of the Wisconsin stage withdrew northward, several lakes of varying outline and outlets evolved as predecessors of the modern Great Lakes, and a fascinating story has been worked out concerning them (see Figures 8–31 and 8–32). In the western United States, especially in western Utah and Nevada almost half the area was covered with magnificent fresh-water lakes, with colder flora and fauna to match. Incident to the impounding of the lakes and the withdrawal of the ice, great floods occurred on the Columbia and Snake Rivers. Lake Winnipeg of Manitoba is the remnant of a late Wisconsin lake larger than all the Great Lakes combined. The Missouri and Ohio Rivers were deranged and deflected from their original courses by the ice. The glaciers thus had a marked effect on our geography.

Ground water

Ground water is water that fills, or partially fills, the openings in the rocks below the surface.

There are vast stores of water underground, and now, more than ever before, man is making good use of them. The rapidly growing communities in the arid west are partly or largely dependent on ground water, and many cities and towns in the more humid parts of the

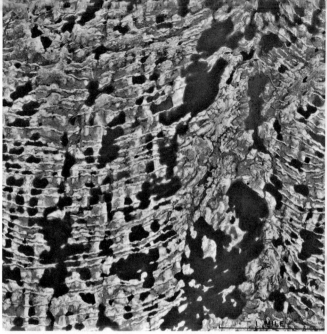

Figure 8–32 Recessional moraines in western Quebec, within 50 miles of Hudson Bay. The once-great ice cap had nearly vanished by this stage and the front of the last remnant was withdrawing rapidly. Each moraine possibly represents the deposits of winter and spring while the front held stationary. During the summer and fall it would withdraw to the next position. If so, the general retreat across the area of the photograph occurred in about 50 years. [Courtesy of Geological Survey of Canada and Ed Schiller.]

country have found a convenient source of their water underground. It has been estimated that much more water is in storage underground than in all the surface reservoirs and lakes, including the Great Lakes. The total usable ground water in the United States is on the order of ten years of annual precipitation. How does water occur below the surface, and how much is available for use in any particular place? Can we consider the underground supply a reservoir that can be managed like a surface reservoir? These and other questions will be taken up in the following pages.

The hydrologic cycle

Water that falls to the earth as rain or snow has a varied course. Part of it drains off the surface to form the streams and rivers. This is called *runoff*. Part of it is returned fairly promptly to the atmosphere by *evaporation* or through the agency of plants that lose water to the air through their leaves. This is called *transpiration.* The last part of the water that falls as precipitation seeps into the soil and rocks below to become ground water. Eventually the ground water finds its way back to the surface through springs to become part of the runoff or it seeps into lakes and even into the oceans. The oceans, through evaporation, replenish the moisture in the atmosphere and thus prepare the way for more precipitation on land. This is the *hydrologic cycle.*

Wells that are drilled for oil often tap flows of salt water or brine in porous sandstone. This water is generally marine and has been locked up in the pore spaces between the sand grains since the grains originally accumulated. If so it is called *connate water.* Connate water is distinguished from the ground water of the hydrologic cycle described above, which is called *meteoric water.* Meteoric water is generally in a state of slow circulation and is fresh unless it has mixed with salty connate water, such as exists in some of the basins of the arid west.

In our discussion of volcanoes we said that the release of water vapor from the magmas was the chief cause of explosive eruptions and that large volumes of such water reach the atmosphere in this way. In most eruptions, it is believed, this water has come from within the earth and has not previously been part of the hydrologic cycle. Such water is called *juvenile water.* Much juvenile water is also released by cooling and crystallizing magmas underground. This water has gained metalliferous compounds from the magma and has carried them upward through porous and cooler rocks, where valuable ore minerals have been deposited.

In the present chapter we are concerned only with meteoric ground water as part of the hydrologic cycle.

Water in the pores of rocks

Kinds of pores

Ground water occurs in the cracks, crevices, and pores of the rocks. All consolidated rocks are riven by joints and the amount of water depends on the abundance of the joints and the degree to which they are open. Generally such joints do not provide much room or storage capacity, but the water can move through them fairly well, even if they are only slightly open.

When water moves along joints in limestone it usually dissolves out passageways of considerable size, sometimes to the extent of large subterranean caves. These will be taken up later.

The chief space for water is in granular aggregates, such as sand and gravel or their cemented equivalents, sandstone and conglomerate. In these rocks from 12 to 45 percent of the total volume may be pore space. The maximum pore space is present in well-sorted grains, such as shown in Figure 8–33, top left and top right. If small grains are present along with larger ones they will partially fill the voids and reduce the amount of space that water might otherwise occupy (Figure 8–33, bottom left).

If the grains are bonded together with cement, the pore space is again reduced, so pore space depends chiefly on the degree of sorting and the extent of cementing (Figure 8–33, bottom right). Another factor, less easily perceived, is the degree of compaction. For instance, given the same sand, there may exist a loose arrangement with much pore space or a compact arrangement with less pore space.

Movement of water through pores

Porosity is defined as the percentage of pore space in a rock. This, then, is a measure of the water-storage capacity in a rock. Even though all the pores are interconnected and filled with water, not all the water thus held is available for use; not all the water can be taken from or will drain out of the rock. For instance, if a piece of porous sandstone is submerged in water, in time the water will penetrate and fill all the connected pores. But if the block is taken out of the water and allowed to drain, only a part of the water held in the rock will drain out. The coarser the sandstone, the more water will be lost; the finer, the more will be retained. It is retained as attracted films on the surface of the grains (they are said to be wetted), and as

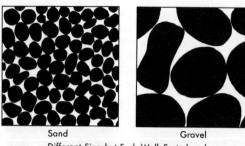

Sand — Gravel
Different Size but Each Well Sorted and of About Equal Porosity

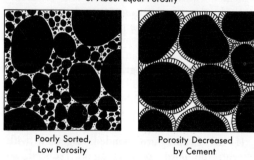

Poorly Sorted, Low Porosity — Porosity Decreased by Cement

Figure 8–33 *Variable porosity in masses of rounded grains.*

filling in the narrow spaces between grains. We have seen water rise in a glass tube with a needle-sized opening and stay there in spite of the pull of gravity downward. The power of small openings to draw a liquid into them is called *capillary attraction*.

Capillary attraction is a very real influence to consider in the movement of water through porous materials. Water will penetrate and moisten a shale or clay, but none of the water will drain out, nor can it be forced out except under high differential pressure. About the same is true with silt and siltstone. But water does move through sand or poorly cemented sandstone, and under a little pressure the flow is appreciable. Since springs and wells are fed by water moving through pores, we are much concerned with the ease or difficulty with which the flow occurs. The measure of the facility of flow through porous rock is called *permeability*. The capacity to hold water is determined by the rock's porosity, but its capacity to yield water is determined by its permeability.

Commonly, good porosity goes along with good permeability, especially if coarse sediments are involved, but we must appreciate that a rock like pumice or scoria has a high porosity, but poor permeability, because the pores are not connected. Also certain clays have moderately high porosities but very low permeabilities.

If a sandstone, for instance, has a porosity of 20 percent but will yield only half of this water, then the rock is said to have a *specific yield* of 10 percent. A specific yield of 10 percent, however, represents a great volume of available water in storage.

Take, for example, a valley or basin 10 miles long and 5 miles wide. It has an area of 50 sq miles. Each square mile has 640 acres, so we are dealing with an area of 32,000 acres. If the porous sediments are 500 ft thick and have a specific yield of 10 percent, the available water content is equivalent to a layer 50 ft thick. Now, $50 \times 32,000 = 1.6$ million acre-feet of water, or 422 billion gal. The equation used is

Area in acres $\times$ depth of porous alluvium $\times$ specific yield $\times$ 326,000 (gal per acre-foot) = gallons of available water in storage below the surface

All we need to do is drill wells and pump out the water. We should know a little more about ground water than this, however, before expensive wells are drilled, or before we misuse a valuable natural resource. In the next paragraphs we will consider the features that geologists have found significant.

Ground water in rocks without barriers to movement

Ground water table

In porous and permeable rocks, in which water can move from the surface downward and laterally without restraints other than that imposed by the capillarity of the pores, a certain distribution always occurs. There is the zone immediately below the surface in which the pores are filled with both air and water, and below this is a second zone in which all the pores are completely filled with water. The upper one is called the *zone of aeration* and the lower one the *zone of saturation*. Rain and stream water filter into the soil and downward through the zone of aeration and become part of the water of the zone of saturation (see Figure 8–34). The upper limit of the zone of saturation is the *water table*. If an open hole or well is dug through the zone of aeration into the zone of saturation, such as shown in Figure 8–34, then the well will fill with water up to the water table. The ground may be moist in the zone of aeration, but the chances are that very little water will ooze out of its pores into the well. In the zone of saturation, on the other hand, water will flow out of the pores into the well due to hydrostatic pressure, and it will fill the well up to the water table. Thus the water level in a series of wells in an area marks the upper limit of the zone of saturation.

In detail we see in places and at times a thin *zone of soil moisture* derived from a recent rain. Any excess water beyond that necessary to fill the small capillary openings will work on down through the zone of aeration. Then there is the *capillary fringe* just above the water table, in which water is drawn up into the capillary openings (see Figure 8–34 again).

If the water table is studied in a hilly terrane it will be found to rise and fall, reflecting the topography, but in a subdued way. It will rise under the hills, but not so much as the surface, and fall under the valleys. The zone of aeration will be thickest under the hill tops. If there is a lake in a topographic depression, the water table will meet the lake level (see Figure 8–35). If the water table were below the lake, then the water of the lake would settle through the permeable sediments and rocks and become part of the zone of saturation, and the lake would disappear. If, on the other hand, the water table rose, then the lake level would rise. During seasons of high precipitation, the water table rises, and during seasons of low precipita-

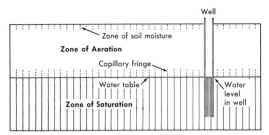

Figure 8–34 Distribution of ground water in permeable homogeneous rock or unconsolidated sediment.

tion it falls. This fluctuation is readily observed in wells and is a result of recharge and discharge of the underground reservoir. Before proceeding further, let it be emphasized that ground water flows slowly through the openings in rocks because of the friction between the water particles and the rock surfaces (capillary attraction), and it always moves under the influence of gravity toward the lower area or point of natural or artificial discharge.

Discharge

Ground water is discharged through various artificial and natural means: (1) water may be pumped from the reservoir through wells—this is an artificial discharge; (2) the ground water may leak out into stream beds and lakes; (3) it may emerge at the surface as a spring, which, in a homogeneous rock, is much the same situation as in the bed of a stream; (4) plants with long roots may reach to the zone of saturation and use some of the ground water.

Recharge

In general, a reservoir is recharged by infiltration from surface precipitation and from

Figure 8–35 Relation of water table to land surface.

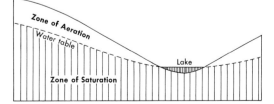

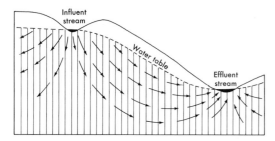

Figure 8–36 Movement of ground water in permeable homogeneous rock. The stream at high elevation loses water, and the lower stream gains water. The flow is treated by ground-water specialists mathematically like an electric current. Permeability may be compared with resistance and gradient with voltage.

streams. The amount of stream recharge may be measured in some basins by gauging a stream's flow before entering the basin and then again as it leaves the basin. If the flow is less upon leaving the basin, then the difference is the amount that has filtered into the rocks or sediments of the basin.

A stream which receives water from the underground and whose flow is thus increased is said to be *effluent*, whereas one that loses water to the underground is said to be *influent* (see Figure 8–36).

Recharge may also be figured by recording total rainfall in a watershed and measuring the runoff, transpiration, and evaporation. Then, by the equation

$$\text{Precipitation} = \text{runoff} + \text{transpiration} + \text{evaporation} + \text{infiltration}$$

the amount of water that seeps into the ground may be calculated. Of course, it is quite a problem to measure transpiration and evaporation, but techniques to do this have been developed.

It should now become evident that a ground-water reservoir may be managed like a surface reservoir if the geologic framework of the reservoir is understood and all the aspects of recharge and discharge are known. For instance, consider a surface reservoir. We determine its size or capacity, we measure the water that the stream brings into it, and we compute the increase in storage as its water surface rises. Finally, the level reaches the spillway, and water overflows and is lost. Only that which is held back of the dam remains to be used. Likewise we assess the capacity of the underground reservoir; we measure the recharge and watch the water level rise. Finally we see springs start up and streams enlarge or dry stream beds start to flow water. This is the overflow that is the equivalent of the water in the dam that goes over the spillway and is lost.

Now, in the case of the dam, if it only took one year to fill to overflowing, we would judge that we could use all the water, say for summer irrigation, and that the supply would be replenished during the fall, winter, and spring. However, if it took five years to fill the reservoir, then we had probably better use only one-fifth of its capacity in one year; otherwise it would not fill up in the next year. Likewise with underground reservoirs, we should not take out each year more than nature puts back in during the following year. But we do not know right off how long it took nature to fill up the underground reservoir. If, however, the water table does not return to its former elevation, then more water has been pumped out than has been put in. It may be possible in this event to augment the recharge. This again requires professional geologic knowledge of the basin, the recharge area, and the management of runoff.

Considerable water is added to the underground by irrigation, because more water is generally applied to the soil than is needed to meet the demands of evaporation and crop growth. This is necessary to keep the soil flushed of salts that would otherwise accumulate, especially if the irrigation water were slightly saline. The "excess irrigation water" seeps downward to the water table and adds to the supply there.

In the impounding of water back of dams, considerable surface water may be fed into the underground, and in certain dams this has been deliberately planned. In others, without careful planning and geologic work beforehand, most of the water has disappeared into the underground, and the surface storage facility has been a failure. This happens particularly in cavernous limestone regions.

The Water Supply Papers of the U.S. Geological Survey are the great store of information on the principles of ground water occurrence and the many ground water regions of the country and should be referred to for information about local ground-water conditions, and particularly about recharge.

Effect of well on water table

If a well is pumped heavily the water table is drawn down around it in the shape of a cone. The *drawdown* is called the *cone of depression* (see Figure 8–37). If a city or industrial plant has a well with a wide-diameter casing, such that a large flow can be pumped from it, and if the well is pumped heavily, the cone of depression of the water table in time will spread hundreds of feet away from the well. Perhaps the area affected will be a mile in diameter, with the big well in the center. All other wells within the cone of depression will experience a lowering of the water level, and possibly in some of the shallow wells the water table will drop below their bottoms. In this event they will have gone dry (see Figure 8–37 again). This situation is a common cause of lawsuits. Since ground water moves and is not fixed in place like a mineral deposit, who owns it? Various interpretations have been made in the courts, but basic to any worthwhile settlement is an understanding of the geology and hydrology of ground water.

Ground water in stratified rocks

Pervious and impervious beds

We have been considering ground water in rock bodies that are homogeneous so far as porosity and permeability are concerned. When ground water in sedimentary rocks is investigated, we deal with some beds that are pervious and transmit water freely and some that are impervious and serve as barriers to movement. Sand and sandstone beds are generally good carriers of water, but clay and shale are bar-

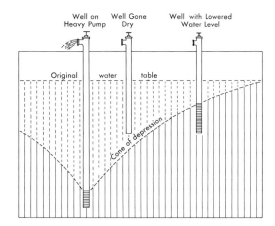

Figure 8–37 *Cone of depression.*

riers. When a sandstone bed, for instance, is underlain and overlain by a shale bed then it is clear that we are dealing with *confined water*, much like in a conduit. In confined water hydrostatic pressures may build up and cause some surprising conditions, as we shall see presently.

A stratum that transmits ground water well, that is, a stratum that has sufficient permeability to pass commercially valuable amounts of water through it, is called an *aquifer*. A bed that prevents the flow of ground water through or across it is called an *aquiclude*. Since sediments and sedimentary rocks are extremely variable in permeability, it follows that there are many variations between very permeable aquifers and completely impervious aquicludes. Not only is the permeability of a bed above or below different, but the permeability of a given bed may change from place to place. Considering these conditions and also that sedimentary rocks are in places tilted, folded, and faulted, we can see that nature has fashioned some rather complex "plumbing systems."

Artesian pressure

Let us take a structure consisting of inclined pervious and impervious beds cropping out at the surface, as shown in Figure 8–38. The sandstone bed is assumed to be a good aquifer between two shale beds or aquicludes. Streams draining from the mountainous area on the left flow across the beveled edge of the sandstone bed, and considerable water filters into it. This is the recharge area. If wells are drilled into the sandstone bed as shown, we find that water will rise in them above the sandstone bed. This is because the water is confined in the sandstone and under pressure. The well serves as one arm of a U tube and the sandstone bed, from the surface down to the well, as the other. This illustrates the well-known principle that water in a U tube seeks a common level on both sides.

Water that rises in a well above the aquifer from which it comes is called *artesian*. If it rises to the land surface where the well is drilled, as it does in a good many places, then a flowing well exists. In Figure 8–38, one well flows and one well does not, but both are artesian wells.

If the pressure is gauged in artesian wells in a certain area, or in other words if the level is noted to which the water rises in the well, then we have defined the artesian *pressure surface*. Where the pressure surface is above the land surface, flowing wells will occur, but where it is below the land surface, the wells will be nonflowing. The artesian pressures in wells in a basin can be contoured, and with such a map in hand, the level can be determined to which water will rise in a new well drilled anywhere in the area.

If there were no resistance to flow through the pores of the aquifer and if there were no leaks in the underground plumbing system, then the pressure surface would be horizontal. But since both these conditions always exist, the pressure surface is never level but slopes away from the recharge area. Its gradient varies from basin to basin and also changes as water is artificially withdrawn from the aquifer.

Nature works out a balance of recharge and discharge in these confined water systems, but man upsets the balance by adding new discharge in the form of wells. If more water is withdrawn than is recharged, then the pressure surface falls, and where it falls below the land surface, the wells cease to flow. They must then be put on pump. This is common in many of our artesian basins. It is possible that recharge increases as the pressure surface drops and that with a steady rate of discharge a new but lower surface will become established. However, if the pressure surface continues to drop, the users of ground water in the area are due for a shortage eventually.

Advantages of ground water over surface water

Now that the main elements of ground water occurrence are understood, it may be emphasized that certain advantages accrue to the users of ground water over surface water.

1. Ground water may be reached within a few hundred feet of the place where it is used, and on the same property, whereas surface water may require pipelines and rights-of-way over stretches of several miles.

2. Ground water may be available for use in areas where the water in streams and lakes has already been appropriated by other users.

3. Yield from wells and springs generally fluctuates less than stream flow in alternating wet and dry periods.

4. Ground water is more uniform in temperature and soluble mineral load than surface water and is generally free of turbidity and pollution by pathologic organisms. Commonly the ground water is cold and clear.

5. There is little or no evaporation loss from underground reservoirs, whereas surface reservoirs lose a tremendous amount in arid regions.

6. Underground reservoirs are ready-made.

On the other side of the ledger, however, it must be said that many ground waters are rather hard, and need to be run through softening plants before the water is suitable for human consumption, particularly for washing.

Ground water in carbonate rocks

Solution effects

The regimen of ground water in limestone and dolomite terranes is somewhat unusual and needs special consideration. This is because the

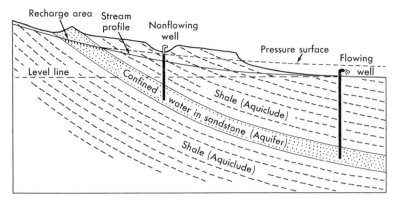

Figure 8–38 Structural and hydrological conditions necessary to produce artesian pressures. The nonflowing well, as pictured, is not in the main valley labeled "stream profile," but some distance away in a tributary valley.

carbonate rocks are fairly soluble in water charged with carbon dioxide. It has been pointed out in Chapter 6 that carbon dioxide gas gets into the ground water chiefly as rain water passes through the soil where plants are decaying. Then as the water continues downward through the joints in the limestone and dolomite, it dissolves away the surfaces along which it passes and creates irregular passageways. Circulation follows the joints that cut across beds and along bedding surfaces between beds. The rock may be badly broken or shattered along a fault, and this zone particularly would be one of circulation and consequently of solution. In time, cavernous openings are dissolved out of the rock, some of considerable height and some of long and devious horizontal extent. Some of the cavernous passages are large and continuous enough to conduct streams of water that can truly be called underground rivers. In the Ozark Mountains of Missouri and in parts of the Allegheny Plateau and Appalachian Mountains certain rivers flow into the underground and others emerge from it.

Limestone caves

Most caves are in limestone and there are literally thousands of them in the United States. Among a few well-known caves that might be listed are the Mammoth Cave of Kentucky, the Shenandoah Caverns of Virginia, the Carlsbad Caverns of New Mexico, and the Lehman Cave of Nevada.

Limestone caves may develop as single- or multiple-story structures. The plan of a single-level cave may show a long winding and branching opening with enlarged chambers along it (Figure 8–39) or a maze of openings controlled by intersecting sets of joints (Figure 8–40). Where solution ways are developed on several levels, vertical or near-vertical shafts connect the levels. In mountainous regions, where thick limestone formations occur, the solution channels may have more of a vertical direction than horizontal. There are three main levels in the Carlsbad Caverns so far explored, and these extend to 1320 ft below the surface. Several caves in the Rocky Mountains are known to extend to depths greater than 1000 ft. Some persons find it exhilarating and fascinating to explore and map caves, and they are called speleologists or spelunkers. They are generally more interested in the techniques of descent, ascent, mapping, and survival than in the geologic aspects.

Karst topography

A limestone terrane in Yugoslavia (the Karst region) is pitted with depressions that drink up all the precipitation. These are solution and caved structures that lead downward into systems of solution channels like those shown in Figure 8–40. These pits are called sink holes. In wet seasons small ponds may exist in their bottoms, and in dry seasons the pond water flows away underground. Many of the sink holes are the result of roof falls of underground chambers. Eventually the caving of the roof rock reaches to the surface, and a pit with broken rock in its bottom results. A landscape pitted with sink holes is said to have a *karst* topography. There are a number of such areas in the United States, especially in Florida, southern Indiana, Kentucky, and Missouri. Figure 8–41 is a photograph of a karst region.

Depositional forms

It seems that caves in limestone are no sooner dissolved out than the chemical process is reversed and calcium carbonate is precipitated. The depositional forms are some of nature's fancy work, but in total, they contrive to fill up the cave openings. What causes the drastic change?

The change from solution to precipitation of calcium carbonate is now known to be caused by a lowering of the water table. The main solution process occurs in the zone of saturation, where the openings are filled with water. Solution etchings are just as pronounced on the ceilings as on the floors of some caves, because the entire passageway or chamber was full of water when solution was occurring. With a lowering of the water table, some or even all of the solution channels come to be in the zone of aeration and are therefore drained of most of their water. Only that water which seeps from the surface downward and passes through the zone of aeration drips into the caves, generally entering from the roof. With air in the caverns, and generally circulating a little, the dripping water loses some of its carbon dioxide, and some of it evaporates, and consequently calcium carbonate is precipitated. Considerable water runs as thin sheets or films down the walls and is thus particularly susceptible to evaporation and loss of CO_2.

Two principal deposits form as a result of dripping. Iciclelike forms grow downward from the ceilings. They maintain an internal tube through which the drip water flows, but some water may run down the outside as well. They are called *stalactites* (see Figure 8–42).

The water that falls from the ends of the stalactites builds mounds on the floor. These are generally stockier than the stalactites and are slightly cupped at the top. They grow by accretion of layers of $CaCO_3$ on their surfaces, where the drip water splashes and runs. They are called *stalagmites*. Stalactites and stalagmites grow in a great variety of forms and sizes. If, by chance, a stalactite and its complementary stalagmite grow until they meet, the resulting form is called a *column* (Figure 8–43).

Water running out of cracks sometimes builds fluted forms called *draperies* (Figures 8–44 and 8–45). On inclined slopes various calcium carbonate dams are built that retain beautiful pools of water back of them. This produces a terraced effect (Figure 8–43). Nearly every cave has unique forms, and many people are fascinated by the endless variety of depositional forms and see in them such varied shapes as heads, faces, dogs, broncos, deer antlers, cathedrals, queens, angels, cowboys, and other strange phenomena.

Springs and seeps

Causes

Ground water finds natural outlets to the surface in several basic geologic settings (examine Figure 8–46). In unconsolidated sediments, a typical situation is provided by an impervious clay overlain by a pervious sand or gravel deposit, with each cropping out on a hillside. In consolidated rocks, the setting might be such that a pervious bed of sandstone or limestone conducts water from a higher place of intake to a lower place of discharge (see second diagram from top).

If the discharge emerges as a distinct flow in a local place it is called a spring, but if it emerges along a zone or in an appreciable area

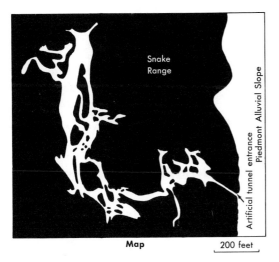

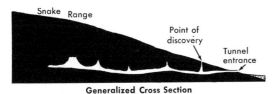

Figure 8–39 Lehman Caves, Nevada. The large north-south caverns have been dissolved along faults from badly shattered limestones.

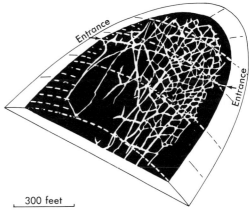

Figure 8–40 Cavernous system in limestone developed by solution along intersecting sets of joints. Part of Cameron Cave, Missouri. The level of the cave system is midway up a hill. [After Bretz, Caves of Missouri, 1956.]

only to produce a moist or swampy condition it is called a seep.

Faulting produces shatter zones in the rock. If confined water, such as the spring depicted in the third diagram from the top, is tapped by the shatter zone, the water will rise to the surface. Some faults are marked by a series of springs along them. Large faults that cut deeply into the crust sometimes bring hot mineral waters to the surface, part of which may be juvenile.

The Snake River in southern Idaho has cut a valley through lava flows, as illustrated in the bottom diagram of Figure 8–46. The basal part of some flows is porous and freely conducts water—from mountain streams, from canal water used for irrigation, and from direct precipitation—to the Snake River valley wall. Along a certain stretch of the valley wall the water gushes out in what is called Thousand Springs, whose combined flow has been gauged at 600 cu ft per sec.

Deposits

Springs that issue from limestone formations are heavily charged with calcium carbonate, part of which is promptly dropped where the spring emerges. This builds sizable, and in places picturesque, deposits called *travertine* (see Figure 8–47). The term is particularly appropriate to those deposits which are quarried and used as building stone. Travertine is usually a porous rock, cuts easily, and has a soft cream, tan, or rusty color. The ancient coloseum in Rome is built of a travertine that was quarried nearby. It is a very popular interior decoration stone in public and industrial buildings today.

Examples of ground-water occurrence

Alluvial fans

Much ground water is obtained from alluvial fans in eastern California, Nevada, Utah, Arizona, and New Mexico. The setting of an alluvial-fan piedmont was shown in Figure 7–53.

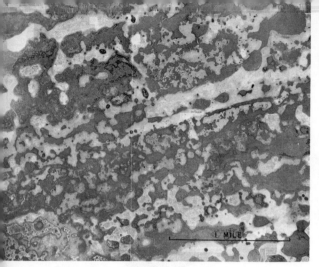

Figure 8–41 *Vertical aerial photograph of a region in Orange County, Florida, where precipitation goes entirely underground, and many solution depressions or sinks have developed. The light areas are the grass lands and the dark areas, the swamps and sink holes. [Courtesy of Docringsfeld, Amnedo, and Ivey, Denver.]*

Figure 8–42 *Onondaga Cave, Crawford County, Missouri, showing particularly beautiful terraced pools. [Courtesy of Massie-Missouri Chamber of Commerce.]*

Figure 8–44 *Draperies in Lehman Cave, Nevada.*

The idealized structure of an alluvial fan is given in Figure 8–48, with the position of confined water in the coarse and porous wedges noted. The water table will be deep in the upper part of the fan, and drilling and pumping costs would be too high to be profitable. The intermediate slopes will be those under which confined water will exist, and here the wells will have artesian pressures. Some of the wells may flow. Farther out in the valley the sediments are

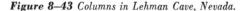

Figure 8–43 *Columns in Lehman Cave, Nevada.*

too fine and the permeability too low to make good wells. Nearly every alluvial-fan piedmont at the foot of a high mountain range has a zone of flowing wells. Owens Valley at the east foot of the Sierra Nevada is a notable example of the yield of large quantities of cold, clear, and potable water. Much of it is piped in aqueducts to Los Angeles and the surrounding area.

Glacial deposits

Many of the cities, villages, and farms of the Great Lakes states are supplied in part or whole by well water from glacial deposits. These are unconsolidated sands, gravels, boulderclay, and clay deposits in an array of forms and arrangements. Some of the porous material was deposited along lake shores, some exists as outwash sheets, and others as river channel sands (see Figure 8–49). Some water is confined and rises in the wells, but generally the wells must be pumped. Some yield excellent flows without much draw-down of the pressure surface. Wells in other places may not penetrate very porous material, and their yield is small.

Sandstone aquifers

The Great Plains of South Dakota, east of the Black Hills, has been one of the most prolific flowing-well regions on the continent. The Dakota Sandstone is upturned along the east front of the Black Hills but flattens out under the plains. It is overlain and underlain by impervious shales and hence serves as an aquifer to conduct water many miles to the east. Probably the sandstone becomes silty and impervious toward its easternmost extent, and hence the confined water in it is sealed off (see Figure 8–50). Streams draining from the Black Hills flow across the beveled edges of the sandstone and presumably supply it with water. In the early farming and ranching days wells drilled into the sandstone spouted over 100 ft in the air. Many of them were left open and flowed continuously, forming small lakes in low areas around. After a few years the pressures began to drop, and eventually many wells ceased to flow. This was a critical situation for the farm-

Figure 8–15 Peculiar drip and flow stone deposit in Lehman Cave, Nevada.

ers, but what could be done? The U.S. Geological Survey made a thorough study of the geology and water wells and instituted a program of controls with implemented recharge, but reservoir pressures have not increased much to date. One reason may be that this particular artesian reservoir was not very well understood. A recent study has shown that an underlying aquifer, the Madison Limestone, plays a significant role. It is a cavernous limestone, and is so porous and permeable that it conducts water 120 miles underground to the east of the Black Hills with little loss of head. It has been estimated that on occasion it receives (recharge) over 35 million gallons of water per day in the area of outcrop. Formerly investigators were at a loss to find an adequate recharge in the Dakota Sandstone belt. The Madi-

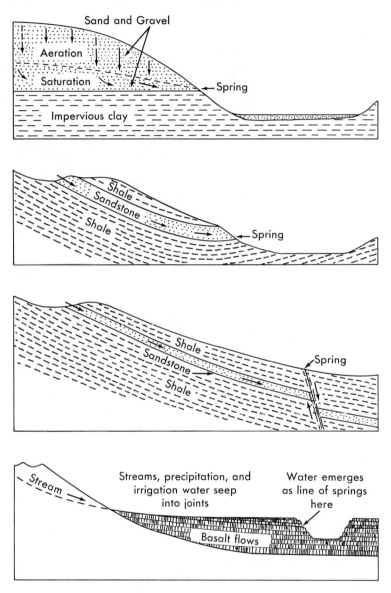

Figure 8-46 *Typical geologic settings in which springs occur.*

son aquifer was differentiated when it was recognized that the Dakota Sandstone from east of its outcrop about to the Missouri River contains its own water which may be identified as the sodium chloride type; east of the Missouri the sandstone contains water of the calcium sulfate type, which comes from the underlying aquifer, the Madison Limestone. This good geologic and chemical study permits a much better attack on the recharge problem. The depletion of ground water in the Dakota Sandstone is an example, however, of wasteful use of a great natural reservoir that might have been better managed. The moral is that, except for very pressing and expedient reasons, more water should not be taken from an underground reservoir than nature, with our help, perhaps, can put in.

Coastal plains

The Atlantic and Gulf Coastal Plains are the most productive ground-water province in the United States. This is the margin of the continent over which the Atlantic Ocean and Gulf of Mexico have spread and, with fluctuations, have gradually withdrawn, leaving bottom de-

Figure 8–47 *Travertine spring deposits, Thermopolis, Wyoming.*

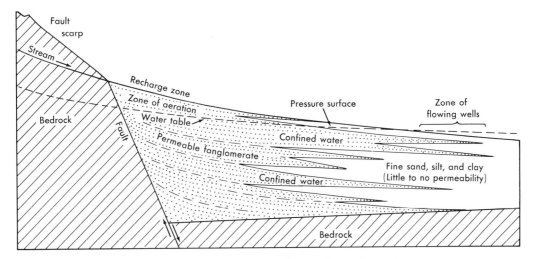

Figure 8–48 *Idealized structure of alluvial fan, and distribution of ground water in it. The fan here is built on the down-faulted block, and the sediments have been derived from the erosion of the up-faulted block. A fanglomerate is the coarse flood debris of the fan.*

Figure 8–49 *An old filled river valley and the conditions necessary for artesian or flowing wells.*

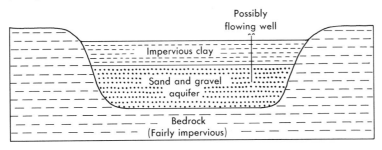

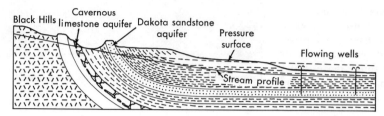

Figure 8–50 The Great Plains east of the Black Hills and an example of a sandstone aquifer and flowing wells.

posits as a testament. These deposits consist of sand, silt, clay, and limestone, and only the limestone is consolidated. The beds dip gently toward the ocean, as shown in Figure 8–51, and because of the favorable structure of coastward-dipping permeable sand and limestone beds alternating with aquicludes of clays, artesian conditions prevail throughout most of the vast region. Flowing wells can be obtained almost everywhere along the coast and in the valleys extending back from it. Since the sediments were deposited in sea water they were originally charged with salt water, but following the retreat of the seas much rain has fallen on the sediments during the last era of geologic time, and they have been flushed fairly well of any salt water. Only in a few places is somewhat salty water found. Heavy pumping of wells could bring salt water up the aquifer.

The Tamiami Limestone of southeastern Florida, part of the Coastal Plain sediments, is the most permeable aquifer ever investigated by the U.S. Geological Survey, and the Edwards Limestone of Texas supplies the largest known flowing well in the world at San Antonio, yielding nearly 25 million gal per day.

Legal aspects of ground-water use

Because ground water is mobile and can be drawn from beneath one person's property by wells in an adjacent person's property, it is easy to see how lawsuits can result. More basically, the intimate relation of surface waters to ground waters make it necessary to consider both in adjudicating the rights of the owners of each. Many court decisions have been based on erroneous concepts or a lack of understanding of ground water, and hence much confusion exists in the laws pertaining to rights and ownership.

Some states have made significant advances in the management of ground-water reservoirs by establishing rules and enforcing them. This has come about by the recognition of the courts of the work and understanding of geologists and engineers on the specific reservoirs in question.

Geysers in volcanic regions

Geysers are hot-water springs that erupt by the blowing off of steam. They occur rarely, but because of their attractive character in Yellow-

Figure 8–51 Idealized section of Atlantic Coastal Plain showing ground-water conditions.

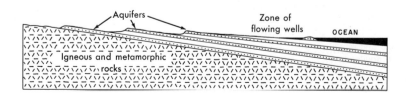

stone National Park most people are aware of them. They occur in Iceland and New Zealand in lesser number and in milder eruptive form, but are practically nonexistent elsewhere. Some erupt mildly every few seconds or minutes, others jet water and steam over a hundred feet skyward regularly every hour or so, and others burst into activity at irregular and longer intervals. They all occur in volcanic regions, and it is clear that the heat that turns the water to steam derives from the still hot volcanic rocks.

The prerequisites for geyser eruptions are first, ground water circulating in the openings of heated volcanic rocks and, second, a narrow irregular conduit to the surface. The rocks must be hot enough to turn the water to steam even at the pressure that a column of water over a hundred feet high exerts. The vent to the surface must be so narrow, so crooked, or so restricted in places that convection circulation cannot take place and bring the heated water to the surface. Under these conditions, the water at depth will reach the boiling temperature before the water in the upper part of the conduit does, and when this occurs the steam lifts the column of water, water spills over at the surface, the pressure on the water below, which is at the boiling temperature, is lessened (perhaps even superheated), and the geyser springs into a major eruption (see Figure 8–52).

The geysers and associated hot springs of Yellowstone National Park build cones and terraced deposits around their orifices. These deposits are not travertine ($CaCO_3$), but *geyserite*, that is, composed of silica (SiO_2). Chemi-

Figure 8–52 Old Faithful geyser in eruption, Yellowstone National Park, Wyoming.

cal studies have demonstrated that meteoric waters percolating through the volcanic rocks in Yellowstone National Park become alkaline or basic and in such condition take up appreciable silica in solution. Then, when they cool upon reaching the surface, the silica is dropped as a spongy amorphous material.

References

Flint, Richard F. *Glacial and Pleistocene geology*. New York: Wiley, 1957.
Glacial map of North America. Geol. Soc. Amer., Special Paper 61, 1945.
Péwé, T. L., and Church, E. *Glacial regimen in Antarctica*. College: University of Alaska, Department of Geology, Reprint Series, n. 15, 1962.
Sayre, A. N. Ground water. *Scientific American*, Nov. 1950.

Motion pictures

Glaciers, understanding our earth. Planet Earth Series, US-42, color, 16 mm, 10 min.
The story of underground water. Encyclopaedia Britannica, FSC-480.

9
winds and waves

Wind as a geologic agent

Wind, like water, erodes, transports, and deposits. Wherever clay, silt, or sand particles are dry and loose and exposed at the surface, they are subject to wind transportation. These conditions are general in deserts, but in more humid regions they are locally present. For instance, along ocean and lake shores sand is left by storm waves, and there it dries and drifts into sizable accumulations by the force of the wind.

Rivers are confined to channels; the wind sweeps across the entire countryside. Rivers erode and transport in downhill courses; wind may transport particles uphill and hence erode out depressions. Water, besides transporting discrete particles of matter, can take compounds into solution; wind is limited as a geologic agent to physical action expended on particles or grains of rocks, minerals, and plant, and animal matter.

Wind erosion and transportation

Bed load and suspended load

Particles are moved by the wind—first along or near the surface, and then considerably above the surface in temporary suspension. The former is called bed-load transport and the latter suspended-load transport. Sand grains particularly within the size range of 0.15 to 1.0 mm are confined in normal winds to within a few feet of the surface, chiefly below 18 in., whereas silt and clay particles are carried in suspension and make up the dust clouds or storms.

The following example of suspended dust is often cited. During the 1930s a prolonged drought occurred in the Great Plains, and a large area was characterized as a dust bowl because the grassy sod had been ploughed under, and the silty loam, now exposed and dry and loose, was picked up by the winds. The dust rose in dense clouds to heights of several thousand feet and obscured the sun for hours at a time (see Figure 9–1):

During the great dust storm of March 20 to 22, 1935, the geology department of the University of Wichita, Wichita, Kansas, "weighed" the atmosphere on Wednesday, March 20, and estimated that at 11:00 a.m. 5,000,000 tons of dust were suspended over the 30 square miles of the Wichita area within one mile of the ground. The dust extended upward to a height of more than 12,000 feet, so the above estimate appears to be very conservative. The source area for most of the dust of this storm was in southeastern Colorado, more than 250 miles west of Wichita. Dust from this area seems to have been deposited, at least to some extent, at most places across Kansas and eastward, and the great dust cloud eventually reached the Atlantic coast, nearly 1,500 miles from the source area.[1]

Much loam was removed from certain areas and dropped many miles away.

During certain dust storms a sharp boundary is sometimes visible between sand bed load and suspended dust load. Bagnold records the following:

In an erosion desert, the only free dust consists of these fine rock particles which have been loosened by weathering since the last wind blew, and have therefore not been carried away. In such country the wind produces for the first hour or so a mist consisting of both dust and sand. Later, although the wind may have shown no signs of slackening, the mist disappears. But the sand still continues to drive across the country as a thick, low-flying cloud with a clearly marked upper surface. The air above the sand cloud becomes clear, the sun shines again, and people's heads and shoulders can often be seen projecting above. . . .[2]

When the wind subsides, the sand movement disappears with it.

In order to understand the fairly sharp boundary between dust storms and sand bed load with little intermediate material, consider an experiment in which the rate of fall of grains of various silt and sand sizes through the air is measured (see Figure 9–2). The rate of fall of a suspended particle is in effect the same as the velocity of an updraft eddy of air that might pick up a particle from the surface and keep it

[1] A. L. Lugn, *The origin and sources of loess,* Lincoln, University of Nebraska Press, 1962.
[2] R. A. Bagnold, *The physics of blown sand and desert dunes,* New York, Morrow, 1942.

Figure 9–1 Front of severe dust storm coming over a town in northeastern New Mexico.

in suspension. We can conclude that, when the velocity of fall is less than the upward eddy currents within the average surface wind, the particle will be picked up and carried in suspension, but if the velocity of fall is greater, then the grain will be driven only in a horizontal direction.

A wind having a horizontal velocity of 5 m per sec (11 mi per hr) will have maximum upward eddies of about 1 m per sec, and these will hardly lift the finest of the sand grains, but

Figure 9–2 Relative size and rate of fall of small particles. The arrows indicate size of atmospheric dust particles and dune sand generally observed. See Chapter 3 for discussion of silt and sand. [After Bagnold, The physics of blown sand and desert dunes, *New York, Morrow, 1942.*]

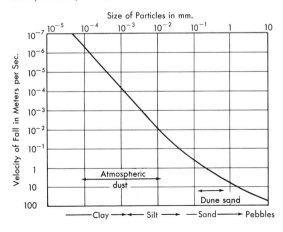

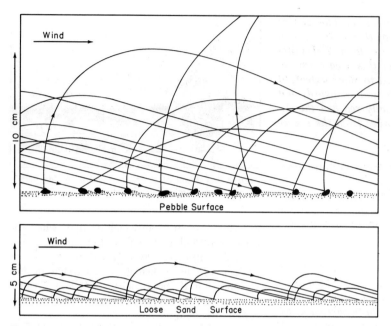

Figure 9-3 Paths of bounding sand grains over surface strewn with pebbles and over loose sand surface. [*After Bagnold, 1942.*]

the particles of silt and clay will be lifted and transported away. Faster winds, of course, will lift the sand grains, but a modest wind will separate, fairly effectively, the sand and dust.

It has been generally observed that when winds reach velocities of about 10 miles per hr near the surface, a few sand grains start to roll, and when the grains have gathered enough speed they begin to jump. The paths of the bouncing sand grains are pictured in Figure 9-3. Upon striking other grains, these in turn are set in motion, and soon a sand cloud develops and moves to the leeward. Most of the bounces or jumps are less than 10 cm (4 in.) high in normal winds, and hence the greatest density of sand movement is close to the surface. Wind velocities necessary to start sand moving are somewhat greater than those required to keep it moving once the bouncing and impact of falling sand grains has started.

We recognize that sand grains in general are too large to be carried in true suspension by the wind. The great bulk of them moves by *saltation* (jumps), but some never leave the ground because they are too large. Other grains that are slightly larger than sand may creep along because they are driven by the impact of smaller grains.

The wind is a remarkable sorter of grains. The silt- and clay-sized particles are removed entirely and blown many miles to leeward. Often they do not settle until brought down by rain. The medium and fine sand particles are concentrated into dunes, as we shall see, and the coarser particles that are left form a residue or protective cover of the surface and are referred to as *desert pavement* (Figure 9-4).

Deflation

Deflation is the process of removal of rock waste from a land surface by the wind. If the material underlying the surface is a mixture of silt, sand, and gravel, then sufficient silt and sand will be transported away until a cover of

Figure 9–4 Desert pavement. Pebbles of limestone and dolomite cover soft tuffaceous shales. Silver Island Mountains, western Utah. [Courtesy of Walter Sadlick.]

gravel remains, and this about ends the cycle of deflation. But if the area is underlain by friable sandstones, siltstones, and shales, the process of deflation may go on for a long time until a basin of considerable proportions has been sculptured out of the rocks. The lower limit of deflation will be the capillary fringe of the water table, and in a desert region the zone of saturation will lose water and slowly fall, thus allowing the wind to erode a little deeper.

Wyoming, New Mexico, Colorado, and Texas have wind-carved basins, or pans, a few to several hundred feet deep; some are up to 15 miles across. The oases of Egypt and Libya are situated in large depressions, such as the Qattara of World War II battles, and have been fashioned, most probably, largely by wind erosion. Some have bottoms below sea level.

Abrasion

The surface movement of sand grains is sufficient to act as a blast. Objects in the way are sometimes mechanically worn and etched by the sand blast, and ledges and rock columns may be undercut. Abrasion is greatest at the ground surface and fades out a few inches above ground, owing to the restriction of movement of sand grains to and near the surface.

Sometimes cobbles on the desert floor become etched or faceted by the sand blast, and fairly sharp edges between the etched faces result. The etched cobbles commonly develop three sharp edges, for some reason not clear,

Figure 9–5 Dreikanter (three-edger) abraded on desert floor by sand blast. Top, vertical view; bottom, end view. Width of base about 4 in.

and they are known by the German word *dreikanter* (see Figures 9–5 and 9–6).

Wind deposition

Collecting of sand in dunes

The drifting sand, freed of the finer silt particles, and with the coarse sand granules and pebbles left behind, collects into dunes. The

Figure 9–6 Selected cobbles from desert floor near Lyman, Wyoming. The flatness comes from the thin beds (quartzite) from which the fragments originally were derived by weathering processes, and the smooth rounding is due to tumbling along a stream channel. Later sand blasting by the wind on the desert floor caused pitting, and the polish is desert varnish. The chipping around the edges is of problematical origin; the upper left specimen is edge-chipped completely and is undoubtedly an artifact. The other two specimens on the top row are partly edge-chipped and are also regarded as artifacts.

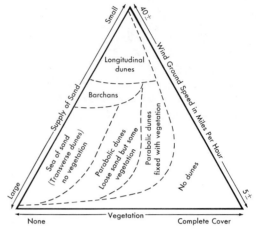

Figure 9–7 Dune forms in relation to supply of sand, wind velocity, and aggressiveness of vegetation. [Modified from J. T. Hack, "Dunes of the western Navajo country," Geogr. Rev., v. 31, 1941, p. 240.]

reasons for the accumulations are (1) obstacles that cause deterrent eddies; (2) the rise of moisture into the sand if the grains are driven into a swamp or onto moist ground; (3) the mastery of vegetation over the drifting sand, holding and fixing it in place; (4) the flow of turbulent air that tends to sweep the sand grains together and build them into streamlined forms.

Dunes take a number of characteristic forms that are repeated in many places the world over. Their shape and size are determined chiefly by (1) supply of sand, (2) wind speed, (3) amount of vegetation. In Figure 9–7 these three factors are related with the common forms of dunes.

The basic dune form

The cross section of a dune in the direction of wind blow reveals a gentle slope to windward up which the sand grains are rolled and a steep slope to leeward down which the grains tumble when drifted over the crest. This is the steepest angle at which the sand will stand without

Figure 9–8 Basic sand dune profile.

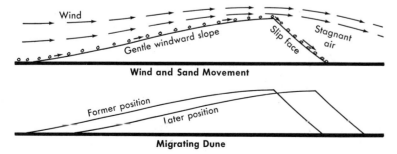

Figure 9–9 Barchan dunes in Monument Valley, northern Arizona. Note the ripple marks. [Photograph by Hal Rumel, Salt Lake City.]

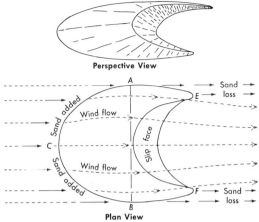

Figure 9–10 Barchan dune, ideally developed.

slumping and is called the *angle of repose slope* or the *slip face* (see Figure 9–8).

The size and ground plan of dunes vary considerably, but the longitudinal section remains fairly constant, in shape at least. A basic design is exemplified by the *barchan* shown in Figures 9–9 and 9–10. A barchan is a crescent-shaped dune with slip face and "wings" pointing downwind. The crests of measured barchans range in height from 12 to 100 ft, and their breadth and length are about ten to twelve times their height. Observation confirms the obvious, that the sand grains are propelled up the gentle slope by the wind and then slide down the steep slip face. The wind movement over the dune is as an elliptical mound, with the space within the bite of the slip face an almost perfect wind shadow. The air is stagnant there. Even in storm periods there is hardly enough movement to cause the sand grains to stir.

It is interesting to examine the sand budget of a barchan. Sand is added and sand is lost. Is the mass of the dune held constant or is it growing or shrinking? Sand is added to the periphery of the dune windward of the transverse profile *AB*. Since it is trapped on the slip face, the only places of loss are the outer wing surfaces downwind, from *A* to *E* and from *B* to *F*. This means that the sand, if any, streams away downwind from the wing tips. In places where the sand supply is abundant, according to Bagnold,[3] the barchans tend to pile up or merge, but as the sand-forms migrate downwind and the supply becomes less abundant, separate barchans take shape and become progressively farther separated from each other (see Figure 9–11). Theoretically the size of the barchans should diminish as they migrate away from the sand source, but commonly other circumstances enter the picture, such as new sand supplies or topographic hindrances, and the diminution is not clearly recognizable.

The rate of "march" of the barchans of the Kharga Oasis, Egypt, has been measured to range from 35 to 60 ft per year. Bagnold relates that belts of barchans trail out from this and other depressions of southern Egypt and the northern Sudan for 50 to 175 miles and that, with the velocity of march given above, it has taken about 7000 years for the front-running dunes in a 50-mile belt to reach their present positions. "This brings us to 5000 B.C., which is a not unreasonable date for the abandonment of the Libyan Desert by the pre-Dynastic folk who are known to have migrated to the Nile Valley."[4]

Perfectly developed barchans seem to form on featureless plains with moderate winds, and most important, with one-directional winds. As soon as the wind shifts, a new streamlining of the previously built dunes will start, and in the transition, many complex shapes will appear.

[3] Ibid.

[4] Ibid.

Figure 9-11 *A belt of barchans. The source of greatest sand supply is to the left. [From Bagnold, 1942.]*

Figure 9-12 *Sand sea of the Sahara Desert in southern Morocco. [Photograph by Cecil B. Jacobsen, U.S. Bureau of Reclamation.]*

Sand seas of northern Africa

The deserts of the earth are scattered along two belts, one in the Northern Hemisphere approximately along the Tropic of Cancer and one in the Southern Hemisphere along the Tropic of Capricorn. In one place the desert extends almost to the equator and in other places to 40° away from it. These regions, characterized by less than 10 in. of rainfall annually, occupy 14 percent of the earth's surface. The largest desert is the Sahara in North Africa. It is continuous with the Arabian Desert, and together they form an arid region larger than the 50 states of the United States. One tenth of the Sahara Desert is covered with dunes and about one third of the Arabian Desert is blanketed with sand. The rest of the region is bare bedrock, either sedimentary, igneous, or metamorphic. The "seas of sand" are almost totally wanting in vegetation. Gentle windward slopes and leeward slip faces are everywhere in evidence (see Figures 9-12 and 9-13). The slip faces are commonly crescent-shaped, reminiscent of the barchan form. Some of the individual dunes rise nearly 1000 ft above the bedrock floors. The great desert regions are broadly related to wind circulation (Chapter 14) and ocean currents (Chapter 10). The winds generally come from the west in the latitudes of the deserts but are cooled first by blowing over cold ocean currents, thus releasing much of their moisture before reaching the western margins of the continents. In places, mountain ranges along the coasts extract the moisture further, so that by the time the winds reach the parched land little precipitation occurs.

Dunes of the White Sands National Monument

In many of the closed arid basins of the West, which have intermittent saline lakes,

Figure 9-13 *The giant dunes of the Murzuch, southwest-central Libya, are among the largest in the world. They approach 300 m (1000 ft) in height and appear as great triangles 1 to 2½ mi along a side, with the desert floor swept clean in between. The desert "washes" grade from south to north, and the present march of the dunes is from north to south. No vegetation impedes the drifting of the sand. Such dunes, not described in geologic literature, do not readily find a place in current classifications. The details of the dunes indicate that the wind has changed direction and that a prevailing direction probably does not obtain. [Courtesy of William E. Humphrey and American International Oil Company.]*

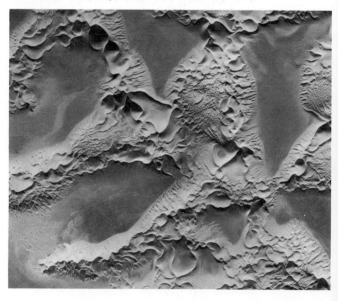

Figure 9–14 Crystals of gypsum that grow in the upper 1 in. of clays of the Great Salt Lake Desert also appear the same after some wind transportation and rounding, as found in dunes.

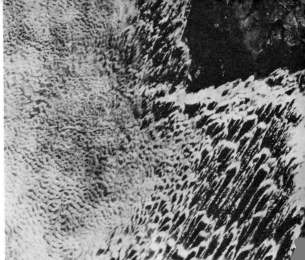

Figure 9–15 Vertical aerial photograph showing the association of transverse, barchan, and parabolic dunes in White Sands National Monument, New Mexico. [Courtesy of E. D. McKee.]

small gypsum crystals are generated in the upper inch of the moist clay and silt (see Figure 9–14). In the hot summer days this upper layer dries out and the gypsum crystals are released, drifting into nearby accumulations commonly called gypsum dunes. Their most noted occurrence is in the White Sands National Monument, southern New Mexico, where four principal types are recognized.[5] Being derived from an intermittently dry lake, Lake Lucero, the sand accumulates in the downwind direction in successive stages from the source: (1) as dome-shaped dunes, (2) as a transverse dune complex, (3) as clusters of barchan dunes, (4) as parabolic dunes. Figure 9–15 is an aerial view showing the transverse, barchan, and parabolic associations. Figure 9–16 is a closer view of parabolic dunes. Parabolic dunes are described as of U-shape or V-shape (Figure 9–17), with the middle part having moved forward with respect to the sides or arms. The arms are somewhat anchored by vegetation, and the central part has migrated ahead.

Vertical cuts of the dunes were made in order

[5] E. D. McKee, Structures of dunes at White Sands National Monument, New Mexico, *Sedimentology*, v. 7, New York, Elsevier, 1966, p. 69.

to study their cross stratification (Figure 9–18) and thus to determine the process of growth and migration. The internal structure, or cross stratification, is as shown in the upper sketch of Figure 9–19. The principle of sand drifting up the gentle windward side and coming to rest on the slip face on the downward side is exemplified, but in a somewhat complicated way (see the second sketch in Figure 9–19). As expected, nearly all cross-laminae dip downwind at high angles (30° to 34°). Numerous bounding surfaces of the sets of cross laminations indicate shifting winds with consequent blowouts and irregular sand accumulations. The bounding surfaces are nearly horizontal on the windward side but steepen progressively downward. Note the bounding surfaces and sets of cross laminations in Figure 9–20. This is a photograph of a hard sandstone face—there can be no doubt that this rock was once a sand dune.

If the wind shifts direction after imparting a streamlined form to the dunes, then remodeling of the forms begins, and in the process irregular and perhaps unrecognizable shapes appear.

Dunes of the Colorado Plateau

In the Colorado Plateau, and particularly in the Navajo country of northeastern Arizona and northwestern New Mexico, dunes are profuse, and three varieties have been recognized:

229

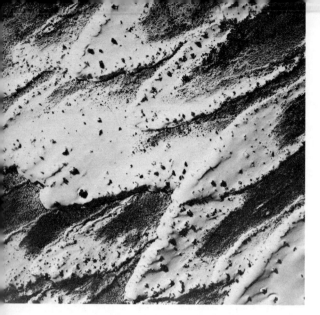

Figure 9–16 Aerial oblique view of parabolic dunes. [From E. D. McKee, Structures of dunes at White Sands National Monument, *New York, Elsevier, 1966.*]

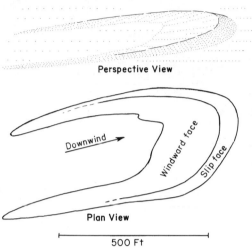

Figure 9–17 Parabolic dune. [Generalized after McKee, 1966.]

Figure 9–18 Vertical cut of a transverse dune parallel to dominant wind direction, White Sands National Monument. [From McKee, 1966.]

Figure 9–19 Cross section in the direction of dominant wind through a barchan. Bottom view shows cross-stratification as found by McKee.

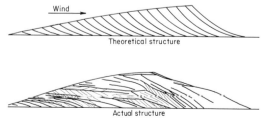

transverse, parabolic, and *longitudinal*.[6] The sand is derived from the old friable sandstones that form many of the cliffs and escarpments of the plateau, and in places it is blown up the escarpment faces onto the plateau surfaces above and also from the high flat surfaces down over the escarpments. The sand trains are scenic in places, for they inherit in delicate hues the colors of the sandstones from which they were derived. The coral sands, the red sands, and the white sands are common appelatives. Their shapes and relation to wind direction are the result of three interacting influences: amount of sand, extent of vegetation, and wind speed.

Transverse dunes are fairly extensive, barren sand zones transverse to the wind direction. There are sand-free places in the zones, and some forms resemble barchans (Figure 9–21). They develop where the source of supply of sand is abundant and the winds are low to moderate. The sands drift, and the dunes are said to be active.

Parabolic dunes, as the name implies, have a parabolic shape, with the wings of the parabola windward, just opposite to the barchan, but the

[6] J. T. Hack, Dunes of the western Navajo country, *Geogr. Rev.*, v. 31, 1941, p. 258.

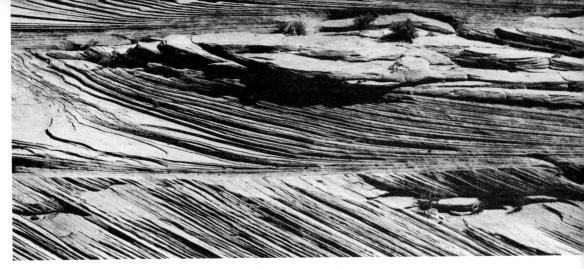

Figure 9–20 Sets of cross laminations and near-horizontal bounding surfaces of Navajo Sandstone, Zion National Park, Utah. [Photograph by Allen Hagood and courtesy of National Park Service.]

gentle windward slope and the steep downwind slip face prevail, as expected. They form by the wind's removal of sand from the windward hollows and the sand's transport to the crest and over to the slip face. They are always associated with vegetation cover, and it is a constant struggle for the plants to prevail against the wind, which blows away the sand in which their roots are embedded. Where the wind gets the upper hand, deflation hollows form. Elsewhere, the sand is somewhat held in place by the vegetation, and low ridges between hollows develop (refer again to Figure 9–21).

Where the wind is strong and the supply of loose sand scarce, long parallel lanes of sand string out over the plateau surface. The lanes or *longitudinal dunes* are only 6 to 30 ft high and 25 to 100 ft wide, and some extend for several miles. They are about 300 ft apart. The zones between the sand ridges, as well as the flanks of the ridges, are covered with vegetation; only the long ridge tops are bare. Longitudinal dunes may also be completely stabilized by the encroachment of vegetation (see Figure 9–22).

Dunes along shores

Source of sand

The sand produced by the waves along shores is a good source of dune sand in many places. The extreme desert coast of northern Chile and southern Peru as well as the temperate but humid coast of the Bay of Biscay of south- western France supply large quantities of sand to the wind, and in either climate, large dune complexes develop. A source of loose sand is unusual in humid climates, because moisture bonds silt and sand grains sufficiently that winds are generally not able to move or pick them up. With moisture comes vegetation. Dune grass particularly mounts an immediate attack on any wind-driven sand and usually does an effective job of preventing further drift. The vegetation is hardly a factor, however, in such intensely dry deserts as those of northern Chile (Figure 9–23), southern Peru, and the south- west coast of Africa.

Coastal dunes of humid climates

The coastal areas of the Great Lakes are marked by dunes in a number of places, and particularly the east side of Lake Michigan and the south side of Lake Superior are noted for their impressive dune complexes.

The waves derive the sand by erosion of glacial deposits, and then it is carried by along shore currents to indentations, where it accumulates in large amounts. During storms with large waves beating on the shore, sand ridges are thrown up, called storm beaches. The sand in a storm beach is above and beyond the reach of the normal-sized waves and hence has a chance to dry out after the storm. Without a storm beach there would be no supply of loose dry sand for the wind. Another requisite for coastal dunes are onshore winds, because off-

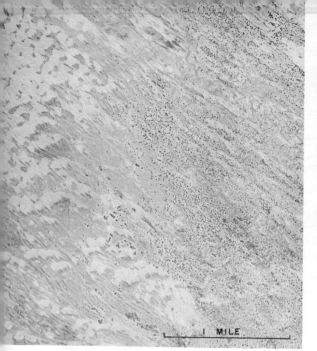

Figure 9–21 Sand dune complex in southwestern Utah. Left is north, and wind is northeasterly. The light gray dunes are vegetation-free and tend to assume barchan forms, but where closely packed they build a transverse fabric. The juniper trees (clots) and desert grasses and brush (medium gray areas) have fixed the dunes fairly well, and in these areas longitudinal and parabolic dunes predominate.

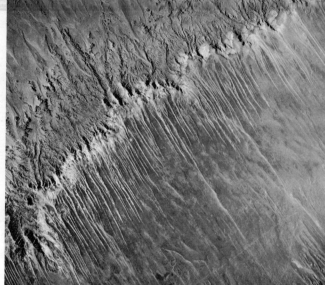

Figure 9–22 Longitudinal dune ridges on the Moenkopi Plateau in north-central Arizona. Sand is blown up the escarpment and strings out on the plateau.

shore winds simply return the sand to the waves.

The onshore winds drift the sand immediately back of the storm beach to build the first true dune, which is sometimes referred to as the *foredune*. It is a ridge 10 to 30 ft high parallel with the storm beach (see *m*, Figure 9–24). Dune grass immediately starts to grow but it generally fails to stabilize or fix the sand completely, and spots develop where the sand is rapidly removed in a second step inland. These spots grow into pockets, or *blowouts*.

In the second stage of transport the sand builds and drifts into parabolic dunes, similar to those of the Colorado Plateau. Here, too, a partial stabilization of the sand by vegetation is the key to the parabolic form (refer again to Figure 9–24). It will be noted that some parabolic dunes are labeled *Al*, some *N*, and some *Aa*. These were built when the lake stood at higher levels than at present. The letter *m* indicates the modern foredune ridge, which is younger than the rest and is in the process of formation. Some of its sand undoubtedly is blown back into the adjacent parabolic dunes.

The coast of Washington and Oregon is one of extensive dune development. A narrow strip of lowland adjacent to the ocean contains the dunes, back of which is the mountain front. Stabilized older dunes of complex habit and history are succeeded in places by active dunes whose dominant element is transverse ridges. Here the summer wind, while the sand is dry, is from the north and northwest, and the sand is driven southward in waves whose crests are normal to the wind with gentle slopes to the north and steep slip face to the south (see Figure 9–25). The dune waves range from 100 to 120 ft from crest to crest, and in plain view they are sinuous in some places. The subparallel sand ridges form where there is a heavy accumulation of sand unencumbered by vegetation and the wind is constant in direction (study also Figure 9–26).

Loess

Nature and origin

Deposits of silt, called *loess*, that mantle the uplands are very extensive in the Missouri and Mississippi River valleys of Nebraska, Kansas, Iowa, Missouri, Illinois, Indiana, Tennessee, and Mississippi. Geologists early recognized

Figure 9–23 Sand-dune chain or ribbon in the Copiapo area of northern Chile. The sand originates here in an uplifted marine strand (terrace) just inland from the present beach. The view is toward the high Andes. [Courtesy of Kenneth Sagerstrom and U.S. Air Force.]

that the loess was made of dust falls because of its position generally above the river courses, its lack of the bedding characteristic of river flood-plain deposits, and because it consists of particles that are usually found in suspended wind load. Some geologists contended that the silt originated in our Western deserts and that the prevailing westerly winds brought it to the more humid climate of the lower Mississippi valley, where the rains helped clear it out of the atmosphere. However, later mapping of the silt deposits showed most of them to be closely tied to the old courses of the Mississippi-Missouri river system, and it has thus been concluded that the loess was derived from the dried-out flood-plain deposits of the nearby rivers. During the last ice age, the Mississippi and its tributaries were much swollen by the melt waters of the glaciers and had spread sand and silt widely over their flood plains. This, when dried out here and there, afforded the wind a fine opportunity to create a major dust storm, and the surrounding silt deposits are the result. Figure 9–27 represents the process and Figure 9–28 the result.

Loess generally develops columnar joints as it compacts and consolidates and erodes into vertical banks or bluffs.

Parts of Alaska, especially the Yukon Valley, are blanketed with silt that has settled out of the air. We observe here today real dust storms on the braided river flood plains (Figure 9–27) and recognize clearly the source and mode of origin of the silt blankets on the hills and mountain slopes up to 1000 ft or more in elevation. In places the silt blankets are 50 ft thick. Sand dunes also originate from the river flood plains even in this arctic climate (Figure 9–29).

Loess soils

Soils formed on loess are some of our very best. Witness the wonderfully productive soils of Iowa, Illinois, and Indiana. This is due to two characteristics: (1) the silt particles are made up of a number of different minerals and hence furnish the elements necessary for plant growth; (2) the silty or loamy nature of the soils holds water yet renders them loose enough for easy plowing. The soils do not clod up like clay, nor are they like sand, from which the soil moisture drains (see Figure 9–30).

Transport of ocean salt to continental areas

Salt, like water, follows the hydrologic cycle. By salt is meant the assorted compounds that are in solution in sea water and which may be precipitated out upon evaporation. It is picked up by the winds as spray from white caps on the ocean surface and transported many miles inland over the continents. It falls to the land surface in the rains and then is drained off by the rivers to the sea again. Part, of course, gets into the ground water, but since the amount is very small, we hardly recognize it in either surface waters or ground waters. However, in those parts of the continents where interior

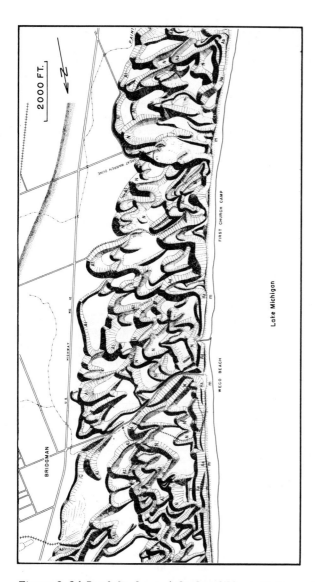

Figure 9–24 *Parabolic dunes of the Grand Marais Embayment, Michigan. Parabolic dunes are denoted by A1, N, and Aa. Lake Michigan stood at a high level toward the end of the Pleistocene, at which time the A1 parabolic dunes were formed; then the lake lowered and the N dunes were built, and at the present level the Aa dunes were built. Many of the parabolic dunes of these three ages have been initiated or augmented in their growth by blowouts. The foredune is noted by m. [Reproduced from G. C. Tague, Post glacial geology of the Grand Marais Embayment in Berrien County, Michigan, Occasional Papers for 1946, Michigan Geological Survey.]*

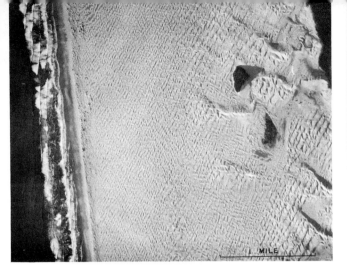

Figure 9–25 Transverse dune complex along the coast of Oregon. Note breaker surf. [Courtesy of *W. S. Cooper.*]

Figure 9–26 Dune advancing on forest along the coast of Oregon. [Courtesy of *W. S. Cooper.*]

drainage occurs, such as the Great Basin of the United States, the salt collects year after year and has contributed to our numerous salt pans and lakes. The Bonneville salt crust and Great Salt Lake are the most noted examples.

Weathering of rocks in the form of chemical decay undoubtedly yields some salt to the streams, and volcanic emanations are important in places, but we are now disposed to think that most of the salt comes from the atmosphere. Sodium and chloride ions compose the chief salt, but other common ions are carried in water droplets in the air such as K^+, Mg^{++}, and SO_4^{--}. The reasons for the atmospheric source are (1) measurable amounts of salt are contained in the atmosphere and rain waters; (2) the rivers gain consistently in salt content as they flow toward the sea regardless of the rocks over which they flow; (3) the salt content of the rivers is similar to the salt content of the atmosphere.[7]

Waves and shoreline activity

The waves of the oceans and lakes are the fundamental force operative along shores. They expend a tremendous energy in eroding, transporting, depositing, and in general, fashioning the exquisite landscape of our lakeshores and ocean coastlines. Extensive sand deposits are there one season and gone the next. Beaches change so rapidly that maps of one decade are of little use the next. Man's efforts to stay the effects of the waves on certain shores have been inconsequential. The short-time changes are conspicuous, but if the long view is taken, we perceive a profound remaking of the shore topography. This will be the theme of the following pages.

Waves in the open water

Wave form

Waves in the open water represent a transfer of form but not mass. This means that, as a wave sweeps by, there is no rush of water but simply a passage of form. If you are in a small boat on a lake, say where 2-ft waves are rolling, your boat will simply rise and fall as each crest and trough passes.

The ideal wave is symmetrical, and its parts are called the *crest* and the *trough*. The distance from successive crest to crest is the wave length, and the vertical distance from trough to crest is the wave height (see Figure 9–31).

The period is the time it takes the wave form to move the distance of one wave length. In deep water the wave length, velocity, and period are related by the simple equation $L = VP$, where L is the length, V is the velocity, and P is the period. For any one velocity in deep water (deeper than ½ wave length) there is only one appropriate wave length and period. If the

[7] For further information see A. J. Eardley, *Salt Economy of Great Salt Lake*, Third Salt Symposium, Salt Institute. Published by Northern Ohio Geol. Soc., Cleveland, Ohio.

Figure 9–27 Silt being caught up and transported by wind from Delta River in Central Alaska. [Courtesy of Troy Péwé and U.S. Geological Survey.]

period, therefore, is measured, which can be done easily, the wave length and velocity can be obtained. On the other hand, length and velocity cannot be measured easily in the open sea. Velocity and length are related by the formula $V = \sqrt{gL/2\pi}$ where g is the force of gravity. The relation of wave length in feet to the period in seconds is given by the formula $L = 5.12P$. The height of the wave is not considered in these formulas, but it can be shown that the ratio of height to length (H/L) does not affect the velocity much. Even the steepest waves travel only 10 percent faster than low waves.

Some wave lengths and their velocities in relatively deep water are given below according to the formulas $L = 5.12P$ and $L = VP$.

WAVE LENGTH IN FT	PERIOD OF SUCCESSIVE WAVES IN SECONDS	VELOCITY IN FT/SEC
5.12	1	5.12
82	4	20.5
184	6	30.8
328	8	41.0
512	10	51.0
1150	15	75.0

When the height of a wave becomes greater than $1/7$ of its length, then the wave becomes unstable and breaks, but this condition is rarely reached in nature, for a steep wave in the open ocean is 10 ft high and 184 ft long. This is an H/L ratio of $1/18$.

The ideal wave, perhaps only theoretically possible, is either sinusoidal or trochoidal (see Figure 9–32). The sinusoidal form has equal curvature in the crest and trough, whereas the trochoidal has a sharper crest and flatter trough.

Waves in the open sea are more complex, especially in the area where they are being generated by the wind. It is the frictional drag of the wind over the water surface that creates the waves, and part of the wind energy is imparted to the waves. In this generating area a number of wave systems are usually superimposed on one another, not only in different lengths but in different periods and directions, and this gives the water surface a most confused appearance. It is an area known as the *sea* to mariners, but as the waves move out of this area into one of relative calm they resolve into *swells*. Here the pattern and form assume some measure of order. The shorter waves have been left behind, and we have to deal with regular, long-crested waves.

Pyramid or short-crested waves and deep, short troughs are common in a sea. To help explain their origin it should be imagined that two trains of waves, traveling in different directions, are superimposed (refer ahead to Figure 9–40 for a view of several superimposed sets of waves). Where a crest of one wave passes across the crest of another a high pointed form results, and where a trough of one wave crosses the trough of another a particularly deep depression results. Crest amplifies crest and trough depresses trough. Where crest intersects trough the two nullify each other. Imagine a third set of superimposed waves, and further regard each as having different wave lengths and heights, and you begin to realize why a sea is so complex in wave form.

Figure 9–28 Loess exposure, U.S. Highway 61 bypass, 5 mi east of Vicksburg, Mississippi. This is a very fresh road cut, and weathering has not yet revealed the vertical joints. [Courtesy of Harold Fisk.]

Water particle movement

As the wave form passes by, what motion does each water particle describe? In the ideal wave, such as described above, the water particles follow the paths of orbits that are nearly circular. Note in Figure 9–33 the movements of a boat as a large wave passes by. Also by reference to Figure 9–34 it can be seen how a wave form is transmitted while adjacent water particles move in circular orbits. The crest of the wave moves ahead as each succeeding water particle rises to the top of its orbit. Each particle at the surface completes one orbit during the wave period. Also the diameter of the orbit is equal to the height of the wave.

Beneath the surface the orbits decrease in diameter progressively downward. For every $1/9$ of the wave length the orbit in depth is halved. For practical purposes, therefore, the water can be considered still at a depth equal to about $1/2$ the wave length.

The drag of the wind on the water surface causes the water particles to advance slightly in the direction of wave advance, and the circles are said to be open (see Figure 9–35). This amounts to a slight forward movement of the water, which is known as *mass transport*. Such slow movement of the surface layer is more significant in high waves than in low and results in the piling up of water on certain leeward shores during wind storms and the complementary return of water as a bottom current. An easterly wind on Lake Erie may cause a 4-ft rise of the lake at the west end; similarly, a strong northwesterly wind over Great Salt Lake will cause the level on the southeast end to rise about 2 ft. Combinations of running waves, wind, and tide all working strongly together cause the disastrous high-water storms that irregularly strike certain coasts.

Size of waves

The size of waves depends on three factors: wind speed, the wind duration, and *fetch*. The fetch is the length of the body of water over which the wind blows. For instance, no matter how hard or how long the wind blows, if the fetch is short, large waves cannot develop.

As the waves move from sea to swell, their height decreases, the sharp crests of the storm area become rounded, and the wave length increases. The swell becomes almost sinusoidal. Although the waves are said to decay as they

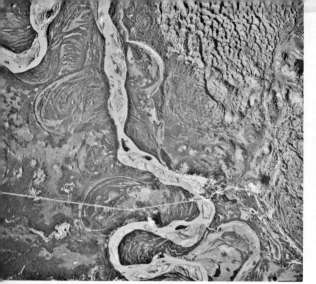

Figure 9–29 Sand dunes adjacent to Tanana River at Tetlin Junction, Alaska. The Tetlin River, frozen over in winter, is generally low in spring and high in summer during melting of glaciers and rains. In low periods it presents a braided appearance. [Courtesy of Helen Foster.]

Figure 9–30 Soil developed on Peoria loess in Doniphan County, Kansas. [Courtesy of John Frye.]

travel away from the storm, they transmit great energy long distances and, as they approach a shore, become the dominant, and sometimes very destructive, waves.

Wave changes in shallow water

Length, height, and steepness

As waves run into shallow water, as depicted in the top diagram of Figure 9–36, two changes occur. The wave length becomes shorter, and the height increases. Wave shortening begins when the depth to bottom equals $\frac{1}{2}$ wave length (wave length as measured in the deep water). By the time the depth is $\frac{1}{20}$ of an original wave length, the wave has shortened to $\frac{1}{2}$ of what it was (see Figure 9–37). The period, however, remains the same. This means that the velocity of the wave motion decreases.

Wave height does not start to increase until the depth of water is about $\frac{1}{20}$ of an original wave length; thereafter the height and the sharpness of the crest increase rapidly (Figure 9–37). At the same time the trough becomes flatter.

Breakers

As the wave height increases, the orbital paths of the water particles become open ellipses with the long axes horizontal. In the sharp crest there is a rapid landward acceleration, while under the flat trough there is a retarded seaward movement. A wave will break when the increased velocity of the water at the crest exceeds the decreased velocity of the wave form. The top of the wave falls forward in a picturesque circle without enough water to make up the central part, and the wave collapses, or is said to break. Refer to top diagram of Figure 9–36 again, and to Figure 9–37. Breaking will occur when the wave height is a little less than the depth to the bottom.

There are two types of breakers: the *plunging* and the *spilling*. The type just depicted, in which the crest describes an arc while falling forward and enclosing a pocket of air, is the plunging breaker. It occurs when a fairly low wave approaches a steep beach. The form of the wave is entirely lost upon breaking, and the energy is converted into a current of tumbling water that rushes up on the sandy beach. This is called a *swash* (see Figure 9–38). The falling breaker water and the swash sweep the bottom sand up the beach 10 to 40 ft, depending on the size of the breaker. With the return runoff or backwash the sand is carried downslope, only to be caught up again in a new breaker swash (see Figure 9–39).

The spilling breaker advances with a foaming or tumbling crest. The wave does not lose its identity but gradually decreases in height

WINDS AND WAVES

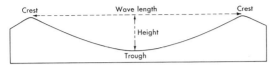

Figure 9–31 *Parts of a wave.*

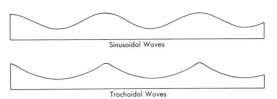

Figure 9–32 *Ideal wave forms.*

until it becomes swash on the beach (see Figure 9–40). Such waves are often fairly steep in deep water and advance over a gently sloping, usually sandy beach. It is these waves that produce several rows of breakers advancing shoreward simultaneously. These are the surf waves down the advancing front of which the surf board riders glide. The best place for surf waves on Waikiki Beach, Honolulu, is where the water shoals over a coral reef about a half mile from shore. The waves increase in height upon approaching the reef area, and it is here that the surf riders rendezvous to catch a wave for the ride in. The waves on occasion may reach heights of 30 ft. At first, the wave is steep-fronted and generally glisteningly smooth; then, as it rolls shoreward, its crest begins to foam, first here and then there, and the surf rider steers obliquely down the front to avoid the foam. The crest actually plunges somewhat at times, but the wave holds its identity and rolls on in. If the surf rider is caught in too much foam, he usually capsizes. If successful, he rides the wave to within 100 ft or so of shore. The wave decreases in height from the time the foaming starts until it is a mere 4 or 5 ft high, when the rider abandons it, or vice versa. The spilling represents a loss of the wave's energy, and hence the decrease in height. Finally, within 20 ft or so of shore, the wave may finally dissipate in a small plunging breaker.

Waves as a geologic agent

Definitions

At this point it might be well to define a few terms. By *beach* is meant the loose material upon which the waves play. The zone extends from the upper limit of storm wave wash on land to a point seaward under water to a depth where the waves have little or no effect on loose particles. On small lakes this depth is a few feet, whereas along ocean shore lines it may be 50 to 100 ft.

By *coast* is meant the general terrane landward of the ocean and includes the vegetation, climate, topographic features, and any human culture related to the water body. Coast is generally not used in connection with lakes.

Shore, like coast, is a general term for the belt of land bordering on the water body, and it has applicability to both lakes and oceans. Some writers limit shore to the zone between low tide and the upper reach of storm waves.

Shoreline is the line, as nearly as it can be defined, between water and land.

Figure 9–33 *Successive positions of crest, orbital movement of water particle in the transfer of wave form, and the positions of a boat as a wave passes by.*

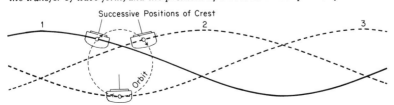

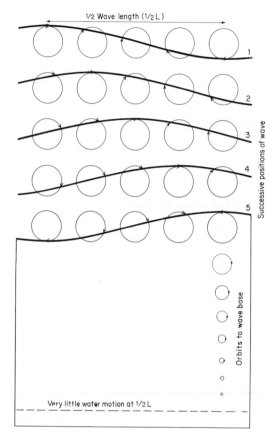

Figure 9–34 *Illustrating water particle motion for sinusoidal wave passing from left to right.*

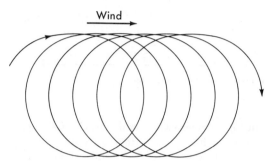

Figure 9–35 *Open circles and mass transfer of surface water.*

Factors affecting the beach

The type of coastal features in any place, with emphasis on the beach, depends on four factors:

1. The waves, their size, their direction of impact on the shore, and the frequency of storms.
2. The tides. A beach with small tides (2 or 3 ft rise and fall of sea level) will have the beach features that a fairly constant play of waves will build, such as spits, bars, and wave-cut cliffs, but one with large tides (20 to 40 ft), will probably have wide mud flats over which the waters advance and retreat twice a day as well as wave-cut cliffs. The cause and nature of tides is discussed in the next chapter. High tides are characteristic of estuaries and inlets, such as in coastal Alaska, and in some of these the tide becomes a vigorous current rather than a play of waves. Islands near the middle of a large tidal basin are apt to have tides of little more than a foot, but funnel-shaped estuaries where the incoming tidal crest has no place to go but up the bay are apt to have tides up to 40 ft. New Brunswick's Bay of Fundy is an outstanding example. Each time the tide comes in, a 4-ft wall of water rushes and foams up the bay and its tributaries, carrying an estimated 3,680 billion cubic feet of water. This is the famous Bay of Fundy tidal bore.
3. The topography of the coast upon which the waves and currents come to act. There are several basic kinds of coast topography, and the waves and currents develop different features on each in their erosional, transportational, and depositional activity.
4. The material of the coast upon which the waves spend their energy. It is quite a different matter whether the waves work on a loose sand and gravel or on a mass of granite.

Breaking on walls and cliffs

Under the right conditions, storm waves impacting on cliff faces produce what is called *shock pressure*. Three variations of waves impinging against steep faces are shown in Figure

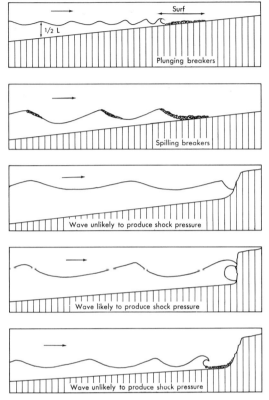

Figure 9–36 *Waves transgressing the shore under varying conditions.* [Partly after C. A. M. King, Beaches and coasts, London, Edward Arnold, 1959.]

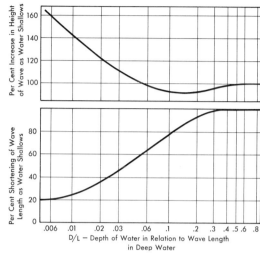

Figure 9–37 *Increase in wave height and decrease in wave length as the wave moves into water less than one-half wave length in depth.* [After King, 1959.]

9–36. The central sketch shows a large plunging breaker enclosing a pocket of air against a cliff face. Undercut cliffs with caves provide a very favorable situation for the compression of a pocket of air. The force will be approximately proportional to the wave velocity where the wave breaks, and thus the enclosed air is compressed considerably. The water, air, and spray are driven into the cracks and joints repeatedly, leading to considerable damage. In the case of abutments or break-waters, the depth of water should be controlled so as to cause the wave to break before striking the structure, if possible. This is one of the ways in which waves erode hard rock in places along steep shores.

Refraction

The energy that waves spend in eroding shores is attenuated in some places and concentrated in others. Oncoming waves at right angles to the general shore line are shown in Figure 9–41. They bend around the headlands and drape shoreward around the bay. The headland, therefore, undergoes a concentrated beating, whereas the bay receives the effect of stretched or weakened waves. The bending of the waves as they run into shallow water is called *refraction*. As waves move into shoaling water the friction at the bottom causes them to slow down, and those in the shallowest water move the slowest. Since the different segments of the wave fronts are traveling in different depths of water, the crests bend and wave direction constantly changes. Thus, the wave fronts tend to become roughly parallel to the underwater contours.

The vertical aerial photograph of Figure 9–42 shows the refracted waves at the east end of San Nicolas Island, California. West of the island the waves approach in a single system, but are refracted around the small island, as

Figure 9–38 Plunging breaker and swash on Waikiki Beach, Hawaii.

Figure 9–39 Spilling-type breakers and swash on the coast of Oregon. [Photograph by Ralph Carlston, Salt Lake City.]

shown in Figure 9–43. The refracted waves are dragged in opposite directions on the north and south sides of the island, and then intersect off the east end. A third small system of waves may be noted in a narrow zone south of the central part of the spit. These may be due to a small, steep-walled island rock from which the waves are reflected.

An interesting analysis of the damage caused by large breakers off the tip of the Long Beach, California, breakwater was made by the coastal engineer M. P. O'Brien.[8] During the period April 20 to 24, 1930, the waves breaking against the tip of the breakwater dislodged stones ranging in weight from 4 to 20 tons. Observers had estimated the period of the waves at 20 to 30 seconds, a very long period for the swell. This incident was unusual because (1) the breakwater had withstood the waves for years; (2) the winds were light and the sea was calm at the time, at least beyond the three-mile limit where gambling ships were anchored; (3) it was noted that at the San Pedro breakwater, only a few miles to the north, there were no breakers, and on the beaches to the south the lifeguards noticed no unusual wave activity. O'Brien's analysis was delayed for 17 years, until refracted waves were better understood.

A refraction diagram must be constructed to analyze the wave conditions. If the deep-water wave front, which depends largely on the submarine topography, is divided into equal parts, and rays are drawn through these perpendicular to each wave front, the distribution of energy is readily seen. Such rays are called

[8] See Willard Bascom, *Waves and beaches*, Garden City, N.Y., Doubleday, Anchor, 1964, p. 267.

orthogonals (see Figure 9–44). The length of the waves, after they begin to drag bottom, becomes progressively less toward shore, and the ratio of the length of the wave in open water to that at the beach is called the *refraction coefficient*. The practical coastal engineer must construct a refraction diagram for each storm (direction of advancing waves in the open sea and the period of the waves), and such information must be sought in the weather records. He must begin with an accurate contour map of the offshore topography out to a depth of at least half a wave length of the maximum wave that will be considered.

After 17 years the nature of the very localized breakers on the tip of the Long Beach breakwater was still an enigma, but O'Brien set out to solve it. He consulted weather maps and found that at the time two storm centers existed that might have been capable of producing long-period waves; one lay to the west and one to the northwest. No weather data were available for the southern hemisphere. Waves from the northwest storm center would have bypassed the San Pedro and Long Beach shores because of the protection afforded by the westward jutting coast and the channel islands, so this possibility could be discounted. Also, a refraction diagram for the storm from the west, using a wave period of 20 seconds as observed, showed that such waves could not possibly have been concentrated at the tip of the Long Beach breakwater. Thus the only possibility was waves generated to the southeast from which direction the channel islands offered no protec-

Figure 9–40 Swash (upper left corner) and rilled sand of the beach incident to swash runoff on the coast of Oregon. The coarse sand grains give an idea of scale. [Photograph by Ralph Carlston, Salt Lake City.]

tion. The diagram of wave fronts, orthogonals, and submarine topography as O'Brien prepared it is given in Figure 9–44, using a 20-second period. A hump in the underwater topography, 180 to 600 feet deep and 10 miles away, acted as a lens to focus waves coming from S 20° E on the breakwater tip. In these special refraction conditions the wave energy of several miles of crest was caused to converge on a point, and the energy of waves only 2 feet high in deep water were concentrated in a narrow point in breakers over 12 feet high, which caused the damage.

Alongshore transport

In the case of waves approaching the shore at an oblique angle, the waves are bent or refracted as shown in Figure 9–45. Breakers play against the beach but also sweep along it. The swash rushes the sand and silt particles obliquely up the beach. They return in the backwash down the steepest slope, or normal to the beach. These same particles are caught up by the next swash and again carried obliquely up the slope, to return downward again in the backwash. This process, repeated many times, although affecting only a few pounds of sediment on any small stretch of beach, and transporting it only a foot or two along the beach in each swash cycle, results in time in the movement along shore of enormous volumes of sediment. Many beach features, as we shall see, are built by the transport of material along shore by the process described.

Alongshore currents are set up by waves rolling in, obliquely to the shore. In the process of breaker swash and backwash there is a slow transfer of water alongshore, and also in the zone of breakers there is a pulsating propagation of water in the same direction.

In the mass transfer of water by wave action, water piles up along the shore. Sometimes there is a gentle return bottom current to compensate (the *undertow*), but at other times a local, restricted seaward current carries the hydrologically unstable water back to the sea. This is called a *rip current*, particularly because it may attain speeds of 2 to 3 miles per hour where

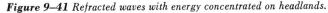

Figure 9–41 Refracted waves with energy concentrated on headlands.

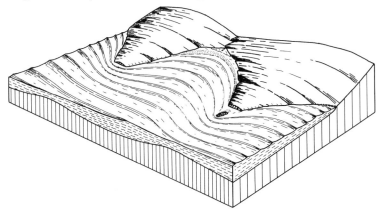

Figure 9–42 Crossing sets of refracted waves off the east end of San Nicolas Island, California. [Courtesy of U.S. Geological Survey and John G. Vedder.]

concentrated in the surf zone. Once through the surf it tends to spread out and dissipate (see Figure 9–46).

Tides

Whereas the variation of wave intensity is irregular and unpredictable, the rise and fall of the tide takes place approximately every 12 hours and can be predicted by a series of accurate calculations. Sometimes meteorological conditions conspire to compound tides with high waves and prolonged high winds; these are generally not predictable but at the same time they are very destructive.

On a tideless coast the beach coming under wave activity is limited by the size of the waves and is generally not very broad. The constancy of position of the breaker activity will produce a certain beach profile and also a certain distribution of materials. Now, if the tidal range is great (30 to 50 ft), then the break point of the waves is never fixed, and a different distribution results at high tide than at low. This has a definite bearing on wave-cut and wave-built features, which will now be described.

In regard to tidal currents it is recognized that *flood current* flows in one direction and the *ebb current* in the other, thus somewhat negating each other. On the other hand, the flood currents affect the shore at higher elevations than the ebb currents, and therefore some changes in beach forms may occur.

Profiles of beaches

The waves deposit and erode and are thus said to be constructive and destructive. On a steep shore the waves fashion a terrace, part of which is generally erosional and part depositional. In a gently seaward dipping plain the wave-cut platform is continuous with the deposit (see Figure 9–47). It has been conjectured that the waves strive to build a stable beach profile that is neither too steep nor too gentle and that, once attained, it remains. This is called the profile of equilibrium. But because of the varying height, steepness, and length of the waves and because of the general progress of erosion, certain beaches in particular are characterized as mobile. In Figure 9–48 it will be seen that with the progress of erosion the beach becomes broader and flatter, with more of the wave energy being spent in crossing the platform and hence becoming progressively less effective in further erosion. In this sense a profile of equilibrium is attained.

In detail the beach may develop a bar at the break point and another at the head of the swash (see Figure 9–49). The break point bar forms at the point of breaking of the waves, and therefore its position depends on the height of the wave. Two bars may be built if small waves follow large waves. Such shallow ridges are often experienced by swimmers on lake beaches. With a falling water level (ebb tide) the break-point bar is destroyed.

Swash bars may be built above the mean water level. They form particularly under the influence of flat waves, which have been found to move sand at depths both inside and outside the break point toward the land. During storms such bars may be built as sand and flotsam is thrown landward several feet above the swash of normal waves. The bars are called *storm beaches* by some, and *berms* by others. The leeward face is in places an angle of repose slope (steepest slope at which the rock particles will stand without slumping). The storm beaches dry out after the waves that built them subside and then furnish the sand of which onshore winds build the coastal sand dunes.

The waves not only remove loose material from the shore but attack the solid rock. In doing so, wave-cut platforms (Figures 9–50

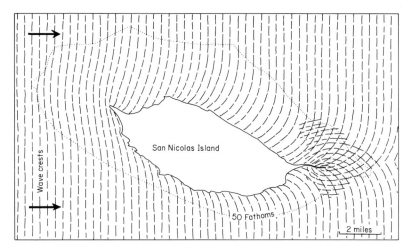

Figure 9–43 *Probable wave pattern around San Nicolas Island at the time that the aerial photograph of Figure 9–42 was taken.*

Figure 9–44 *Refraction diagram for the destructive waves at Long Beach, California. The effects of a salient or nose between 30 and 200 fathoms deep and 12 mi offshore in focusing the waves' energy are clearly shown. [After Willard Bascom, 1964.]*

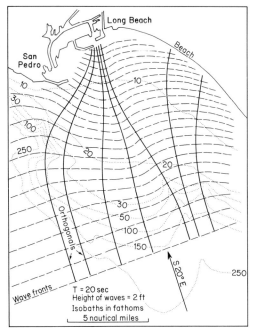

and 9–51) and cliffs are formed. The retreat of the cliffs is intimately related to the cutting of the platform, and both processes are dependent on the waves reaching the rocks.

The processes involved are solution, corrasion, attrition, and hydraulic action. Solution by water is particularly significant in limestone, and where the cliffs are eroded in limestone various solution etching effects are prominent. Corrasion implies the direct striking and rubbing of the rocks by sand grains, pebbles, and boulders. During a storm huge blocks of rock are hurled around with great force by the waves. Fine sand and silt exert little corrasive action, however. At the same time that the platforms and cliffs are being corraded, the particles themselves are being worn down. They become rounded and smaller.

After the cliff is formed, other erosional processes may become active, such as frost action, chemical decay, and gravity, and fragments fall to the base of the cliff. These are broken up by the wave action and are used as tools for corrasion; they eventually are carried away.

We have mentioned the hydraulic pressures and shock pressures of pockets of air enclosed in breakers striking a steep cliff face. For a

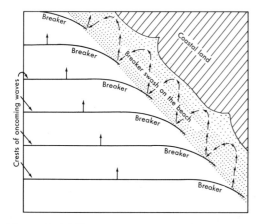

Figure 9–45 Transportation of sand along the beach by waves approaching at an oblique angle. Note also the bending (refraction) of the waves as they roll into shallow water. Material is moved by the breaker swash on the beach and in the turbid surf zone.

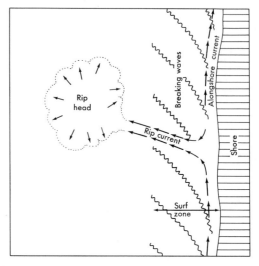

Figure 9–46 Waves, alongshore current, and rip current. [After R. S. Arthur, "Dynamics of rip currents," Jour. Geophys. Res., vol. 67, 1962.]

wave 10 ft high and 150 ft long the hydraulic pressure against the cliff is about 1200 lb per sq ft. A very high shock pressure of 12,700 lb per sq ft was measured in a cliff cave at Dieppe, France. These pressures seem sufficient to pry apart and dislodge rock fragments. In badly jointed rock formations such effects are undoubtedly great, and the process is possibly more significant than corrasion.

In this regard, the wave-cut platform at Santa Cruz, California, is instructive (Figure 9–52). It slopes seaward at an angle of less than 1° and is carved out of the Monterey Formation, which consists of bedded porcelainites and shales. Porcelainites are hard shales that have the texture and breaking proportion of porcelain. At times the platform is bare, but at other times it is mantled by a thin layer of loose sand. The platform extends seaward to a depth of 65 ft and is over 1½ miles broad. W. C. Bradley[9] of the University of Colorado has concluded that:

[9] Submarine abrasions and wave-cut platforms. *Bull. Geol. Soc. Amer.* v. 69, 1958.

1. *Abrasion (corrasion) is restricted to the surf zone, which has a maximum depth of 30 feet.*

2. *Sediments are transported to depths beyond the surf zone but are too fine and the water movement is too slow to accomplish any abrasion there.*

3. *Since the wave-cut platform extends out to a depth of 65 ft, it is concluded that the coast has been sinking, or the water level rising, and that the deeper outer part of the platform was eroded during a lower water level.*

4. *If the sea level is stationary relative to the land, erosion of a platform about one-third of a mile wide is all that may be expected. The*

Figure 9–47 Wave-cut and wave-built terrace.

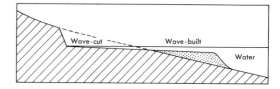

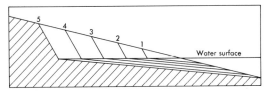

Figure 9–48 Progressive cutting of rock platform.

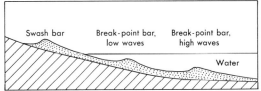

Figure 9–49 Kinds of bars along the beach.

shaping of a continuously smooth platform of greater breadth than this requires the slow submergence of the land.

The maximum depth of effective erosion is called the *wave base*.

Headlands are particularly susceptible to platform and cliff cutting because the waves are refracted around them and the energy is concentrated on them. They appear in many variations similar to that shown in Figure 9–53. Incidental to such erosion, pedestal-like remnants are often left for a while. They are called *stacks* (Figure 9–54). They are soon worn away, however. Other forms are caves and arches, as the irregularities in resistance and fracturing of the rocks allow the erosional processes various irregular avenues of penetration.

Deposits along shores

Bars, spits, and barrier beaches

So far we have thought of deposits along shores in cross section. Now, when we take the map view and consider the significance of alongshore transportation, some of the most striking features come into focus. The deposits generally produce straight or gently curved beaches. In the attainment of this kind of beach, the waves often trim off certain prominences as they fill inbetween with sand (see Figures 9–53 and 9–55).

The alongshore currents commonly build sand and shingle (gravel) projections out from eroded headlands. These are called spits (see Figures 9–53 and 9–56). If spits from headlands on either side of a bay are joined, the deposit is a bar, which would be called a *bay-mouth bar*. In the course of spit building the end may become hooked, which always occurs toward land. Possibly strong onshore waves during a storm accomplish the turn. Successive new hooks may form as the alongshore currents continue to unload sand at the end of the spit (Figure 9–57). There are many irregularities in shore topography, and thus the erosional and depositional forms incident to the work of the waves and currents are quite varied, but almost always the waves and currents conspire to smooth out the shore. This we should look for on our next visit to a large lake shore or the ocean coast.

In the large plan certain coasts for hundreds of miles are remarkably straight or smoothly curved. Spits and bars across bays are generally smaller features, but much of the coast of the eastern and southeastern United States has

Figure 9–50 Wave-cut platform across inclined limestone and chert beds, near St. Jean de Luz, Bay of Biscay, France.

Figure 9–51 *Wave-cut cliff at Acapulco, Mexico. [Photograph by Hal Rumel, Salt Lake City.]*

been remade by the waves on a scale dwarfing all engineering projects of man. The straightening of an originally irregular coast has been accomplished chiefly by the building of great bars in shallow water occasionally some distance from the shore line. These are called *barriers, barrier bars,* or *barrier beaches.* They enclose a swampy lagoon between them and the mainland. Examination of a large and detailed map of the eastern seaboard reveals that long stretches of the shore are formed by the barrier beaches beginning with Long Island and extending to the Florida Keys. There are also long and smooth barrier beaches west of the Mississippi Delta along the coast of Texas and Mexico. At Atlantic City, New Jersey, one of several cities built on the barrier beaches, a southward movement of 400,000 cu yd of sand per year has been estimated. The barrier ends southward at Great Egg Inlet to Great Egg Bay only to take up on the south in another barrier on which Ocean City is located. Here, however, the sand is drifting northward along the shore at the rate of about 50,000 cu yd per year and is building the barrier northward into the inlet (see Figure 9–58).

It will probably be surmised by now that some of our harbor breakwaters have interfered with the beach-forming process, and there are a number of noted examples where this interference has resulted in unexpected deposition and erosion in others. We have succeeded, at least temporarily, in some of our engineering projects along the coasts but have lost in the struggle against the sea in others. It is no wonder then that wave action is considered a very important phenomenon and is the subject of ongoing investigation.

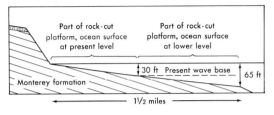

Figure 9–52 Wave-cut platform at Santa Cruz, California. [After W. C. Bradley, Bull. Geol. Soc. Amer., v. 69, 1958. See also Robert S. Dietz, Bull. Geol. Soc. Amer., v. 74, 1963, p. 971.]

The origin of the long barrier beaches is a controversial subject, but the theory that they are built by sand thrown up from the bottom seems to have preference. A slope or profile is presumed to have existed that is gentler than the equilibrium gradient for the waves that play upon the beach. The waves rectify the situation by throwing up sand from the bottom to form a bar, thus making the slope greater and more agreeable (Figure 9–59). There is no doubt that alongshore transportation of sand occurs after the barrier is built, but alongshore currents are not considered very important at the beginning of building of the barrier.

The tidal outlets of barrier bars make fascinating views from the air. Figure 9–60 shows plumes or underwater sand deposits both on the lagoon side and on the ocean side as the tide surges through and withdraws. Generally the deposit is mostly on the lagoon side. Hurricane storms are very destructive to barrier beaches. The series of three aerial photographs of Matagorda Peninsula speak for themselves (Figures 9–61, 9–62, and 9–63).

Cusps

Cusps are evenly spaced, crescent-shaped depressions that are built by wave action along sandy and gravelly beaches. They vary greatly in shape and size (note those of Figure 9–64). They are made up of sand or gravel and occur both on exposed beaches and projected bays. They have been measured from horn to horn to range from 14 to 1200 feet. Why they occur in the various sizes, or why they occur at all, is still unknown. Yet we may stand on the beach where they have formed and think we can see the reason why. We will probably note that (1) the swash bearing its usual load of suspended sand rushed up the beach where it splits into two segments by the apex point. The two segments are deflected along the sides of the bay until their force is spent, or until they meet the

Figure 9–53 Typical features of a headland undergoing wave erosion.

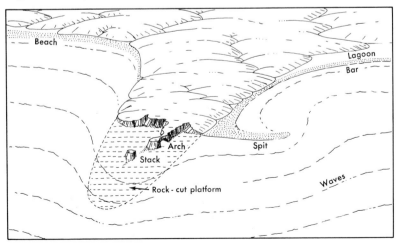

Figure 9–54 Sea cliffs and stack near Cheticamp, Nova Scotia. [Courtesy of Geological Survey of Canada and R. F. Block.]

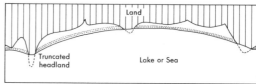

Figure 9–55 Smoothly curved beach formed by trimming off headlands and deposition of sand in recesses or across embayments.

water of the opposite side. Then (2) the centralized swash returns seaward by the steepest path. In doing so (3) it leads to or scours a channel in the center of the bay. We will note then (4) that a small submarine deposit (delta) builds at the bottom of the channel. The effect of this channel flow of water and the delta deposit is to (5) impede the next oncoming wave in the center of the cusp, but the wave swash opposite the horn is unobstructed. From these observations it appears that the bays are deepened and the horns built out by every wave, but we must keep in mind that if the wave height and period vary, the waves will get out of phase and tend to undo each other. This is the reason why most students agree that regularly spaced and well developed cusps are formed by regular waves. They also tend to agree that (1) conditions are best for cusp formation if the waves approach perpendicular to the shore and are unconfused by local currents and winds, (2) some irregularity in the beach is necessary to start the cusps forming, and (3) the spacing of the cusps is related to wave height. But the relation between wave height and cusp size is not answered, and particularly why the cusps start on a clean sandy beach in the first place remains an enigma.

Shores of emergence

The crust of the earth is restless, and in some places geologists note uplift and in others subsidence. Particularly along shores, it is easy to recognize these movements. For instance, if a coast has been rising, we should note a succession of wave-fashioned beaches above the present one (Figures 9–65, 9–66, and 9–68). On the west side of Ben Lomond Mountain near Santa Cruz, California, there are five main terraces and several indistinct intermediate ones that give the appearance of a crude gigantic staircase from sea level up to an eleva-

Figure 9–57 Hooked spit at Cape Kellett, Banks Island, Canada. [Courtesy of U.S. Navy and James A. Whelan.]

Figure 9–56 Spits, bars, and truncated headlands.

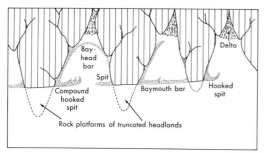

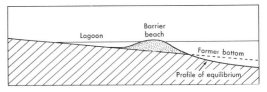

Figure 9–59 *Origin of barrier beach.*

tion of 850 ft. They are all wave-cut platforms and attest to the rise of this section of the coast of 850 ft in rather recent geologic time.

Shores of emergence are usually fairly steep and straight and have only a few harbors.

Figure 9–60 *Red Fish Pass, southwest Florida, is a tidal inlet in a barrier bar. The lagoon, called Charlotte Harbor, is fairly quiet, and incoming tidal currents have built in it a birdsfoot delta of sediments brought by longshore currents to the pass. The Gulf of Mexico is also rather calm here, and a smaller tidal delta has been built into the gulf by the ebb tide currents. [Courtesy of Harold R. Wanless and the University of Illinois collection of aerial photographs.]*

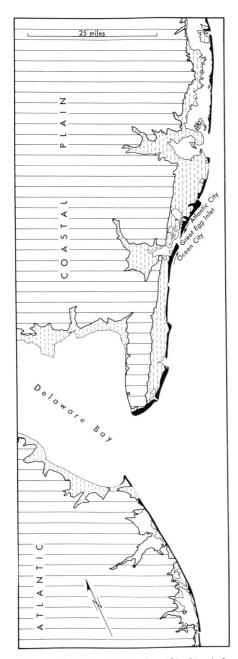

Figure 9–58 *Barrier beaches (black) of the Atlantic seaboard, near Atlantic City, and Delaware Bay. Lagoonal areas are in dashed lines.*

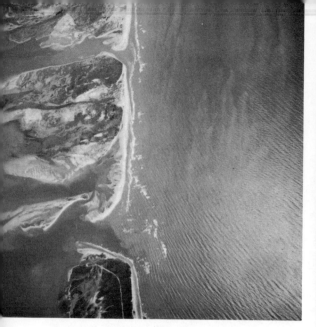
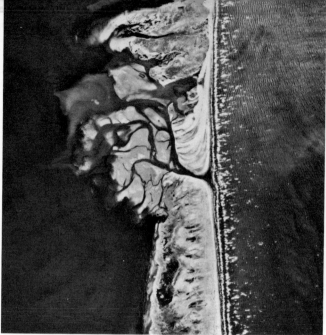

Figure 9–61 This aerial photograph and the following two views depict the modification of a barrier beach by alongshore currents and hurricane storms in Matagorda Peninsula, off the Texas coast. A hurricane in 1942 sent the sea surging across the barrier and ripped open three inlets. A spit has already been built across the central inlet. [Courtesy of the University of Illinois collection of aerial photographs.]

Figure 9–62 Between 1943 and 1957, shore currents transported much sand from the north, closing the northern inlet and nearly closing the southern inlet by building a series of hooked spits. Note the road at the south end of the barrier in Figure 9–61; erosion on the south side of the inlet has removed the point to the road as the spits built southward. [Courtesy of Harold R. Wanless and the University of Illinois collection of aerial photographs.]

Shores of submergence

If a coastal area subsides, the lower part of the valleys of rivers that flow into the sea will be flooded, with the development of estuaries like those of Chesapeake and Delaware Bays. Such valleys are said to be drowned. In fact, the entire coast from North Carolina to Newfoundland is one of fairly recent submergence, geologically speaking in point of time. The Gulf of St. Lawrence is another great drowned river valley. Needless to say, the strong waves of the Atlantic have worked over the irregular relief features as the seas have advanced on the land but have not had time to straighten out much of the shore from Long Island northward. There are thousands of truncated headlands, bars, spits, and curved beaches, but all on a much smaller scale than the barrier beaches to the south.

In a submerged beach of headlands and bays the process of erosion of the headlands has been imagined from the early inception to an advanced stage, where they have all been worn back to a straight coast line. Such may be the case along the Chalk Cliffs of England (see Figure 9–68).

The shoreline environment

The past pages have been chiefly devoted to the changes in shorelines by erosion and deposition. Much is being done in harbor development, protection, and maintenance, and several examples of this have been cited. Bathing beaches are also receiving much attention. Until recently the problem here was to keep the sand in place and to prevent the waves and currents from transporting it up or down the shoreline. The new element in beach maintenance is that of oil pollution. First in the news was a large tanker, the Torrey Canyon, that went aground on the south shore of England in March 1967, and spilled its cargo of 30 million gallons of crude oil on the ocean waters, much of which promptly drifted to the shores of Cornwall and across the English Channel to Brittany, where it soaked the sand grains of the beaches there.

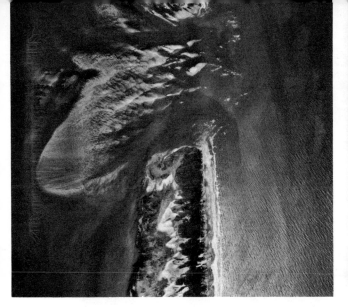

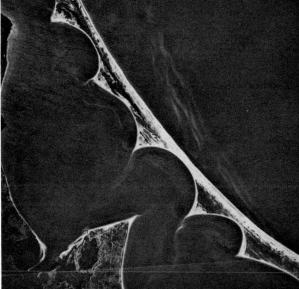

Figure 9–63 In the early fall of 1961, Hurricane Carla struck the Texas coast and removed most of this section of the barrier. Because of the surge of sea across the barrier and into the lagoon, the previous tidal inlet delta deposits were reworked into a complex of current ripple forms. Now the road is almost on the beach.

Figure 9–64 The tip of Massachusetts' Nantucket Island, Coatue Point, may be recognized by roads. A lagoon and a barrier bar or island lie beyond, and the remarkable cusps occur on the lagoonal side of the barrier. The seaward side is also remarkably smooth and curved. Could this also be a cusp? [Courtesy of Harold R. Wanless and the University of Illinois collection of aerial photographs.]

The flight of aquatic birds was impaired when their feathers became coated with oil. They lose their buoyancy in water and many perish. The possible effect on marine life is alarming but hardly yet explored. Three-quarters of a million gallons of detergent used on the Torrey Canyon oil spill apparently did not hurt the deep-water lobsters and crabs or the commercial catch of fish, according to one report, but according to another, the detergents and dispersants are unjustified in the clean-up operation, and are far

Figure 9–66 Elevated platform cut below water near Santa Cruz, California. [Courtesy of W. C. Bradley.]

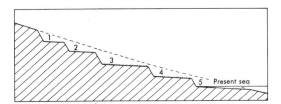

Figure 9–65 Platforms cut as coast emerged.

Figure 9–67 Seven Sisters of the Chalk Cliffs of England, showing shoreline straightened by erosion.

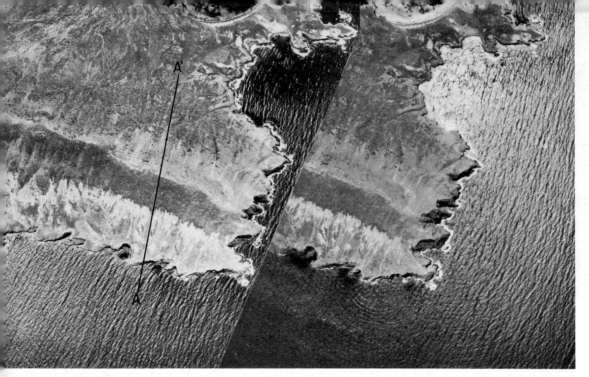

Figure 9-68 Frazer Point, Santa Cruz Island, California. Note the sea cliffs, headlands, coves, stacks, arcuate beach, and cusps. Study with a stereoscope to recognize the raised beaches.

more damaging to marine life than oil.

It has been estimated that leaky tankers and sloppy handling measures release 1 million barrels of oil per year into the ocean waters. Tankers discharging oily ballast waters off the coast of Kodiak Island are reported to have killed 10,000 waterfowl and coated 1,000 miles of shoreline with oil. This has brought about a rebuke of the oil companies by the Secretary of the Interior.

Widespread public concern was also directed to the oil spill in the offshore Santa Barbara shelf area where oil wells were being drilled. An unexpected oil accumulation which was under considerable pressure was penetrated. Unfortunately it leaked up to the sea floor around the outside of the casing and also in a zone running 800 feet eastward from the drilling platform. This posed an uncontrollable situation, at least for the several weeks ahead. Much oil emerged to the water surface, a huge oil slick formed, and thousands of barrels of oil drifted to the beach, ruining it for bathing at least temporarily. The drilling was discontinued and very difficult grouting operations were affected (cement was pumped down the sites of the oil seeps). After about two months most of the oil leakage had been checked, but even after 10 months some oil was still exuding.

Wells drilled from a platform 30 miles off Venice, Louisiana, caught fire on February 10, 1970, and burned for a month, when a blast of dynamite put out the blaze. The wells then spewed dark brown crude oil like a geyser 100 feet high above the drilling platform. It is reported that 11 wells had been drilled from the same platform and that 8 of them had caught fire. It is estimated that between 600 and 1000 barrels of oil per day were released into the Gulf of Mexico waters. Within a few hours an oil slick had formed which drifted toward Breton Island, 10 miles away. In order to retain most of the oil, the oil company that owns the wells had a string of barges locked together with floating booms surrounding the well. Facilities for sucking and lapping up the oil were also at hand. But 10-ft waves developed, the boom system broke apart, and the great oil slick spread toward the mainland. The biggest and last of the flowing wells was finally capped seven weeks after the oil first broke loose. The great worry was the welfare of rich shrimp and oyster beds nearby—it takes two or three years to evaluate the damage to marine life from the effects of an oil spill.

The Santa Barbara and Venice incidents dramatized the potentially disastrous effects of oil-well blowouts, no matter how rare they may be. Because the continental shelves now produce about 20 percent of the world's oil and gas, and are estimated to contain a reserve as great as that of the land areas, there will surely be an expanded effort in off-shore drilling. We already have the capability for drilling oil wells in 1600 feet of water, and offshore accidents are an uncomfortable reminder that exploitation of the sea bed poses serious hazards to the shoreline environment. We must underscore the need for further improvement of offshore drilling technology and for further studies of means to handle such accidents, some of which might be almost catastrophic.

References

Bagnold, R. A. *Physics of blown sand and desert dunes.* New York: William Morrow, 1942.

Bascom, Willard. Ocean waves. *Scientific American,* Aug. 1959.

Bascom, Willard. *Waves and beaches.* Garden City, N.Y.: Doubleday, Anchor, 1964.

Bradley, W. C. Submarine abrasions and wave-cut platforms. *Bull. Geol. Soc. Amer.,* v. 69, 1958.

Dietz, Robert S. Wave-base, marine profile of equilibrium and wave-built terraces: a critical appraisal. *Bull. Geol. Soc. Amer.,* v. 74, 1963, p. 971.

Hack, J. T. Dunes of the western Navajo country. *Geogr. Rev.,* v. 31, 1941, p. 258.

King, C. A. M. *Beaches and coasts.* London: Edward Arnold, 1959.

Lugn, A. L. *Origin and sources of loess.* Lincoln: University of Nebraska Press, 1962.

McLellan, Hugh J. *Elements of physical oceanography.* New York: Pergamon, 1965.

McKee, E. D. Structures of dunes at White Sands National Monument, New Mexico. *Sedimentology. Vol. VII.* New York: Elsevier, 1966.

Schultz, C. B., and Frye, John C. Loess and related eolian deposits of the world. *INQUA,* v. 12, 1968, pp. 3–21.

Thorp, James, and Smith, H. T. U., eds. Pleistocene eolian deposits of the United States, Alaska, and parts of Canada (map). 1952. Sold by Geological Society of America, New York.

Motion Picture

The beach: a river of sand, Encyclopaedia Britannica, color, 16 mm, 20 min.

part two
the oceans

10

the chemistry, dynamics, and origin of the oceans

Uniqueness of the oceans

Two special conditions are needed to provide a planet with oceans: (1) a planetary body sufficiently large to have enough gravity to hold water or water vapor in an atmosphere, and (2) the right temperature. Mercury, the planet nearest the sun, has no free water at all, because it lacks sufficient gravity to hold an atmosphere or water vapor. Any gases that may have existed in its atmosphere have long since escaped into space, leaving the planet dry and lifeless. Mars, which follows the earth in order from the sun, is believed to have a very thin atmosphere, probably mostly carbon dioxide, and possibly a small amount of water. White patches come and go on its poles like a thin frost and drifts at certain seasons, and astronomers think they see a faint green at other times suggesting a sparse vegetation, but there is nothing on Mars resembling an ocean. (It looks like the sterile moon on the remarkable photographs taken by the U.S. space probes of Mariners 4, 6, and 7.) Venus, the last of the four inner planets, is large enough to hold gases and a dense layer of clouds. Spectroscopic examination, however, reveals only a trace of water in Venus's atmosphere, which is mostly carbon dioxide. The surface temperature of Venus is much higher than the boiling point of water.

The giant planets that lie far beyond the inner planets—Jupiter, Saturn, Uranus, Neptune—are too cold to have oceans, and that brings us to the second condition for the existence of oceans, *the right temperature*. Water is a liquid only between 1° C and 100° C (at atmospheric pressure), and although the giant planets contain a great deal of H_2O, it is in the form of ice. Ice comprises most of the outer layers of Jupiter and Saturn, and probably of Uranus and Neptune. Saturn's rings are believed to consist of a swarm of ice and grit particles. Even in areas that are warmed by the sun, surface temperatures on the outer planets range below $-150°$ C.

The earth's oceans, then, are unique among the sun's family of planets. They exist only because the earth has a temperature in the exceedingly narrow range within which water remains a liquid. In the known universe temperatures generally tend toward extremes, either toward the near absolute zero of interstellar space or the tens of millions of degrees found within some of the stars. Nearly all the matter of the universe is either flaming gas or frozen solid.

The earth's great oceans are the Pacific

(equal in size to the other three oceans combined), the Atlantic, the Indian, and the Arctic. These, with the associated and interconnected gulfs and mediterraneans, occupy 70.8 percent of the earth's surface and contain 330 million cubic miles of water. The earth's tallest peak, Mount Everest, 29,028 ft high, compares with the ocean's deepest abyss, the Mariana Trench, 35,800 ft deep. If all the irregularities were smoothed out both on land and under the oceans, the sea would cover the entire earth to a depth of 12,000 ft.

Chemical nature of sea water

Ionic constituents in sea water

Inasmuch as the salts in sea water are ionized, their amounts should be stated in terms of ionic concentrations. Table 10–1 lists the

TABLE 10–1 *Principal Ions in Sea Water*

IONS	G/KG	SALINITY: WEIGHT OF TOTAL IONS DISSOLVED (%)
Cations		
Sodium	10.47	30.4
Magnesium	1.28	3.7
Calcium	0.41	1.2
Potassium	0.38	1.1
Strontium	0.013	0.05
Anions		
Chloride	18.97	55.2
Sulfate	2.65	7.7
Bromide	0.065	0.2
Bicarbonate	0.14	0.4
Borate	0.027	0.08

Adapted from Hugh J. McLellan, *Elements of physical oceanography*, p. 16.

principal ions both in grams per kilogram of solution, and in percentage of salinity.

The composition of the earth's crust is similar to that of sea water, with one notable exception. There is much more chlorine in sea water than in rocks; otherwise, every element for which an adequate method of chemical analysis has been devised has been detected in sea water. Most are present in low or extremely

TABLE 10–2 *Less Abundant Elements in Sea Water*

ELEMENTS	CONCENTRATION (MG/M^3)
Fluorine	1400
Silicon	1000
Nitrogen (NO, NO^2, NH^3)	1000
Rubidium	200
Aluminum	120
Lithium	70
Phosphorus	60
Barium	54
Iron	50
Iodine	50
Arsenic	15
Copper	5
Manganese	5
Zinc	5
Selenium	4
Uranium	2
Cesium	2
Molybdenum	0.7
Cerium	0.4
Thorium	0.4
Vanadium	0.3
Yttrium	0.3
Lanthanum	0.3
Silver	0.3
Nickel	0.1
Scandium	0.04
Mercury	0.03
Gold	0.004
Radium	0.0000001

Adapted from McLellan, *Elements of physical oceanography*, p. 17.

low concentrations. Table 10–2 lists the less abundant elements.

Salinity

Normally, salinity means the amount of solids in percent contained in a given volume of solution, but in oceanographic research it is defined as the total amount of solid material in grains contained in one kilogram of sea water.

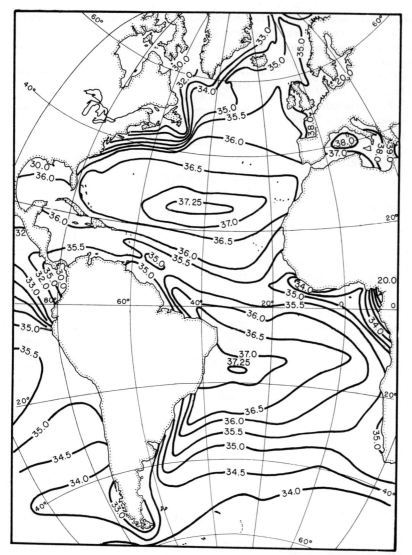

Figure 10–1 Average surface distribution of salinity (in ‰) in the Atlantic Ocean.

The units used in reporting salinity are grams per kilogram, or parts per thousand.

The normal range of oceanic salinity is 33 to 37 parts per thousand, or as generally expressed, 33‰ to 37‰. Although the oceans could hold much more salt in solution, and are thus somewhat weak solutions, it has been computed that the salt present amounts to 4.8×10^{16} tons. In terms of dry volume this would equal 21.8×10^6 cu km and would cover the area of the oceans with a layer 60 m (about 200 ft) thick. It has also been shown that, although the salinity varies widely, the relative proportion of the major ions remains almost constant. Since the proportion of the constituents of sea water remains almost constant, it was early realized that in a salinity analysis the quantity present of only one element would suffice for all. Thus, some 77 samples collected in 1884 by the British *Challenger,* of Darwin fame, were analyzed for chlorine as a measure of total salinity, and chlorine has since come to be the standard element for analysis. It is the most abundant, and a titration technique with silver nitrate, using potassium chromate as an indicator, is simple and quick. With this technique the small amounts of bromine and iodine as well as chlorine are measured. Thus *chlorinity* is defined as the total amount in grams of chlorine, bromine, and iodine contained in a kilogram of sea water, assuming the iodine and bromine to be replaced by chlorine.

Samples of sea water for analyses are collected at specific depths by means of Nansen bottles, which have become standard throughout the world. They are attached in series to a line lowered from a ship to any depth. Each bottle is open at both ends as it is lowered so that the water passes freely through it, but is capable of being closed at will. A metal messenger is allowed to slide down the line and trip the catch of the uppermost bottle. The closing of this bottle releases a similar messenger previously suspended below the uppermost bottle, and this slides down the line, closes the next bottle, and so on.

The distribution of salinity at the surface of both the Pacific and Atlantic oceans is marked by a narrow equatorial belt of about average salinity (34.5 to 35.0‰). This belt is bordered north and south by broad regions of fairly high salinities (36.5 to 37.25‰), which are due to excessive evaporation, and this in turn is due to a combination of prevailing winds (the trade winds) and to warm temperatures (see Figure 10–1). The North Atlantic region of high salinity extends from about lat 20° to 40° N, whereas the high-salinity region of the South Atlantic extends from about the equator to lat 30° S. Salinity decreases from these broad regions generally toward the polar regions. The Mediterranean Sea is mostly land-locked and is a region of fairly high evaporation; consequently it shows exceptionally high salinities (38–39‰). Salinity has been measured above 40‰ in the Red Sea. The Baltic Sea, which is fed by many rivers in a region of good precipitation, has salinities not much above 20‰. In the Gulf of Bothnia the salinities are regularly less than 10‰, and in some of the fjords less than 1‰.

The distribution of salinity with depth has been found to vary from place to place. These variations point to significant movements of the oceans and to the source of some of the ocean layers or masses. For instance, examine the five plots of salinity-depth curves of Figure 10–2. The curve for the Sargasso Sea shows a well-mixed layer to a depth of about 750 meters; in contrast, the northeast Atlantic has a thin surface layer, and a deeper high salinity layer, with a maximum at 1200 m depth. This high-salinity water is believed to come from the

Figure 10–2 Curves showing salinity at depth at five selected stations. [*After Hugh J. McLellan*, Elements of physical oceanography, *New York, Pergamon, 1965, p. 23.*]

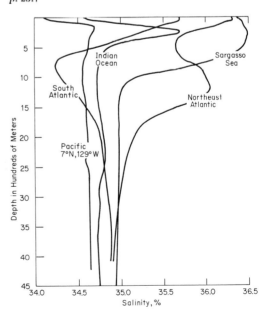

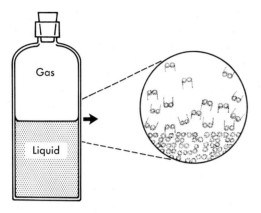

Figure 10–3 *Schematic representation of a liquid in equilibrium with its vapor in a closed container. The rate of evaporation is equal to the rate of condensation.*

Mediterranean, where excessive evaporation produces salinities greater than normal. A South Atlantic profile shows a salinity minimum at 750 m, and indicates "Antarctic Intermediate" water. The Pacific is characterized by generally lower salinity, at about 250 m. The great ocean currents that are deduced in part from the salinity profiles will be discussed presently.

Dissolved gases

Possibly the most important process involving the oceans and the atmosphere is the interchange of gases between them. One thinks immediately of the supply of water vapor to the atmosphere from the ocean, but this is only part of it. The transfer is a two-way street, with the atmosphere giving and taking gases from the oceans. The continents also participate in the exchange, and it thus becomes a complex process. The atmosphere will be discussed in Chapter 13, so our concern here is only with the dissolved gases of the oceans, their nature, distribution, and significance.

The surface of contact between the atmosphere and ocean is an interface across which gases pass in both directions. For any one gas the exchange ceases, or reaches equilibrium, when the vapor pressure of the water equals the vapor pressure of the atmosphere[1] (see Figure 10–3). The temperature of the water is the controlling factor. The gases that concern us most in oceanographic research are oxygen and carbon dioxide. Nitrogen, the most abundant gas in the atmosphere, and also abundantly dissolved in ocean water, is nearly inert and so far has not been of much interest.

Oxygen dissolved in sea water may be reported in millimeters per liter (ml/liter). This refers to the volume in millimeters that the oxygen dissolved in a liter of sea water would occupy at standard conditions, 20° C temperature and 760 mm of mercury pressure. The amount of oxygen in sea water ranges from 4.5 to 9 ml/liter and varies according to salinity and temperature. In the normal range of salinity, 27‰ to 36‰ about $\frac{1}{10}$ more of the gas is held at the lower salinity than at the higher. However, in a temperature range of −2° C to 30° C, slightly more than twice as much oxygen is dissolved at the lower temperature than at the higher (refer to Table 10–3).

TABLE 10–3 *Saturation Values of Oxygen in Sea Water (ml/l) in Contact with Normal Dry Atmosphere*

TEMP.	CHLORINITY ‰ —15 SALINITY ‰ —27.11	20‰ 36.11‰
−2° C	9.01 ml/l	8.39 ml/l
30° C	4.74 ml/l	4.46 ml/l

Taken from McLellan, *Elements of physical oceanography*, p. 20.

The oxygen concentration may either exceed or fall short of the above amounts. If there is plant growth beneath the surface of the water, the oxygen content may exceed these values. Where organic material is decaying oxygen is

---

[1] The pressure exerted by the gaseous molecules of a substance in equilibrium with the liquid molecules of a substance is known as the *vapor pressure* of a liquid.

consumed, and the values decrease. A distinct oxygen minimum at a depth of a few hundred meters occurs in most parts of the ocean, and it is believed to be due to a balance of oxygen consumption and water renewal. In basins where circulation is impeded and new water is not brought in, all oxygen may be used up and *anaerobic* (no free oxygen) *conditions* exist. Such is the situation in the lower waters of the Black Sea.

In the case of carbon dioxide, its transfer back and forth from water to atmosphere is complicated by the fact that it enters into a reaction with water, forming carbonate ions $(CO_3)^{--}$ and bicarbonate ions (HCO_3^-). The reaction is represented as follows:

$$CO_2 + H_2O \rightleftharpoons H_2CO_3 \rightleftharpoons H^+ + HCO_3^- \rightleftharpoons 2H^+ + CO_3^{--}$$

The HCO_3^-, CO_3^{--}, and H^+ ions are all present, and their relative proportions depend on the hydrogen ion concentration, called the pH of the medium. Transitions from one form to another take place in directions that tend to maintain a constant pH.

To the depth of light penetration, where plants live and photosynthesis takes place, the plants use up carbon dioxide and bicarbonate in the synthesis of hydrocarbons, and oxygen is liberated. On the other hand, respiration of animals or the oxidation of dead organic material consumes oxygen and releases carbon dioxide.

It is apparent now that the constancy of sea water composition, regardless of salinity variations, pertains to the inorganic ions and not to the gases that are constantly exchanged with the atmosphere. The chemicals that enter into biologic and geologic processes can vary greatly and as such are called *nonconservative constituents*. Examples are dissolved oxygen and carbon dioxide as well as inorganic nutrients such as nitrate, nitrite, phosphate, and silica. The concentration of these nutrients or plant foods is high in deep water and low in the surface layers where the plants grow. Where deep waters well up and mix with the surface waters, more vigorous plant growth occurs, and since the plants furnish the food for the animals, these are also well nourished and abundant.

In recent years refined analytical methods have revealed that a number of *organic* compounds exist in sea water. These must be classed as nonconservative constituents. At least nine fatty acids, which are the result of both the growth and decay of life forms, have been identified. In addition, literally hundreds of individual organic compounds, including hydrocarbons, which lead to the valuable oil accumulations of the world, have been recognized. The oceanographer is interested in the distribution of nonconservative constituents because they provide clues to the history of sea water.

Temperatures of the oceans

Of all the measurements made of the ocean waters, temperature is the most common, and of these, the surface variations were first noted and are best understood. Subsurface temperature profiles are being taken at more places and are becoming better known as a result of modern research, particularly because of keen interest in deep circulations. Ocean temperatures range from $-2\,°C$ to around $35\,°C$. The lower limit is set by the freezing of sea water, and the upper limit occurs in certain shallow tropical areas. The full range of temperatures, therefore, occurs at the surface. Below a depth of 1500 m there is little variation, ranging from $1°$ to $3°\,C$. This points up the fact that almost half of all the water in the oceans is cooler than $2°\,C$, and only a small fraction is warmer than $6°\,C$.

The recording of valid temperatures, both at the surface and at depth, is a difficult job—the procedures and equipment involved will not be described here. The results are significant, however, and in brief are as follows.

The surface temperatures of the Atlantic Ocean for August are shown in Figure 10–4. A

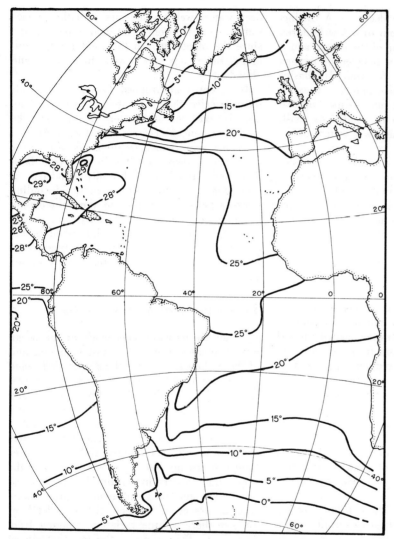

Figure 10–4 *Surface temperatures for August of the Atlantic Ocean.*

broad equatorial zone ranges from 25° C to 28° C, and from this zone northward and southward the temperatures decrease to 0° C and less. Approximately the same condition holds for the Pacific Ocean. The effect of the Gulf Stream across the North Atlantic is evident in the northward shift of the isotherms (zones of equal temperatures). Both Atlantic and Pacific oceans show higher temperatures in the summer than in the winter. Isotherms are closely spaced along the east coasts of Asia and North America, where the western boundary currents of the warm subtropical and cold boreal gyres (vortexes, or circular currents) interact. Also noteworthy is the steep temperature gradient off the Peruvian coast, where the

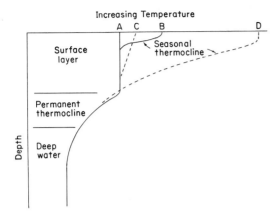

Figure 10–5 *Temperature layers in the ocean. A, extreme winter; B, after vernal warming with light winds; C, B after vigorous wind stirring; D, extreme summer condition. [After McLellan, 1965.]*

cold Humboldt Current turns westward. Periodically the north front of this current system is displaced southward, and all life in the area suffers. The marine plants, accustomed to the high CO_2 content of the cold waters, find themselves in warmer waters that are deficient in CO_2. Their mortality is tremendous, and the fishes, lacking plant food and now in an unnaturally warm environment, die in countless numbers.

The warm waters of the equatorial regions tend to drift to the polar regions, and at the same time certain north and south sea areas, because of cooling and evaporation, produce heavy waters that dive and spread widely under the oceans. One of these areas is south of Greenland and another is in the Weddell Sea of Antarctica. The cold heavy water sinks and fills the North Atlantic deep even to the South of the Atlantic. The Weddell Sea water sinks and moves northward partly under the South Atlantic, where it forces itself as a wedge under the cold North Atlantic bottom waters.

Temperature profiles for a temperate belt with pronounced winter and summer seasons are shown in Figure 10–5. Three layers are indicated:

1. A surface layer, which is usually less than 200 meters thick. It is that portion of the column that serves for temporary storage of large quantities of thermal energy from the sun. Its temperature changes in response to the exchange of energy with the atmosphere and the absorption of solar radiation.

2. A permanent thermocline layer, which lies below the surface layer and marks a rapid decrease to deep water temperatures.

3. A deep water zone, which exhibits a gradual decrease in temperature to the bottom.

In many coastal regions of polar seas a cold, low-salinity layer lies over warmer oceanic water, and in the summer this low-salinity layer develops a thin somewhat warmer upper layer, then a layer of distinct minimum temperature. Below these two thin temperature layers occur the slightly warmer waters down to the bottom.

Ocean currents

The surface waters of the Atlantic are involved in two great circling currents, the one in the North Atlantic turning clockwise, and the one in the South Atlantic turning counterclockwise as shown in Figure 10–6. There are similar but larger currents in the Pacific Ocean, and there, also, the circling in the north is clockwise and in the south counterclockwise. Each pair of turning wheels is separated by a major westerly flowing current called the Equatorial Current.

The Equatorial Current of the Atlantic is compounded of the North Equatorial Current and the South Equatorial Current between which a counter current exists toward Africa. The Pacific Equatorial Current is made up of a central westward-flowing current flanked north and south first by counter currents and then by another North Equatorial Current and South Equatorial Current.

Another circular current in the Indian Ocean adds a fifth great gyre of circulation to the oceans' ceaseless motions. Still another great current must be recognized to complete the

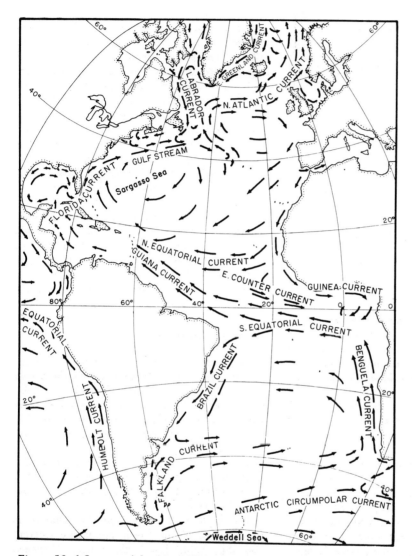

Figure 10–6 *Currents of the Atlantic Ocean.*

major circulation patterns of the oceans, and this is the Antarctic Circumpolar Current. It moves easterly around the entire continent of Antarctica approximately at lat. 60° S. Many other smaller circulations exist incident to the shore-line configurations, and several are shown in Figure 10–6. The great currents north and south of the equator probably have irregularities as well.

Benjamin Franklin was the first to recognize the Gulf Stream, which he named. He noted that American ships commonly took about two weeks' less time to cross the Atlantic than English vessels, and through a cousin, a Nantucket whaling captain, he learned that American captains were steering a more northerly course to England, taking advantage of a 3-knot current (3 nautical miles per hour). The Gulf Stream is part of a great gyre that encompasses the whole North Atlantic. The southern section of the gyre, the Equatorial Current, sweeps westward to the West Indies. Part is deflected northward past the Bahamas but most of it flows between the islands of the Lesser Antilles into the Caribbean Sea. Thence, funneling between Cuba and Yucatán, the water piles up with a head of more than 7 inches before spilling into the Gulf, then spins around and eastward, and converges through the narrow passage between Florida and Cuba to spurt into the Atlantic. Along the east coast of Florida it is now known as the Florida Current, but farther to the northeast it is called the Gulf Stream proper. Flowing past Miami it has a surface speed of about 5 knots and is 50 mi wide. Recently new survey methods at the surface, mid-depth, and bottom have indicated that the current extends in places to the bottom, more than 1500 ft deep. The submarine *Ben Franklin*, a 50-foot, 140-ton research craft, drifted 1650 miles at various depths in the Gulf Stream from Palm Springs, Florida, to a point 300 mi south of Halifax, Nova Scotia, and recorded speeds of 3 knots at depths of 1400 to 1500 ft. The Gulf Stream is thus a current with a volume a thousand or two thousand times greater than that of the Mississippi River.

Off the Grand Banks of Newfoundland the warm blue waters of the Gulf Stream collide with the cold green waters of the Labrador Current. Long fingers of the cold waters are thrust into the warm, and the mixing results in the billowing banks of fog, for which Newfoundland and the Banks are famous. The Labrador Current brings down hundreds of icebergs, but on reaching the warm waters of the Gulf Stream they melt in 10 days or less.

The northern edge of the Gulf Stream leading into the Atlantic from the North Carolina Capes is now being tracked by an airborne infrared radiometer. Flying at an altitude of 500 feet above sea level the temperature of the upper film of water can be measured to within $0.4°$ C. The flights indicate that much is to be learned not only of the change in position of the interface of the warm Gulf Stream and the cold Labrador waters but of the wave conditions on either side. On one flight 8- to 10-ft waves were noticed in the Gulf Stream and relatively calm water beyond the interface. The color difference is reported as startling, with the Gulf Stream deep blue and the Labrador waters murky gray-green. The interface is a great collector. On one flight a tremendous amount of lumber was seen strung out and held at the interface. On other flights large schools of fish including tuna and blues were noted repeatedly just inside the warm blue water. At one place a herd of some 200 sea turtles, each as large as a washtub, was spotted bucking the current swimming southward, also just inside the Gulf Stream. The interface is also an area harboring thousands of birds.

Recent observations demonstrate unsuspected large and swift currents underlying the surface currents. The westerly flowing South Equatorial Current of the Atlantic is countered by an easterly flowing undercurrent. An equatorial undercurrent has been extensively measured 4000 nautical miles in the Pacific flowing toward the Galápagos Islands.

What are the causes of the circulations as we have reviewed them? This problem absorbs more time of the oceanographers than possibly any other of the numerous aspects of the oceans. In brief, the causes are threefold: (1) the concentration of the sun's heat-energy in the equatorial regions, (2) the Coriolis force due to the earth's rotation, (3) the prevailing wind belts.

The seas near the equator receive the direct rays of the sun, and thus much more heat than the polar seas. The warming of the surface layer of the equatorial regions results in instability and motion. The water expands, and the level actually stands a few inches higher than to the north or south. This is not much, but enough to produce in a broad way a downhill

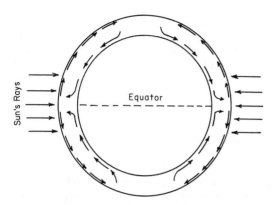

Figure 10–7 *Concentration of the sun's heat-energy in equatorial regions contributes to circulation of water.*

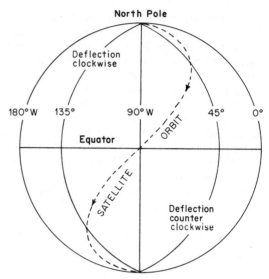

Figure 10–8 *The straight path of the satellite is apparently reflected because of the rotation of the earth. This is the Coriolis force.*

streaming toward the poles. The cold and heavier waters of the northern and southern seas sink below the warm surface water. The circulation pattern is shown in concept in Figure 10–7. Such a pattern is complicated by the sweep of other forces; nevertheless, it is a guiding concept in understanding the ocean currents. Density variations due to variable evaporation rates and temperature changes are also profoundly important in the production of currents.

The second cause of ocean currents is the Coriolis force. As the earth rotates around its axis it has a speed at the equator of 1000 miles per hour, and it tends to spin out from under its fluid oceans and atmosphere and leave them behind. Since the rotation is eastward, the waters tend to pile up on the western shores. But suppose the current direction is originally northward or southward from the equatorial regions, or southward or northward from the polar regions, what effect does the earth's spinning have? The result is that the ocean currents and the winds are deflected clockwise in the northern hemisphere and counterclockwise in the southern hemisphere. The effect is better understood by considering an earth satellite in a polar orbit (see Figure 10–8). If you were observing the satellite from some point in space, say as far out as a weather satellite, it would describe a closed circle, always oriented the same relative to the stars, and the earth would be seen rotating within the closed circle. But now, if you were located on the earth, say at the Greenwich meridian at the equator, you would observe that each succeeding pass would be farther to the west, because the earth is rotating eastward. The earth would turn completely inside this satellite circle in 24 hours. As a result, it would *appear* that there is a force that causes the satellite to be deflected. Assume the satellite takes a course due south along the zero meridian from the North Pole. The earth will have rotated by the time the satellite crosses the equator, but since it is above the earth's atmosphere and is free of the influence of rotation (not of gravity) it will cross the equator west of the zero meridian. Actually, the path of the satellite does not curve, but the path it follows across the earth curves. After crossing the equator it will appear to curve more and more to the left until it approaches the South Pole. This deflection was called the Coriolis force after the nineteenth-century French mathematician Gaspard G. de Coriolis, who first discussed it. (Some prefer to call it the Coriolis effect.)

In the case of an ocean current with a velocity of 1 knot, the deflection is computed to be 0.186 nautical miles, or about 20 percent of the distance traveled. Thus, even at low current speeds the Coriolis force is important in changing the direction of the currents.

Another analysis of the Coriolis force is shown in Figure 10–9. Here, a plate of some light material floats on water and is given an initial impulse by the wind in a northerly direction. The Coriolis force acts on the plate at a right angle, and if the plate glides without friction on the water it will be accelerated in a circular course. If the plate is retarded by friction it will slow down, spiraling toward the center. The North Atlantic gyre with its central Sargasso Sea is an example not only of circular motion but of inward spiraling motion.

Another effect of the Coriolis force on sea water is the Ekman spiral, shown in Figure 10–10. The previous figure illustrated the Coriolis effect in the horizontal directions, but in depth the direction of water movement is rotated in decreasing amounts in a spiral converging to a point, which is believed to be about 100 m below the surface. Thus a layer of water about 100 m deep is affected. In the Northern Hemisphere and the North Atlantic ocean particularly, the flow is pushed to the right toward the Sargasso Sea. Thus the Ekman spiral helps deflect the water into a central collecting region, where the waters are piled up perhaps a meter above normal sea level.

Added to the effect of the Coriolis force is another force, that of the winds. The steadiest winds on earth are at the fringes of the tropics, where the Trade Winds blow from an easterly direction diagonally toward the equator from both hemispheres to produce the Equatorial Easterlies (see Figure 15–23). The friction or pressure of the air on the water drives the surface waters unrelentingly westward. Since the atmosphere is just as much subject as is water to the sun's warming influence and the deflecting influence of the Coriolis force, the winds circle too.

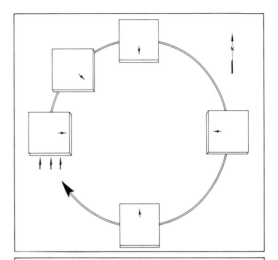

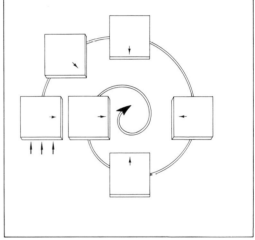

Figure 10–9 Upper diagram represents Coriolis acceleration on slabs floating in water if not affected by friction. The slab is set in motion by an impulse toward the north (heavy arrows) and then the Coriolis force accelerates it in a clockwise circle in the Northern Hemisphere (small arrows show the direction of the Coriolis force). In lower diagram, friction, if present, causes the slab to slow down and move in a converging spiral. [After R. W. Stewart, The atmosphere and the ocean, Scientific American, Sept. 1969.]

Solar heat and evaporation plus the Coriolis force probably start and accelerate water movement in the oceans, but it is evident that the winds provide the most significant force in

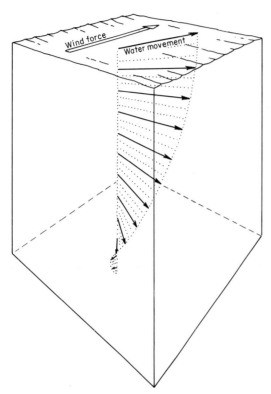

Figure 10-10 Ekman spiral of water movement in the Northern Hemisphere. The average water movement is at right angles to the wind direction, and the bottom of the Ekman layer is about 100 meters below the surface. [After R. W. Stewart, 1969.]

starting and driving the water currents. As shown in Figure 10-10, the initial water movement at the surface departs only a few degrees from the wind direction, to the right in the Northern Hemisphere and to the left in the Southern, but the *average* departure from the wind in the Ekman spiral is 90 degrees. The Ekman spiral layer is believed to adjust to surface wind changes within a day or so. The Ekman spiral also drives the water generally downward, and thus at depth water is forced to move horizontally. The downward and horizontal movements of the water also participate in the general rotary circulation, and as the water is forced horizontally outward its radius of gyration increases and so its rate of rotation must slow. This means that the rotation no longer fits the path or curve that it should have, because for every latitude there is a different radius of circulation, greater at the equator and becoming smaller northward and southward. As a generalization there is a water movement from the midlatitudes to the equator below the Ekman layer, where its circulation fits the radius of rotation of the earth better. In the North Atlantic gyre this sub-Ekman layer current flows westerly in the equatorial regions and turns northerly upon contact with the North American shoreline, where it is helped in this direction by the Coriolis force. There it runs into a strong current, the Gulf Stream, which is in effect the return flow of the water squeezed downward and outward from the North Atlantic gyre.

The combination of the three great forces—solar heat, Coriolis force, and the prevailing winds—produces the major circulations of the oceans, as shown in gross outline for the Atlantic in Figure 10-11.

Deep circulation and water masses in the Atlantic Ocean

Figure 10-12 is a longitudinal section of the Atlantic Ocean from south (left) to north, and is a composite schematic interpretation on the basis of temperature, salinity, and oxygen analysis. At the far south is the shelf water of the Weddell Sea at near-freezing temperature. It mixes with the Antarctic Circumpolar Current water ($AACP$) to form the heaviest waters in the oceans, and sinks along the continental slope to form the Antarctic Bottom Water (AAB). Its characteristics are temperatures of less than 0° C, salinity over 34.6‰, and dissolved oxygen over 5 ml/liter. This water wedges into the deep parts of the South Atlantic and has been detected as far as the north coast of Brazil. The great mass of deep water is believed to originate around Greenland, and perhaps in the Norwegian Sea, where evaporation and winter cooling contribute to its characteris-

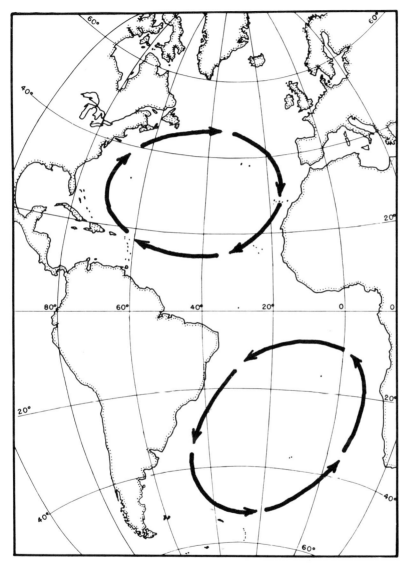

Figure 10–11 *Three forces—solar heat, Coriolis force, prevailing winds—combine to produce circulation patterns in the Atlantic Ocean.*

tics. It is called the North Atlantic Deep and Bottom Water (*NAD&B*) and has a salinity of 34.9‰, temperatures from 2° to 4° C, and oxygen content greater than 6 ml/liter. It mixes with the Antarctic Bottom Water and probably spills into the South Pacific and Indian oceans to form the deep Common Water.

At lat. 50° to 60° S. the water becomes somewhat lighter and warmer than that of the Weddell Sea. It is labeled *AAI* (for Antarctic Intermediate Water) on Figure 10–12 and has a salinity of 34.2‰, temperature of 4° C, dissolved oxygen content greater than 6 ml/liter, and flows as a broad sheet to the north at an

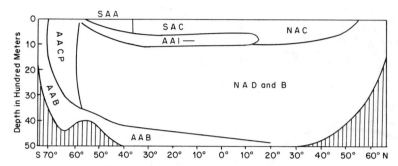

Figure 10-12 Idealized longitudinal section of the Atlantic Ocean with water masses recognized on the basis of temperature, salinity, and oxygen analysis. Bottom irregularities not shown. [Taken from McLellan, 1965.]

intermediate depth of 500 to 1000 m. This is detected well north of the equator, even in the Gulf of Mexico.

In the central part of the North Atlantic is the Central Atlantic Water (NAC), which has a salinity of 36.5‰ to 36.1‰ and a temperature of 18° C to 8° C. It forms a sheet from the surface to a depth of about 1000 and is characteristic of the subtropical gyre of the Sargasso Sea. The corresponding water in the South Atlantic is labeled SAC. Not depicted is the wedge of heavy Mediterranean water, previously described, that spreads out into the Atlantic from the Strait of Gibraltar at a depth of 1500 m. It is characterized by a salinity of 36.5‰ and a temperature of 12° C.

Inside the North Atlantic gyre is the strange Sargasso Sea, measuring 1000 miles wide and 2000 miles long. It is littered with patches of seaweed, which terrified the sailors of the old sailing ships. Mariners long believed they were running aground or that the weeds could trap the unwary voyager. The impressive show of the giant algae is, indeed, misleading. Modern expeditions have discovered that the waters are not only deep but leaner in plankton than any other known body of surface sea water, and thus scarcer in all other life that depends on the plankton for food. The Sargasso is a warm, blue, but rather barren sea except for the abundance of floating brown seaweed. It has been concluded that because of the rotating currents and the Trade Winds little new water enters the system of the Sargasso Sea, and because of the high surface temperatures no cold water wells upward from the deeps. Thus the sea lacks the necessary nutrients to support an abundance of life.

Tides

In addition to tides caused by lunar gravitational attraction there are those caused by solar attraction. Lunar tides are two to three times as high as those caused by the sun. The lunar tide ebbs (falls) and flows (rises) twice in each lunar day, which is 24 hr and 51 min long (see lunar orbit period in Chapter 1). The two daily tides are caused by two tidal bulges, one on the side of the earth facing the moon and one on the opposite side. The bulge on the opposite side is due to the weaker gravitational attraction of the moon there and the rotation of the earth. There are times when the moon and the sun pull in the same direction and produce especially high tides, known as the *spring tides*. When their attractions are at right angles, low tides result, which are known as *neap tides*. The relation of sun, earth, and moon with the resulting tides are shown in Figure 10-13.

Unlike the winds that disturb and agitate the sea's top layer, the tides move the entire ocean. The gravitational pull of the moon and the sun

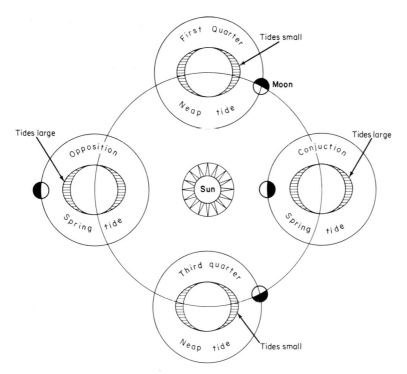

Figure 10–13 *Gravitational attractions of the sun and the moon affect the rotating earth and produce various types of tides.*

also affects the solid earth and the atmosphere. Where there is a 10-ft tide in the water the crust rises about 6 in., and the atmospheric blanket bulges out several miles.

Tides are a complex phenomenon because they vary latitudinally and, particularly, in each separate basin. The activity in certain estuaries, like the Bay of Fundy, has been described earlier. When tides of the moon and sun coincide, together with storm waves, frightful damage may result along the coast. The tide at Galveston, Texas, in 1900 coincided with an incoming hurricane; the sea rose 15 ft and nearly 6000 people were drowned.

Earthquakes that disturb the ocean floor suddenly set up low, broad waves that traverse the oceans for thousands of miles, at times even around the world. Ships at sea may not be disturbed by these low waves, yet they move at the incredible average speed of 450 mph. The waves are separated by about 15 min, and the first is not necessarily the highest. The very broad waves rise to overwhelming heights when they run into shallow water along a sea coast and they may smash the coast with pulverizing force. They are properly called *tsunamis*, but popularly they are known as tidal waves. They really are not related to tides, but to earthquakes and volcanic eruptions. In 1883 the explosion of the East Indian volcano Krakatoa and the resulting tsunamis killed 36,380 people as the waves overwhelmed village after village on neighboring islands. They went all the way around the world, leaving their marks on tide gauges in many places, particularly in the English Channel. Tsunamis are common in the

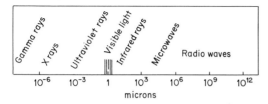

Figure 10–14 *Relative wave-length distribution.*

Pacific Ocean, originating in Chile, the Aleutian Islands, and the volcanic arcs of the western Pacific. With waves traveling at 450 mph not much time is available for people in coastal towns and cities to prepare or evacuate. In 1946 the U.S. Coast and Geodetic Survey set up a network of stations to flash warnings of undersea earthquakes that may send tsunamis racing toward one shore or another. Other Pacific nations are now cooperating.

Heat budget

Solar radiation

It is clear that the moon and Mars receive abundant energy from the sun, but lacking an atmosphere and extensive oceans, such as those of the earth, their surfaces experience wide ranges of temperature, whereas the earth's surface has relatively mild temperatures. The temperatures of the earth's land, oceans, and atmosphere provide hospitable media for an abundance of life. The oceans are tied inseparably to the atmosphere in receiving, retaining, and distributing the energy of the sun, but the relations are complicated and difficult to reduce to simple calculations, such as we would desire in formulating the heat budget of the earth or any part of it.

The energy that we receive from the sun is in the form of electromagnetic radiation, or waves of different lengths that travel with the speed of light (186,000 m per sec). When electromagnetic waves are arranged according to their wave lengths they form a continuous spectrum. The shortwave end of the spectrum begins with waves known as cosmic waves, followed by gamma rays, X rays, and ultraviolet rays. Next come the rays that comprise visible light, the light rays, and still longer are the infrared rays. Finally comes the broad band of radio waves. The diagram in Figure 10–14 shows the relative wave length distribution. The infrared waves are heat waves and are emitted by warm or hot bodies. The sun's rays include not only the visible light rays but extend throughout the entire spectrum.

The radiant energy from the sun passes through the vacuum of space, while light waves pass through air, water, and glass, as well as space, X rays pass also through flesh and other substances opaque to light rays. In either case the rays are said to be *transmitted*. As we shall see, most substances exhibit selective transmission.

The portion of the radiant energy that enters a substance but is not transmitted is *absorbed*. It is changed into some other form of energy, often into heat, but sometimes into the energy used in evaporation or in chemical changes. Selective transmission implies selective absorption, because the wave lengths not selected for transmission are absorbed.

Some of the waves reaching a material may be *reflected*. They are turned back without entering the substance, and the only change is the direction of propagation of the waves. Reflection may be regular as from a mirror, or diffuse, as from the surface of the ground. That part of the sun's radiation that reaches the earth's surface is called *insolation*.

Somewhat less than 20 percent of the sun's radiation is absorbed by the atmosphere and its clouds, and the remainder is stored temporarily by the earth and its oceans. The release of this temporarily stored energy powers the circulation of the atmosphere and oceans, which redistributes the energy and maintains the climatic conditions that we observe. The capacity to absorb and hold energy, at least for a time, makes them most important in the global energy system, or in what we call the *heat budget*.

Elements of the heat budget

Figure 10–15 illustrates three conditions of the earth's receipt of solar energy, and what happens to it.

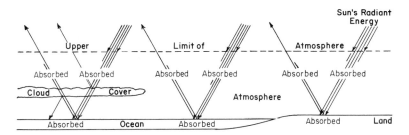

Figure 10–15 Three conditions in the earth's receipt of solar energy.

1. At the right side is radiation passing through the clear atmosphere where part of it is absorbed, the remainder reaches the ground, where it is partly absorbed and partly reflected, and in reflected rays it is further partly absorbed, but finally a fraction is lost into space;

2. In the center, radiation passing through the clear atmosphere where part of it is absorbed, then striking the ocean surface where part of it is absorbed, but part reflected. The reflected rays are partly absorbed by the atmosphere but a small part is transmitted into space;

3. At the left, solar radiation through an atmosphere with a cloud cover—the clouds selectively absorb somewhat more energy than the clear atmosphere, and in turn do the same to the reflected rays (in addition they reflect some of the rays back to sea or to the land).

Solar radiation at the outer boundary of the atmosphere, measured on a plane normal to the sun, is about 2 calories/cm²/min. But the amount that reaches the earth's surface and the oceans and is absorbed depends on (1) the angle of the sun, which depends on the latitude and the season, (2) the amount of absorption in the atmosphere, which depends on the variables of water vapor, dust, and a cloud cover, (3) the amount immediately reflected from the surface of the ocean. The amount reflected depends mostly on the sun's angle and the state of the sea.

The sea surface, by virtue of its temperature, emits long-wave radiation to the atmosphere. Most of this long-wave radiation is absorbed efficiently by carbon dioxide and water vapor in the atmosphere, but in the particular band of wave lengths from 8 to 14 microns, which is within the long wave lengths emitted by the sea water, the air is almost transparent. This is known as Simpson's window. The earth is heated by short-wave solar radiation that passes effectively through the atmosphere, and as the sea is heated builds up to an emission of the longer wave lengths. In fact, the wave lengths increase to the band of the window, and then a balance of incoming and outgoing radiant energy is effected, and the skin temperature of water is stabilized.

Liquid water in clouds is opaque to terrestrial (ocean) radiation, including the window range, and since the clouds radiate back to sea, they decrease the amount of radiation that gets back into space. At any rate, the difficulty in computing the many variables of the insolation that reaches sea water must be apparent. Table 10–4 shows the amount in cal/cm²/min that

TABLE 10–4 *Amount of Radiation that Reaches the Sea*

LATITUDE	JAN.	JULY
60° N	0.002	0.267
30° N	0.146	0.301
0°	0.239	0.188
30° S	0.380	0.145
60° S	0.213	0.003

SOURCE: H. H. Kimball, Amount of solar radiation that reaches the earth, *Monthly Weather Rev.*, v. 56, 1928, p. 373.

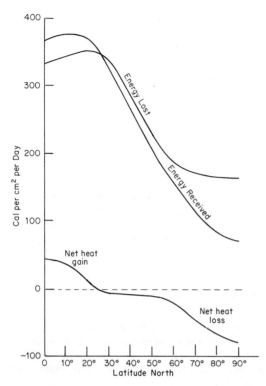

Figure 10–16 *Heat budget for oceans of the northern hemisphere.* [After Defant, 1961.]

is believed to reach the sea, which should be compared with the 2 cal/cm²/min that reach the outer limit of the atmosphere. The observations and calculations vary at different longitudes.

The difference between the incoming radiation from the sun and the effective back radiation from the surface is called the radiation surplus. The relative stability of the temperature on a global basis leads to the belief that the total surplus is zero. The surplus is positive in the tropics where the sun's angle is high, and the daylight hours long, and negative where the sun is low and the days short. At about lat 38° the radiation surplus is zero on an annual basis.

As a result of this unequal heating of polar and tropical regions heat has to be transported poleward by the oceans and by the atmosphere.

The currents of the oceans account for only about 10 percent of this transport. The rest is by winds.

The greatest factor in heat loss from the oceans is evaporation. About half of the radiant energy absorbed by the oceans is returned to the atmosphere by evaporation. Obviously, the tropics, which have a large radiation surplus and a great water area, have high evaporation rates. Evaporation is augmented by the winds, and thus, in the temperate zone of the stormy westerlies evaporation is also high.

The heat budget, in all its complications, can be summarized as follows.

Energy received = energy lost + energy stored

Energy lost is made up of radiation from the sea surface, heat lost by evaporation, and heat conducted across the water–air interface. Energy stored is in the form of heat associated with the rise of water temperature. The average annual latitudinal variations in the heat budget of the oceans of the Northern Hemisphere have been evaluated and, with certain deletions, are reproduced in Figure 10-16.

Origin of the oceans

Problems and aspects

The puzzle of origin of the oceans is inseparably connected with the origin of the atmosphere, and with the oceans must be included all the earth's waters. Any lasting solution will hinge on evidence from a surprising range of scientific disciplines. Foremost are the fields of geology, geochemistry, meteorology, oceanography, and hydrology, but indirectly also must be listed astronomy, seismology, nuclear physics, biochemistry, and others. Theories of the early development of the hydrosphere and the atmosphere certainly are hypothetical, but the history of the oceans, at least from about 600 million years ago to the present, is known substantially. From the geologic record it is clear that the oceans have existed this long, and with them an abundance of life.

The seaways were the receptacle of voluminous sediments which provide us with the uncontestable fossil record of past marine conditions. But how long before this were the oceans in existence?

Hypotheses of the origin of the primeval oceans and atmosphere fall into two categories: (1) all air and water of the earth are residual from a dense primitive atmosphere that once enveloped a molten earth; (2) air and water have accumulated at the earth's surface by leakage from the interior. These hypotheses will be briefly reviewed.

Another problem is the origin of the ocean basins. Have they always been in major outline as they are now, or have material changes taken place in the course of geologic time? This problem is considered in Chapter 11.

Constancy of salinity and composition

The sedimentary rocks of the past 600 million years bear evidence of deposition in extensive bodies of water, and the fossils entombed in them, because of their affinities with the tens of thousands of living marine species, indicate that the waters were marine. Zoologists know that many varieties of animal life, particularly the invertebrates, inhabit only the saline ocean waters, and have never taken to fresh-water lakes or rivers; thus it is generally not difficult to identify marine fossil forms. Not only have hundreds of thousands of marine animal species been identified and named from the fossil record, but representatives of all the major phyla are present. Already in the early Paleozoic era all of the phyla except the vertebrates existed, and they all lived in the oceans. In fact some of them have continued with little change to the present. Considering the many multicellular and complex invertebrates (sponges, corals, crustaceans, gastropods, pelecypods, cephalopods, and brachiopods) that we find in the modern marine environments as well as in the fossil record, it is difficult to believe that these environments were much different 600 million years ago. We can thus conclude that the salinity and chemical composition of the oceans were little different then than they are now. We may also conclude that the oxygen and carbon dioxide content of the oceans and the atmosphere varied within the same range as now. The only essential difference is that there is no evidence of cold polar seas in the early Paleozoic. All the waters in which plants and animals lived were fairly warm, indicating warm, equable climates everywhere.

Water and air from within the earth

The question of the original source of the water, air, and salts of the oceans hinges ultimately on the origin of the sun and planets, and this takes us into the realm of astronomy. A number of hypotheses for the origin of the solar system have been proposed. The most acceptable one at present begins with a nebula like the Lagoon Nebula of Sagittarius, which is a vast swirling cloud of gas and dust, the so-called cosmic material. Over a long period of time the cosmic material of the cloud was pulled towards its center by gravitational attraction, and there the growing ball of gas and dust became the sun. A number of satellitic nuclei developed and grew by their own gravitational attraction, among them the earth. But the sun has over 99 percent of the mass of the solar system, and eventually under its own gravitational field a heat build-up resulted until it began to radiate electromagnetic waves. Whether the earth passed through a high-temperature stage and a molten condition is problematical. Harold C. Urey, who proposed the theory, further concludes that after a central metallic core separated from the mantle, gradual cooling began, with the mantle solidifying from the bottom upward until finally the crust firmed into the first solid primeval surface. This was about 4.6 billion years ago and was the beginning of the Archean era.

As stated previously, two hypotheses of the origin of the atmosphere are current—that the earth in a molten condition had a dense envelope of gas, or that the earth slowly released from within, through geologic time, gases and solutions to form the atmosphere and hydrosphere. The problem was ably attacked by the American scientist William W. Rubey,[2] who compared the quantities of rocks that have been weathered and decomposed with their products that have been transported to the sea (the water-soluble constituents and sediments) during geologic time, on the theory that the major rock-forming elements—silicon (Si), aluminum (Al), iron (Fe), calcium (Ca), magnesium (Mg), sodium (Na), potassium (K) in sedimentary rocks and all the dissolved cations (Tables 10–1 and 10–2) in sea water—have been derived from the weathering of earlier rocks. But weathering is *not* an adequate source of other constituents, namely, water (H_2O), carbon dioxide (CO_2), chlorine (Cl), nitrogen (N), and sulfur (S), all of which are much too abundant in the present atmosphere and oceans and in ancient sedimentary rocks to be accounted for solely by rock weathering. Some other sources are required. Rubey calls the products of these sources "excess volatiles" (see Table 10–5).

TABLE 10–5 "*Excess Volatiles*" *Present in Atmosphere, Hydrosphere, and Buried Sedimentary Rocks* (*units are in* 10^{20} *grams*)

H_2O	16,600
Total C as CO_2	910
Cl	300
N	42
H	10
B, Br, A, F, etc	4

In terms of the first hypothesis, these excess volatiles are considered part of the hot and dense atmosphere which later cooled. The water vapor condensed and formed the primitive ocean, and the atmosphere is that part which remained gaseous. In terms of the second hypothesis of degassing of the interior of the earth, two variants may be recognized: (1) the water and other volatiles emerged within some brief period of time very early in the earth's history, (2) these volatiles have been escaping gradually and at about the same rate throughout geologic time.

Now, if all the excess volatiles were present at one time in an early ocean and atmosphere, regardless of whether they were inherited from a primitive atmosphere or erupted suddenly from the interior, the predominant gases would be CO_2, N_2, and H_2S. The ocean water and the rain would be intensely acid, and, in contact with bare rock, they would vigorously decompose it. In so doing some of the cations or the rocks would dissolve, and the acidity of the water would decrease. When the right hydrogen ion concentration was reached limestone ($CaCO_3$), pyrite (FeS_2), and other compounds would be precipitated on the sea floor. Under the postulated conditions the atmosphere would have been made up largely of CO_2 at the outset, but as carbonates continued to precipitate from the ocean more and more of the original CO_2 would leave the atmosphere. Eventually the concentration of this gas would be decreased to the point at which the most primitive forms of life might conceivably have existed.

However, the amount of rock that would have to be decomposed under this concept before life could begin would be very large; in fact, larger than the amount decomposed in all geologic time before and after the advent of life. Also an amount of Na equal to that now dissolved in sea water would have been contributed to the oceans before life could begin, and thus no room is left for the weathering of rocks and the contribution of Na in all the time since (approximately the Proterozoic, Paleozoic, Mesozoic, and Cenozoic eras). Furthermore the Archean Era should have been a time of excessive $CaCO_3$ precipitation, and this is contrary to geologic findings. Thus the chemical evidence does not bear out the postulate of an original gaseous envelope with the excess volatiles that must have been in it.

If the excess volatiles had accumulated gradually, according to the second hypothesis, the

[2] Development of the hydrosphere and atmosphere, in *Study of the earth*, J. F. White, ed., Englewood Cliffs, N.J., Prentice-Hall, 1962, p. 363.

results would be significantly different. Almost from the beginning of ocean accumulation from the emerging volatiles, the CO_2 content would not have exceeded the tolerance limit of primitive life and $CaCO_3$ would start to be deposited. And as CO_2 continued to be gradually released into the system at a relatively constant rate with rock decomposition proceeding, $CaCO_3$ precipitation would continue at a more or less uniform rate. All this means that the second hypothesis does not lead us to expect embarrassingly large quantities of rocks decomposed, Na dissolved, and $CaCO_3$ deposited in early Archean time, as we were forced to believe by the first hypothesis. The geologic evidence thus distinctly favors the postulate that the atmosphere and hydrosphere developed gradually over the ages.

Perhaps you have been wondering how the volatiles emerged from the earth, how the interior is degassed? Well, we are very familiar with the phenomenon in the upper crust, at least, because here it is manifest in many phases of volcanic activity. The very materials that Rubey calls the excess volatiles are erupted in large amounts with the molten rock. See Chapter 2.

Certain observations, partly by other workers cause Preston E. Cloud of the U.S. Geological Survey to raise questions about the time in earth history that the atmosphere and hydrosphere originated and what changes they have undergone.[3] Judging from the abundance of the noble or inert gases, helium, neon, argon, krypton and xenon, of the sun and the planets of the solar system, the earth's atmosphere should have much more of these elements than it now contains. Either the earth originated in some way without an original atmosphere, or the atmosphere was lost mainly by a crustal melting episode and then began accumulating again by the continued degassing of the interior, as Rubey has proposed.

In addition to the above observation, we are left with a gap of about one billion years, about 4.5 billion to 3.5 billion years ago, without

[3] Atmospheric and hydrospheric evolution on the primitive earth, *Science*, v. 160, 729–736, 1968.

knowledge of the earth. Isotope data from meteorites and uranium ratios in lead have led to the proposal that the solar system and the earth had their beginning about 4.6 billion years ago. Isotope data have also shown that the oldest rocks of the continents are from 3.4 to 2.7 billion years old. It is not particularly the gap in the record but the recognition that these oldest rocks were once sediments laid down in water that concerns us. The earth had a substantial hydrosphere, and undoubtedly an atmosphere by 3 billion years ago, but how did it evolve? What was it like at this time? Certain minerals in about 2 billion year old sedimentary rocks, such as pyrite (Fe_2S) and uraninite (a black mineral mostly of uranium) indicate by their presence that the atmosphere could not have contained much oxygen; also, the rarity of carbonate rocks (Ca and Mg carbonates) and the great abundance of chert or silica in these ancient sediments indicate that there could have been little ammonia (NH_3) in those times. Ammonia would have raised the pH of the water to favor the precipitation of the carbonates, but a low pH, on the contrary, favors the precipitation of much silica. We need only turn to the cherty iron ores of the Lake Superior region of this age to realize the abundance of water-precipitated silica. Thus, the early atmosphere would seem to have been composed of the gases that were occluded in igneous rocks (once molten) or that reached the surface as components of volcanic eruptions and hot springs. These are mainly H_2O, CO_2, CO, N_2, SO_2, and hydrochloric acid (HCl).

But whence came the oxygen of the atmosphere and oceans? Degassing of the interior does not produce oxygen, and in the preceding discussion it was concluded that the earth was left with no atmosphere, let alone one containing oxygen. Oxygen is most necessary because it is essential for life, and thus the origin of oxygen in the atmosphere is a most basic question confronting planetary scientists today. It appears that there are only two oxygen-producing mechanisms possible; (1) oxygen resulting from the photodissociation of water vapor caused by ultraviolet rays, in which case the hydrogen produced would escape into space and the oxygen atoms would combine into O_2; (2) oxygen produced by photosynthesis. This second mechanism requires an abundance of highly evolved plant life. The present-day level of oxygen is the result of a balance between photosynthetically-produced oxygen and oxygen consumed by respiration, decay, fires, and inorganic oxidation.

The paucity of oxygen in the early atmosphere produced by degassing of the interior is a problem in any theory about the origin of life, because oxygen is of vital importance to most life forms. Preston Cloud postulates that life arose in the early earth atmosphere and oceans as follows. It was presumably anaerobic and dependent on external food sources; thus it could not have lasted long enough to give rise to the evolutionary succession. Life may have started many times unsuccessfully, but finally one start endured, which was possibly a kind of organism that manufactured its own substance and thus probably its own food by photosynthesis. It is also reasoned that oxygen-mediating enzymes arose simultaneously, otherwise the oxygen-releasing green plant would have burned up. The acceptor for the oxygen is believed to have been ferrous iron that took up the oxygen with the conversion of the ferrous to ferric iron. This resulted in the build-up of the tremendous silicious iron-ore deposits in the various continents. The oxygen presumably became locked up in these biochemical sediments and at first did not appear in the atmosphere except in small quantities. Eventually the green-plant photosynthesizers, equipped with very efficient enzymes, could spread widely through the seas, and the oxygen of the atmosphere began to build up. With an excess of free oxygen, the stage was set for the evolution of higher life forms.

The first possible method suggested, namely by dissociation of water vapor by ultraviolet radiation, was believed until lately to be capable of yielding only $1/1000$ as much oxygen as is now present, and this would hardly suffice to nourish life. But now it has been calculated that oxygen may have reached an appreciable fraction of that now present even in the absence of biological activity.

A consideration of the evolution of strontium and lead isotopes in rocks of the earth's crust has led R. E. Armstrong of Yale University to the conclusion that the oceans, and the continents, have remained nearly constant in bulk volume at least for the past 2.5 billion years. This supports the biological considerations of Cloud, namely, that the oceans in which voluminous sediments accumulated were in existence 2.7 to 3.4 billion years ago.

References

Brinkmann, R. T. Dissociation of water vapor and the evolution of oxygen in the terrestrial atmosphere. *Jour. Geophys. Res.*, v. 74, 1969, p. 5355.
Cloud, Preston E., Jr. Atmospheric and hydrospheric evolution on the primitive earth. *Science*, May 17, 1968.
Defant, A., *Physical oceanography*, v. 1, New York, Pergamon, 1961.
Engle, Leonard. *The sea.* New York: Time Inc., Life Nature Library, 1961.
Leopold, Luna B., and Davis, Kenneth S. *Water.* New York: Time Inc., Life Nature Library, 1966.
McLellan, Hugh J. *Elements of physical oceanography.* New York: Pergamon, 1965.
Mero, John L. *The mineral resources of the sea.* New York: Elsevier, 1965.
Rubey, William W. Development of the hydrosphere and atmosphere, with special reference to probable composition of the early atmosphere. In *Study of the earth*, ed. J. F. White, Englewood Cliffs, N.J.: Prentice-Hall, 1962.
Stewart, R. W. The atmosphere and the ocean. *Scientific American*, Sept. 1969.
Urey, Harold C. The origin of the earth. In *Study of the earth*, ed. J. F. White, Englewood Cliffs, N.J.: Prentice-Hall, 1962.

11
topography and sediments of the ocean floor

Research on the ocean floor was sporadic and of little consequence until 1923. A few soundings by wire line had been made in the several oceans, and a few samples had been dredged from the deep bottom sediments.

The perfection of the bathometer, or continuous depth sounder, in the years 1923 to 1930 started a surge in oceanographic research that is at its peak today.[1] Amazing detail of the topography is recorded in all depths of water, and the profiles are accurately positioned at great distances from land by means of radio devices. Topographic features have been discovered by the thousands, and researchers seeking to explore them soon perfected a number of new and remarkable instruments for that purpose (see Figure 11-1). Core barrels were built that procure an undisturbed core of the deep-sea sediments, 75 ft long under favorable circumstances, and this in places is a 10-million-year-long record. Compression waves, artificially excited in the water by dynamite explosions, penetrate into the bottom sediments and even into deeper crustal layers. They are reflected and refracted and return to the surface to be recorded for study. The difference between the oceanic crust and the continental crust was largely determined by this method. It also permitted recording of the layering of sediments and rocks in the oceanic crust. Gravity-recording meters for use at sea were perfected, and some unexpected large anomalies in the earth's gravity field were uncovered. The anomalies are now fairly well understood, but only after seismic data were related to them.

Magnetometers have been trailed behind ships over hundreds of thousands of miles, and the results were contoured to produce some surprising patterns in the earth's magnetic field. What do they mean?

The rate of heat flow or loss from the interior may now be measured by lowering temperature measuring instruments into the bottom muds. We have thus learned that there are zones on the ocean floors where more heat is escaping from the interior than at others, and these zones are now being related to seismic, gravimetric, magnetic, and volcanic data for an explanation. The major inquiries pry into such problems as, Why do ocean basins and continents exist? Have the continents really shifted about? What causes the great belt of earthquakes, volcanoes, and crustal deformation around the Pacific

---

[1] See Willard Bascom, Technology and the ocean, *Scientific American*, Sept. 1969.

Figure 11–1 *The* Vema, *oceanic research vessel of the Lamont Geological Laboratory of Columbia University, with which many pioneering explorations were made, uses instrumentation and techniques developed by the scientists of this institution.* [*Courtesy of Marie Tharp and Lamont Geological Laboratory.*]

Ocean? In what ways and why is the Atlantic basin different from the Pacific? In the present chapter the ocean-floor topography and the sediments will be treated.

Major geomorphic divisions

Three extensive geomorphic divisions of the ocean floors are recognized: the *continental shelves*, the *ocean-basin floors* or *abyssal plains*, and the *midoceanic ridges*. These divisions are particularly clear in the Atlantic, where each accounts for about one-third of the area (see Figure 11–2). Projecting upward from each are the conspicuous seamounts, some of which rise even to the surface and appear as islands. The seamounts are numerous and widespread throughout the oceans. The shelves are trenched by deep canyons whose origin has provoked much thought. Deep trenches and associated volcanic island arcs represent another fascinating geomorphic division of ocean floors.

Continental shelves

The typical form of a continental shelf is shown in Figure 11–3. It has a gently sloping upper "tread," known as the *shelf* proper, a steep slope down into deep water, the *continental slope*, and a gentle slope at the foot, the *continental rise*, that leads to the *abyssal plain*. There are many variations in relative width, slope, and depth of these divisions, but it is common for the shelves to grade out to a depth of about 100 fathoms (600 ft) and then break into the steep slope, even though some shelves are several hundred miles broad and others only a few miles. The varying width can be seen plainly on almost all maps of the oceans and continents. The shelf proper and continental slope are fairly uniform and relatively smooth-surfaced, but the continental rise is likely to be irregular and studded with seamounts.

The continental shelves are continuous with, and part of, the emerged coastal plain. The present position of the shoreline, which marks the beginning of the continental shelf, is a transient feature that depends closely on changing sea levels. During the past ice age, the sea level fluctuated up and down about 400 ft, and the shoreline moved in and out, in some places hundreds of miles.

Submarine canyons and fans

The continental shelves are dissected by many valleys, called *submarine canyons*. They are of two general classes, those which furrow the shelf slope and those which cut the shelf and extend in places almost to the present shore. These latter are the typical, deep, submarine canyons, and their origin has been a subject of much controversy. In Figure 11–4, the Hudson submarine canyon is clearly related to the Hudson River and, at least down to a depth of 300 ft, it may have been partially eroded as the shelf was emerging during the last glacial stage. A submarine canyon like the Monterey (Figure 11–5), which extends seaward from Monterey

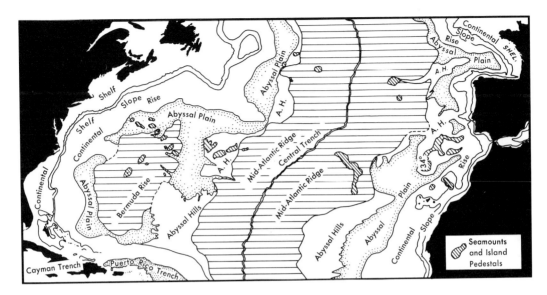

Figure 11-2 *Physiographic provinces of the North Atlantic Ocean. [After Bruce C. Heezen et al., The floors of the oceans, 1959, p. 20, pl. 20.]*

Bay on the California coast, does not have a complementary river adjacent on land.

The large submarine canyons have continuous downhill longitudinal profiles, V-shaped transverse profiles, and tributaries much like subaerial stream valleys. It is, therefore, little wonder that some pioneer researchers proposed that the sea level was once thousands of feet lower than now and that the canyons were cut by subaerial rivers. With echo sounders tracing some of them to 10,000 ft and more in depth, this proposal became untenable. No one could conceive of the oceans shrinking to such an extent or that so much water could be piled on continental ice caps during a glacial stage. Some suggested that certain coasts were locally elevated thousands of feet and dissected, that they then subsided, and that deep canyons became drowned. But investigation of the geologic history of the adjacent land did not bear out this idea. It must also be borne in mind that sampling from the walls of the submarine canyons showed they had been cut through partially and well lithified sediments, and in places even through granitic rocks, so that some energetic abrasive process must be active.

The present most widely accepted theory for the origin of submarine canyons is still not entirely satisfying, probably because we do not know enough about the process proposed. According to this theory sediment-laden and hence heavy currents, called turbidity currents, pour down the bottoms of the canyons at appreciable velocities and affect the erosion. These currents may be vividly demonstrated in laboratory tanks and artificial canals. They are heavier than clear water because of the rock and mineral particles they carry, and they flow along the bottom even at low gradients attended by little mixing with the surrounding clear water. They most probably can move fair-sized par-

Figure 11-3 *Common divisions of the continental shelf. Slopes are exaggerated.*

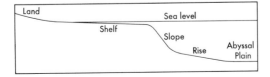

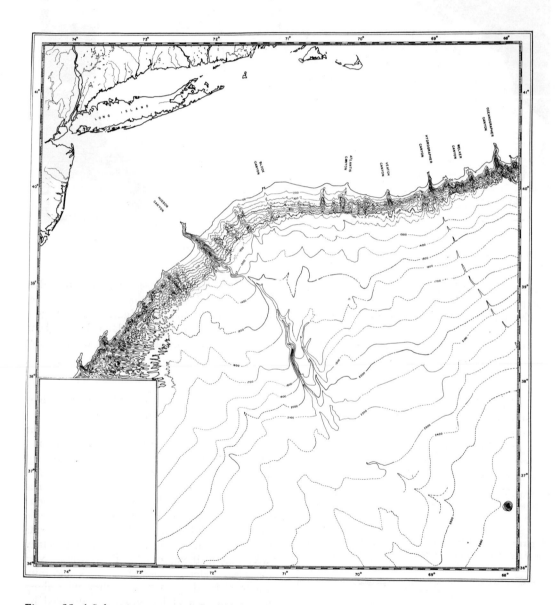

Figure 11–4 Submarine canyons of the Atlantic continental shelf, slope, and rise off Long Island. Topographic contours, or isobaths, are in fathoms. [From Heezen et al., 1959, p. 30.]

ticles, such as pebbles, cobbles, and boulders, and thus by the abrasive activity of these large particles considerable erosion occurs. Turbidity currents may be generated by muddy rivers flowing into the ocean but more likely they are caused by submarine landslides which occasionally cave from the canyon walls and discharge suddenly large volumes of heavy particles into the canyon bottoms. Submarine landslides also account for the widening of the valleys.

There can be no doubt that the canyons are the lines of transit for great volumes of clastic sediment, because the detailed topography at the lower end of the canyons is commonly one of fanlike deposits (Figure 11-5). These have been described as *deep-sea fans*, *levied deep-sea channels*, and *abyssal cones*. They build the continental rise, which is undoubtedly due largely to the deposits of turbidity currents.

Abyssal plains

Aside from the trenches, the abyssal floors are the deepest parts of the ocean basins. They consist of the abyssal hills and the remarkably flat abyssal plains. The abyssal hills are small relief features that rise slightly above the plain. The smooth abyssal plain is thought to be a turbidity current deposit that has flooded and buried an irregular terrane. The abyssal hills at first glance appear to be unburied parts of the old floor, but their texture is so sharp and rugged that a pattern of faulting seems to have emphasized them after most of the turbidity current deposition of the plain has taken place. The abyssal hills have a cover of clay, which is probably thin and has a very slow rate of deposition.

Seamounts

A *seamount* is any isolated elevation that rises more than 3,000 ft above the sea floor. Some rise as high as 12,000 ft above the abyssal floor and reach almost to the surface. These may be as much as 35 mi in diameter at the base. Seamounts occur on the continental rise, on the abyssal floor, and even in the trenches.

Most rise abruptly from the general bottom, and it is thus reasoned that their bases have been covered by bottom sediments. This supposition is based further on the theory that seamounts are volcanic piles or cones built on the ocean floor. Their shape, the many samples dredged from their sides, large magnetic anomalies, and actual submarine eruptions all support the volcanic theory of their origin.

Some seamounts have flat tops and appear as if the volcanic cones had been decapitated. These decapitated cones are called *guyots* (Figure 11-6). The flat tops generally range in depth in both the Pacific and Atlantic oceans from 3000 to 5000 ft. Dredgings from the flat tops yield sand and gravel, like beach deposits, and shallow-water microfossils. Photographs (Figure 11-7) show ripple marks and, on one guyot top at least, single corals. These characteristics have led to the belief that guyots are in reality wave-truncated volcanic cones. The cones were built above sea level, then planed off at sea level by the waves, and then the ocean floor or crust on which the cones were built subsided. The amount of subsidence is measured by the present depth of the cones below sea level. No fossils older than late Cretaceous (fairly late in geologic history) have yet been recovered from the sediments of the flat tops, and hence we are left with the conclusion that in both the Atlantic and Pacific, where the guyots occur, the subsidence has taken place since the Cretaceous period. This is now understood in terms of sea-floor spreading, discussed in the next chapter.

Rises

Rises[2] are broad relief features above the abyssal floors and stand surrounded by deep water. For instance, the Bermuda Rise is an oval-shaped uplift over 300 miles across and about 3000 ft high above the abyssal plain. From it rises a cluster of volcanic cones some of which emerge above the ocean surface to

[2] The terms rise and ridge are used interchangeably by different authors, with some confusion.

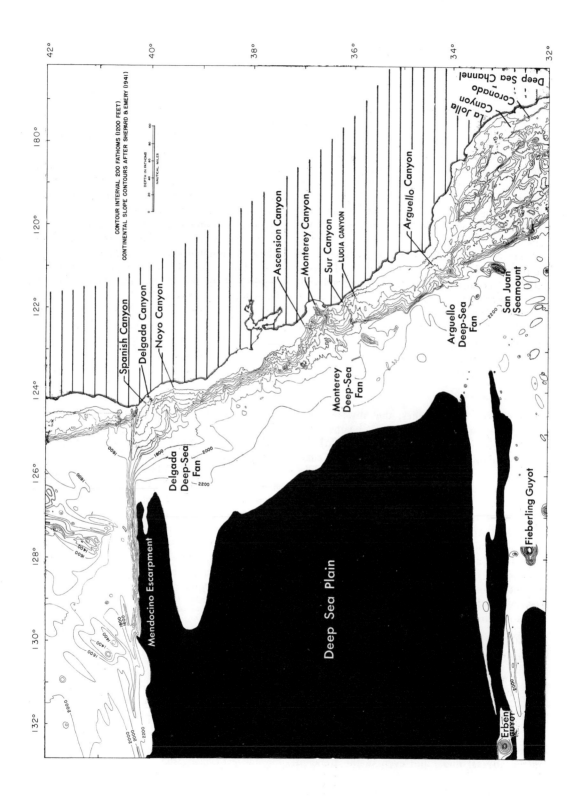

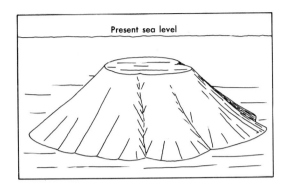

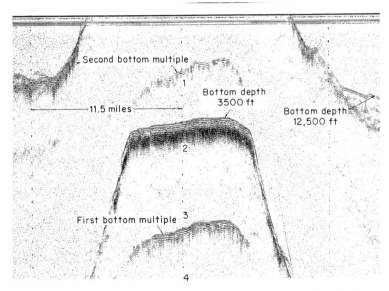

Figure 11–6 Above, diagram of a guyot, which is a truncated volcanic cone with top below sea level. Below, seismic profile of a guyot in the Atlantic. The form is approximately 9 mi wide across its top, and it rises 9000 ft above the floor. A thin veneer of sediments on top overlies the volcanic cone. [Courtesy of Teledyne, Inc.]

form the Bermuda Islands. The Bermuda Rise has a fine-textured relief that contrasts with the smooth slopes of the continental margin and abyssal plain. From its topography, the rise, but not the volcanoes, is thought to be an old ocean-bottom feature. It is bounded on the east by a bold escarpment over 3600 ft high in places and 600 mi long—probably a fault scarp of younger age.

An example in the Pacific Ocean is of the Hawaiian Rise. It is a mound-shaped feature 450 mi long and 175 mi wide and supports

Figure 11–5 Bathymetric chart of sea floor off California, showing submarine canyons and fans. Isobaths in fathoms. [After H. W. Menard, 1955.]

Figure 11–7 Camera for deep-sea photography being lowered from the Vema. [*Courtesy of Lamont Geological Laboratory.*]

many volcanic piles that reach to the ocean's surface and above to form the Hawaiian Islands (Figure 11–8). The entire rise may be of volcanic origin and is probably more recent than the Bermuda Rise. The Hawaiian Rise is surrounded by a sag and then an arch in the ocean floor. These are thought to be due to the weight of the great volcanic mass depressing the crust.

Midoceanic ridges

Probably the most significant discovery in oceanographic research has been that of the midoceanic ridges. These are typified by the Mid-Atlantic Ridge, which is a broad swell running lengthwise along the Atlantic basin and about midway between the continents of the Eastern and Western Hemispheres (Figure 11–9). As shown in profile, Figure 11–10, the ridge is topographically rough, with numerous small ridges and seamounts on either flank and a spectacular trench or rift in the center. The broad arch occupies the central third of the ocean floor, and its central "backbone" ranges are less than 9,500 ft below sea level, whereas the abyssal hills are approximately under 16,500 ft of water. The Mid-Atlantic Ridge, thus, has a relief of 7,000 to 8,000 ft.

On an average profile of the rift valley the floor lies at about 12,000 ft, and the adjacent bounding peaks at about 6,000 ft below sea level, but the depth of the valley ranges from 2,000 to 12,500 ft. The width of the valley between crests of adjacent peaks ranges from 15 to 30 mi.

The walls flanking the rift valley are steep and appear to be fault scarps of uplifted and tilted blocks like those of the Great Basin of the western United States. The immediate outer slopes of the mountains adjacent to the rift seem to be minor tilted fault blocks as much as 3,000 ft high and 10 mi wide.

The Mid-Atlantic Ridge is also marked by many transverse fracture zones (shown conspicuously in Figure 11–9). These fractures have been called transform faults, and a number of them have displaced the central rift. The great fractures and the rift and its parallel side ridges build a pattern that provokes much thought and analysis.

A worldwide midoceanic rift system has been recognized and partially explored. It is shown in Figure 11–11. The midoceanic ridges are of salient interest and are treated appropriately in Chapter 12.

Fracture zones

A number of linear relief features mark the bottom of the northeastern Pacific Ocean. Three of these extend westward from the coast of North and Central America and range up to 3300 mi long and 60 mi wide (Figures 11–12 and 11–13). The northern two, the Mendocino and Murray fracture zones, stretch across the Pacific floor to the Hawaiian Rise. All follow great-circle courses and are approximately parallel. Many others are now known. These linear zones are characterized by parallel narrow troughs, asymmetrical ridges, escarpments, and scattered seamounts. The block of crust between the Mendocino and Murray fracture zones is $\frac{1}{4}$

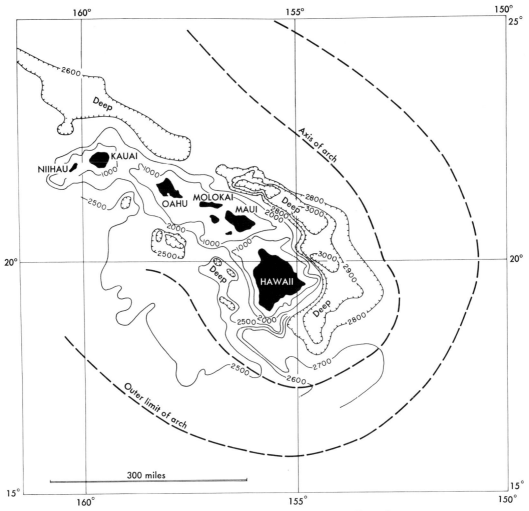

Figure 11–8 Generalized topography around southern end of Hawaiian Rise, showing peripheral deep and arch. Contours in fathoms. [After Hamilton, 1957.]

to ½ mi lower than the adjacent blocks lying north and south. Because of the several characteristics listed above, the linear features are postulated to be zones of fractures or faults in the crust of the earth.

Movements along the fracture zones opposite North America are not now taking place because few earthquake foci occur along them. The magnetic field reflects the fracture zones strongly, as shown in Figure 11–14, and comparison of the unusual anomalies along the north and south sides indicates that many miles of horizontal displacement along them has occurred.

Island arcs and trenches

The margins of the Pacific Ocean are marked in places by arcuate rows of volcanic islands, such as the Aleutians. Each island arc is bordered on the convex side, usually the side to-

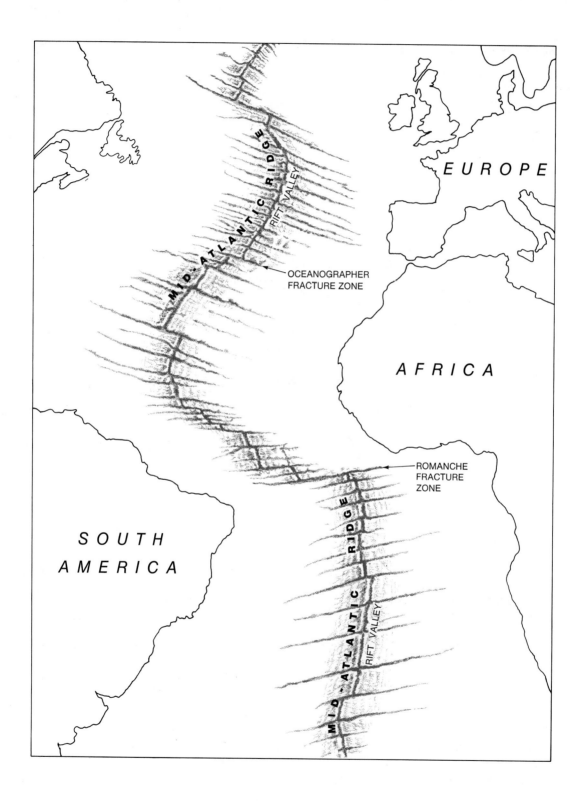

ward the ocean, by a deep trench. Whereas the ocean floor may be 12,000 ft deep, the trenches have been sounded to depths of 35,000 ft. Although much restricted in area, these are truly the profound depths of the ocean floors. The western margin of South America is marked almost continuously by a trench, and so is that of Central America. Rows of active volcanoes lie adjacent to part of the length of the trenches on the border of the mainland (Figure 11–15). The best graphic maps showing these features are probably those of *Life* magazine's publication *The Sea*.

The trenches are everywhere the sites of greatest seismic activity, and hence, it is believed, the crust is actively being deformed in these belts and is buckling downward under compressive stress. A great convective cell in the mantle is probably dragging the oceanic crust against the continent and plunging it downward and under the continental margin. The volcanoes and trenches are presumably surficial manifestations of this deep-seated activity.

The volcanic arcs and trenches have received much study by geologists and geophysicists because they are undoubtedly the regions of active mountain building today, and an extensive bibliography exists on them. They are further discussed in Chapter 12.

Deposits on the ocean floors

Sampling methods

Samples of the ocean bottom may be procured either by dredges or by coring devices. Drag-type dredges have nets fastened to triangular or rectangular frames. The lips of the frames are beveled or have teeth to scrape and plow the bottom muds and stir up the burrowing worms and mollusks. Another type of dredge is the clam-shell bucket type, in which the clam-shell scoops are dropped to the bottom in open position, a messenger is sent down the line that trips a catch, and powerful springs close the heavy scoops. They bite into a few inches of the mud at best, hopefully grabbing up something of interest. The material brought up is mixed, and the fine particles get somewhat washed out.

Coring devices, on the other hand, cut deep into the bottom sediments and generally bring up samples showing the undisturbed layering, worm holes, and other features of interest (see Figures 11–16 and 11–17). A steel tube, the core barrel, is forcefully driven by an explosive charge into the sediments. Cores as long as 70 ft may be procured (see Figure 11–18).

You can appreciate the significance of the core barrel technique when you realize that a 50- to 70-ft core might transect sediments that range in age from modern to several million years. The change of temperatures, salinities, and ocean life have been chronicled over this long time by analyses of the cores.

Cameras have been developed for deep-water photography of the physical forms and of the life at the ocean bottoms, with amazing results. We now search the bottoms with television cameras in manned submersibles. Finally, a decade of dreams has become a reality. A $12.6-million oceanographic ship, replete with all modern devices, especially the capability to drill a hole over 2500 ft deep in the bottom sediments and rocks under 20,000 ft of water, has started an 18-month odyssey. It is the *Glomar Challenger*, a 400-ft craft, and it carries a crew of 75, including 25 scientists. It has already traversed both the Atlantic and Pacific over a course of 40,000 nautical miles and is producing much new evidence for the origin of the ocean basins and their history. Perhaps some valuable "spin-offs" will result, such as the discovery of new foods, drug sources, and mineral resources. The *Glomar Challenger* will be guided at sea by four polar-orbiting satel-

Figure 11–9 *Topography of the floor of the Atlantic Ocean showing especially the Mid-Atlantic Ridge and transverse fractures or transform faults.*

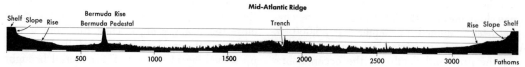

Figure 11–10 *Profile across the North Atlantic from Cape Henry to Rio de Oro. Horizontal scale is in nautical miles and vertical scale in thousands of fathoms.* [*After Heezen et al., pl. 22.*]

lites, backed up by the customary celestial navigation aids.

The new craft cannot drop anchor in 20,000 ft of water in order to hold position on the bottom, so a positioning system has been devised to keep her in place during drilling operations, even in rough water. A combination of hull-installed thrusters, the ship's regular engines, four hydrophones below the hull, and an ocean-floor sonar homing beacon team up to keep the *Glomar* in position while the drill cores the bottom sediments. All are operated automatically by a computer, and it is expected that no more than 300 ft of drift in any direction will occur. The most evident reason why 2500-ft holes in the bottom will yield most valuable results is that so far we have only been able to penetrate a few tens of feet and thus our knowledge concerns only the very late Cenozoic Era.

The *Glomar Challenger* has more than lived up to expectations, and many cores have been taken through the ocean floor sediments into the basaltic crust, throughout the Atlantic and the Pacific. As a result many exciting new data about the evolving ocean floor are known. In a recent voyage in the Caribbean Sea a worn-out bit was exchanged for a new one and the very difficult job of reentry in the original hole was accomplished. Previously a worn-out bit meant the termination of coring in that particular hole.

Classification of bottom sediments

A very broad classification of sea-floor sediments has been used for a long time and still seems satisfactory. Two groups are recognized: *pelagic* and *terrigenous*. Terrigenous sediments derive from the continents, and pelagic sediments originate beyond the direct influence of the land. Pelagic sediments come mostly from the sea itself, although they contain some dust borne by the winds to the oceans or from interplanetary space. They are mostly deep water sediments, but some terrigenous sediments have also been deposited on the abyssal floors.

Pelagic sediments

The pelagic sediments are described as four kinds of "oozes" and as "red clay" (see Figure

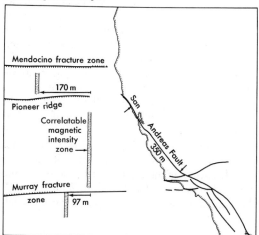

Figure 11–12 *Horizontal displacements along fracture zones indicated by the offset magnetic intensity field. Horizontal displacement along San Andreas fault also shown. Distances are in miles.* [*After H. W. Menard, private map.*]

Figure 11–11 *Midoceanic ridges with central trench, and the deep-sea trenches associated with volcanic archipelagoes.*

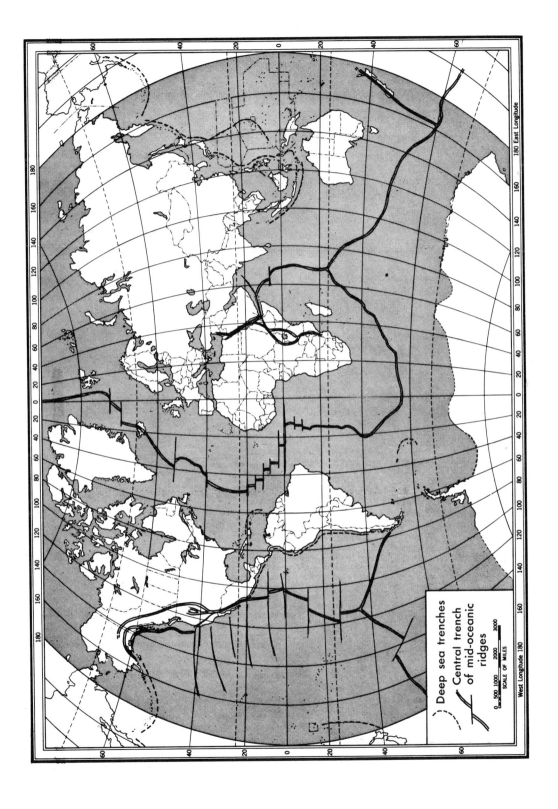

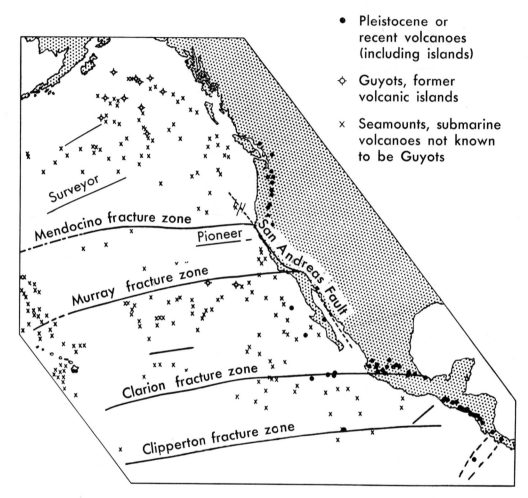

Figure 11–13 *Fracture zones and seamounts of northeastern Pacific. Also volcanoes of adjacent coastland.* [From Menard, 1955.] *Surveyor fracture zone added.* [From Peter, 1966.]

11–19 for their distribution). *Diatom ooze, globigerina ooze, radiolarian ooze,* and *pteropod ooze* are the common varieties. Red clay is mostly inorganic. In the open sea most falling organic material is dissolved or decomposed in the course of a lingering descent from the surface waters, where life is prolific, to the ocean floor. Only the hard parts of plants and animals arrive at the bottom. The microscopically small diatoms, globigerina, and radiolaria require a long time to settle through the water. The calcareous skeletal parts are more soluble and do not reach as deep water as those that are siliceous. In very deep water no microscopically small shell fragments survive, but only the hard parts of larger animals occur, such as shark teeth. These are mixed with mineral-dust particles. The ooze and clay is thus not an ooze in

TOPOGRAPHY AND SEDIMENTS OF THE OCEAN FLOOR   295

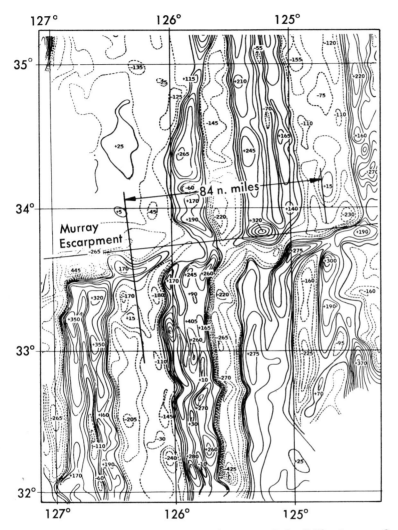

Figure 11–14 *Total magnetic intensity of an area off the California coast. Contour interval is 50 gammas.* [*From Menard and Vacquier, 1959.*]

the sense of the gelatinous mud of a pond, but rather a thin blanket of the hard remains, generally microscopic, of animals and plants, and tiny dust particles.

Diatoms are minute plants that grow everywhere in the plankton, but most profusely in the cold waters of the oceans. Sometimes and in some places they are present in inconceivable numbers to make the water soupy and give it a yellowish tint. Each tiny plant has a pair of shells or covers of silica and under the microscope appears like glass. Diatoms constitute a major source of food for such small animals as copepods and larvae. They are the "grass" of

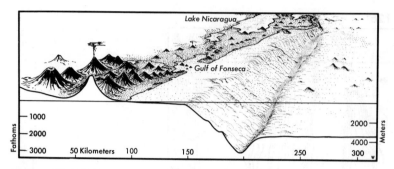

Figure 11–15 *View of the southeastern end of Middle America Trench. [After Fisher and Shor, ms., 1959.]*

the sea. Diatom ooze is found principally in the cold regions of the Antarctic and in the far northern Pacific (see map, Figure 11–19) at depths of 600 to 2000 fathoms. Here the diatom skeletons are so profuse that they obscure other skeletal remains.

Globigerina ooze originates from the foraminifera in the plankton. These single-celled animals move by extensions of their living protoplasm known as pseudopodia—false feet. In this respect they are like amoebae, but they differ in having a calcareous shell with one or more chambers. The shell is porous and has larger openings through which the pseudopodia stream. When a globigerina outgrows its tiny shell, it forms another and larger shell, which remains attached to the first shell. Successively larger shells form a tapering linear series or chambered spiral aggregate suggestive of a nautilus shell but on a minute scale. With needlelike spicules the foraminifera become very ornate and elaborate, and are of many species. About two-thirds of the floor of the Atlantic, and more than one third of all the ocean floors, is covered by globigerina ooze. It is a mixture of shell fragments of the globigerina and coccoliths or coccolith-spores, and is about 65 percent calcareous matter. It occurs at depths of 1000 to 2500 fathoms.

Radiolarian ooze consists of remains of radiolarian shells in a matrix of red clay and occurs at depths of 2500 to 5000 fathoms in isolated areas of the tropical Pacific and Indian oceans. Radiolaria, like globigerina, are protozoans, but their skeleton is siliceous. It is also internal, with the body in two divisions, one within the skeleton or capsule and one without. The capsule is elaborately perforated to provide continuity between the two body portions. The shells may thus be extremely complex and beautiful.

Pteropod ooze is a deep-sea deposit that comes from a distinctly higher type of animal; pteropods are mollusks—thin-shelled, snail-like creatures that spend their entire life in the open water. They swim by finlike structures on the head. They are large enough to be visible, some slender ones being half an inch long. Not closely related generically but living together are the heteropods, which are minute conchs with transparent conical, spiral, or disclike shapes. Some of the heteropods are larger than the pteropods.

Like the diatoms, globigerina, and radiolaria, the pteropods become very abundant under favorable conditions and dominate large areas of the sea. Such waters attract the whalers because the pteropods are food for whales, and give promise of the proximity of "right whales," or those that have whalebone sieves in their jaws. Pteropods, as well as copepods and small shrimplike crustaceans, are strained from the water by the great whales. Pteropod ooze composed mostly of pteropod skeletons but

Figure 11–16 *Core barrel being lowered from the Vema. [Courtesy of Lamont Geological Laboratory.]*

Figure 11–17 *Core barrel being retrieved.*

mixed with heteropods and globigerina is calcareous and occurs in tropical regions at depths of 500 to 1300 fathoms.

Red clay is particularly noted in the great depths. There, it is not obscured by the organic deposits. The red clays constitute more than half of the floor of the Pacific but are not widespread in the Atlantic, and thus we note a distinct difference between the two oceans. The material that makes up the red clays is of several origins:

1. *Volcanic dust*, either from continental eruptions or from oceanic basin volcanoes.
2. *Dust from the lands* that is carried long distances by the winds.
3. *Interplanetary particles* which do not bulk large in marine deposits but may be recognized by their dark color, generally, and by being magnetic. The particles are chiefly black and brown metallic spherules.
4. *Rock fragments*, some of large size, dropped in specific areas by floating icebergs.
5. *Concretions and nodules*, some dominantly of manganese, some of iron and manganese, some of barium, some phosphatic. Con-

Figure 11–18 *Core containing sand and mud layers on continental rise off New York.*

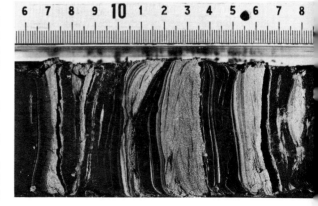

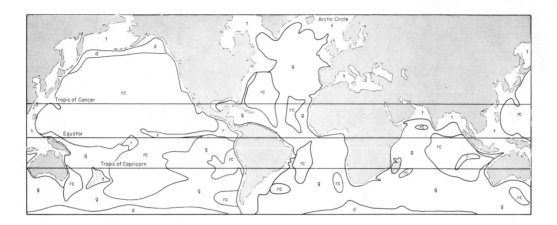

Figure 11-19 Distribution of pelagic sediments. rc, red clay; r, radiolarian ooze; d, diatom ooze; t, terrigenous sediment. Small areas of pteropod ooze not shown. [After McGraw-Hill Encyclopedia.]

siderable interest has recently centered on the possibility of commercial dredging for these concretions. Some of the concretionary patches appear to be associated with terrigenous bottom sediments.

The terrigenous deposits are black, bluish, green, and gray. The pelagic deposits are red, brown, yellow or white (volcanic ash). It has been suggested that the calcareous oozes, in order to deserve the name, should contain more than 30 percent $CaCO_3$, and the silicious oozes should contain more than 30 percent silica and less than 30 percent $CaCO_3$. It should be understood that all oozes are mixtures, and that the varieties here described are simply those with a dominance of one kind of material or another.

By means of isotope dating the rates of sediment accumulation have been determined in several places. Measured by the ionium method, abyssal red clays accumulate at a rate of 0.07 to 0.20 cm per 1000 years. Globigerina ooze off equatorial Africa accumulate at a rate of about 1.6 cm per 1000 years. Deposition of terrigenous sediments is much faster and will be recounted later.

It should be noted that the deep-water pelagic sediments, especially the red clays, have been searched for in the mountainous regions of the continents and the volcanic archipelagos like those of Indonesia, but have never been reliably identified. If found, they would indicate that a certain deep ocean bottom has been elevated and transformed into a mountain range. Deep water terrigenous sediments have been identified in certain mountain ranges, but not the red clays.

The midoceanic ridges have come to be the sites of intensive oceanographic research, and the sediments found on them are of particular concern in this chapter (see Figure 11-20). The East Pacific Ridge west of Mexico, Central America, and South America has been the subject of a recent study (see Figure 11-21). Twelve crossings were made by vessels of the Lamont Geological Observatory and 200 cores were taken. Fifty of these penetrated Pliocene and Miocene sediments. The cores ranged in length from less than 1 m to over 12 m, with the sediments varying from biogenous oozes to red clay. In general, the oozes were made up of foraminifera, coccoliths, radiolaria, and diatoms and occurred on the crest and upper flanks. On the lower flanks adjacent abyssal

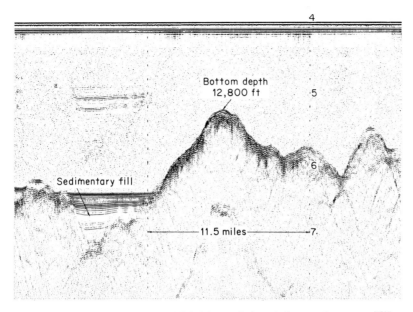

Figure 11-20 A portion of the Mid-Atlantic Ridge. Sediments having a maximum thickness of 2500 ft have accumulated in the depression shown on the left. [Transoceanic seismic profile courtesy of Teledyne, Inc.]

plains were the red clays. In several cores from the lower flanks the red clay overlies a calcareous ooze containing Miocene microfossils. The distribution of the microfossils—with those of Pleistocene origin near the axis, those of Pliocene and Miocene origin down the flanks, and eventually Eocene fossils on the lower slopes and abyssal plains—is most significant. It is seen as a demonstration of "sea floor spreading," a process leading to the rise and widening of the ocean basin. In the case of the Atlantic Ocean, the Mid-Atlantic rise has been widening from a central fracture zone (Figure 11-9), and the abyssal plains have been spreading farther apart. The details of sea-floor spreading with its fascinating implications of ocean basin and continental evolution will be taken up in Chapter 12.

Terrigenous sediments

It is not evident yet that any oceanographer has attempted a serious classification of terrigenous ocean floor sediments, but from topographic occurrence the sediments may be described as follows:

1. Various continental shelf types, including the shoreline, estuary, and delta sediments.
2. Deposits of the continental shelf slope and at its base.
3. Abyssal plain and trench deposits.
4. Sediments on guyots.

Inasmuch as sediments of past seas, especially shallow ones, have hardened into sedimentary rocks that are now exposed in the mountains of our continents, they have been a great concern of geologists.

Sediments on continental shelves

The continental shelves are composed of layer upon layer of sediments. One of the ways in which they are built is by rivers forming deltas in the sea such as the one being formed

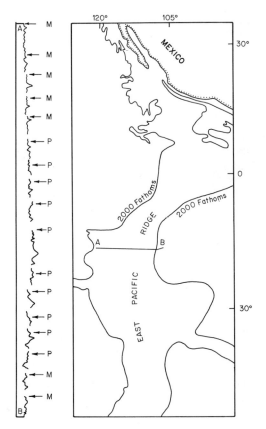

Figure 11–21 Track V 19 (A-B) across the East Pacific Ridge shows distribution of Miocene (M) and Pliocene (P) microfossils. [After Riedel, 1967.]

by the Mississippi River in the Gulf of Mexico. Each year the river transports literally cubic miles of sand, silt, and clay into the waters of the Gulf, where the sediments settle in layers. As time goes on the deposits grow both seaward and upward, as shown in Figure 11–22. The crust of the earth is not strong enough to support the tremendous load of the delta and sags to produce the lens-shaped deposit illustrated. The maximum thickness of the delta is 25,000 to 30,000 ft, and it extends as a gently sprawling fan almost across the entire Gulf of Mexico to Yucatán and Cuba. Such are the deltas of many great rivers; they are the result of the incessant wearing away of the rocks of the continents and the transportation and deposition of the weathered products, the sediments, in the bordering seas. The marginal shelf deposits are characteristic of many of the sedimentary rocks of geological antiquity, and thus the careful study of the modern shelf sediments and the conditions under which they form helps immensely in understanding the old and hardened sedimentary rocks.

Along the Atlantic shelf slope from Cape Hatteras to Cape May, minor benches and slopes have been followed lengthwise and interpreted to mean that formations composing the shelf extend out to the shelf front (see Figure 11–23). Cores and dredgings substantiate the conclusion that the beds continue nearly horizontally to the shelf front. This then is another type of continental shelf, and is believed by some oceanographers to be a common occurrence around the world.

Borings and seismic-reflection profiles on the shelf east of Florida indicate that sedimentation has built the shelf oceanward (prograded) about 15 km since the Eocene (Figure 11–24). This outward building of the shelf has occurred continuously. The sediments on the shelf are porous, well-sorted clastic carbonates composed

Figure 11–22 Delta type of continental shelf.

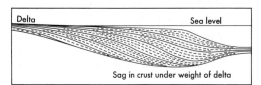

Figure 11–23 Continental shelf built by deposition on shelf top, or by erosion on slope. Typical of the Atlantic Shelf between Cape Hatteras and Cape May. Benches along slope reflect beds. [After Heezen et al., 1959, p. 48.]

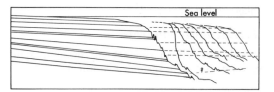

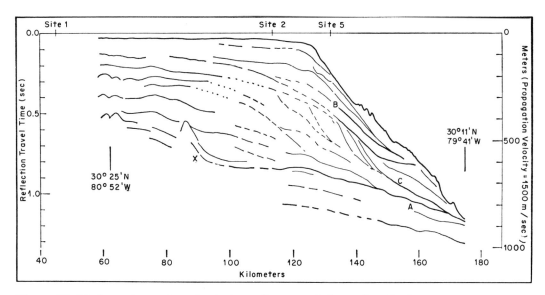

Figure 11-24 *Composite tracing of seismic profiles east of Fernandian Beach, Florida, at latitude 30° 39′ N, longitude 81° 27′ W. Kilometers are east of the beach, and the vertical exaggeration is X67. [After JOIDES Program, Science, 1965.]*

of molluscan shell fragments, foraminiferal shells, algae, and coral debris, all cemented by calcium and magnesium carbonate. Offshore in deeper water the sediments are finer. This is a sediment assemblage more of pelagic than terrigenous affinities. The rate of sediment accumulation for the Eocene epoch of the shelf is 1.6 cm per 1000 years and 0.3 cm per 1000 years at the base of the shelf on the adjacent Blake Plateau.

Many shelves have a dominant terrigenous sediment cover, made up of detrital mineral and rock grains from the land. The fluctuations of sea level during the ice age have left several strand lines, with the shore-line sediments now under variable depths of water down to 100 m.

Still other types of shelves may be erosional platforms or complex deformational, erosional, and depositional features (see the profiles of Figure 11-25).

Certain shelf slopes are unusually steep. These are known to be due to reef building (see Figure 11-26). The periphery of a modern reef is a buttress of limestone built principally by coral and algae that resists the onslaught of the waves. It protects the loose or less resistant material landward. The reef may be an almost vertical cliff facing the ocean, as is the Great Barrier Reef of Australia. The platform of the Bahama Islands is a continuation of the Atlantic Coastal Plain but is now a region of almost pure calcium carbonate deposition. On the Atlantic side it slopes steeply into great depths, and the steep slopes appear to be held up by the reef limestone. A well drilled on Andros Island penetrated carbonate sediments to a depth of over 12,000 ft, which suggests that the region has been subsiding slowly for the past 100 million years. It is also evident that carbonate sediments have been deposited at the rate of subsidence, keeping the region at sea level. Thus a steep continental slope over 12,000 ft in height could have been built.

Continuous seismic reflection profiles have recently been run from Florida to Yucatán across the Gulf of Mexico (Uchupi and Emery, 1968), especially from the Florida escarpment to the Yucatán escarpment. These are steep shelf slopes that have been thought to be faults, but the profiles and cores indicate that they

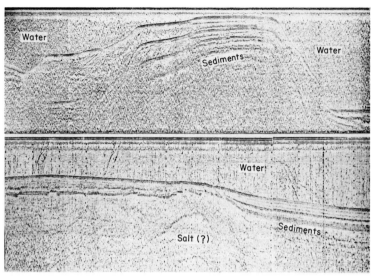

Figure 11–25 Upper, electrosonic record off the southeast end of Santa Catalina Island, California, showing eroded edges of beds at bottom of water on left, general folding of beds, and later sediments in basin on right (1 sec indicates a depth of about 2400 ft). Lower, portion of continental shelf off Corpus Christi, showing probably a salt intrusion and small-scale attendant faulting of the sedimentary layers. [Courtesy of Raflex Exploration Company.]

were built up by reefs of early Cretaceous age (100 million years ago) and subsequently still more by fragmental calcareous deposits.

Observations on the origin of the continental shelves

The several pulses or stages of the ice age were accompanied by an equal number of fluctuations of sea level, each on the order of 400 ft. This has produced a seesaw battle between erosion and deposition on the continental shelf, but along the Atlantic shelf deposition and permanent accumulation has generally won out as an upbuilding process. The Pacific Coast of California, Oregon, and Washington, however, is a belt of active crustal deformation and exhibits many surf-cut benches. In places, the postglacial rise of sea level has resulted in a recent drowning of the wave-eroded, narrow shelf.

The average relief of the continental-shelf slopes above the sea floor is 12,000 ft. Along almost half of the continental shelf slopes of the world are deep trenches, and in such places the slope descends to as much as 30,000 ft below sea level. The trenches will be described in later paragraphs. The grade of the shelf slope to a depth of 1 mi averages 4° 17′, and somewhat less at greater depths. The mountain fronts of the Sierra Nevada, Wasatch Range, and Teton Range of late geologic age in our western Cordillera average greater than this.

Continental shelf slopes are remarkably straight for hundreds and, in places, even thousands of miles. Protruding bulges are rare. If one presumes that the shelves are due to a series of coalescing deltas, then the almost total lack of bulges argues against this premise. The form of the Mississippi delta is unusual, because it has an average slope of only 1° 21′ for the first 6000 ft.

From seismic profile surveys K. O. Emery[3] emphasizes that the continental shelves in many

[3] The continental shelves, *Scientific American*, Sept. 1969.

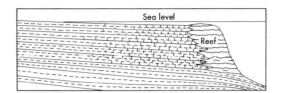

Figure 11–26 Continental shelf due to reef build-up.

places owe their straight or smoothly curved fronts to horstlike uplifts of the older rocks of the continental margin. The uplifts have served as dams to sediments washed off the continents, and in most places the sediments have built up to the crest of the dams and have spilled over them to form the continental shelf slope. Such, according to Emery, is the situation along most of the Pacific Coast of North America.

Deposits on the shelf slope and at its base

The shelf slope off the east coast of Florida, as indicated in Figure 11–24, consists of the same materials as the deeper part of the shelf, and thus its sediments are mostly pelagic. The deepening front of the Mississippi delta is more terrigenous in nature with silt and clay from the river being admixed with pelagic calcareous microfossils.

Where the shelf and slope are trenched by submarine canyons, a number of the canyons have proved to be chutes down which shore sands are flowing and cascading. Long-shore currents transport the beach sands to the heads of the valleys. Manned submersibles have photographed the process in graphic clearness. Of 93 submarine canyons studied 34 appear to have a present-day source of shoreline sediments or a direct river source. Forty-three others have heads so far out from the present shore that their shoreline sources are no longer evident, yet sediments may have been dumped into them at the lower sea levels that were obtained during the ice age.

The sediments that pour down the submarine canyons have built large sprawling, fan-shaped deposits (study especially Figure 11–15, on which you will see the Delgada deep-sea fan, the Monterey deep-sea fan, and the Arguelo deep-sea fan). From what is known, these fans are composed of alternating sand and mud layers, containing many foraminifera that have been identified as coming from shallow waters.

An interesting story has been discerned in the shelf slope, deep-sea fans, and abyssal plain deposits off Oregon and Washington. The volcanic field that now shelters beautiful Crater Lake, Oregon, was a mighty cone 6600 years ago, much like Mount Rainier today. The old volcanic mountain is referred to as Mount Mazama. At that time a cataclysmic eruption spread volcanic pumice and ash to the north and northeast and the winds carried it over all of Oregon and Washington, except the Coast Ranges, and much of Idaho and western Montana. Mount Mazama was destroyed. The ash fall has been much studied and rather accurately dated by radiocarbon analyses. Now, off the mouth of the Columbia River, at the great submarine Astoria Canyon, over 80 cores of the bottom sediments have been procured and were found to contain Mazama ash (see Figure 11–27). The optical and chemical properties of the deep water ash relate it securely to the Mazama ash on land, and radiocarbon analyses support the correlation. The ash appears to have been washed down the Astoria Canyon and to have spread out on the Astoria deep-sea fan and abyssal plain. The turbidity currents carrying the ash passed through a gap in the Gorda ridge onto the west-lying Tufts abyssal plain. The cored sediments are regarded for the large part as terrigenous and to have been derived from the extensive drainage of the Columbia River after the eruption. The layers in the cores, as indicated by their graded bedding, were deposited by turbidity currents. Some of the material of the sediments, however, is pelagic, containing shallow water foraminifera. This suggests that at the time of the great ash flood or floods carrying the volcanic ash down the Astoria Canyon, there were major landslide slumps on the walls of the canyon. Thus the resulting spread of debris on the deep-sea fan was a mixture of pelagic silts, calcareous microfossils, and volcanic ash.

The above research is significant, because in

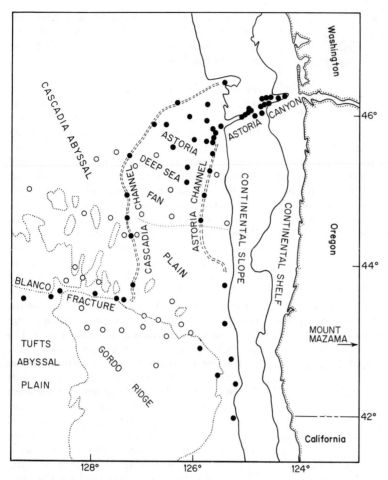

Figure 11-27 *Mount Mazama ash off the mouth of the Columbia River. Black dots show cores that contained the ash layers, and open circles are cores without ash. [After Nelson et al., 1968.]*

many other ocean-floor localities volcanic ash is being cored, and these ash beds prove invaluable in relating one core with another. Such correlations are necessary in working out the sequence of events that comprise the geologic history of a region.

Terrigenous sediments of abyssal plains and deep trenches

Figures 11-28 and 11-29 are seismic profiles of the abyssal plain of the Atlantic Ocean. Both show a very rough basement surface which has been covered and leveled by a fill of sediments up to 3500 ft deep. The sediments, at least in the upper part, are probably of turbidity-current origin.

A striking example of turbidity-current transport and deposition on an abyssal plain is the Grand Banks turbidity current. Here an earthquake triggered an underwater landslide on the Grand Banks of Newfoundland. The head of the slide is at the 100-fathom contour, and the mass that broke away was possibly 200 km from side to side (see Figure 11-30). The great volume of sediments from the upper shelf slope moved down-slope and broke in succession several transatlantic cables lying across its path. The exact time of each cable break was noted, and thus the exact velocity of the submarine sediment current was calculated. At the top and down a slope of 1 to 170 the current moved at a speed of 55 knots, but farther down the shelf slope toward the abyssal plain, at a gradient of 1 to 2000, it slowed to 12 knots. Undoubtedly, the sediments of the shelf mixed with the sea water to produce a highly mobile, turbid suspension. A great tongue spread out on the abyssal plain covering an estimated 80,000 square miles with 3 feet of sediment. Some of the sediment may have flowed southward along the abyssal floor to the Puerto Rican Trench, which, as shown by seismic profiling, contains a fill of unconsolidated sediments several thousand feet thick. Much of this material is terrigenous in origin.

Mineral resources of the marine environment

Petroleum (oil and gas)

Subsea petroleum produced offshore in 25 countries presently amounts to 17 percent of the world's output. This is 90 percent of the total value of all minerals produced from the sea floor. Petroleum will remain the chief fuel for the rest of this century and perhaps longer. It is estimated that by 1980 offshore production will increase to 30 or 35 percent from the present 17 percent, and soon the annual value of subsea petroleum production will exceed that of all marine resources, including chemicals and fish.[4]

The subsea petroleum resources are largely confined to the continental shelves, shelf slopes, and small ocean basins, such as the Gulf of Mexico, the Caribbean Sea, the Mediterranean Sea, the Black Sea, the Caspian Sea, the Bering Sea, the Sea of Okhotsk, the Sea of Japan, the South China Sea, and the seas of the Indonesian Archipelago. Current offshore petroleum production is confined to water depths of less than 105 m within 120 km of the coast. The present drilling capability is about 500 m of water, but this may be extended to 1800 m by 1980, although at much higher costs. It is estimated, because of the high costs of deep-water production plus the available oil and gas at shallower depths, that production from areas beyond the 200 m isobath will be very restricted during the next decade, but by the end of the century it may reach a few billion barrels per year. Sea-floor geophysical exploration (not drilling) is relatively inexpensive because large areas can be covered quickly, generally, with very meaningful structural results. Exploration is not the deterrent to production; it is the high cost of drilling and the installation of production facilities in water beyond the 200-m depth that limits the extension of production areas.

Geophysical exploration is already under way off the coast of more than 75 countries and drilling is in progress off 42 of them. Wide shelves where a possibility of oil and gas exists, and where the oil and gas could be recovered economically, occur off the coasts of Greenland, Norway, the United Kingdom, Canada, Mexico, Trinidad-Tobago, Venezuela, Guyana, Surinam, French Guiana, Brazil, Uruguay, Argentina, Australia, New Zealand, China, Korea, Taiwan, and the Soviet Union. The United States has potential sites along the Atlantic, Gulf of Mexico, and in several places off Alaska.

Manganese oxide nodules

Manganese oxide nodules, a fraction of an inch to 3 inches in diameter, occur on the surface of the ocean floor, generally at depths of 3500 to 4000 m. In addition to nodules, crusts or pavements of manganese oxide have been noted. The composition of the nodules varies considerably, iron being a dominant constituent. In one analysis of nodules from the East

---

[4] These paragraphs recount material contained in V. E. McKelvey et al., *World subsea mineral resources*, U.S. Geological Survey, Misc. Geol. Investigations Map I-632, 1969.

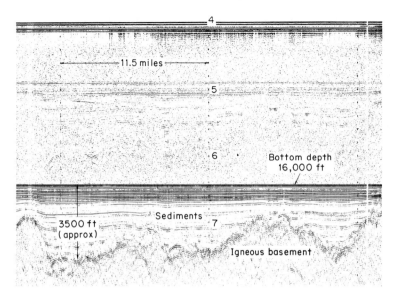

Figure 11–28 *Abyssal plain of the Western Atlantic area showing thick sediments overlying a rough basement surface. [Courtesy of Teledyne, Inc.]*

Indian Ocean, the amount of iron was found to exceed that of manganese. But of most interest commercially is the nickel, cobalt, and copper content. Although small, it is significant. Nickel runs from 0.1 to 1.1 percent, cobalt from 0.1 to 1.1 percent, and copper 0.1 to 0.7 percent.

As noted, most nodule occurrences are on the abyssal floors of the oceans, but they have also

Figure 11–29 *Western Atlantic abyssal plain area showing a basement outcrop. [Courtesy of Teledyne, Inc.]*

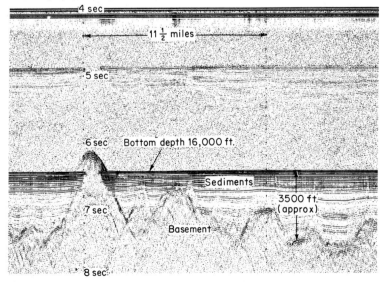

been sampled from guyot tops, and in shallower waters nearer land—notably on the Blake Plateau off the eastern United States, off Baja California, and near some of the islands in the large ocean basins. Three dives by the manned submersible *Deepstar* on the Blake Plateau revealed that the nodules are concentrated at the crests of sandwaves and gradually grade to solid pavement on gentle slopes. The near-shore occurrences are low in nickel, cobalt, and copper. Cores and photographs are as yet so scattered over the wide ocean floors that the continuity of the deposits cannot be reliably reported, but it is evident that the nodules are extensive in many areas. One theory of their origin envisages hydrothermal solution in connection with volcanic activity as adding the metals manganese, iron, copper, cobalt, and nickel to the bottom muds.

The nodules constitute a huge resource. It has been estimated that they aggregate 1.7 trillion tons and contain 400 billion tons of manganese, 16.4 billion tons of nickel, 8.8 billion tons of copper, and 9.8 billion tons of cobalt. The Russians estimate the amounts to be a twentieth of these figures. At any rate the reserves must be great. But production is not economically feasible now. The reasons are several: (1) the cost of dredging in the great depths is very high, and even in the shallow depths dredging would still not be competitive with land mining; (2) the metallurgy of separation of the several metals in the nodules is costly; (3) the ratio of nickel, cobalt, and copper is out of line with the present consumption ratio, and it appears that if all the copper were sold, there would be a large surplus of nickel and cobalt.

Other minerals

Nodules containing phosphorus in the form of phosphate, and muds containing copper and zinc, have been discovered, but at the present time these finds only suggest that the ocean-bottom sediments contain valuable concentrations of these minerals and metals, and possibly others. Heavy mineral concentrates, which on land we would call placer deposits, contain

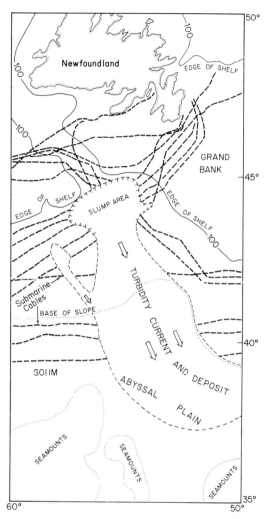

Figure 11–30 Grand Banks submarine slide of 1929 and turbidity current showing the several cables cut or buried. Edge of shelf is noted at 100 fathoms. [After Newman and Pierson, after Heezen and Ewing.]

magnetite, zircon, titanite, gold, platinum, and perhaps the tin mineral, cassiterite, have also been noted and to a limited extent prospected for.

The phosphorite concentrations occur in low latitudes on continental shelves where cold, upwelling deep-water currents coming up the shelf slopes mingle with the warmer surface waters.

The subsea gold placers have been found and are undoubtedly ancient beaches that have been overlapped by the waters of a rising sea level, and hence occur approximately parallel with modern shorelines and generally not far out from the shoreline.

The Woods Hole research vessel *Atlantis II* in 1965 charted and explored three small deeps in the Red Sea that contained unusual water below depths of 2000 m. The bottom brines contain 257 ppt of dissolved solids and have temperatures of up to 56° C (133° F). This is about double the temperature of the surface waters, and the brine is almost 10 times saltier than normal Red Sea water, which in turn is somewhat saltier than normal ocean water. But the significance of the deep Red Sea brines and the bottom muds to the present discussion is the fact that they contain appreciable zinc, manganese, iron, and lead. The amounts of these metals in the deep brines in the Red Sea pockets over the amounts in normal sea water are as follows: zinc, 1500×; iron, 5,000×; manganese, 26,000×; and lead, 30,000×. The bottom muds are very hot to touch, are black, and flow like tar. They consist of metalliferous sulfides, oxides, sulfates, carbonates, and silicates, principally of iron, manganese, zinc, and copper. A fascinating study concludes that the brines are neither evaporation products nor juvenile (magmatic) waters, but brines that originated near the surface at the south end of the Red Sea and found their way by subsurface circulation to the deep pockets of the central area. On the way they became hot and dissolved the metals that they contain from the rocks as well as more sodium chloride.[5]

Production of all of the metals and minerals in sea-floor deposits has not yet reached the feasibility stage. There might be some justification for the occasional optimistic reports in the newspapers, but because of the low-cost operation on land and the high cost of retrieving the sea-floor deposits, not many attempts have been made.

[5] See Egon T. Degans and David A. Ross, The Red Sea hot brines, *Scientific American*, Apr. 1970.

References

Bascom, Willard. Technology and the ocean. *Scientific American*, Sept. 1969.
Degans, Egon T., and Ross, David A. The Red Sea hot brines. *Scientific American*, Apr. 1970.
Emery, K. O. The continental shelves. *Scientific American*, Sept. 1969.
Engle, Leonard. *The sea.* New York: Time Inc., Life Nature Library, 1961.
Fisher, R. L., and Revelle, Roger. The trenches of the Pacific. *Scientific American*, Nov. 1955, pp. 36–41.
Heezen, Bruce C., et al. *The floors of the oceans.* Geol. Soc. Amer., Special Paper 65, 1959.
Hill, M. N., ed. *The sea.* 3 vols. New York: Wiley, 1963.
King, C. A. M. *An introduction to oceanography.* New York: McGraw-Hill, 1963.
Menard, H. W. The deep-ocean floor. *Scientific American*, Sept. 1969.
Munk, Walter H. The circulation of the oceans. *Scientific American*, Sept. 1955, pp. 96–104.
Shepard, Francis P. *Submarine geology*, 2d ed. New York: Harper & Row, 1963.
Shepard, Francis P., and Dill, Robert F. *Submarine canyons and other sea valleys.* Skokie, Ill.: Rand McNally, 1965.
Uchupi, E., and Emery, K. O. Structure of the continental margin of the Gulf Coast of the United States. *Bull. Amer. Assoc. Petrol. Geol.*, v. 52, 1968, pp. 1162–1193.

12

evolution of the continents and the ocean basins

Constitution of the North American continental crust

Tectonic units of North America

A tectonic unit is a structural division of a continent of considerable size. A typical tectonic unit is a mountain system, or cordillera, and on the new *Tectonic Map of North America*[1] such a unit is called a fold belt. The fold belts were originally the sites of geosynclines (Chapter 3). These thick sedimentary rock sequences then became faulted and folded to various degrees and in places invaded by batholiths and metamorphosed. Fold belts are of various ages, and thus North America has been divided on the new map into eight fold belts ranging in age from Early Precambrian (2.5 billion years) to the present.

The three earliest fold belts have been recognized chiefly in the Precambrian rocks of central and eastern Canada (see Figure 12-1). The early fold belts have been inactive or stable since Precambrian time and have thus been subject to extensive erosion. Nothing is left but the stumps of the folds and fault blocks; the

[1] U.S. Geological Survey, 1969.

granitic batholiths have been unroofed and extensively exposed. All this had occurred before the beginning of Paleozoic time, because then shallow seas began to spread over the Precambrian fold belts and deposit sediments unconformably over the deformed rocks. It seems probable that all parts of the fold belts were eventually covered, but then another erosion episode began stripping away the nearly flat-lying sedimentary layers, so that by now an extensive Precambrian rock region, which we call the Canadian Shield, is exposed. Figure 12-2 shows the relatively thin veneer of sedimentary rocks over the Precambrian fold belts, which are sometimes referred to as the crystalline basement.

The region of the nearly flat-lying sedimentary veneer may be called a *platform*. It comprises all areas between the shield, the Appalachians and the Rocky Mountains. The platform includes the Great Plains of western Canada and the United States, the Missouri, Mississippi and Ohio drainages, and most of the continent southward to the Gulf Coastal Plain. This vast area of sedimentary rocks thickens to form local basins and thins to form local arches and domes, but only in very small

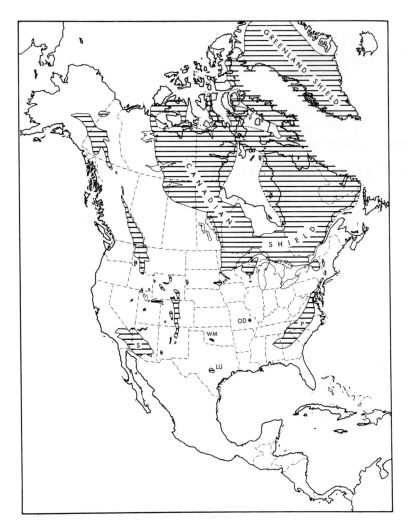

Figure 12–1 *Exposures of Precambrian rocks in North America, showing particularly the Canadian Shield. S, area of scattered outcrops; P, partly Precambrian and partly Paleozoic; OD, Ozark Dome; WM, Wichita Mountains; UL, Llano Uplift.*

part has it been sharply deformed by folds and faults such as we studied in Chapter 3. It is a rich hunting ground for oil and gas and coal and many other mineral products. The maps of Figures 12–3 to 12–6 depict the changes in the shield and sedimentary cover during Paleozoic, Mesozoic, and Cenozoic time.

Geosynclines and fold belts

The Western Cordillera

The vast continental platform was bordered by geosynclinal basins and fold belts. Geosynclines are basins in which very thick sequences of sediments and layered volcanic rocks have

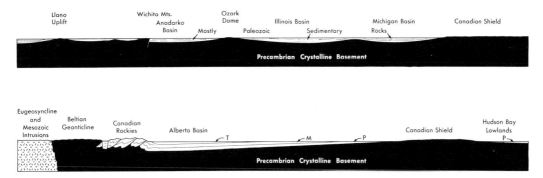

Figure 12–2 Cross sections showing the Precambrian basement of North America and the sedimentary veneer. Lower section, from the Hudson Bay Lowlands westward across the Shield, the Alberta Basin, the Canadian Rockies, and the Beltian uplift or geanticline. No attempt is made to show the different divisions of Precambrian rocks. Top section, from the Canadian Shield along a crooked course through the Michigan and Illinois basins to the Ozark Dome, and thence to the Wichita Mountains of Oklahoma and the Llano uplift of Texas.

accumulated. They consist of two divisions: (1) an inner division composed of sedimentary rocks—conglomerates, sandstones, shales, limestones and dolomites, (2) an outer division composed of interlayered volcanics and various sedimentary rocks (see Figure 3–22, Chapter 3). The inner division is called the *miogeosyncline*, and the outer the *eugeosyncline*. During its evolution, the eugeosyncline was subject to crustal disturbance and mountain building. This was accompanied by earthquakes, volcanic eruptions, uplifts above sea level and subsidences below sea level. The uplifts were vigorously eroded, and the debris was deposited in adjacent marine troughs. All this can be deciphered by the nature and distribution of the sediments.

The crustal deformation spread to parts of the miogeosyncline and, in the case of the Rocky Mountains of the western United States, even to adjacent parts of the shelf. The classical anticlines and synclines of the Appalachians were formed by deformation of part of the miogeosyncline that developed along the eastern margin of North America.

The climactic deformation of the eugeosyncline was marked by the invasion of the deformed sediments and volcanics by tremendous volumes of silicic magmas. In cooled and solidified state, and now exposed by erosion, these magmas make up large parts of the crystalline piedmont (Acadian and Taconic fold belts, Figure 12–3), and the Sierra Nevada of California (Nevadan fold belt, Figure 12–5). They form very large batholiths of granite and quartz diorite. But the Sierra Nevada batholith is just part of a continuous belt of granitic rocks that extends from Baja California to Alaska. Further, the granitic belt extends southward through Mexico to South America and also through the Andes from Colombia to Tierra del Fuego. The batholiths extend into Asia from Alaska, and thus the batholithic belt is an intricate part of the continental margins around the Pacific. It is associated with eugeosynclines, ocean deeps, active volcanoes, and incessant earthquakes (see Figure 12–7).

The Sierra Nevada batholith has been found to be made up of a number of separate intrusions, which have been studied chemically and dated by isotopic analysis. The separate intrusions occur in discrete masses in sharp contact with each other, only separated by thin dikelike masses of metamorphic rocks or mafic igneous rocks. They range in size from 1 to 1000 sq km, and the larger ones generally are elongate in a northwesterly direction. Where two intrusive masses meet it is usually possible to determine

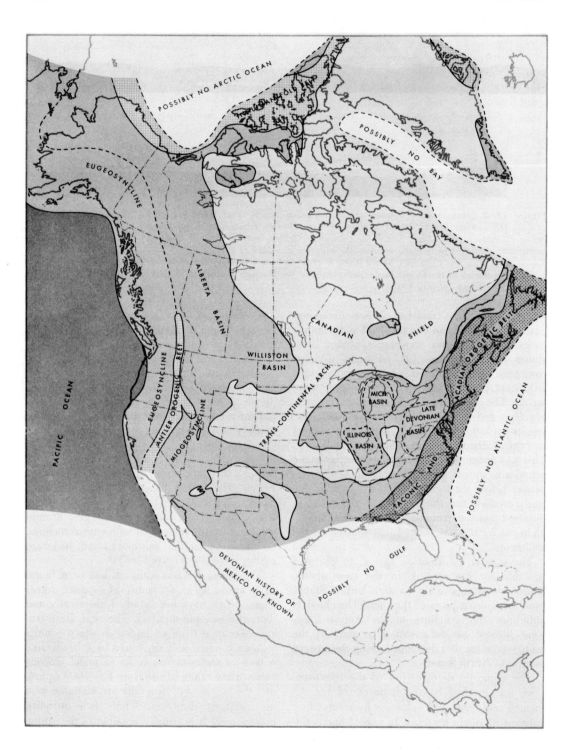

which is older by means of truncated structures and from fragments of the older mass included in the younger. On the basis of isotope-age determinations those along the east side of the great batholith are 170 to 210 million years old, those in the western part 125 to 145 million years old, and those just west of the range crest 80 to 90 million years old. This has led to the conclusion that three episodes of intrusion in the Sierra Nevada have occurred, the first on the east, the second on the west, and the last in the middle.

During the past 10 years several seismic refraction profiles have established that the crust under the Sierra Nevada is conspicuously thick; in fact, the range has a thick root under it. The seismic velocities of the several "layers," as well as the presumed densities of the layers, are shown in Figure 12-8. These velocities and densities are compatible with granite at the surface gradually changing to diorite at depths of about 25 to 30 km, and with a rock similar to gabbro in the lower crust. Peridotite probably composes the upper mantle. The downwarp of the geosyncline and its filling with sediments and volcanics occurred during Paleozoic and Mesozoic time, but in Early Mesozoic time (200 million years ago) the lower part of the silicic crust began to melt and granitic magmas began to form. The melting is postulated to be due to the presence of mineral grains in the sediments and volcanics containing radioactive elements, which yield heat upon decay. Heat is also acquired by the depression of the eugeosynclinal rocks to depths where higher temperatures prevailed. The dry sediments and volcanics of the eugeosyncline melt at temperatures above 1400° C, but the presence of much water in them reduced the melting or partial melting temperature considerably, possibly to 900° C. Certain refractory rocks and minerals were not melted, and these, with early crystallized mafic minerals, settled and thus helped thicken the lower crust. Because granitic magma is significantly less dense than granitic rock of the same composition or the silicic rocks being melted, it would tend to move upward in the manner of a salt intrusion, exploiting lines of structural weakness on the way up. Some magma may have broken through to the surface in volcanic eruptions, but most crystallized at depth to form the batholith we see today. Figure 12-9 is a series of diagrams showing the evolution of the Sierra Nevada as thus conceived.

Erosion followed each magmatic episode; 10 to 12 km of rock may have been eroded away since the Jurassic period (130 million years ago) and 7 to 10 km since the early Late Cretaceous (100 million years ago).

The magmatic cycle as represented by the Sierra Nevada batholith, although truly the climactic event of the western marginal orogenic belt, is only part of the entire story. The Coast Ranges to the west of the Sierra Nevada along the outermost margin of the continent were elevated chiefly in Cenozoic time after the great batholithic intrusions. They are composed of Cretaceous and Cenozoic sediments and volcanics and are noted for the San Andreas fault with great horizontal displacement. The Coast Ranges and the San Andreas fault will be referred to again in this chapter, in connection with sea-floor spreading.

Turning to the western cordillera east of the Sierra Nevada, we recognize a belt of folding, overthrusting, and uplift with scattered smaller intrusions (stocks). This developed in Late Paleozoic time (Figures 12-3 and 12-4). It lay at about the border of the eugeosyncline and the miogeosyncline in central Nevada. Folding and thrusting continued on either side in Nevada during the Mesozoic, and then a Late Mesozoic and Early Cenozoic belt involved the eastern part of the miogeosyncline in Utah and the shelf in Wyoming and Colorado, and the Rocky Mountain states in general. This fold belt—the Laramide—is noted for great thrust

Figure 12-3 *Tectonic map of the Devonian period. Gray, seaways and regions of sedimentation; light gray, gently emergent land areas; cross-hatched, fold belts; stippled, Taconian (Taconic) and Acadian belts of folding, large intrusions, and metamorphism. See discussion of ocean regions beginning on page 319.*

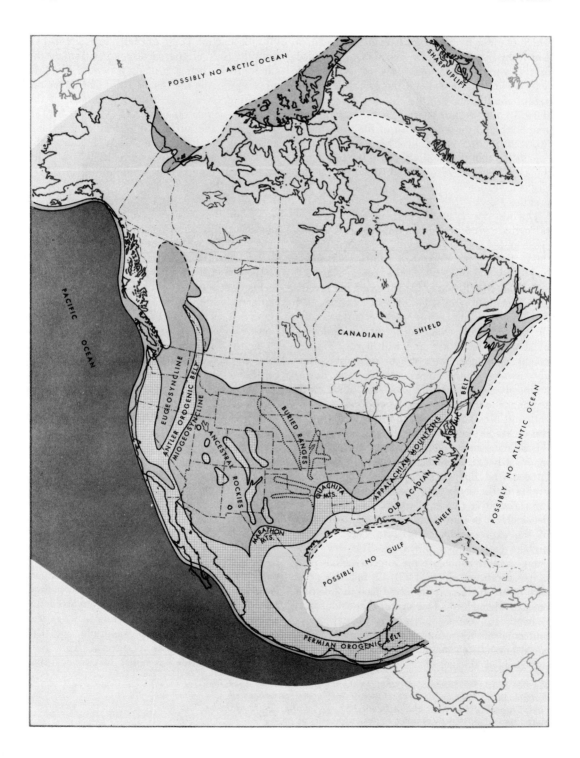

sheets in the miogeosyncline. It spread through western Canada to form the Canadian Rockies and through much of Mexico to form the folded structures of the Mesa Central.

Finally, as shown on Figure 12–6, many extensive volcanic fields developed through Cenozoic time, and some of the erosional forms on the mountains that were elevated during the Mesozoic and Early Cenozoic times were buried. The extensive black basalt flows of the Columbia River volcanic field were erupted, the Sierra Madre Occidental of western Mexico grew to great areal extent and thickness, and a voluminous silicic volcanic field spread over Nevada and western Utah, where the so-called Basin and Range orogeny occurred. This kind of crustal deformation by normal faulting and presumed east-west extension of the crust created the Great Basin of internal drainage and extended from the eastern escarpment of the Sierra Nevada in California across Nevada to the Wasatch Mountains of central Utah.

Seismic layering

Refractory seismic surveys have been made across the Great Basin and Wyoming and Colorado to the Great Plains, and the generalized results are shown in Figure 12–10. The velocity layering as interpreted does not compare too closely with that of the Sierra Nevada crust of Figure 12–8, but what is important is the thick intermediate layer between crust and mantle, the 7.6 km per sec layer or "layer of low density and low velocity." It has been thought of as a transitional layer between crust and mantle. Because it has intermediate velocities between those of the mantle and the lower crust, it has been conceived of as a changing upper mantle—changing both physically and chemically. In the physical sense it might be a solid-state transformation of the high-density silicate minerals to a somewhat lighter density state, or it may be a state of partial fusion, with accompanying decrease in density and decrease in seismic velocities. With the partial melting of a complex ferromagnesium silicate, such as peridotite, the first melt products would be of basaltic composition, which would contain a little more SiO_2 than peridotite. Thus there may be a chemical change taking place.

The idea of large volumes of basalt generation in the upper mantle is appealing, because we can see a need for much basalt in generating the Cenozoic volcanic rocks and the oval-shaped uplifts of the Rocky Mountains. Many careful microscopic and chemical studies of the volcanic rocks in Montana, Wyoming, Colorado, New Mexico, Arizona, and Utah have resulted in the conclusion that an olivine-rich basalt rose to the silicic upper crust and partially mixed with, mobilized, and fused the silicic rocks and produced the many interesting but somewhat exceptional light-colored volcanic rocks that occur in these states. Here and there the black basaltic magma found a line of weakness directly through to the surface, perhaps along a deep-cutting fault, and produced a basalt eruption with related basaltic sills and dikes.

The major mountain ranges of the Rockies, the Big Horn Mountains of Wyoming, the Uinta Mountains of Utah, and the Front Range of Colorado, for instance, are oval-shaped uplifts (Figure 12–5) and presumably have been produced by vertical doming of the upper part of the crystalline basement and the overlying sedimentary veneer (Figure 12–11). Figure 12–12 illustrates a basaltic intrusion from the mantle that has blistered up the overlying rocks, perhaps asymmetrically, with a fault on one side. This fault is conceived to be vertical at depth but to turn into a thrust fault at the surface, chiefly due to mass movement of uplifted rock toward the lower, adjacent basin.

It is also likely that the transitional layer has

Figure 12–4 Tectonic map of the Pennsylvanian period. Gray, seaways and regions of sediment accumulation; light gray, gently emergent lands; cross-hatched, ranges and mountain systems of energetic uplift. The nature of the Pacific crustal region had not been defined by Pennsylvanian time. Africa then lay in the area of the Atlantic Ocean. The fold (orogenic) belt across southern Mexico is of Permian age.

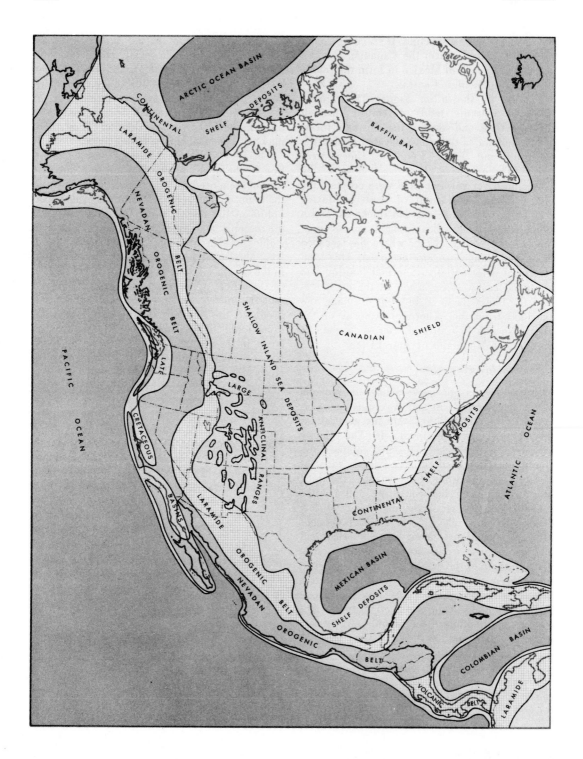

something to do with the Basin and Range faulting and the great volume of associated silicic volcanics. In the Sierra Nevada crust a root was deeply depressed, and thus the elevated temperatures were brought about that resulted in the fusing of the lower part of the silicic crust and the creation of the granitic magmas. No such deep root appears under Nevada and western Utah, yet the silicic crust must have been copiously fused to generate the silicic magmas of the volcanics. The process of magmatic differentiation (Chapter 2) can create silicic magma, but the amounts would be quite inadequate. Thus, the idea has been suggested that the liquid basalt, rising from the transitional layer, has brought up enough heat to partially fuse the lower part of the silicic layer. This is the magma that rose to eruption producing the vast fields of silicic volcanic rocks in the Great Basin. Conceivably the mobilized basalt and silicic layer became so extensive as to result in an unstable crust above, and with a small component of gravity, effective toward the Pacific basin, the continental crust broke into a system of generally north-south orientated blocks. Some of the blocks sank, some rose, but many rotated (see Figure 12-13). In general the crust was here extended, possibly 25 to 50 miles between the Wasatch and the Sierra Nevada. An alternate hypothesis, to be discussed later, is that a rising convection current exists under the Great Basin.

The Greater Appalachian Mountain System

The Atlantic margin of the continent also has a fold belt, called the Greater Appalachian Mountain System. It stretches from Newfoundland, through the Acadian Provinces, New England, and the classical Appalachians to Georgia and Alabama. It has many of the geological features of the western cordillera, but differs in one important way—it is older. Whereas the climactic, magmatic, and structural episode occurred in Mesozoic time in the west, it occurred in Paleozoic time in the east. The Appalachians have been passive or quiet, from the geological point of view, since Early Mesozoic time and have suffered erosion ever since. The once-high mountains have been reduced many thousands of feet, and if a great root was ever existent it has vanished as an isostatic response to erosion and release of weight. No appreciable root is recognized by seismic surveys today.

The White Mountains of New England are as classical in the east as is Sierra Nevada in the west, because each mountain mass is a most instructive locality of magmatic intrusional succession. The magma injection occurred in the White Mountains in Devonian and Mississippian time, with folding, faulting, and metamorphism before and after (see Figure 12-3). The belt of batholiths extends from central Newfoundland through Nova Scotia, New Brunswick, and New England to the Piedmont region of New Jersey, Maryland, Virginia, the Carolinas, and Georgia.

The Appalachian Mountains of Pennsylvania, West Virginia, eastern Kentucky, and Tennessee are a fold and thrustbelt of Late Paleozoic age, and bear the same relation to the batholithic belt as the Rockies do to the Sierra Nevada. The analog of the faulting of the Great Basin is the so-called zone of Triassic rifting in the eastern fold belt. Great extension faults created a zone of basins and uplifted blocks in Late Triassic time from Nova Scotia to South Carolina. Since then no disturbances have occurred, other than broad and gentle arching.

Figure 12–5 Tectonic map of the late Cretaceous and early Tertiary periods. Light gray, seaways and regions of sediment accumulation; white, gently emergent land; cross-hatched, regions of vigorous uplift and deformation, the Laramian (Laramide) orogeny; gray, the Nevadian (Nevadan) orogenic belt of strong deformation, metamorphism, and batholithic intrusions. Late Cretaceous seas spread widely across the Laramide belt before deformation in Alaska, the Yukon, and Mexico. They spread also over parts of the Nevadan belt in Mexico. The Laramian belt in eastern Nevada and western Utah is now called the Sevier.

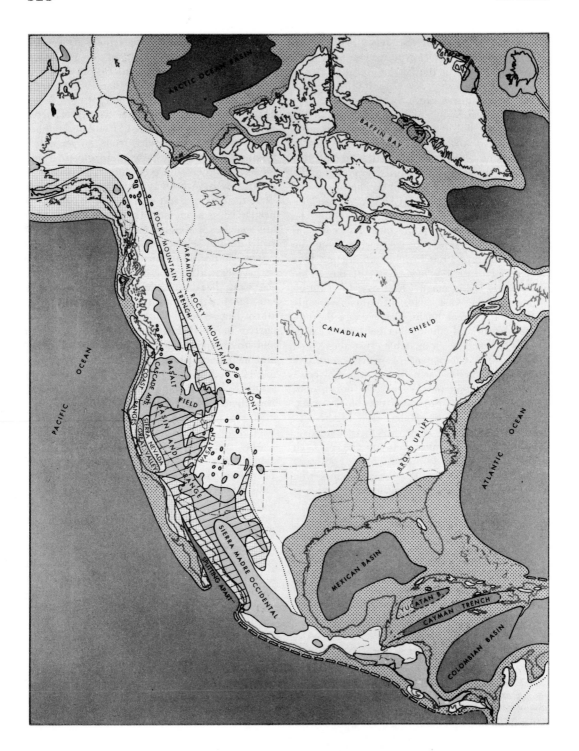

Skin-deep deformation and deep-seated causes

For many years geologists were convinced that the great folds and thrust faults of the Alps and the Appalachians indicated horizontal crustal compression and shortening. Then came the realization that the folds might be only skin deep, like wrinkles on the top of your hand when a finger of the other hand pushes the skin sidewise. The striking folds of the Jura Mountains, which demonstrated that a succession of sedimentary layers had been skidded and folded on an undeformed crystalline rock basement, had been known for a long time (refer again to Figure 3–47). This was clearly an example of "skin-deep" deformation. The crust in its entire thickness had not been folded or compressed. The cause of skin-deep deformation is attributed by some to down-slope gliding of rock masses in a variety of forms and sizes. To create a down-slope gradient we must invoke the rise of major uplifts of parts of the earth's crust. Upward surging of magma seems to be a reasonable cause of the uplifts. And with this hypothesis we arrive at the conclusion that the crust in places has not been primarily compressed horizontally but primarily elevated. We should not, however, swing summarily to the thesis that all crustal deformation is originally or primarily of a vertical nature. Careful studies leave some leading geologists with the conviction that parts of the fold belts represent the compression of thick sections of the earth's crust, not just the sedimentary veneer. And they find a cause of such compression in the new theory of plate tectonics, to be discussed on following pages.

Convection currents in the upper mantle

In trying to envision the cause of horizontal crustal compression, geologists and geophysicists some 50 years ago conceived of convection currents in the mantle. In simple form and in cross section, a convection cell that would deform the crust is imagined to be like that shown in Figure 12–14. A central rising current would reach the crust, divide, spread horizontally, drag the overlying crust with it, and pile up the crust where the currents descend. The zone of deformation would not only be one of badly folded and thrust-faulted strata but also one of crustal thickening, with a deep root extending into the mantle. Above the rising convection would be a zone of crustal tension, with the development of rift valleys or graben structures. The hypothesis gained wide acclaim for a while, but then the concept of gravity tectonics (down-slope mass movements) gained attention and convection currents lost some of their attraction. This was before geologists and geophysicists started to learn much about the ocean floors. Their attention was centered on deformation of the continental crust, not of the ocean floors, but now the amazing structures of the ocean floors seem to demand the existence of convection currents in the upper mantle. From the concept of sea-floor spreading came the concept that major plates—not continents—are involved in translation, and with it the hypothesis of great convection cells is now being questioned. Gravity again seems to be the cause of movement. The concept of major plates (plate tectonics) is explained in later paragraphs.

Constitution of the oceanic basins

Atlantic Ocean basin

Mid-Atlantic Ridge, magnetic anomalies, and sea-floor spreading

The topography of the Atlantic Ocean floor has been presented in Chapter 11, and especially in Figure 11–9. The figure represents the

Figure 12–6 Tectonic map of the middle and late Tertiary period. White, gently emergent land; stippled, seaways over the continental margins; gray, regions of volcanism; dark gray, oceanic crust; cross-hatched, fold belt; lined areas, block faulting and trench faulting. Pacific fold belt is the result of the collision of the North American plate and the eastern Pacific plate.

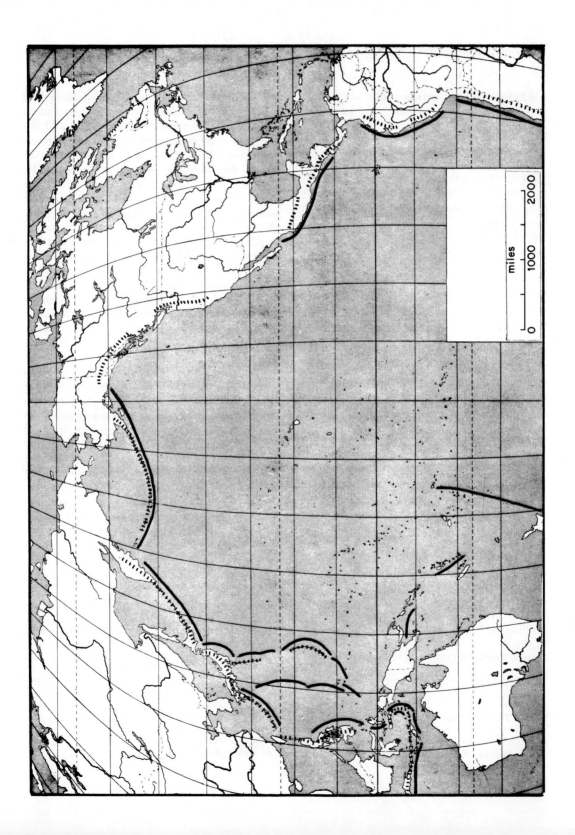

culmination of much remunerative research by the staff of the Lamont Geological Laboratory. What the detailed topographic picture means in terms of the evolution of the basin is the concern of the following paragraphs. The detailed mapping of the relief features is surely a noted achievement; however, another outstanding accomplishment is the mapping of the magnetic anomalies over the Mid-Atlantic Ridge and their analysis. Figure 5–13 is a magnetic intensity anomaly map off the California coast, but at the time the survey was made no one could guess what structures or rocks produced it. It was not long, however, before the strips of anomalies, especially those paralleling the Mid-Atlantic Ridge, were demonstrated to represent basalt flows paralleling the central trench or rift (Figure 12–15). Further, the basalt flows were dated by the potassium-argon isotope method and were found to be youngest at the rift and successively older away from it, thus demonstrating that the rift was the source of the lavas. The rift was also recognized to be widening with each eruption as the crust was transported away on either side possibly by the underlying convection currents. The spreading produced a new fissure each time for the further emission of basalt, and thus each time the rift was healed with freshly crystallized basalt. This is the sea-floor spreading theory (see Figure 12–16).

Emphasizing the stripped anomaly picture was the discovery that some strips had reversed magnetic polarity. It was known from various studies of remanent polarization in the rocks that the earth's magnetic field had reversed itself a number of times, but now with the anomaly maps and the ages of the basalts causing the anomalies, it could be demonstrated exactly when the most recent reversals had occurred. The reversals and origin of the basalt flows are indicated in Figure 12–16. With the dating of the anomaly strips the rate of sea floor spreading was determined.

As the Mid-Atlantic Ridge was further charted, the magnetic anomaly pattern proved to be symmetrical and agreed with the relief features, including the transform faults. The theory of sea-floor spreading was thus confirmed in the Atlantic Ocean and soon also in the Indian Ocean, as well as in parts of the Pacific, and the rates of spreading were determined in several places. This rate was first determined as approximately 1 cm per year. Recently, using the fission-track dating method of age determination on the ocean-floor basalts, the local rate within a few kilometers of the central rift has been set as 2–3 cm per year whereas farther away it is only 0.8 cm per year.

The anomaly pattern with its reversed strips is, in itself, of interest, because now that the pattern has become established, as shown in Figure 12–16, one need not procure basalt samples from the ocean floors for potassium-argon isotope of fission track age determinations. One simply reads the anomaly pattern and recognizes the age of individual strips. Each reversal has been accorded a name, as is shown in Figure 12–16.

Sediments that settle to the ocean floor preserve the existing polarity of the earth's magnetic field in the same way as crystals do in crystallizing magmas. Thus a chronology of time cycles may be recognized in cores taken from the ocean floor sediments.

The magnetic anomaly pattern of the North Atlantic on profiles from one side of the ocean to the other show a boundary between disturbed and undisturbed magnetic regions. By a disturbed magnetic region is meant the striped anomaly picture that we have been discussing. This extends 2000 to 2500 km on each side of the central rift. Beyond the disturbed region, and roughly equidistant from it on either side, are the undisturbed regions, which are believed to reflect a long period of no reversals that occurred in the pre-Tertiary time (before 80 million years ago). Since these regions of undisturbed magnetic fields are approximately at the

Figure 12–7 The Pacific Ocean showing the bordering trenches (dark lines) and active volcanoes (rows of dashes).

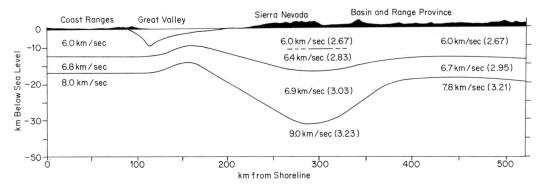

Figure 12-8 *Crustal layering under the Sierra Nevada, California and Nevada.* [*After Paul C. Bateman and Jerry P. Eaton,* Science, *December 15, 1967.*]

continental edge, it may be concluded that the process of sea-floor spreading, at least for the Atlantic, started in Cretaceous time.

Wegener and continental drift

We have just concluded that the topography of the Atlantic Ocean floor, the volcanic rocks of the ocean floor with their isotope ages, and the magnetic anomaly pattern together build a strong case for the drifting apart of the Western Hemisphere from Europe and Africa. We call it the theory of continental drift. The fit of South America and Africa was so conspicuous that as long ago as 1620 Francis Bacon discussed the possibility that the two were once joined, but it was not until the close of the 19th century that real scientific evidence was brought to bear on the concept. Edward Suess, of Austria, noted that formations and many fossils of the continents of the Southern Hemisphere had a close resemblance, and to explain this similarity he postulated that the continents were once connected. The single continent he called "Gondwanaland" from a province in India. In 1910 Alfred Wegener brought together much more geological and paleontological evidence indicating a common record on the two sides of the Atlantic. He dealt with the common kinds of rock on either side where the fit seemed most evident, with common belts of crustal deformation, with common plants and animals that needed land connections to spread from one region to another, and with common temperature zones, such as warm, coal-forming swamps and regions of continental glaciation. He built up a convincing argument, except for one thing; he could not propose a plausible cause of continental drift. But now that we have sea-floor spreading, we seem to understand the cause better.

Figure 12–17 is a recently proposed arrangement of North America, South America, Africa, Europe, and Greenland. It has a calculated "misfit" that does not exceed 1 degree of longitude in any place except the Gulf of Mexico. This is the most obvious grouping of the continents. It is also generally accepted that Eurasia was once bridged fairly solidly over the North Polar region to Greenland and North America before the Arctic Sea Basin evolved. The region was warm, moist, and equable in Late Paleozoic time. However, other continents give us much more trouble, as they did Wegener. He postulated that India, Antarctica, Africa, Australia, and South America were all together—a greater Gondwanaland—but it was

Figure 12-9 *Schematic drawings showing the three stages of magmatic invasion of the crust in the Sierra Nevada.* [*After Bateman and Eaton, 1967.*]

EVOLUTION OF THE CONTINENTS AND THE OCEAN BASINS

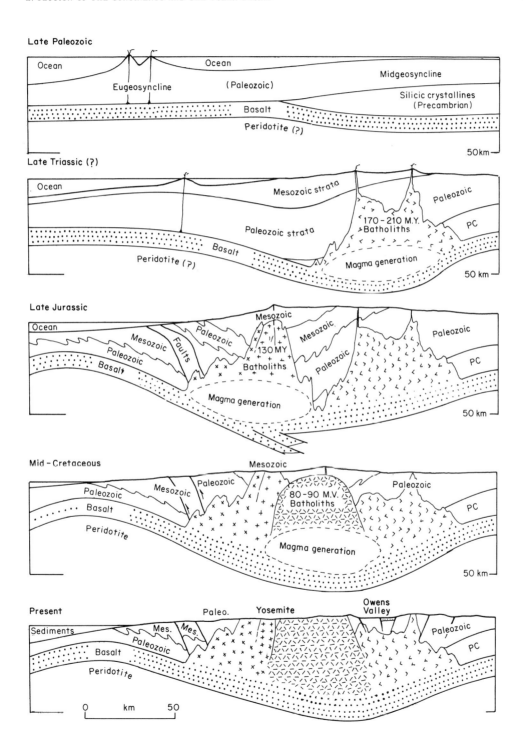

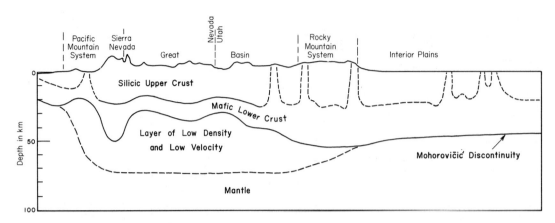

Figure 12–10 Approximate extent of crustal and upper mantle layering of the western United States along the 38th parallel. [After Paliser and Zeitz, 1965.]

awkward to fit Australia and Antarctica into the supercontinent without rotating them inordinately (see Figure 12–18). However, a good fit of Weddell Sea–Princess Martha Coast of Antarctica with the Southeast Coast of Africa has been recognized. The most surprising thing about Wegener's arrangement was the way India broke away from Africa, leaving

Figure 12–11 Cross sections of the Uinta and Wind River Mountains showing nature of the border faults.

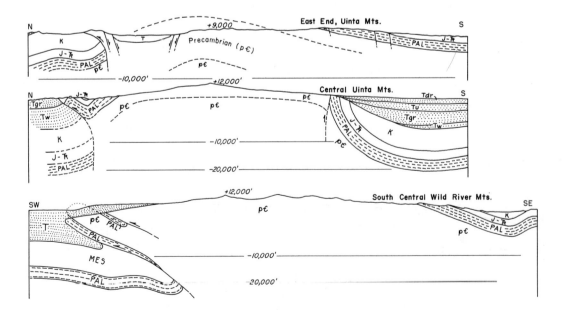

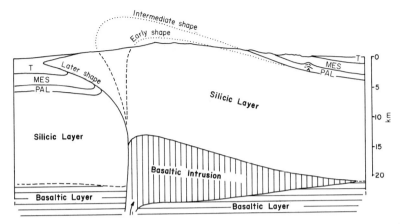

Figure 12–12 Postulated origin of the uplifts by deep-seated basaltic intrusion. The early shape of the uplift is attained after about half of the basaltic pluton is intruded. The intermediate shape is attained after all the pluton is intruded. The final stage is attained by mass movements under the duress of gravity.

Madagascar as a chunk of crust in the widening Indian Ocean and acting as a ram into the continent of Asia piling up the Himalayan Mountain system with its great folds and thrust faults. Wegener proposed that all the continents once composed a single great continent, including Gondwanaland, which he called Pangaea. When we understand the makeup of each of the continents today with their platforms and fold belts, then we find the postulates of Wegener rather plausible in the light of sea-floor spreading.

Pacific Ocean basin

Midoceanic ridges

The Pacific basin is much larger than that of the Atlantic. In fact, it covers half of the earth. There is no apparent match-up of the continents across the Pacific Ocean. There may be a Mid-Pacific Ridge where the Hawaiian Islands lie, but it is as yet not well understood and is probably not now an active zone of spreading (see Figure 12–22). An East Pacific Ridge and a South Pacific Ridge, like the Mid-Atlantic, have all the characteristics of zones of sea floor spreading, but what of the vast stretches of the Pacific from the Hawaiian Islands to the volcanic arcs of Asia?

The old crust of the northwestern Pacific

A summary of the Deep Sea Drilling Project, Leg 6 of the western and northwestern Pacific Ocean has been published[2] in which 17 cores of the deep-sea sediments to depths of penetration up to 348 meters were obtained. Microscopic

[2] *Geotimes*, v. 14, n. 8, Oct. 1969.

Figure 12–13 Schematic representation of fault blocks across the Great Basin from the Sierra Nevada to the Wasatch Mountains. The vertical scale is exaggerated. The uplifted blocks are now much eroded and the depressed blocks covered with waste products. The faults probably extend to depths of 15 km or more.

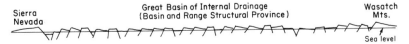

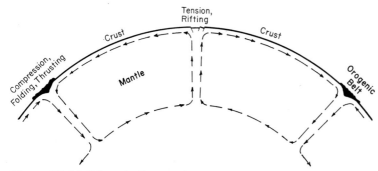

Figure 12-14 *Schematic diagram of convection currents in the mantle and the nature of crustal deformation caused by them.*

Figure 12-15 *Magnetic anomaly pattern over the Mid-Atlantic Ridge south of Iceland. Parallel strips on either side of ridge axis are strikingly symmetrical. Stronger-than-normal magnetic fields are stippled; weaker areas are white. [After Cox.]*

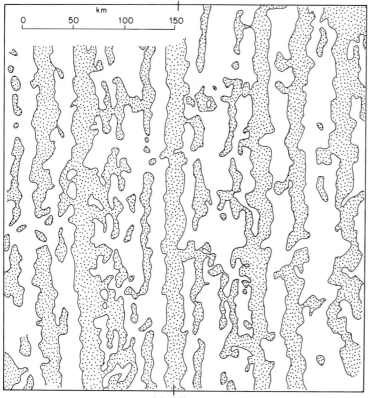

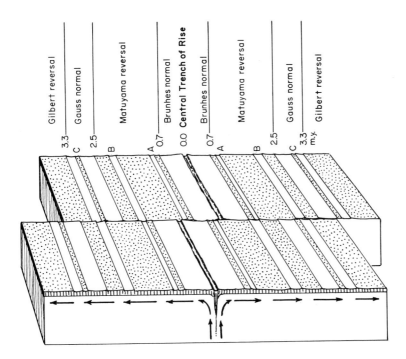

Figure 12–16 Parallel strips of magnetic anomaly across the Mid-Atlantic Ridge showing their symmetry and a transform fault. Stippled lanes are strips of reversed polarization, white strips are normal. The established chronology over the past 3.5 million years is given above the block diagram. A, B, and C are times of short-lived reversals within the designated epochs.

fossils reveal nearly continuous sedimentation from late Cretaceous to the present except in places where a notable unconformity occurs between overlying late Miocene strata and underlying beds as old as late Cretaceous. In two cores in the northwest Pacific late Jurassic sediments were sampled. By means of seismic soundings, the bottoms of these sedimentary deposits can be determined, along with the reasonable certainty that they rest on basalt flows, taken to represent the top of the true oceanic crust. An analysis of the cores and related seismic data results in the following conclusions:

1. The northwest Pacific basin is the oldest part of the greater Pacific basin with the last basalt outpourings occurring in Late Jurassic time (about 150 million years ago), and the western Pacific in the region of the Philippine Sea and the Carolina Ridge is now basaltic crust of Oligocene age (about 35 million years).

2. The Upper Jurassic crust of the northwest Pacific is the oldest known remnant of the oceanic crust anywhere, and the Oligocene crust adjacent to the volcanic arcs, if it once was part of the older crust, has been thoroughly remodeled by volcanic processes or is completely new volcanic crust of Oligocene age.

3. The last basaltic outpourings produced comparatively smooth floors for later sediment

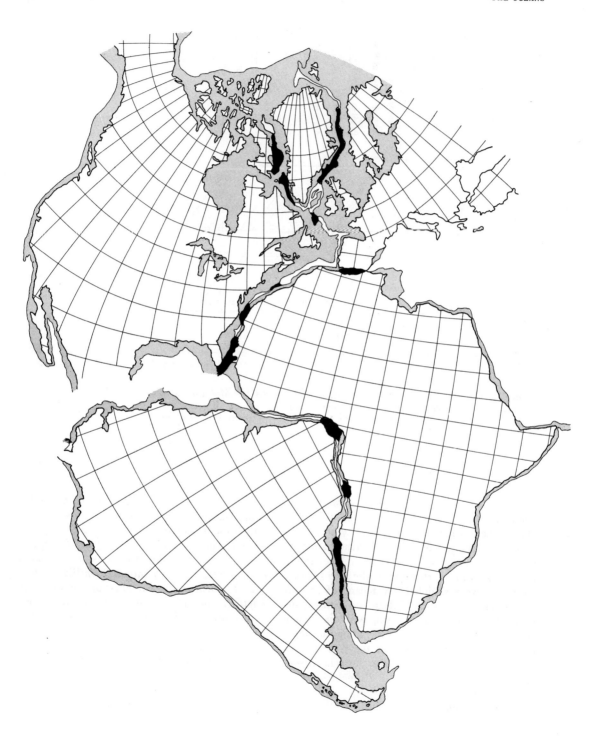

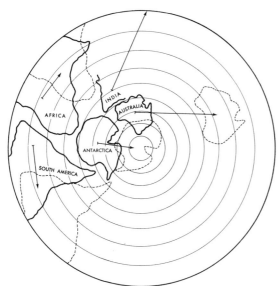

Figure 12–18 Attempted reconstruction of the Southern Hemisphere for mid-Mesozoic time. Dashed lines show continents in present positions, and solid lines show postulated mid-Mesozoic positions shortly after breakup. [After Runcorn, 1962.]

accumulation, unlike those of the midoceanic ridge system, and thus there is cause to think that new oceanic crust may not originate solely by sea-floor spreading as explained on previous pages. This is also cause to question the current concept of island-arc dynamics with the oceanic crust being carried under the margin of the continental crust by convection currents in the mantle (see below).

The Pacific Ocean basin is imagined by some to be the primitive ocean basin of the earth. It would seem to be as old as the primitive continent of Pangaea. The Atlantic, Indian, and Arctic Ocean basins are supposedly younger than the Pacific and caused by the breakup of Pangaea and the drifting apart of the fragments.

Volcanic arcs

The northern, western, and part of the southern Pacific is ringed by a series of majestic volcanic archipelagoes and deep trenches (these are shown in Figure 12–7). Even along the eastern Pacific margin there are deep trenches bordering South America and Mexico, and also adjacent rows of volcanoes on the continental margin. The rows of volcanic islands are convex toward the ocean basin, and the deep trench lies on the convex side of the island arc. The deep-seated earthquakes are also shown in relation to the convection currents in the mantle (under the ocean crust and under the continental crust), as illustrated in Figure 12–19. Basically, in the theory of sea-floor spreading, the suboceanic current must sweep the thin oceanic crust under the continent, and as the crust plunges downward along a fairly steep incline (the Benioff zone) to depths which approach 700 km (those of the deep-seated earthquakes), magma is generated as shown. The depth of magma generation is believed by some to be at the intersection of the Benioff zone and the low-velocity zone in the upper mantle. Earthquakes occur during the generation of the magma. The swept-down crust is presumed to add some silica to the magma, which otherwise would be of basaltic composition. The andesitic magma, thus formed, then rises to form the volcanoes along the inside of the trench. Much has been written on this subject, but the illustrations given must suffice to present the best thinking to date.

The lava erupted from the volcanic arcs is andesitic for the most part. Thus we have a border of andesite around the Pacific. Oceanward from the arcs is the vast domain of the Pacific floor where the volcanoes eject almost only basalt. The line that separates the two volcanic provinces has been called the *andesite line*.

Figure 12–17 Fit of the continents bordering the Atlantic Ocean. Stippled areas indicate continental shelves. White areas indicate gaps in the fit, and dark areas indicate overlap. [After Patrick M. Hurley, The confirmation of continental drift, Scientific American, 1968.]

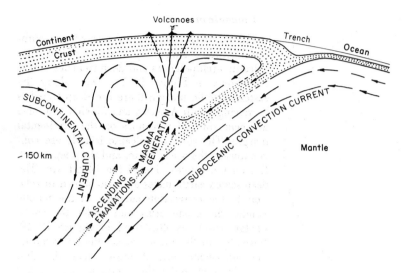

Figure 12–19 *Diagrammatic cross section of the crust and mantle showing suboceanic convection current plunging under the continent and producing trench and volcanic arc.* [*After Sillen,* Science, *1967.*]

Fracture zones of the eastern Pacific

A number of fracture zones up to 3300 miles long have been described in Chapter 11, stretching westward from North and Central America. One of these, the Murray Fracture Zone, intersects the Hawaiian Rise. These great fracture zones are now regarded as transform faults of the East Pacific Ridge.

Eastern Pacific Ridge and the San Andreas Fault

The East Pacific Ridge was recognized and traced along a northeasterly course until it seemed to terminate against the continent of North America at the mouth of the Gulf of California. It has been postulated that it continued under the continent and was responsible for the block faulting or rifting of the Great Basin. Later it was thought that it was offset by the San Andreas Fault of California, and finally by a series of transform faults in staggered arrangement, as shown in Figure 12–20. The San Andreas Fault thus becomes the northern member of the series of transform faults.

Geologists had already concluded that Baja California had been rafted away from the mainland of Mexico, and that the oceanward sliver of the land beyond the San Andreas Fault had been transported as much as 350 miles northwestward. The segment of the fault between Tejon Pass and the Salton Sea shows that similar formations have been offset up to 150 miles. As a result of the analysis of the Tamayo fracture zone, which shows that the crest of the rise there has been offset 75 km (45 mi), it is now concluded that the rafting of Baja California is due to sea floor spreading. Further, the ribbons of magnetic anomaly indicate that the drifting of Baja California away from the mainland started about 4 million years ago, and that it has slid along the several staggered faults a total of about 260 km. Movement in excess of 260 km, as postulated by geologists, could be explained if the excess had occurred before the opening of the Gulf of California.

To the north off Oregon, Washington, and Vancouver Island the ridge has been recognized, where again it is offset by transform

EVOLUTION OF THE CONTINENTS AND THE OCEAN BASINS

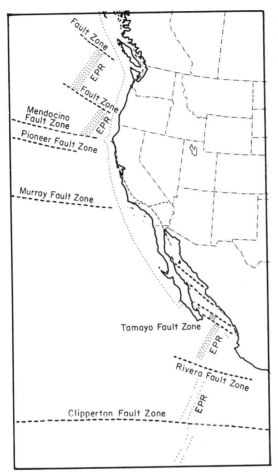

Figure 12–20 *East Pacific Rise and transform faults off the United States and Mexico. EPR, Eastern Pacific Ridge; FZ, fracture zone. [After R. L. Larson, H. W. Menard, and S. M. Smith, Science, August 23, 1968, p. 790.]*

faults, called fracture zones on Figure 12–20. From the mouth of the Gulf of California to the sea floor off Oregon the zone of spreading is now believed represented by the San Andreas fault, a transform fault.

A recent interpretation of the Late Jurassic and Cretaceous strata of the Coast Ranges making up the Franciscan Group is of interest in connection with the East-Pacific Ridge and sea-floor spreading. The Franciscan Group of strata is very thick, displays characteristics of shelf-slope and deep-trench deposition, and is intricately folded and faulted. All this constitutes the compressed, down-dragged continental margin as the oceanic crust was thrust under the continent.[3] Figure 12–21 shows the theory in the context of sea-floor spreading. The deformation of the Franciscan Group occurred approximately at the same time as the procession of batholiths were being intruded to form the Sierra Nevada.

The encounter of the Pacific crust and the North American crust as represented in Figure 12–21 is an organization of crustal movements that occurred in late Mesozoic and early Cenozoic time. But beginning in about mid-Cenozoic time a reorganization of crustal motions set in, and the Pacific crust changed from a collision course to one of lateral movement. The lateral movement is characterized by the northwest translation of the San Andreas block, which has become locked to the Pacific crust or plate. The Pacific plate now seems to be moving to the northwest and carrying not only the San Andreas sliver of continental crust with it, but also a good slice of coastal Alaska.

Major plates of the crust involved in drifting

As a consequence of sea-floor spreading, geophysicists have turned to the study of continental drift in precise terms. That is, the direction and amount of drifting may be specified. Previously, the continents had been fitted together on the basis of outline and internal geology, helped out by concepts of polar migration. The latest effort presumes movement of a few large, fairly rigid segments or plates, rather than of the individual continents (see Figure 12–22). These plates are defined by the mid-

[3] W. G. Ernst, Tectonic contact between the Franciscan melange and the Great Valley sequence, *Jour. Geophys. Research*, v. 78, pp. 586–902. See also Warren Hamilton, Mesozoic California and the underflow of the Pacific mantle, *Bull. Geol. Soc. Amer.*, v. 80, 1969, p. 2409.

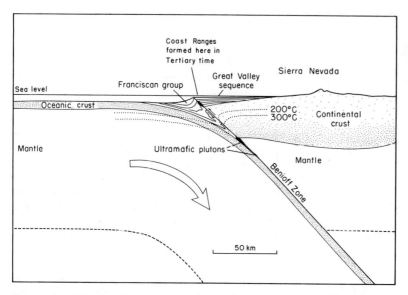

Figure 12–21 *Model diagram showing the theory of sea-floor spreading and development of the Benioff zone under the western margin of the North American continent in California. This is an explanation of the origin and deformation of the Franciscan group of strata. See also Figure 11–9. [After W. G. Ernst,* Earth materials, *Englewood Cliffs, N.J., Prentice-Hall, 1969.]*

oceanic ridges and the deep trenches and associated volcanic island arcs. As previously stated, the midoceanic ridges are regarded as sites of upwelling of convection currents in the mantle, and the trenches and island arcs as the lanes of downwelling. The horizontally flowing convection current between the two complementary components transports the plate of crust along the earth's surface.

Defining the lanes of downwelling is the first problem. The great belts of modern mountain building that exhibit island arc characteristics as well as the island arcs themselves are believed to be underlain by downwelling currents. This is attested to by belts of high earthquake activity which are continuous from the island arcs to the modern mountain systems. The next problem is to match a zone of downwelling with a zone of upwelling, and this is not everywhere apparent. The midoceanic ridges are zones of earthquake activity also, but in a far less frequent and severe way than the island arcs and belts of modern mountain building.

In attempting to recognize the major plates and their boundaries, it has been discovered that the ridges and their transform faults can be treated as longitudes and latitudes, respectively, in global projection, and when plotted define *poles of spreading*. For instance, the ridges and transform faults of the North Atlantic point to a pole of spreading near the southern tip of Greenland. For the Southern Atlantic and the East Pacific Ridge the poles of spreading appear to be close to the earth's present magnetic poles. For the Indian Ocean the pole of spreading would lie in North Africa. Furthermore, the

Figure 12–22 *Major and minor plates recognized to date to be involved in sea-floor spreading. The African plate is assumed to be stationary and the arrows show the directions in which the other plates have moved. [After Sir Edward Bullard, "The origin of the oceans,"* Scientific American, *September 1969, p. 66.]*

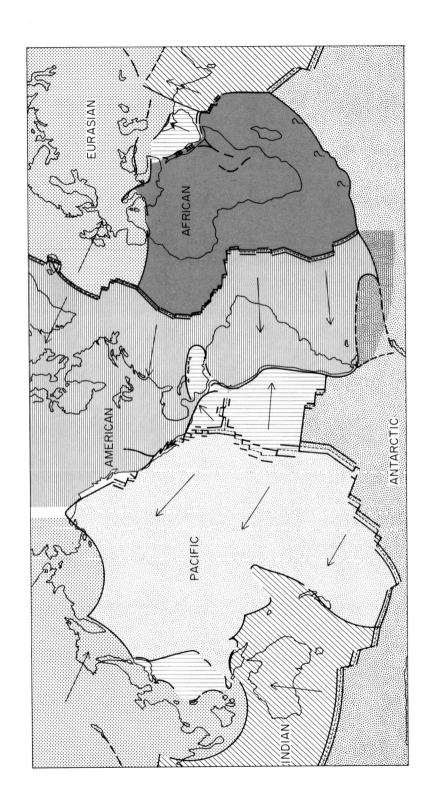

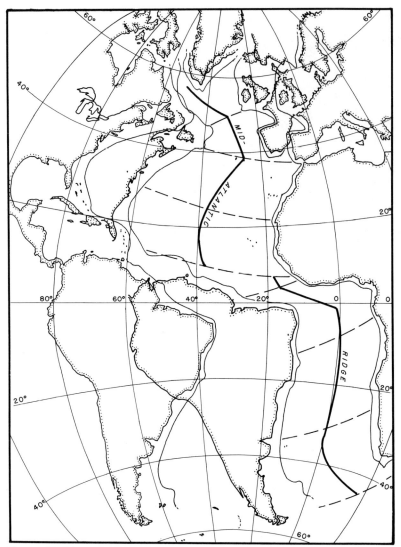

Figure 12–23 *Positions of the Americas and Europe now (hatched) and 80 million years ago (plain) in relation to Africa. Africa is assumed to have remained fixed while Europe and the Americas drifted away. The continental shelves, which are the true edges of the continent, are shown. North and South America began separating from Europe and Africa 120–150 million years ago. The dashed lines are the general directions of the transform faults and indicate the direction of separation of the several continents. [After Heirtzler, 1968.]*

rates of spreading are at a minimum at these poles and increase from about 1 cm per year to a maximum of about 10 cm per year at the spreading pole's equator. By means of these poles of spreading, the ridges and the transform faults, the direction and rates of separation of the major plates in relation to each other have been visualized; at least, the former positions and possible rotation of the plates for Cenozoic time have been suggested. The relations of the continents on either side of the Atlantic and their directions of drifting seem especially clear (see Figure 12-23).

From the study of the ages of the Precambrian fold belts of the several continents, Hurley and Rand[4] show that these very old rocks are clustered together into two great regions in the ancient continent of Pangaea (refer to Figure 12-17). These regions of old rocks are circumscribed and transected by younger fold belts. Further, it appears that this coherent clustering with related fold belts obtained until about 200 million years ago when the continents first began drifting apart. Thus, the origin of fold belts older than about 200 million years may possibly be sought in a regimen of mantle convection circulation and sea floor spreading quite different from that of the present.

The theory of sea-floor spreading and global tectonics is so new and so far-reaching that some of the many ramifications pose questions, and these will represent subjects of continuing research for many years. Several writers[5] have already listed the problems as they see them, and although problems have arisen, none of them seem to indicate that the theory in substance will not survive.

[4] Patrick M. Hurley and John R. Rand, Pre-drift continental nuclei, *Science*, June 13, 1929.

[5] See, for example, Egon Orowan, The origin of the oceanic ridges, *Scientific American*, Jan. 1969, pp. 102-119.

References

Bateman, Paul C., and Eaton, Jerry P. Sierra Nevada batholith. *Science*, Dec. 15, 1967.
Biehler, R. L., and Kovach, C. R. Allen. Geophysical framework of northern end of Gulf of California structural province. In *Marine Geology of the Gulf of California*, ed. T. H. Van Andel and G. G. Shor, Jr. American Association of Petroleum Geologists, Memoir 3, 1964.
Bullard, Sir Edward. The origin of the oceans. *Scientific American*, Mar. 1969.
Cox, A., Dalrymple, G. B., and Doell, R. R. Reversals in the earth's magnetic field. *Scientific American*, Feb. 1967.
Cox, A., Doell, R. R., and Dalrymple, G. B. Reversals in the earth's magnetic field. *Science*, June 26, 1964.
Deep-sea drilling project: Leg 6. *Geotimes*, v. 14, n. 8, Oct. 1969. This is one of the Legs of the *Glomar Challenger*, which is managed by the Scripps Institution.
Dietz, Robert S., and Sproll, Walter P. Fit between Africa and Antarctica: a continental drift reconstruction. *Science*, Mar. 20, 1970.
Eardley, A. J. *Structural geology of North America*. New York: Harper & Row, 1962.
Geological map of North America. U.S. Geol. Survey, 1965.
Hurley, Patrick M. The confirmation of continental drift. *Scientific American*, Apr. 1968.
Hurley, Patrick M., and Rand, John R. Pre-drift continental nuclei. *Science*, Nov. 1968.
King, Philip B. *Tectonic map of North America; Tectonics of North America*. U.S. Geol. Survey, Professional Paper 628, 1969.
Larson, R. L., Menard, H. W., and Smith, S. M. Gulf of California: A result of ocean-floor spreading. *Science*, Aug. 23, 1968.
Menard, H. W. *Marine geology of the Pacific*. New York: McGraw-Hill, 1964.
Menard, H. W. The deep-ocean floor. *Scientific American*, Mar. 1969.
Orowan, Egon. The origin of the oceanic ridges. *Scientific American*, Jan. 1969.
Stirton, R. A. *Time, life and man: The fossil record*. New York: Wiley, 1959.
Takeuchi, H., Uyeda, S., and Kanamori, N. *Debate about the earth*. San Francisco: Freeman, Cooper, 1967.
Talwani, M., LaPichon, X., and Heirtzler, J. R. East Pacific rise: The magnetic pattern and the fracture zones. *Science*, Nov. 26, 1965.
Tectonic map of North America. U.S. Geological Survey, 1969.

part three
the atmosphere

13
characteristics of the atmosphere

Nature of gases

Matter exists in three states of aggregation: gaseous, liquid, and solid. Gases are noted for the following characteristics:

1. They have neither definite shape nor definite volume.
2. They have low density under ordinary conditions of temperature and pressure.
3. They can be compressed markedly.
4. They exert pressure.
5. They diffuse spontaneously one through another, and also through certain liquids and solids.
6. If a gas is confined to a certain volume, and the temperature is increased, the pressure of the gas will be increased.

These characteristics are explained in terms of the atomic theory, which considers gases in terms of its constituent molecules or atoms separated by a vacuum. The distance between molecules or atoms is large compared with the almost negligible size of the atomic or molecular particles, which are always in very rapid motion. They travel in straight lines until their direction is changed by collision with other molecules of the gas or with the walls of the container. As the temperature rises, the number of molecules with high velocity increases. Thus there is a greater number of collisions with the walls of the container and this is manifested by increased pressure. The molecules are perfectly elastic, and collisions do not result in loss of energy. When a gas is compressed, its molecules are simply forced closer together.

The large distances between gas molecules and their continuous random movement result in rapid diffusion without stirring when two different gases are placed together. Since the atmosphere is made up of several kinds of molecules and atoms they remain perfectly diffused, and thus the air people breathe wherever they are has the same composition (polluted air excepted).

Gas molecules have weak electrical forces of attraction that are normally negligible when the molecules are far apart. Their rapid motion carries them away from one another, but as a gas is compressed and the molecules are brought closer together, their forces of attraction overcome the normal tendency to move apart, and suddenly the liquid state of aggregation occurs. A decrease in temperature slows down the motion of the particles and assists in bringing about the liquid state.

Composition of the atmosphere

Permanent gases

The atmosphere is a stable mixture of a number of gases. First, there are the compounds that remain permanently in gaseous condition under all natural conditions. Second, there is gaseous water or *water vapor*, which is a variable part of the mixture. Under certain conditions, liquid and solid forms of water also occur in the air. And third, the air always contains a great number of solid particles of various kinds, known collectively as *dust*.

When all water vapor and dust are removed from the air, we have to deal with the *permanent gases*, 99 percent of which by volume is made up of the elements nitrogen and oxygen. Nitrogen forms about 78 percent and oxygen about 21 percent. Of the remaining 1 percent, the greater part is argon. Only about 0.04 percent remains, of which approximately ¾ (0.03 percent) is carbon dioxide. The components of the atmosphere are shown in Table 13–1.

TABLE 13–1 *Components of the Atmosphere*

GASES	PERCENTAGE OF ATMOSPHERE BY VOLUME	BOILING POINT, °C
Nitrogen	78.09	−195.8
Oxygen	20.95	−183.0
Argon	.93	−185.8
Carbon dioxide	about .03	− 78.0
Neon	.0018	−245.9
Helium	.0005	−268.9
Krypton	.0001	−152.9
Hydrogen	.00005	−252.7
Xenon	.000008	−109.1
Oxone	.000002	−112.0
Radon	Minute traces	− 61.8

The best method of separation of the above gases is by liquefaction of the air itself. The separation takes advantage of the fact that each of the gases has a different boiling point, as shown in the above table. As the liquid air is evaporated each gas is drawn off at its own boiling point. The production and effects of these supercold temperatures is a new frontier of science and is called *cryogenics*. Because of the many industrial uses for pure gases, this process has grown into a billion-dollar-a-year industry. For instance, helium is invaluable in producing corrosion-resistant zirconium, argon is used for welding, krypton for long-lasting light bulbs, and xenon for an experimental anesthetic that brings near-natural sleep. Additional uses of atmospheric gases are as follows: oxygen as an oxidizer in steel-making furnaces and as a rocket fuel; a combination of hydrogen and fluorine as a powerful rocket fuel; and liquid nitrogen for refrigeration, such as flash freezing of fruits and vegetables and preservation of human blood.

Since the days of the Italian chemist Amedeo Avogadro (1776–1856) and the English chemist and physicist Michael Faraday (1791–1867), nitrogen, oxygen, carbon dioxide, and water vapor are known to occur in the atmosphere in the molecular state; that is, in the gaseous state they consist of combinations of two or more atoms of the elements. Two atoms of oxygen are held together by bivalent bonds to form oxygen molecules, two atoms of nitrogen to form nitrogen molecules, two atoms of oxygen are bonded with one carbon atom to form carbon dioxide molecules, and two hydrogen and one oxygen atom bonded together in a V-shape form water vapor molecules (see Figure 13–1). The gases helium, neon, argon, krypton, xenon, and radon, also known as the noble gases, do not naturally react with other substances, nor do two atoms of the same noble gas react with each other. These gases are therefore said to be inert, and hence their fundamental unit is an individual atom. Ozone, in the gaseous state, consists of molecules of three oxygen atoms, and as a natural gas it exists only where lightning strikes and at elevations between 12 and 21 miles above sea level. Ozone will be given special treatment later.

Water vapor

As we have learned in the earlier chapters, water vapor is contributed to the atmosphere chiefly by evaporation from the vast water surfaces of the oceans. The water surfaces of lakes and the soils and plants of the land add appreciable water to the atmosphere, also. Unlike the other gases of the air, however, the amount of water vapor is quite variable and ranges from minute amounts in the air of the deserts and polar regions to a maximum of 4 percent in the warm and humid tropics. The amount of water vapor that may be mixed with the air is small at low temperatures compared with that at high temperatures. In any event, the role that water vapor plays in the atmosphere is all-important. First, it is important because it is necessary to all forms of life, and second, because it affects the temperature, density, and humidity of the atmosphere. Thus it determines in a large measure our climate and environment.

Carbon dioxide

The amount of carbon dioxide in the atmosphere is about .03 percent, and is fairly steady, although it may vary a little from place to place. The atmosphere loses and gains carbon dioxide constantly by exchange with that dissolved in the ocean waters. The atmosphere also loses carbon dioxide owing to photosynthesis in plants. The same is true for the carbon dioxide in ocean waters where marine plants live. The total loss by photosynthesis is calculated to be about 60 billion tons per year. But it is believed that the same amount is returned to the atmosphere by the processes of decay of organic material and by respiration. Thus the exchange is believed to be in balance, and the total amount of carbon dioxide in the atmosphere and the oceans is steady, at least for the short term, geologically speaking.

In the geologic past, during the times of great coal and oil formation some carbon dioxide may have been retained by the earth. Also in the past much carbon dioxide was extracted from the atmosphere and precipitated on the shallow sea floors combined with calcium and magnesium as calcium carbonate (limestone, $CaCO_3$) and calcium magnesium carbonate [dolomite, $CaMg(CO_3)_2$]. On the other hand, carbon dioxide may have been added to the atmosphere by hot springs and volcanoes in extra large amounts at certain times, as recounted in Chapter 9. These two factors, one adding and the other subtracting, may upset the balance of exchange momentarily, and change the amount of carbon dioxide in the atmosphere. Until recently the CO_2 gain and loss were calculated to be about equal, one contributing 0.1 billion tons and one subtracting 0.1 billion tons annually. But now, man has upset the balance by adding a new factor. The combustion of coal, gas, and oil has added about 6 billion tons of carbon dioxide in 1954 alone, and the amount is doubling every 10 years. This is larger by far than any other contribution from the organic or inorganic world, and thus it is calculated that man is increasing the carbon dioxide content of the atmosphere by 30 percent a century.

Does an increase from .03 percent to .04 percent (30 percent increase per century) make any difference to life on the earth? Some think it does, materially! They argue as follows.

1. Because of the relatively low temperature of the earth's surface and atmosphere, nearly all the outgoing radiation is in the infrared region of the spectrum.

2. The three most abundant gases of the atmosphere—nitrogen, oxygen, and argon—do not appreciably absorb the infrared radiation. If these were the only gases in the atmosphere our climates would be considerably colder.

3. Three other gases, although occurring in small amounts, do absorb strongly over a portion of the infrared spectrum. These are carbon dioxide, water vapor, and ozone. Carbon dioxide is fairly uniformly mixed in the atmosphere, but water vapor and ozone are variable from time to time and place to place. The action of these three gases, and carbon dioxide in particular, may be compared to that of a greenhouse where the transparent glass admits the heat rays of the sun but prevents the escape of the outgoing heat waves emanating from the ground and plants.

4. If the carbon dioxide content of the atmosphere should increase, the outgoing radiation would be trapped more effectively near the earth's surface, and the temperature would rise. The latest calculations show that if the carbon dioxide content of the atmosphere were doubled, the average annual surface temperature would rise 3.6° C, and if the amount were cut in half it would fall 3.8° C.

Various authorities estimate that if the annual temperature rose by 4° C, a tropical climate would envelop most of the earth, and conversely, if the temperature dropped 1.5° C to 8.0° C, another glacial period would result. Several natural signs, such as the melting back of many glaciers around the world, tell us that the climate has become warmer, particularly in the past 50 years. Is man causing this change by adding to the carbon dioxide content of the atmosphere, and could it finally bring harmful, if not catastrophic, results to much of the life of the earth?

Dust

Suspended in the gases of the atmosphere is a great amount of dust, which consists of discrete mineral particles plus seeds, spores, pollen, and bacteria. We use the term mineral particles here rather loosely and mean particles that can be identified as true minerals such as mica, quartz, and feldspar, plus shards of volcanic glass, plus soil particles (which includes both mineral and organic materials), and lastly, salt crystals that have been derived from the ocean spray. Carbon particles from smoke also contribute to atmospheric dust. All of these particles, called *aerosols*, are lifted and diffused by the winds and the rising air currents. They are naturally more numerous in the lower atmosphere but some are carried to heights of several miles. The burning of meteors which enter the upper

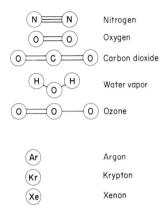

Figure 13–1 Molecules and atoms in the atmosphere. Relative sizes are not indicated.

atmosphere at great speeds contributes dust to the air at great heights.

Large amounts of dust are continually exchanged between the earth and the atmosphere. Dust particles are picked up by the wind from the desert floors and arid plains and later carried down to earth by rain droplets, for which they serve as nuclei of condensation. The wind also picks up salt spray from the oceans, which evaporates, leaving tiny salt crystals as dust. These also serve as condensation nuclei for raindrops. Thus the dust in the atmosphere is an important ingredient of the weather.

The dust in the atmosphere also intercepts some of the heat coming from the sun. When there is an unusually large amount of atmospheric dust, as in times of great volcanic activity, the result is to reduce the amount of heat that reaches the ocean and land surfaces, thus reducing the surface temperatures.

Dust also plays a role in the creation of the colors seen at sunrises and sunsets. For three years after the violent explosion of the volcano Krakatoa between Java and Sumatra, brilliant twilight colors were seen around the world as the dust spread around the globe and then gradually settled.

Over midocean the air has been found to contain 500 to 2000 microscopic and submicroscopic particles per cubic centimeter, and over

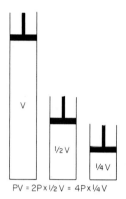

Figure 13–2 *Relation of volume to pressure on a mass of gas at constant temperature.*

large cities more than 100,000 particles. The greater the number of aerosol particles the lower is the visibility.

General characteristics of the atmosphere

Mobility

It is obvious to everyone that air has great mobility. Several times each year we read of hurricane winds that reach speeds of over 100 mph, and of the jet stream at elevations of about 6.5 miles that attains velocities of 250 mph. Unlike water, which seeks the lowest depressions, air moves vertically with great freedom. This is important not only in understanding the weather but in geological processes. As mentioned before, air is readily compressed but is also capable of indeterminate expansion.

The gas laws

Since gas is compressible, we should like to know the relation of pressure to volume. This relation was discovered by the English chemist and physicist Robert Boyle (1627–1691). He found that doubling the pressure on a confined mass of gas reduces its volume by one-half. He stated the law as follows: the volume of a given mass of gas is inversely proportional to its pressure, provided the temperature remains unchanged. It may also be stated as for a given mass of a gas at constant temperature, the product of the pressure times the volume is a constant, or

$$PV = K$$

where P is the pressure, V the volume, and K is a constant.

Since density means the ratio of mass to volume, density also varies inversely with the volume, and therefore the density of a gas at constant temperature is directly proportional to its pressure, or

$$P/D = K'$$

where D is the density and K' is a constant. This is *Boyle's Law* (see Figure 13–2).

So much for pressure effects, but what happens if the temperature changes? We know already that an increase of temperature causes a mass of confined gas to attempt to expand, but if it cannot expand, its pressure on the walls of the jar increases. The exact relation was established toward the close of the nineteenth century by two French physicists, Jacques Charles and Joseph Gay-Lussac. They found that when the volume remains constant the pressure of a gas increases at a rate of $1/273$ of the initial value at $0°$ C for each centigrade degree increase in temperature. This relationship may be expressed as follows:

$$P_t = P_o \left[1 + \left(\frac{1}{273} \right) T \right] \text{ at constant volume}$$

where P_t is the pressure at any given time, and P_o is the pressure at $0°$ C.

It is also true that at constant pressure the volume of a gas increases at the rate of $1/273$ of its initial value at $0°$ C for each centigrade degree increase in temperature. This may be expressed as

$$V_t = V_o \left[1 + \left(\frac{1}{273} \right) T \right] \text{ at constant pressure}$$

where V_t is the volume at any given time, and V_o is the volume at $0°$ C. This is the Charles and Gay-Lussac law, or commonly called Charles's law.

One of the consequences of Charles's law is that at a temperature of $-273°$ C (also written as $0°$ absolute or $O°$ A, also as $0°$ K, the Kelvin scale) a gas should cease to exert any pressure, and the molecules should cease to move. No lower temperature is believed possible. It has been found that real gases do not follow Charles's law precisely at very low temperatures.

The two laws may be combined as follows:

new volume = original volume
$$\times \frac{\text{new temp °A}}{\text{original temp °A}} \times \frac{\text{original pressure}}{\text{new pressure}}$$

Let us consider a specific problem. If a given quantity of gas occupies 100 ml at 0° C and 750 mm pressure, what volume will the gas occupy at 20° C and 770 mm pressure? (1 ml is a thousandth of a liter, and a mm pressure is a thousandth of a meter of mercury pressure as measured by a barometer.) The solution is as follows.

$$\text{new volume} = 100 \text{ ml} \times \frac{293°A}{273°A} \times \frac{750 \text{ mm}}{777 \text{ mm}}$$

new volume = 105 ml *ans*

Pressures in the atmosphere

The above laws begin to make sense when we consider the atmosphere. The air is held to the earth by the force of gravity. If it were not for the sizable pull of gravity, the earth would lose most of its gaseous envelope into space. The atmosphere, therefore, has weight which is indicated by atmospheric pressure. At sea level it is 14.7 lb per sq in. on the average. As we take readings with increasing elevation the pressure decreases. In other words, a column of air 1 in. sq and, in height from sea level, as high as air extends weighs 14.7 lb. The pressure at sea level is, accordingly, about 1 ton per sq ft. The pressure at sea level or at any land surface fluctuates a little from time to time, and this is

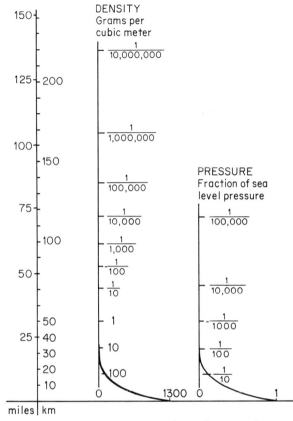

Figure 13–3 *Relation of atmospheric density and pressure to altitude.*

one of our concerns in the study of the weather.

The weight of a cubic foot of air at sea level is about 1.2 oz, or 0.08 lb. The density is therefore said to be 0.08 lb per cu ft. The density and pressure of air decrease with increasing height, at first rather rapidly, and then more and more slowly (see Figure 13–3).

Atmospheric pressure is measured by barometers, which are of two kinds, the *mercurial* and the *aneroid*. To construct a mercurial barometer, a strong glass tube, about 36 in. long and sealed at one end, is first filled with mercury while the sealed end is held down. Then, with a finger held over the open end, the tube is inverted and inserted into a cup or small basin also filled with mercury (see Figure 13–4).

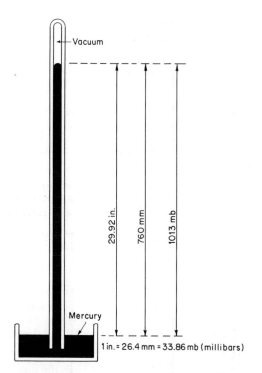

Figure 13–4 *Construction of a mercurial barometer, with scales showing height of mercury at sea level.*

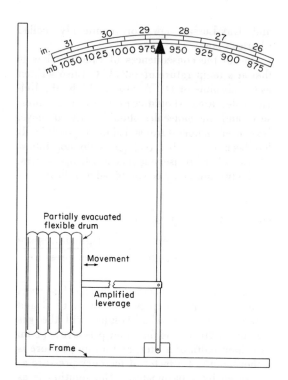

Figure 13–5 *Principle of the aneroid barometer.*

Care must be taken to prevent air from getting into the tube. In the inverted position you will see that the column of mercury settles in the glass tube leaving a vacuum in the top. At sea level the height of the column of mercury above the level of the mercury in the cup will be very close to 29.92 in. or 76 cm, which is equal to the weight of a similar-sized column of air from sea level up. A column of mercury one in. sq and 29.92 in. high weighs 14.7 lb, and hence we determine that the sea level pressure per sq in. is 14.7 lb. The mercurial barometer is an accurate instrument although simple in its construction. There may be one mounted on the wall in your laboratory or in the department somewhere.

An *aneroid* barometer is designed to be portable, and is the circular type you see in airplanes and in weather stations. It is the kind hanging on the wall of your home, perhaps.

The principle involved is a small sealed and partially evacuated metal chamber or tank of which one side is a flexible diaphragm. As the air pressure on the outside of the chamber fluctuates, the diaphragm moves in and out. The movement is magnified by a leverage or linkage system, and this, to a hand on a circular dial. Figure 13–5 is an idealized drawing of the construction. Each instrument should be calibrated to a mercurial barometer in a pressure chamber to insure accurate performance.

Three scales may be shown on mercurial barometers, one ruled in hundreds of inches, one in millimeters, and one in millibars. For the inch scale we read simply *inches of mercury*, and for the millimeter scale again simply *millimeters of mercury*. The millimeter scale may be converted to millibars. For instance, 29.92 in. of mercury equals 760 mm of mercury, which equals 1013.2 mb. The millibar is one-thousandth of a bar, and the bar is defined as a force of one million dynes per sq cm of surface. A dyne in turn is the force necessary to give a mass of 1 g an acceleration of 1 cm/sec/sec. The three scales are shown on the sketch of the mercurial barometer. In meteorological work (see Chapter 15) all three scales are used at one time or another.

Stratification of the atmosphere

Divisions

The atmosphere may be separated into layers according to three criteria. One system of stratification is based on compositional variations, one on pressure characteristics, and one on temperature changes. As shown in Figure 13–6, the strata are called spheres, and according to the classification by temperature characteristics they are as follows: the *troposphere* at the bottom in which all weather occurs, next the *stratosphere*, then the *mesosphere*, and at the top and fading out into space is the *thermosphere*. All the gases of the atmosphere that lie above the stratosphere have been classed as the *upper atmosphere*, one reason being that it lies just beyond the reach of conventional meteorological sounding instruments. Another reason is that it serves as a protective umbrella against excessive ultraviolet radiation from the sun and as a shield to the earth against meteorite bombardment. Here most meteors disintegrate and vaporize due to frictional heat and only reach the earth as meteorite dust. Space exploration has had to contend with the upper atmosphere, and as a result we have learned much about it and are still deeply engrossed in research on it.

The upper boundary of the atmosphere must be determined on an arbitrary basis because its gases seem to thin and fade gradually into a *solar atmosphere* or *solar corona*. Even inter-

Figure 13–6 Stratification of the atmosphere according to temperature and composition.

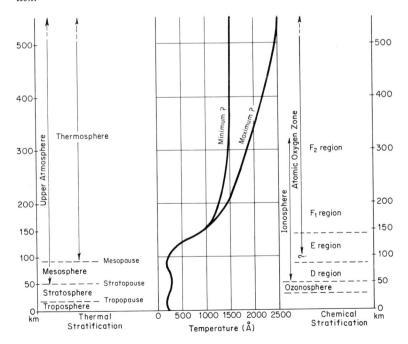

stellar space seems to contain many gas particles per cubic centimeter, which constitute the *solar atmosphere,* and through it the earth and its atmosphere create a wake like a submarine passing through water.

The space-age interest in the upper atmosphere has given rise to a new science called *aeronomy.* It is concerned, we might say, with the physical and chemical properties of the upper atmosphere and the processes occurring within it. Although the term is new, some aspects have been known since the late 1920s, for it was then that radio waves were discovered to be reflected by layers in the upper atmosphere, and this led to further expansion of the science. Needless to say, the efforts to land astronauts on the moon have augmented the study of the upper atmosphere.

Probing the upper atmosphere

The earliest scientific observations of the upper atmosphere were made in the late 1800s on auroras and meteor trails. The next phase of research, 1925–1950, concerned radio waves. Later it became possible to send sensing devices with radio transmitters aloft, by balloons, aircraft, rockets, and finally satellites. The meteorologists and aeronomists, in identifying the kinds of data that should be collected for study, were remarkably able, but the electronic specialists who designed the necessary "hardware" were phenomenal. As a result of the successes of the International Geophysical Year, the most practical and economical means of gathering atmospheric data have come to be sounding rockets.

By the late 1950s a program was inaugurated for regular observations of the lower part of the upper atmosphere through the Meteorological Rocket Network. This is sponsored jointly by the U.S. Weather Bureau, the National Aeronautics and Space Administration (NASA), and the various branches of the armed forces. A dozen or more sites in the United States and Canada have been used to launch meteorological rockets into the upper atmosphere, and soon the schedule reached a scale of multiple launchings per month per site. Within the first four years of the program over 2500 rockets were fired.

Rockets used extensively are the "Loki" and the "Arcas." The Loki carries a payload of 1 pound of 3-band chaff, which after ejection at some predetermined altitude, is tracked by radar as a chaff cloud. The Arcas carries a radio-sonde as its payload. When ejected, it spreads into a 15-foot metallized parachute which is also tracked by radar.

The sun was marked by intense solar activity from July, 1957, to December, 1958, and by international agreement a worldwide coordinated effort was made to observe the solar and terrestrial atmospheres during this interval. The results were highly successful, thus another program was scheduled from January, 1965, to December, 1965, to study the solar and terrestrial atmospheres during a period of minimum solar activity. In this program meteorological rockets played a prime role, and it is encouraging that Argentina, Australia, England, France, India, Italy, Japan, Pakistan, South Africa, the Soviet Union, and the United States have participated.

Compositional variations

The composition of the atmosphere remains uniform, and the laws of gases hold, up to fairly great heights, but then changes occur. In the upper stratosphere from a height of about 20 km to a height of about 50 km, ultraviolet light with wavelengths in the vicinity of 1800 angstroms[1] is absorbed by oxygen to cause the dissociation of the biatomic oxygen into monatomic particles, which then may recombine into triplets called ozone (O_3) (see Figure 13–1). This creates the ozonosphere. Countering the action of 1800-Å light is radiation at 2550 Å which tends to decompose the ozone and maintain a balance. Even so, the amount of ozone in the ozonosphere is

---

[1] An angstrom unit of length is equal to one ten-thousandth of a micron or one hundred-millionth of a centimeter. See Figure 2–17.

only about one part per million of O_2, but the zone is quite effective in shielding the lower layers from the deadly ultraviolet rays, and in absorbing the heat rays.

Above the ozonosphere, radiation of 1216 Å dissociates O_2 molecules into monatomic oxygen giving rise to the "D-Region," which extends to about 85 km.

The absorption of X rays at about 100 km ionizes molecules of hydrogen and nitrogen, and this is characteristic of the "E-Region." Less distinct is the "F-Region," which absorbs radiation at 100 Å and 800 Å, producing positive oxygen ions. Two subdivisions, F, and F_2, develop during the day but fade in the night. The D, E, and F Regions are collectively called the *ionosphere* and have been studied in detail because they affect radio communications. Atomic oxygen is believed to dominate from about 100 km to 1000 km.[2] Atomic oxygen rests above molecular oxygen because it is lighter. Above the atomic oxygen layer hydrogen is the predominant species. Helium ions are present in small amounts in both the atomic oxygen layer and the hydrogen layer. Hydrogen is lighter than atomic oxygen and hence has moved above and decreases in density outward until at a distance of about 10,000 km its density is equal to that found in interplanetary space (see Figure 13–7).

With the very minor but important exceptions noted above, the composition of the atmosphere remains constant up to an elevation of about 100 km, and thus the gas laws relating to pressure, density, and temperature hold true. Beyond 100 km, the rate of decrease in pressure is not as rapid, but it is believed to continue to decrease at least to 800 km (see Figure 13–8).

Temperature stratification

The temperature characteristics of the troposphere are well known, but since this near-surface zone is the one in which weather changes

[2] John H. Hoffman, Composition measurements of the topside ionosphere, *Science*, v. 155, 1967, pp. 322–324.

occur, the entire next chapter will be devoted to it. The stratosphere is by now almost as well known as the troposphere, and of major importance in it is the reversal of the temperature gradient. Whereas the temperature of the troposphere decreases from its value at the ground to about $-55°$ C to $-70°$ C at the *tropopause* (the upper limit of the troposphere), it increases slightly or remains about constant to the stratopause (the upper limit of the stratosphere). Then it decreases fairly sharply again through the mesosphere to a minimum of about $-93°$ C ($180°$ K) at the mesopause, at about 85 km in altitude. This is shown in Figure 13–6. The presence of heat absorbent ozone in the stratosphere causes the temperature to rise.

From the mesopause upward the temperatures rise greatly but the observations have been far less abundant and also less reliable than those below. Due to the attenuation of the molecules and atoms above 100 km the meaning of temperature becomes vague, and the sensing instruments on spaceships are hardly affected by the infrequent collision with gas particles. Little heat is imparted to the instruments, spaceships, or space-walking astronauts by the collisions, although the temperature of the particles themselves may be high. Figure 13–6 is a most recent chart of the temperatures to 700 km, published by the U.S. Committee on Extension of the Standard Atmosphere. Two lines above the 120-km altitude represent the possible variations in interpretations. At any rate, the temperature of the gas particles becomes very high, and these temperatures, when more accurately fixed, will serve as a measure of the speed of the particles, and hence they will help determine the rate at which different gases escape from the atmosphere into space. The variation in the interpretations of the temperatures of the thermosphere are due to variations in the understanding of the amount of photodissociation and recombination of the gaseous molecules at these high elevations.

In summary, it should be emphasized that these characteristics of the upper atmosphere are only explained in terms of energy gained

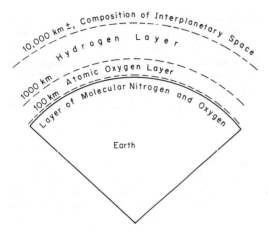

Figure 13-7 Layers of the extremely rarefied outer atmosphere.

through absorption at certain wavelengths of the electromagnetic spectrum. The sun emits a spectrum of energy in the form of waves which travel at the speed of light in all directions, and essentially all the waves at both ends of the spectrum are effectively absorbed at selective levels of the upper atmosphere. Only in the ranges of visible light, infrared radiation, and microwaves do appreciable amounts of solar energy reach the surface of the earth. As the solar radiation spectrum enters and passes through the ever-denser atmosphere, first the gamma and X rays are absorbed. Whereas at the 150 km level the X-ray energy is still close to 100 percent of its original value, the absorption is complete by the 88 km level. Other shortwave bands of ultraviolet light are also absorbed in the 150–88 km region. The effect of the absorption of these energetic rays is the loss of an electron by the molecules and atoms which then become positively charged ions, thus creating the ionosphere.

Other high-altitude phenomena

Winds

Winds in the upper atmosphere are as yet only fragmentarily charted, but very strong winds at a height of 58 km have been recorded over Fort Churchill, Canada, blowing from west to east. These are winter winds and reached a velocity of over 300 miles per hour. Surprisingly, the summer winds at this high altitude blow toward the west. It remains to be seen if winds in the upper atmosphere have any effect on the weather in the lower atmosphere. Recognizing the attenuation of the air at the altitudes of satellite orbits in which only a fraction of an ounce resists the space craft during a complete orbit, it is difficult to imagine that winds of the upper atmosphere do have any influence.

Clouds

Winds and clouds in the troposphere will be our concern in the next two chapters, but clouds in the upper atmosphere are less well known and should be briefly mentioned. They exist at two greatly different elevations. Those at altitudes between 20 and 30 km are called *mother of pearl* clouds, and are frequently observed in Norway. They are believed to be composed of ice crystals. *Night-glowing* clouds, also called *noctilucent* clouds, are sometimes visible at an altitude of 85 km and are very thin and only distinguishable against the night sky while the sun's rays still illuminate them. They are believed to be composed of water vapor, but this is only a guess. Rocket soundings indicate that their temperature is less than 150° K but does not vary between night and day. It does vary seasonally, however, and this probably means circulation of the mesosphere. This circulation brings very small amounts of water vapor during the summer to the high altitudes, and there, saturation or super-saturation at the very low temperatures occurs, producing the clouds.

Both mother-of-pearl and night-glowing clouds have been seen moving rapidly across the sky, indicating high winds. Likewise vapor trails of meteors and rockets confirm high-wind velocities in the upper atmosphere with considerable turbulence and sharp shears.

Auroras and airglow

The normal pattern of absorption and transmission of solar energy is seriously disrupted at

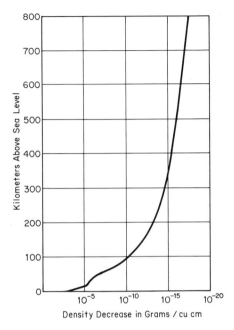

Figure 13–8 *Rate of decrease of atmospheric density with increasing altitude. The gas laws hold true up to about 100 km, beyond which the rate of decrease of density and pressure is not as rapid.*

Figure 13–9 *Photograph of the aurora borealis (multi-arc aurora—College of Alaska). [Courtesy of J. Ohtake, Geophysical Institute, University of Alaska.]*

times of unusual flares on the sun's surface. These solar storms disturb the earth's magnetic field, especially the ionosphere. Radio signals, which are normally reflected by regions of the upper atmosphere, may at these times be either absorbed or allowed to escape into space.

Auroras are similarly caused by highly excited particles attracted into the fields of maximum magnetic intensity (flux) near the earth's magnetic poles. The *aurora borealis*, or northern lights, and the comparable *aurora australis* are concentrated in the high latitudes of the northern and southern hemispheres, respectively. At altitudes of 75 to several hundred km a symphony of colors, particularly green, flash across the northern and southern skies. Sometimes the forms unfold in luxurious drapes, sometimes in radiate bands and streamers (see Figure 13–9). They are most spectacular during solar flares. The light of the auroras apparently is given off by atoms of gas that have been excited by solar radiation. The greenish color is due to the excitation of atomic oxygen. Rarely, reds and purples are produced by excited molecular nitrogen.

Another wave phenomenon of the upper atmosphere is the *airglow*. On certain moonless nights airglow occurs and is due to visible and ultraviolet waves. Astronauts have reported beautiful hues of reds, greens, and yellows at the top of the nighttime atmosphere. Unlike the aurora, airglow does not seem to be controlled by the earth's magnetic field because it is fairly evenly distributed over all latitudes, although not uniformly at any one time. The glow is diffused and its patches move and change shape, and appear to be influenced by air movements. Observations indicate that the visible and ultraviolet waves come from excited particles of oxygen, hydrogen, and sodium, which are concentrated in the lower reaches of the thermosphere.

Van Allen radiation belts

Led by J. A. Van Allen, a team of State University of Iowa scientists equipped satellites to study radiation at extreme distances from the earth (2,000 to 20,000 miles). They found that two intense radiation belts, high above what we have considered the earth's atmosphere, girdle the earth in a doughnut shape around its geomagnetic equator. The inner belt has a maximum intensity exceeding 10,000 counts per sec and is 2,000 to 3,000 miles above the surface. The outer belt also has a maximum intensity that stretches beyond 10,000 miles from the surface of the earth. The radiation particles are primarily electrons and protons and have their own peculiar motions under the influence of the earth's geomagnetic field. They seem to be alien to the atmosphere, which is under the influence of the gravitational field. It is believed that the radiation in the Van Allen belts is produced by bombardment of cosmic rays and not by the rays of the solar spectrum. These cosmic rays are made up possibly of protons but their source is unknown. At least they do not come from the sun because they do not vary appreciably in intensity from night to day in a given place.

References

Blair, Thomas A., and Fite, Robert C. *Weather elements*, 5th ed. Englewood Cliffs, N.J.: Prentice-Hall, 1965.

Miller, Albert, and Thompson, Jack C. *Elements of meteorology*. Columbus, Ohio: Charles E. Merrill, 1970.

Petterssen, S. *Introduction to meteorology*. New York: McGraw-Hill, 1958.

Strahler, A. N. *The earth sciences*. New York: Harper & Row, 1963.

14
the earth's atmospheric environment

In the previous chapter the composition and layering of the atmosphere were discussed. In the present chapter we will be concerned with the atmosphere in motion—the winds, clouds and precipitation, the temperature and pressure variations, and basically, the solar radiation that drives the atmospheric engine. Our aim here will be to describe the weather and analyze the environment with which the atmosphere envelops the earth.

Solar radiation and the atmospheric engine

Heat budget

Energy originating in the sun drives the circulation of the oceans, and likewise it powers the circulation of the atmosphere. The two are intimately related. Meteorologists and oceanographers have tried to understand this tremendous solar engine and its effect on the earth by computing the *receipt* of energy, the amount of energy *stored*, and the amount *lost*. This is known as the *heat budget*, and significant elements of it have already been discussed in Chapter 10, especially as they apply to the oceans. The aspects of the heat budget may be summarized as follows:

1. The energy that the earth receives from the sun is in the form of electromagnetic radiation, or waves that travel at the speed of light. The spectrum of wavelengths, from shortest to longest, is shown in Table 14–1.

2. Radiant energy from the sun passes through the vacuum of space, and upon reaching the outer limit of the atmosphere it has an intensity of 1.94 calories per cm^2 per min. This is known as the solar constant, and it assumes an average orbital distance of the earth from the sun and a measurement on a surface perpendicular to the solar beam. If we assume that in the course of a year half of the energy expressed by the solar constant reaches the earth at lat 49°, the energy received amounts to more than 5 million kilowatt hours per acre.

3. The radiant energy that reaches the outer limit of the atmosphere then proceeds to penetrate the atmosphere, but in so doing, part is absorbed and scattered by the atmosphere, while some reaches the sea or land surface (and) where it is partly absorbed and partly reflected back into the atmosphere. Here again, part of the reflected energy is absorbed by the atmosphere, but a part passes out into space, and is lost. These relations have been illustrated in Figure 10–15. That part of the sun's radi-

TABLE 14–1 *The Electromagnetic Spectrum Coming from the Sun*

	LENGTH IN MICRONS	PERCENTAGE OF SOLAR ENERGY
Gamma rays and X rays (shortest)	1/2000 – 1/100	} 9
Ultraviolet rays	0.2 – 0.4	
Visible light	0.4 – 0.7	41
Infrared rays	0.7 – 3.0	} 50
Heat rays (longest)	3 – 3000	

ation that reaches the earth's surface (land and ocean) is called insolation.

4. The absorption of the sun's energy is preferential. First, the gamma and X rays and a small part of the ultraviolet band are absorbed in the atmosphere. Then the ozonosphere filters out most of the rest of the ultraviolet rays, so

that the chief rays that reach the earth's surface are the visible-light, infrared and heat bands of the spectrum.

5. Somewhat less than 20 percent of the sun's radiation is absorbed by the atmosphere and its clouds, and the remainder is stored temporarily by the land and oceans. The release

Figure 14–1 *Earth positions in its orbit about the sun.*

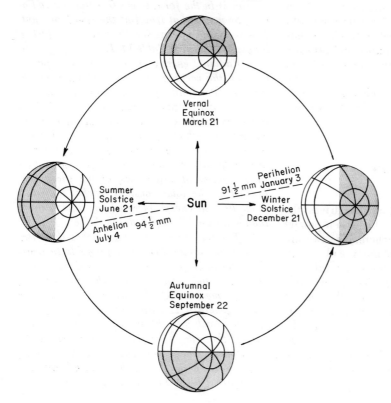

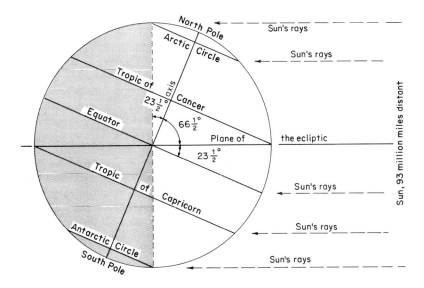

Figure 14–2 *Inclination of the earth's axis to the plane of the ecliptic and the earth's position at the summer solstice for the Northern Hemisphere.*

of this temporarily stored energy chiefly powers the circulation of the atmosphere and the oceans, which is a means of redistributing the energy and producing the prevailing climates.

In the study of our weather, further attention must be given to the *absorption, reflection,* and *transmission* of solar energy. These aspects are outlined below.

Factors determining absorption and reflection

Distance from the sun

The amount of insolation depends on the amount of energy absorbed by the atmosphere, for one thing, and this depends on the clouds, dust, and the humidity. These factors are variable both in time and position and will be taken up presently. Two other factors that determine the amount of insolation received in any one place are regularly recurring, predictable, and subject to calculation. These are the radiation changes that result from the elliptical orbit of the earth around the sun, the angle of incidence of the sun's rays, and the length of the day at any one latitude at the different seasons of the year.

Figure 14–1 shows that on July 4 of each year the earth is at a maximum distance of 94½ million miles from the sun, and on January 3 it is 91½ million miles away. Thus it is about 3 million miles closer on January 3 than on July 4, and consequently receives a little more radiation. The difference is measured to be about 7 percent. However, the effect of this unequal heating in January and in July is not of material difference, as we will see when we consider the most significant factor, the angle of incidence.

Angle of incidence and length of day

During the course of the earth's orbit around the sun its axis of rotation remains inclined to the plane of the orbit (the ecliptic) by 66½ degrees (Figure 14–2). Consequently, the angle at which the sun's rays strike a point on the earth changes from day to day. On June 21 the direct rays of the sun fall on North Latitude 23½°, which is called the *Tropic of Cancer.*

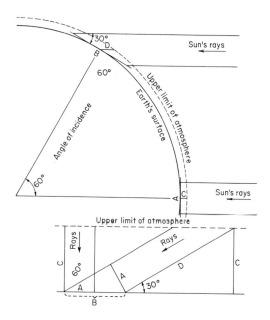

Figure 14–3 *The effect of unequal heating because of variable angle of incidence. The intensity of insolation on the earth's surface when the sun is vertical is shown by the line A, and when its angular elevation is 30° by line B. The ratio of A to B is the sine of 30°. Also the thickness of the atmosphere that the rays must penetrate is shown by C (vertical) and D (inclined) and their ratio is likewise the sine 30°. The relative amount of insolation received at any place on the earth's surface may thus be expressed as the amount received where the rays are vertical times the sine of the angle of elevation of the sun at the point to be determined. In the case illustrated, the angle of incidence is defined as 60°, the angle the sun makes with the vertical.*

Also on this day the sun reaches its greatest noon elevation at all points in the Northern Hemisphere. Six months later, on December 21, the relations are reversed, and the noon rays strike the earth directly at South Latitude 23½°, which is called the *Tropic of Capricorn*. The dates June 21 and December 21 for the Northern Hemisphere are called the *summer solstice* and the *winter solstice*, respectively, and in the Southern Hemisphere they are reversed. By studying Figure 14–1, it will be seen that as the earth moves in its orbit from December 21 to March 21, the rays of the noonday sun move northward and are directly overhead at the equator on March 21. This is called the time of the *vernal equinox*. They continue to shift to the Tropic of Cancer, arriving there on June 21. By September 22 they are directly above the equator again; this date is known as the *autumnal equinox*. At the equinoxes the days and nights are of equal length throughout the world. At the solstices the days are longest or shortest depending on the hemisphere, with either one pole or the other (lat 66½° to 90°) continuously illuminated or in darkness. The dates of the equinoxes and solstices vary by a day from year to year, because the calendar year is slightly out of phase with the solar year.

The greatest variation in insolation is due to the variation of the *angle of incidence*, that is, the angle at which the sun's rays strike the earth's surface. It is defined as the angle which the sun's rays make with the vertical. In viewing Figure 14–3, it is evident that rays coming from directly overhead have the greatest concentration, or energy per unit area, and that inclined rays scatter their energy over a larger area. If the maximum amount of insolation when the rays are vertical is expressed as 100 percent, then the relatively smaller amount of insolation for any angle less than 90° is expressed by the cosine of the angle of incidence. In other words, when the angle of incidence is 0°, the amount of insolation will be 100 percent; at 10° it will be 98 percent (cos 10° = 0.98), at 30° it will be 87 percent, at 45° it will be 71 percent, and at 60° it will be 50 percent.

In Figure 14–3 you will notice that the path through the atmosphere is longer for the inclined rays than for the vertical rays. This means that the amount of absorption and scattering of energy in the atmosphere is more for the inclined rays than for the vertical, and at the same time less energy reaches the earth's surface. The relative thickness is the ratio of C to D, which is the same as A to B, in Figure 14–3. Thus, the factor of absorption in the atmosphere counters absorption by the earth's surface (land and water). The total effect depends on (1) the condition of the atmosphere, chiefly the amount of water vapor in it, (2) the length of the day. It is evident that other factors being equal, the amount of insolation received is directly proportional to the length of time during which the sun shines. With increasing latitude the days increase in length in the summer in the Northern Hemisphere. For instance, the North Pole on June 21 is in continuous daylight, whereas the equator at the same time receives only 12 hours of daylight. The North Pole, on the other hand, receives only 41 percent of the energy received by the equator. Actually, the lines of equal insolation do not follow the latitudes very well because of local influences, principally prevailing cloudiness. Of practical importance is that long daylight hours permit the growing of certain crops, particularly wheat, in the fairly high latitudes, in spite of the short frost-free summer period.

It should also be noted that there is a lag of about one month in effectiveness of high sun in summer and low sun in winter in producing temperature changes; the hottest temperatures in the Northern Hemisphere occur in July, and the lowest temperatures in January. This lag also makes spring on a calendar basis in the Northern Hemisphere colder than autumn.

Variation and effects of solar radiation

Albedo

The percentage of the total incoming light that is reflected light is called the *albedo*. The average reflectivity of the earth's surface, assuming average cloud cover and other atmospheric conditions, is 34 percent. The albedo of the earth is thus 34 percent. The clouds have an albedo averaging 55 percent, while the land averages only 10 percent.

Effect on the atmosphere

The reflectivity of the clouds is strikingly noticeable if one is flying above the cloud cover. There the bright sunshine is made even brighter by the dazzling white clouds below (Figure 14–4). Probably 75 to 80 percent of the light is returned upward and is lost to the

Figure 14–4 View from commercial jet at about 33,000 ft atop a cloud cover over Ohio, January 15, 1967, illustrating the bright reflected light.

earth. But some of the reflected radiation is bounced back and forth from one cloud deck to another, especially when the clouds are scattered, and thence to the earth. The earth radiates not only light waves but heat waves, and these are reflected back to the earth from the clouds. Incoming radiation thus pursues a complex path. It is evident also that a cloud cover, by reflection back to earth, keeps the lower air from cooling rapidly, as it would if there were no clouds present.

The temperature of the air in any one place is due to the solar radiation that is absorbed by it, and the chief cause of absorption is dust and water vapor. Oxygen and nitrogen are practically transparent to the sun's radiation. Dust and moisture absorb largely the long-wave heat rays (infrared), and the ozone is affected by the ultraviolet as well as the infrared. And again the moisture in the air is particularly effective in absorbing the radiation emitted by the earth, because it is mostly in the infrared range of the spectrum.

Dry air in the deserts, if not filled with dust particles, absorbs little radiation, and hence the sun has an intense heating effect on the rocks and soils. The air of high plateaus and mountains is not only likely to be fairly free of dust and moisture but is also less dense than at lower elevations. Both conditions induce rapid warmth when the sun is up, and rapid cooling when it is down. Being in the sunshine in the skiing areas generally means being warm, because of the transparency of the atmosphere to heat waves, which warm the clothing; but note the change the second the sun goes down, and no more heat radiation gets through the atmosphere. Then one feels how cold the atmosphere really is.

Effect on the land surfaces

Forests or dark, freshly cultivated soil will reflect 3 to 10 percent of the radiation that reaches them, but grassy fields reflect 14 to 35 percent. Sandy, dry soil may return 20 percent, and freshly fallen snow 70 to 80 percent. The remaining part of the radiation that reaches the earth is absorbed and changed to other forms of energy. The soil and rocks are good absorbers and heat readily when the sun is shining on them. Because they do not conduct the heat away rapidly, only their surface layer warms up. We note that the daily temperature change is conspicuous at the surface but hardly extends more than 4 inches deep.

The soil and rocks, being good absorbers in the sunshine, become good radiators at night, and thus they cool off rapidly when the source of incoming radiation stops. Moist soil does not heat up as much as dry soil because part of the absorbed radiation is used to vaporize the water.

Snow not only reflects solar radiation upward, it also absorbs and reflects downward any radiation from the soil. Thus, a good snow cover protects the land from large daily changes in temperature and also prevents deep freezing of the soil through a period of extra cold weather.

The rocks and soils of areas outside the tropics may slowly increase in temperature as the longer days and the shorter nights pass by during the summer, and then the temperature may decrease during the longer nights and shorter days of the winter. This is a temperature effect upon which the daily variations are superposed. The reason for this is that the soils are better absorbers than radiators during the times of longer days, and better radiators than absorbers during the time of longer nights.

Effect on water surfaces

We have already mentioned the reflection and absorption of solar radiation on water in Chapter 10, but this was for the purpose of exploring the temperature conditions of the oceans. Now, we are interested in the atmosphere, and the interrelations of the oceans and the atmosphere.

Reflections from water surfaces depend on (1) the smoothness or the roughness of the water and (2) the angle of incidence. The smoother the surface, the greater the reflection. For the angle of incidence the results are a little surprising: the lower the sun the greater the reflection. When the sun is 5° above the horizon 40 percent of the rays is reflected, but when it is at an angle of 50° only 3 to 4 percent is reflected. These observations lead to the conclusion that on the average, water surfaces reflect about as much radiation as the land, and conversely, absorb about as much.

In treating the oceans we have seen that they are the great storehouses of energy, and thus affect our weather and climates immensely. Water differs from land in absorbing and holding energy in four ways:

1. Radiation penetrates deeper in water. More than one-third reaches a depth of 3 ft; more than one-fourth reaches to 30 ft, and very small amounts to depths of 1700 to 1900 ft.

2. The general turbulence (wave action) of the sea distributes heat effectively, at least in the upper 30 ft. Thus, there is a much thicker layer of water to be heated than the 4-in. layer of the soil and rocks, and also it will not rise in temperature as much as the land. It must also be reiterated that the absorbed heat is transported great distances by the currents of various kinds.

3. A large part of the absorbed energy, probably 30 percent, is used in evaporation, and therefore this part does not contribute in raising the temperature of the water. This, together with the penetration and mixing, produce a rather mild temperature effect on the water. Furthermore, evaporation increases the salinity, and thereby sinking of the heavier water and vertical mixing at considerable depths takes place, as we have seen in Chapter 10.

4. Raising the temperature of a certain mass of water 1° C requires three times as much heat energy as heating the same mass of rock or soil 1° C.

For all these reasons, water masses heat slowly, store much energy, and cool slowly. On the other side of the coin, the land areas heat rapidly, and cool rapidly. The oceans are savers, the continents spenders, and therein lies the cause of much of our weather.

Conduction and convection

By *conduction* is meant the transfer of heat in matter without the transfer of matter itself. If an iron rod is heated at one end, the heat passes through the rod slowly to the other end, which gets hot also. Heat is thus conducted by excited atoms which in turn excite adjacent atoms along the iron bar. Now, if the soil or water is warmed by absorption, and is in contact with cooler air, some of the heat of the soil is transferred to the air by direct conduction. If the air is still, the heat reaches only 2 or 3 ft into the air by conduction, but air turbulence picks up the warmed air and carries it to considerable heights. However, meteorologists have calculated that transfer of heat by radiation from the ground or ocean to the air is greater than by conduction.

Convection in the atmosphere is similar to that in liquids, only more free. When air is heated, it expands, becomes lighter, and hence rises, and thus the great commotion in the atmosphere. In meteorology, convection refers to vertical movements of the atmosphere, whereas *advection* refers to the horizontal winds and slow drifting movements of the air.

The winds

Simple convective system

A simple convective system is presented in the four sketches of Figure 14–5. A central low-lying land is flanked by water areas, and the atmosphere consists of horizontal isothermal layers. It is assumed that the barometric pressure is equal at the surface points A, B, and C.

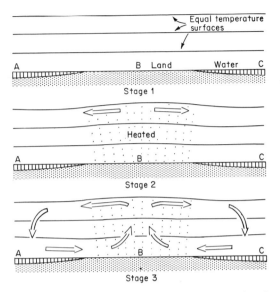

Figure 14–5 Simple convective system of air produced by unequal heating over land and water.

Now, suppose solar radiation warms the land more than the water, with consequently greater warming of the *air* over the land than over the water. The air, thus, expands over the land, as shown in the second sketch (stage 2), and the top of the column of air is elevated over the adjacent columns. This slope or gradient results in the flow of air away from the central elevated column, and with the flow additional air is loaded upon the adjacent columns A and C. With the pressures at A and C now being greater, a near-surface pressure gradient is established between A and B and C and B, and a return circulation ensues (see stage 3).

Atmospheric pressure gradients are basic weather data through which we understand the wind circulations. With modern instruments the pressures at selected elevations can be measured, and thus surfaces of equal pressure can be drawn. The lines connecting points of equal pressure are called *isobars;* an assumed set of isobaric surfaces is shown in cross section in Figure 14–6. The isobars at sea level are shown on the first map below the cross section, and at an elevation of 5000 ft in the second map

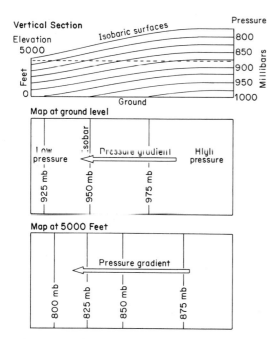

Figure 14–6 Isobars seen in vertical section and on corresponding maps, one on the ground (sea level), and one at 5000 ft.

below. The wind moves in response to the pressure gradient, and *the pressure gradient acts in the direction of lower pressure and is proportional to the spacing of the isobaric contours.* The wind speed will tend to be greater where the isobars are closely spaced, and although the pressure gradient is normal to the isobars and toward the centers of low pressure, the wind tends to be deflected in a direction parallel to the isobars, for reasons which will presently be discussed.

Sea and land breezes

Figure 14–5 outlines the conditions causing the so-called sea and land breezes. On warm summer days, providing no strong winds of regional scale are present, the land and the air above are warmed, the air expands, and a circulation gets started to produce the cooling sea breezes. At Cape Cod, over which the sea breezes blow toward the mainland, the days are generally mild during the summer, while the mainland is warm and muggy. But at night the mainland as well as the Cape cool by radiating heat faster than the sea, and the reverse convection circulation sets in, with the winds blowing from the Cape to the sea.

Valley and mountain breezes

Valleys adjacent to major ranges are commonly cooled at night by breezes coming down the mountain slopes. By day a reverse flow from the valleys up the mountain slopes takes place. At night, the air close to the slopes is cooled rapidly, becomes heavier, settles, and a reverse circulation occurs. The most conspicuous breezes occur in the evenings near the mouths of canyons that funnel the cool air down toward the wide valleys or basins. The cool air moderates the piedmont slopes of the range but is hardly felt at the valley bottom. The flow of cool air often moves downward but over a layer of somewhat stagnant air in the valley bottom, as shown in Figure 14–7.

Laws governing regional atmospheric circulations

Major influences

Wind is an atmospheric circulation in predominantly horizontal directions. The major regional winds with their variable velocities and directions, are due to four influences: (1) the pressure-gradient force, (2) the Coriolis force, (3) the centrifugal force, (4) the force of friction. The simple elements of the pressure-gradient force which cause the atmosphere to circulate have already been discussed. After considering the effect of the Coriolis force, we will consider the combined effect of these.

The Coriolis force

The great gyres of the oceans were explained in Chapter 10 as the result of prevailing westerly winds, the concentration of the sun's energy in the equatorial regions, and the Coriolis force. The effect of the Coriolis force is shown diagrammatically in Figures 10–8 on page 268 and

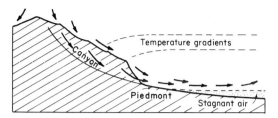

Figure 14-7 *Movement of mountain air to the valley at night.*

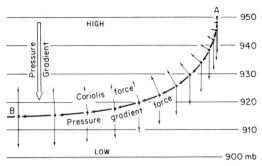

Figure 14-8 *A parcel of air starting at A and moving along the pressure gradient is deflected until it moves parallel with the isobars. It gains velocity until it reaches a maximum at B. [After A. N. Strahler,* The earth sciences, *New York, Harper & Row, 1963; and S. Petterssen,* Introduction to meteorology, *New York, McGraw-Hill, 1958.]*

10-9 on page 269. Since the circulation of the winds is almost unhindered, they are dominated more by the Coriolis force than are the oceans, and hence the aspects of the Coriolis force must be further elucidated. Here are the points to be recalled:

1. *Any body in horizontal motion is deflected by the Coriolis force to the right of its path in the Northern Hemisphere, and to the left in the Southern Hemisphere. The Coriolis force always acts horizontally in a direction that is at right angles to the path of the motion.*
2. *The magnitude of the Coriolis force depends upon (a) the speed of the body and (b) its latitude. It is directly proportional to speed and to the sine of the latitude. By the sine of the latitude is meant that at the equator the Coriolis force is zero, and that it is at a maximum at the poles.*

What is the relation of the Coriolis force to the pressure-gradient force? Let us start with a small mass or parcel of air sufficiently above the ground that frictional drag is unimportant, and say at lat 45° N so that the Coriolis force will be appreciable. Suppose the pressure gradient is southward (low pressure to the south) and that the parcel of air begins to move southward. Refer to point *A* in Figure 14-8 in which the pressure field is shown by parallel isobars. As soon as motion begins, the Coriolis force begins to act, increasing in strength as the speed of the parcel increases, but always directed at right angles to the path of the parcel. In time the veering to the right will have changed the direction sufficiently so that the path of the parcel will be parallel with the isobars (Figure 14-8). At this stage, *B*, the Coriolis force will be opposite and equal to the pressure gradient force, thus the two will be balancing each other. Henceforth the wind will move parallel with the isobars. Arriving at this condition a rule of thumb (using your arms) reads as follows: In the Northern Hemisphere if you stand with your back to the wind, arms outstretched to either side, the region of low pressure is toward your left hand and high pressure to your right hand; in the Southern Hemisphere the relations are just the reverse.

The wind velocity parallel with the isobars depends upon the pressure gradient. In other words, the wind is stronger where the isobars are closely spaced, and milder where they are widely spaced.

In viewing a weather map of North America we may see one, two, or three centers of low pressure and one, two, or three centers of high pressure, and these centers are nicely defined by curved isobars. What is the nature of the circulation around the lows and the highs? The principles of circulation are shown in Figure 14-9 in idealized form. Here the isobars surround a low pressure center and a high pres-

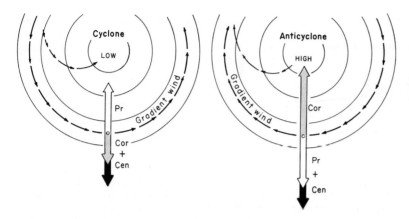

Figure 14–9 Cyclonic and anticyclonic circulations in the Northern Hemisphere, showing the balancing of forces. Note that the pressure gradient force (Pr) and the centrifugal force (Cen) are the same in both diagrams. A theoretical movement of a parcel of air, beginning in the center of each, is also shown. Refer to Figure 14–8.

sure center in the Northern Hemisphere, with the winds moving around the low in a counter-clockwise direction and around the high in a clockwise direction. This is always the rule in the Northern Hemisphere. The circular motion around the low pressure center is called a *cyclone*, and around the high an *anticyclone*, and they should be visualized as normal circulations of the general weather pattern. They have diameters of 1000 to 1500 mi. Cyclones should not be confused with hurricanes or tornados, because the latter are smaller but violent weather spasms, not at all like the cyclones.

Where the wind blows parallel with straight isobars the Coriolis force equals the pressure gradient force, but in circular wind patterns a third force enters the picture. This is the centrifugal force. Recall the example of the satellite in orbit, which would fly off in a straight line if it were not for the pull of gravity downward toward the earth. The pull of gravity is called the *centripetal force*, and, as long as the satellite is in uniform orbit this must equal an imaginary and opposite force, called the *centrifugal force*. Now, a circular wind must be attracted either to the low-pressure center or the high-pressure center by one of these forces, and counterbalanced by the other in the opposite direction, if the wind remains in its circular path. In the case of the cyclone low, the inward force, or the centripetal force, will be the pressure gradient force, and the outward pulling force will be a combination of the Coriolis force and the centrifugal force (study Figure 14–9). In the circulation around an anticyclone high the inward force, or the centripetal force, is the Coriolis force, and the outward or centrifugal force is made up of the pressure-gradient force and the centrifugal force.

Circulation patterns like those described are common 1500 ft above the land surface, and the above principles must govern our thinking when we analyze various weather maps on subsequent pages.

Surface winds

The ground exerts frictional drag on winds at and near its surface. Winds near the ground therefore do not follow the higher regional circulation. The friction may be considered as a force which acts opposite to the direction of the wind, but is proportional to its speed and to the roughness of the land surface. As the friction reduces the wind speed, the Coriolis force is also reduced. But the pressure-gradient force

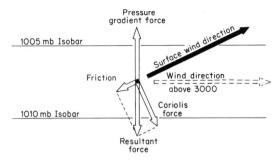

Figure 14–10 *Resolution of forces producing surface winds. [After Petterssen, 1958, and Strahler, 1963.]*

remains the same, and thus comes to dominate the Coriolis force, and effectively turns the winds toward the low pressure. The new direction becomes stabilized when the pressure-gradient force equals the resultant of the force of friction and the Coriolis force, according to Figure 14–10.

Condensation of moisture

Water-air contact

The water vapor content of the air ranges from near-zero to about 4 percent by volume. It not only contributes to the heating and cooling of the earth's surface, but is directly related to the distribution and extent of precipitation over the earth's surface.

We have learned that molecules of water at the interface between water and air are continually escaping the water and entering the air as gaseous molecules. The number that escape in a period time depends solely upon the speed at which they are moving, and this is determined by the temperature of the water. Raising the temperature of the water increases the rate at which the molecules break away and enter the air as water vapor. This process is called *evaporation*. Ice and snow under certain conditions change directly from the solid to the gaseous state, and this is called *sublimation*.

In breaking away from the attraction of the other water molecules in the liquid state, the molecules use heat energy at the expense of the immediate environment. A wet towel in dry air cools off, and the heat that is lost is imparted to the newly formed gaseous water-vapor molecules. This is known as *latent heat of vaporization*. Likewise the transformation from the solid state (ice or snow) to the gaseous state involves absorption of *latent heat of sublimation*. When water vapor returns to the liquid state, which is called *condensation*, the latent heat of vaporation is given off, and in turn it warms the environment.

As the number of water molecules increase in the air, the vapor pressure increases, and likewise the number of water vapor molecules that return to the liquid. When the number that leave the liquid equals the number that return the equilibrium point is reached, and the air is said to be *saturated*. Such a condition of equilibrium is specified by a certain temperature and a certain pressure. If the temperature of the air is now increased, the tendency of the water molecules to return to the liquid will be decreased, the tendency to leave the liquid will be increased, and more water vapor will be added to the air—the molecules will be striving to reach a new state of equilibrium between the water and the air.

Measurement of water vapor

Water vapor is one of the several gases in the air and as such contributes to the total pressure of the atmosphere. Its partial pressure is known as the *vapor pressure* of the air. The amount of water vapor per unit mass of dry air is simply expressed by the ratio

$$\frac{\text{mass of water vapor}}{\text{mass of dry air}} = \text{mixing ratio, or}$$

$$\frac{m_v}{m_a} = r, \text{ where } r \text{ is the mixing ratio}$$

When equilibrium has been established for a certain temperature and pressure, the mixing ratio for such conditions (saturated air) is denoted by r_w. Vapor pressures for saturated

TABLE 14–2 *Saturation Vapor Pressures in Inches of Mercury and Temperature of Dew Point in Degrees Fahrenheit (barometric pressure, 30.00 inches)*

AIR TEMP. (°F)	SATURATION VAPOR PRESSURE (IN.)	DEPRESSION OF WET-BULB THERMOMETER (°F)													
		1	2	3	4	6	8	10	12	14	16	18	20	25	30
0	.038	−7	−20												
10	.063	5	−2	−10	−27										
20	.103	16	12	8	2	−21									
30	.164	27	25	21	18	8	−7								
40	.247	38	35	33	20	25	18	7	−14						
50	.360	48	46	44	42	37	32	26	18	8	−13				
60	.517	58	57	55	53	49	45	40	35	29	21	11	−8		
70	.732	69	67	65	64	61	57	53	49	44	39	33	26	−11	
80	1.022	79	77	76	74	72	68	65	62	58	54	50	44	28	−7
90	1.408	89	87	86	85	82	79	76	73	70	67	63	59	48	32
100	1.916	99	98	96	95	93	90	87	85	82	79	76	72	63	52

From Smithsonian Meteorological Tables.

air at a barometric pressure of 30 in. mercury and at temperatures from 0° F to 100° F are given in the second column of Table 14–2.

The most common way of expressing the amount of water vapor in the air is by the comparison of the percentage of water vapor in the air with the amount that it could carry if it were saturated. This is *relative humidity*. It is expressed by the equation

$$\text{relative humidity} = 100 \, \frac{r}{r_w}$$

A similar term, *specific humidity*, implies the mass of water vapor in a unit mass of air. It would be stated as grams of water in a kilogram of air.

The *dew point* of a given mass of air is the temperature at which saturation occurs when the air is gradually cooled.

A simple device for measuring humidity is widely used by meteorologists. It is called a *psychrometer* and consists of two similar bulb thermometers mounted side by side, but with one having a piece of clean muslin tied around its bulb. The muslin and bulb are dipped in water, and then the air is circulated around the two thermometers. A common way to circulate air is to whirl the two mounted thermometers by means of a cord or chain. The wet-bulb thermometer, because of the evaporation, will usually record a lower temperature than the dry bulb thermometer, which should record the correct temperature of the air. Repeated whirlings will make sure that the wet bulb thermometer has reached a minimum level. The difference between the readings of the two thermometers can be translated into a measure of the relative humidity, as shown in Table 14–3.

Dew points may be determined by reference to Table 14–2. For instance, when the temperature is 50° and the depression is 6°, the dew point is 37°; when the temperature is 60° and the depression is 10°, the dew point is 40°.

Adiabatic temperature changes

If a gas is compressed, its temperature rises and, if allowed to expand, its temperature falls. You have experienced the cooling effect of air being released from an automobile tire, and if you have had to use an old-fashioned tire pump you will know that it gets hot from the repeated

TABLE 14–3 *Relative Humidity, Percent (barometric pressure, 30.00 inches)*

AIR TEMP. (°F)	DEPRESSION OF WET-BULB THERMOMETER (°F)													
	1	2	3	4	6	8	10	12	14	16	18	20	25	30
0	67	33	1											
10	78	56	34	13										
20	85	70	55	40	12									
30	89	78	67	56	36	16								
40	92	83	75	68	52	37	22	7						
50	93	87	80	74	61	49	38	27	16	5				
60	94	89	83	78	68	58	48	39	30	21	13	5		
70	95	90	86	81	72	64	55	48	40	33	25	19	3	
80	96	91	87	83	75	68	61	54	47	41	35	29	15	3
90	96	92	89	85	78	71	65	58	52	47	41	36	24	13
100	96	93	89	86	80	73	68	62	56	51	46	41	30	21

From Smithsonian Meteorological Tables.

strokes of compression. Turning to natural conditions in the atmosphere, we find that ascending air becomes cooler because it expands, and conversely, descending air becomes compressed and warms up. This seems evident, but what is generally not appreciated is that the temperature change is an internal energy affair, and that heat has not been added to or taken away from the air. Mechanical energy has been changed to heat energy and released in the case of the tire pump, and energy, as heat, has been absorbed as the air is released from the tire. So, an ascending current of air cools without any help from the outside, and the process is said to be *adiabatic*. The term implies "without transfer of heat." The significance of adiabatic cooling may be appreciated when we learn that the temperature of dry air drops at the rate of 1° C per 100 m of ascent (5.5° F per 1000 ft), and increases at the same rate in descending.

When considerable moisture is present in rising air, the cooling may cause the condensation of some of the water vapor. The level at which condensation begins is known as the *lifting condensation level*. As the air rises above this level it will continue to cool adiabatically, but now another temperature influence sets in. Latent heat of condensation is given off and tends to warm the air. The adiabatic cooling is dominant, however, although its rate is retarded by the latent heat. The retarded rate is called the *wet adiabatic rate*. It depends on the temperature and pressure and how much of the condensed moisture is carried along with the rising air or how much falls as rain. It varies from 0.4° C to nearly 1° C per 100 m; an average value for warm temperatures is about 0.5° C per 100 meters or 3° F per 1000 ft.

The conditions are reversed when an air current descends. It is warmed by compression, its ability to hold water vapor increases, and there will be no lifting condensation level or no condensation at all. On the contrary, if liquid water in the form of raindrops is carried downward, these will evaporate. Thus we may have the adiabatic rate in reverse. However, usually when condensation has occurred, part of the water will have fallen and the rest disappears soon after the air begins its descent. During the remainder of the descent the air is unsaturated, and follows the normal adiabatic rate, and arrives at its original elevation somewhat warmer than when it left. Figure 14–11 is a chart of the processes involved. A parcel of air (*A*) beginning at 80° F at a pressure of 1,000 Mb rises to 5,000 feet and cools at the adiabatic rate of

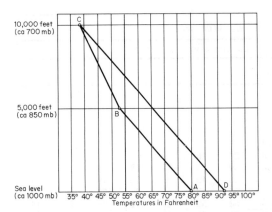

Figure 14–11 Adiabatic changes in temperature before and after becoming saturated. In terms of potential temperature, that of the air of B at 5,000 ft would be the temperature of A; that of C at 10,000 ft would be the sea level temperature of D.

unsaturated air. At 5,000 feet (B) condensation begins and continues to 10,000 feet (C). The rate of cooling from B to C is at the rate of saturated air (dry adiabatic minus the latent heat of condensation). The parcel of air at C, which has lost much of its water vapor, now begins to descend, and since no condensation will occur, its temperature rises at the unsaturated adiabatic rate, and it arrives at the 1,000 Mb level at a temperature of $92\frac{1}{2}°$ F.

The adiabatic chart

The adiabatic chart (Figure 14–12) is a graphical means of solving problems that involve relationships of pressure, temperature, and water vapor in the atmosphere. The ordinate (vertical coordinate) is a function of pressure, as you will see in Figure 14–12, and the abscissa (horizontal axis) represents temperature, increasing to the right. It has two sets of lines, each representing functions dependent on temperature and pressure. They are:

1. The dry adiabats (heavy solid lines) or lines of equal potential temperature. They show a rate of 5.5° F per 1000 ft, or 1° C per 100 m.
2. The wet adiabats (dashed lines), which are numbered in degrees A (absolute) according to the equivalent potential temperature of a saturated parcel of air situated on the line.

The use of the adiabatic chart will be illustrated in some examples described in the next paragraphs.

Lapse rates

Temperature generally decreases with increasing elevation, but not always. The rate of change is not always the expected adiabatic increment, and the reasons are as follows: (1) the air is not always rising or falling, and therefore not always changing adiabatically; (2) air is constantly gaining and losing heat by radiation, absorption, and conduction, and also by evaporation and condensation; (3) horizontal movements bring warm or cold air from other sources. The temperature distribution that you find, may therefore be quite different than that expected from the adiabatic rates. The vertical temperature gradient, as a concept, is expressed by the meteorologists as a *lapse rate*, and it means the actual change in temperature with elevation.

From the many data available meteorologists have realized that the lapse rates have great variability from time to time and place to place, especially up to the first 2 or 3 miles, where they are greatly influenced by air circulations. Higher than this the rates are more nearly uniform. Under certain conditions the temperature increases with elevation instead of decreasing—this is called an *inversion,* or an example of a negative lapse rate. Inversions may be due to rapid radiation from the ground surface at night with consequent conspicuous cooling of the near-surface layer more than higher layers, but inversions are also noted locally at higher elevations. The inversions aloft are usually caused by warmer air advected over colder air. The average observed lapse rate at lower elevations is 3.3° F per 1000 ft, but it increases slightly with higher elevations.

If we find that a mass of air (such as plotted as rate 1 on Figure 14–13) is a little heavier

THE EARTH'S ATMOSPHERIC ENVIRONMENT 365

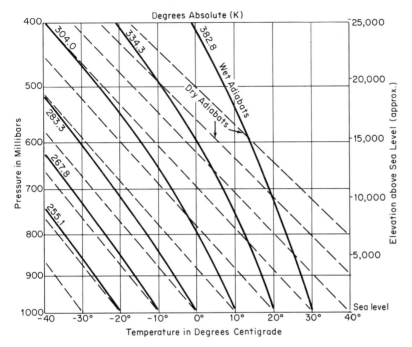

Figure 14–12 *Adiabatic chart.*

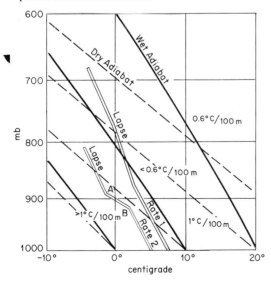

Figure 14–13 *Lapse rate 1 shows a stable air, and lapse rate 2 shows an unstable air for a short interval, plotted on an adiabatic chart.*

than normal, it should be descending. Or if it should be forced upward through the surrounding air it will cool at the adiabatic rate and will continue to have a greater than normal rate, and upward motion will stop. This generally gives rise to fair weather, and as such we call it *stable*. The rule used by the weather experts is: Unsaturated air is stable when its lapse rate is less than the dry adiabatic rate, and saturated air is stable when its lapse rate is less than the wet adiabatic rate. Thermal convection is not possible in these conditions.

Now if part of a column of air is as illustrated by *A-B* in the lapse rate 2 of Figure 14–13, with a rate greater than the dry adiabatic rate, it is warmer than the surrounding air. It is light air, and will favor upward movement and convection. It is said to be *unstable air*. Unstable air is conducive to cloudiness and rain.

Assume that the lapse rate of the air is as

shown by the double line in Figure 14–14. This is a condition between the dry and wet adiabatic curves. Such air is stable when unsaturated but unstable when condensation is occurring. Assume also that a body of air—as represented by the double-dashed line—has considerable water vapor in it, that it is lifting and is being cooled adiabatically. The portion below A cools at the dry adiabatic rate. Condensation begins at A. Above A the rising column cools according to the wet adiabatic rate, and becomes warmer than the air around it, and therefore unstable. B is the level of *free convection*, because it is the height at which instability begins. The instability of the rising current was latent until an outside force lifted the air to the level of free convection. When the lapse rate is greater than the dry adiabatic curve, condensation makes convection more active by increasing the difference between the rising air and the surrounding air.

Fog

Condensation of moisture in free air is the result of cooling and assumes a number of forms, such as haze, fog, clouds, drizzle, rain, and snow. These are the result of variations in the moisture content of the air, its movements and turbulence, rate of cooling, and nuclei of condensation. We should understand, first of all, something about the condensation phenomenon. If air is perfectly free of dust it may be cooled below its dew point without any condensation, in which case it is said to be supersaturated. If physical particles in suspension are present, condensation of the water vapor may begin around these *condensation nuclei*. They are essential for changing the gaseous water vapor into tiny physical droplets of water. It has been determined that physical particles of visible size and of *nonhygroscopic* character do not serve well as condensation nuclei, but on the other hand, particles of microscopic size and particularly of crystalline nature attract the water vapor and stimulate the condensation process. By *hygroscopic* is meant a crystal that will attract moisture. This is why salt crystals in the atmosphere, which are the

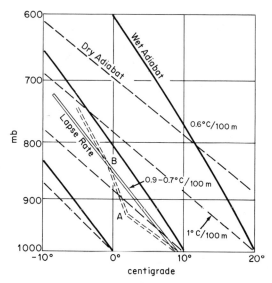

Figure 14–14 Lapse rate lying between the wet adiabats and dry adiabats. Rate shown by double lines is air that does not reach condensation, and rate shown by double dashed lines is for air with considerable moisture.

result of evaporation of ocean water spray, cause immediate condensation when the dew point is reached. Other silicate mineral particles whose atomic crystal structure is somewhat similar to the atomic structure of water or ice may stimulate condensation. The presence of favorable particles in the air may cause condensation even before a mass of air is completely saturated. Nuclei of condensation are present in the atmosphere in large numbers.

Fog and cloud masses consist of microscopically small drops of condensed moisture in sufficient amount to appreciably reduce visibility. The particles range in size from about $\frac{1}{10}$ to $\frac{1}{100}$ mm in diameter, and may remain as liquid droplets appreciably below freezing temperature, surprisingly without freezing themselves. Such conditions cause the rapid icing of aircraft wings. Since the advent of commercial air transportation, fog density is arranged according to visibility, as follows:

Light fog   visibility $\frac{5}{8}$ mi or more
Moderate fog visibility between $\frac{5}{8}$ and $\frac{5}{16}$ mi

Figure 14–15 Smog over Salt Lake City, Utah, is an example of temperature inversion.

Thick fog   visibility between $5/16$ and $1/5$ mi
Dense fog   visibility less than $1/5$ mi

When the moisture droplets grow larger, the fog merges into drizzle, and when the droplets become less numerous and are more widely scattered, they grade into a moist haze. Dry haze results from dust or smoke without moisture.

Most fogs are produced in two ways, by *radiation* and by *advection*. On cold clear nights the earth's surface cools strikingly by radiation, and the adjacent air, if moist, suffers condensation to produce a fog blanket. The cool air tends to collect in the valley bottoms, with the fog appearing as a lake when viewed from above. The fogs disappear when the sun rises, and are said to "burn off," although the earth does most of the warming up.

Advective fog is due to horizontal transfer of air. When warm, moist air flows over a cold dry air, or vice versa, condensation occurs at the interface. The Gulf Stream, with its warm water flowing far northward into boreal waters and adjacent to cold land, is the cause of great fog regions, such as those over and east of Newfoundland, around southern Greenland and Iceland, and around the British Isles. Warm, moist air from the Gulf Stream moves over the cold air of the land, and condensation occurs. Or cold air from the lands of the north moves over a thin layer of warm, moist air, cooling the warm air and again causing condensation of moisture at the interface.

Another type of advective fog is the so-called up-slope fog which occurs commonly east of Cheyenne, Wyoming, and Denver, Colorado. Moist air coming from the east and moving up the Great Plains from an elevation of 500 ft to 5000 or 6000 feet at Denver and Cheyenne is cooled to the point of condensation. Thus a

Figure 14–16 Representative cloud types. *Top, cirrus; middle, stratus; bottom, cumulus.*

broad and fairly durable fog blanket is spread east of the Rockies.

Smog

Smog is the term used to describe the low-visibility air that collects over some of our modern cities. The word "smog" is a contraction of smoke and fog, but sometimes moisture in the air has little to do with poor visibility. Smog is now considered a matter of air pollution, and there is much astir about it. Smog collects where there is poor circulation or where a temperature inversion occurs, or both. Figure 14–15 shows the valley in which Salt Lake City is nestled, with a layer of smog not more than 500 ft thick filling it like a lake. This phenome-

Figure 14–17 A, fair-weather cumulus over Iowa as seen from above; B, band of stratocumulus such as seen in satellite photographs; C, cumulonimbus and associated rain storm. [Courtesy of Harry D. Goode.]

non occurs particularly in the winter when the atmospheric pressures are high, the skies are clear, and the winds are practically nonexistent. Smokestack fumes, automobile exhaust, and other combustion products begin to accumulate as smog, and the more smog accumulates the less insolation reaches the ground to warm up the air at the surface. Soon the air is cold in the city during the day, but up above, particularly in the adjacent mountains, the air is warm. This is then an example of a temperature inversion. If a storm front approaches, the winds quickly remove the smog and the air becomes clear again. Every city has its own peculiar atmospheric conditions that require special study.

Clouds

Some clouds consist of aggregates of small droplets of water, some of ice crystals, and some of both. The particles, as mentioned before, range in size from $\frac{1}{10}$ to $\frac{1}{100}$ mm in diameter and are easily sustained and transported by air movements even as slow as $\frac{1}{10}$ mph. The particles, of course, contain myriads of microscopic nuclei of various inorganic and organic materials.

Much can be determined about the weather by observing the clouds. Since 1803 three principal forms have been recognized (see Figure 14–16). At that time Luke Howard, an Englishman, identified the basic forms and named them *cirrus*, *stratus*, and *cumulus*.

Ten divisions or genera of clouds are described in the standard International Cloud Atlas. Brief descriptions of them are as follows (in reading the descriptions of each you should study the individual cloud photographs and drawings of Figures 14–16, 14–17, 14–18, 14–19, 14–20, 14–21, and 14–22):

Cirrus clouds appear as delicate, thin, fibrous, or silky white tufts, featherlike plumes, or veils. The hooks or filaments on cirrus clouds are termed mares' tails. Because of their great altitude, cirrus clouds sometimes reflect beautiful hues of red or yellow before sunrise and after sunset.

Stratus clouds are more or less uniform cloud layers resembling fog, but not resting on the ground. When such a low layer is broken up into shreds, it may be designated as *stratus fractus*, an indication of bad weather.

Figure 14–18 Fair-weather cumulus over Zion National Park. Sandstone face is part of an ancient sand-dune complex. [Photograph courtesy of National Park Service and Allen Hagood.]

Figure 14–19 Top appearance of stratocumulus over the Appalachians. Jet at about 33,000 ft and top of clouds at about 15,000 ft.

Cumulus clouds are thick and are characterized by their prominent vertical development. The cumulus base is nearly horizontal and the upper surface is dome-shaped with various protuberances. When the light comes from the side, cumulus clouds exhibit strong contrasts of light and shade; against the sun, they look dark with bright edges. During bad weather, cumulus clouds may be shredded by the winds until they have a very ragged appearance and are called *cumulus fractus*.

Cirrocumulus clouds are relatively rare. They appear as small white flakes or small globular masses, without shadows and are associated with cirrus or cirrostratus clouds. The globular masses of cirrocumulus clouds are often arranged in rows.

Cirrostratus clouds are thin, whitish veils which do not obscure the outline of the sun or moon but give rise to halos. Sometimes cirrostratus clouds merely give the sky a milky look. Sometimes they show a fibrous structure with disordered filaments.

Stratocumulus clouds occur in layers or patches of flakes or globular masses. The smallest of the regularly arranged elements appear to be fairly large; they are soft and gray with darker parts arranged in groups, lines, or rolls. Often the edges of the rolls join together to make a continuous cloud cover with a wavy appearance.

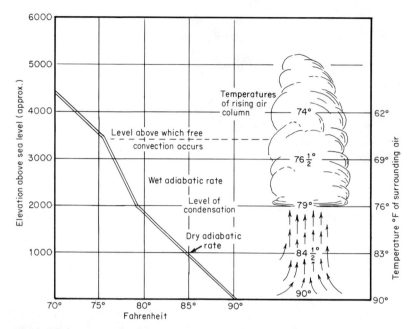

Figure 14–20 *Adiabatic chart showing an unstable air, rising currents, condensation level, and the formation of cumulus.* [*After Strahler, 1963.*]

Cumulonimbus is a heavy mass of cloud with great vertical development whose summits rise in the form of mountains or towers, develop a fibrous quality, and often spread out in the shape of an anvil. Lightning and thunder are associated with cumulonimbus clouds as well as rain, snow, or hail. The base of a cumulonimbus cloud often has a layer of low, ragged stratus fractus below it.

Altocumulus clouds are observed in layers or patches of flattened globular masses, with or without shadows. The globules frequently have definite dark shading. Sometimes they occur in a regular pattern of lines or waves, producing what is called a mackerel sky.

Altostratus clouds are stratified veils of clouds ranging in color from gray to dark blue with a ground-glass appearance. The sun or the moon is visible through thin altostratus, but without the halo phenomenon which is characteristic of cirrostratus.

Nimbostratus clouds are amorphous rainy layers of clouds of dark-gray color. The height of the base of a nimbostratus cloud is usually much lower than that of an altostratus cloud. When nimbostratus clouds produce precipitation, continuous rain or snow results; but the characteristic cloud formation is called nimbostratus even when no precipitation occurs.

Cirrus, cirrocumulus, altocumulus, and cumulus clouds occur in detached masses, usually covering only part of the sky. Precipitation normally does not fall from them. Cirrostratus, altostratus, nimbostratus, stratocumulus, and cumulonimbus clouds form almost continuous layers and often cover the entire sky. Precipitation may occur from any of these except cirrostratus. Cumulus congestus (a species of the cumulus genus) and cumulonimbus develop to great heights, their tops often extending 2 to 5 mi (3 to 8 km) above their bases.

Convection and condensation in cumulus

Cumulus and cumulonimbus clouds develop clearly in updrafts of air. They are the favorite places for gliders (aircraft without engines),

Figure 14–21 *Flat-based cumulus with overlying cirrostratus, Uinta Mountains.* [*Courtesy of U.S. Geological Survey and Max D. Crittenden, Jr.*]

which seek a rising current or thermal that will carry them to higher elevations. As the air rises, it is subjected to adiabatic cooling; if it is moist, it may reach the condensation point. An adiabatic plot of unstable air which starts to condense at 2000 ft or 79° is shown in Figure 14–20. At this altitude the base of the cumulus clouds is defined, but the rising air current may continue much higher, thus continuing to build up the cloud. Figure 14–21 shows a typical development of flat-based cumulus, and Figure 14–22 illustrates three stages, with the last resulting in rain.

Measuring precipitation

Precipitation is any kind of moisture—snow, hail, or rain—that falls to the earth. It is expressed in depth of water, either by inches or centimeters. The word *rainfall* is used synonymously with precipitation, meaning the amount of water in whatever form it has fallen. For instance, the snow that falls in a mountainous region is converted to inches of water, and this is added to the normal rain to give the total rainfall.

A common rain gauge has a cylindrical opening at the top, 8 in. in diameter, which receives the rain. This funnels into a smaller cylindrical column 20 in. long and $\frac{1}{10}$ of the cross section of the 8-in. opening. The depth of the rainfall is thus magnified 10 times, and one inch of rain measures 10 inches in the smaller inner tube.

Snowfall is measured as the actual depth of snow and also as its water equivalent. Snow drifts badly, and the surface of mountainous regions is irregular so that measurements of snow thickness must be taken at several places in each local area and averaged. To measure the water equivalent, the snow may either be melted or weighed. A common practice is to force an 8-in. apperture cylinder downward through a representative layer of snow and cut

Figure 14–22 *Development of convection storm cloud.* [*After Strahler, 1963.*]

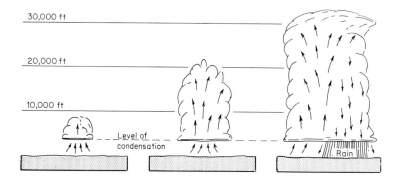

off the section by sliding a thin metal sheet beneath the opening. Then the sample is melted and poured into the small measuring tube, as if it were rain.

A simple balance, supporting a probing tube and calibrated in inches of water, may be carried into the snow field. The tube is then thrust through the snow layer, withdrawn with the core of snow, and weighed by means of the balance. Thus many measurements can be made without bringing the samples of snow into a cabin for melting. With the new snowmobiles a team of two men can measure the snow pack thoroughly after each snowfall. The arid states watch their snow fields anxiously each winter to predict how much their reservoirs will fill up and how much water is going to be available the following summer.

References

Blair, Thomas A., and Fite, Robert C. *Weather elements*, 5th ed. Englewood Cliffs, N.J.: Prentice-Hall, 1965.

Hoffman, John H. Composition measurements of the topside ionosphere. *Science*, Jan. 20, 1967, pp. 322–324.

Miller, Albert, and Thompson, Jack C. *Elements of meteorology*. Columbus, Ohio: Charles E. Merrill, 1970.

Petterssen, S. *Introduction to meteorology*. New York: McGraw-Hill, 1958.

Strahler, A. N. *The earth sciences*. New York: Harper & Row, 1963.

15
air masses and weather prediction

General circulation pattern

Basic model

Figure 15–1 shows the major circulation of the winds of the globe. It is called a model because it illustrates in a simple way the basic surface directions and convections as we believe them to exist. The winds and storms are never quite as ideal as shown, yet the model serves as a base upon which we can superimpose the exceptions and irregularities, and better understand them.

First, the equatorial region receives the greatest amount of insulation and, as we should imagine, is the place where the atmosphere is dominantly warmed and expanded. Hence ascending currents are started in motion and produce a general low-pressure belt. This is the Equatorial Low. The currents rise to the upper troposphere, turn northward and southward as upper air currents, and descend to the lower troposphere a little north of the Tropic of Cancer and a little south of the Tropic of Capricorn (between lat 25° N and 35° S). These become zones of a series of high pressure centers. Now, turning to the cold regions, the Arctic and Antarctic are each high pressure areas, with descending air currents. Such cold currents spread outward to lower latitudes.

The equatorial lows, the subtropical highs, and the polar highs spawn the major wind systems. To understand these systems, we must call upon the Coriolis force, and then the patterns of circulation of Figure 15–1 make good sense. The highs and lows and associated Coriolis-directed winds are almost ideally developed over the wide oceans, but either interfered with or largely modified by the continents.

To the equator side of the subtropical highs lies a belt of the most reliable easterlies. They are best developed over the Pacific and Atlantic oceans, and are called the *trade winds*. They blow more than 98 percent of the time, and fair weather with little rainfall is the rule. The trades converge from the northeast and southeast toward the equator, meeting in a zone called the equatorial trough. Here the air escapes by rising. This low-pressure zone has an axial region of weak pressure gradients, so parts of it have long periods of calms, known as the *doldrums*.

To the north of the subtropical high in the Northern Hemisphere, and to the south of the

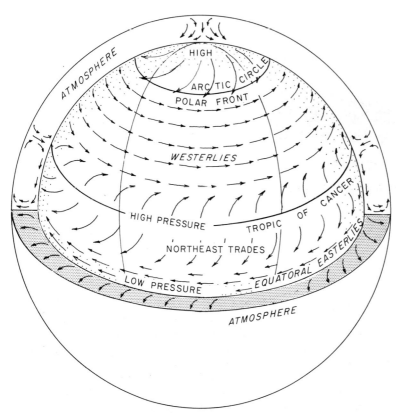

Figure 15-1 *Basic model of global air circulation for the Northern Hemisphere in winter, showing the highs on land and the lows on water. [After Strahler, 1963.]*

subtropical high in the Southern Hemisphere, are the *prevailing westerlies*. The great mass of the continents in the Northern Hemisphere prevents their formation in orderly arrangement, but in the southern oceans, which girdle the earth, they are clearly defined between lat 40° and 65° S. Here the winds frequently reach gale velocities and are known to mariners as the "roaring forties." The westerlies are the result of outflowing air from the subtropical highs which is deflected toward the east by the Coriolis force.

Over the continent of Antarctica is a permanent high, intensified by the extreme cold, and from it and flowing off the great ice cap are winds that spiral counterclockwise. They meet the prevailing westerlies at about lat 65° or 60° S to form a belt of very low pressure, the *subpolar low*. Low-pressure cyclonic storms are generated in this zone and commonly disturb the general wind directions and pressures.

The Arctic region is largely ocean but frozen over, and like the Antarctic, it is very cold, with cold heavy air that settles and spirals outward in a clockwise motion. This polar air meets the winds of the westerlies that are moving in an opposite direction, and complex circulations and convections result. We must investigate them fairly well because they are responsible for the changeable weather of Canada and the United States. The struggle for dominance shifts continually, first moving poleward here and equatorward there. The Norwegians proposed a new theory about this shifting zone

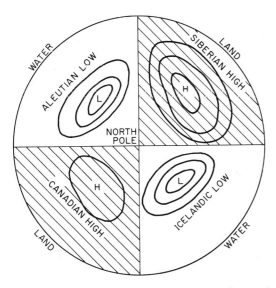

Figure 15–2 Schematic diagram of the quadrants of the Northern Hemisphere.

setting somewhat the latitudinal distributions shown on the model (Figure 15–1). The schematic arrangement of the four quadrants, with lows over the oceans and highs over the land, is depicted in Figure 15–2. In winter the land areas are dominated by large centers of high pressure, induced by the intense cooling. In contrast, the intervening, relatively warm oceans develop large lows. The Siberian high is especially strong and has high pressure gradients. The Canadian high rests over northwestern Canada and is weaker but distinct. The Icelandic low centers over Iceland and southern Greenland in the North Atlantic, and its analog, the Aleutian low, centers in the Gulf of Alaska of the Northern Pacific. In the summer the land areas heat up rapidly while the adjacent oceans lag, thus reversing the pressure relations. Lows develop over southern Asia and North America, and the subtropical high pressure cells over the oceans shift northward, enlarge, and intensify (see Figure 15–3). The Azores high develops in the North Atlantic and the Hawaiian in the North Pacific.

General global barometric pressures and surface winds

The first complete photographic view of the world's weather is shown in Figure 15–4. This is a composite of 450 individual photographs

where polar air meets tropical air, and depicted the meeting place as the *polar front*. It is the interface of cold dry air and warm humid air, and the breeding place of storms.

The middle and high latitudes of the Northern Hemisphere may be divided into quadrants, two mostly of water and two of land, with their own pressure conditions, thus up-

Figure 15–3 Scheme of alternations of surface pressures and winds over a Northern Hemisphere continent in the middle latitudes.

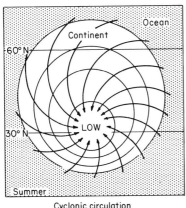
Cyclonic circulation

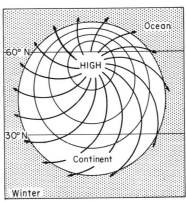
Anticyclonic circulation

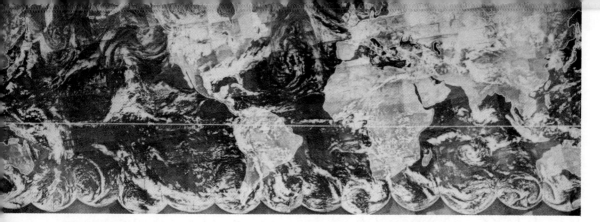

Figure 15–4 A composite of 450 satellite photographs gives a complete view of the world's weather.

taken by the U.S. satellite Tiros IX during 24 hours of February 13, 1965. The white line is the equator, and the clouds seemingly indicate the wind directions, but not necessarily so. Many cloud masses are representative of rising currents associated with air mass fronts, which will be discussed below. It is seen that cyclonic and anticyclonic circulations compose a major aspect of the weather picture. But before trying to analyze this most remarkable photograph, let us take a look at the conventional weather picture. On the same base as the photograph are shown in general form the world's barometric pressures as of January (Figure 15–5) and as of July (Figure 15–6). The heavy solid lines are isobars in inches of mercury.

The January map depicts the major highs and lows and the circulations that we have discussed. The Southern Hemisphere shows the subtropical highs and prevailing westerlies in almost ideal development, because of the great expanse of the oceans. The equatorial low and the prevailing easterly trade winds are interrupted only by South America and Africa, but the large continental mass in the Northern Hemisphere introduces marked changes. The Pacific subtropical high veers northeastward across North America and southeastward across northern Africa. Then it composes itself in a very strong high in central Asia. Note also the Aleutian low and the Icelandic low over the ocean areas. The Aleutian low lies north of the Pacific subtropical high, as the model indicates it should, but the Asian high is so far north as to lie opposite to or at the same latitude as, the Aleutian low. A high appears in northwestern

Figure 15–5 Generalized view of the world's barometric pressure for January.

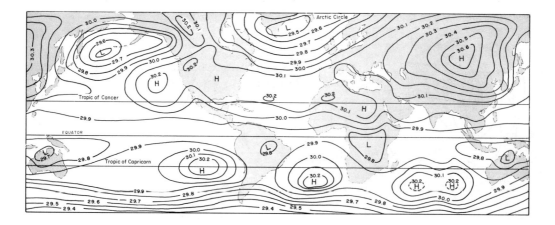

Canada and is an extension of the general Arctic high. The map does not extend sufficiently south to include the Antarctic high and the subpolar low.

The July map of the Southern Hemisphere presents about the same picture as the January map, but for the Northern Hemisphere, because of the influence of much land mass, there are notable changes. The subtropic high in the Pacific has pushed northward and about removed the Aleutian low, and the strong Asian high has disappeared, to be replaced by a weak low. The low is centered in India with a strong pressure gradient to the south or toward the Indian Ocean, thus setting the stage for the *monsoon* system. The development of the summer low, replacing the winter high over the land mass, together with the equatorially warm waters of the oceans, creates an alternation of strongly opposed seasons, dry in the winter and then very wet and warm in the summer. The change bringing the summer monsoon occurs with dramatic suddenness, and it is evident that the change from high to low pressure over India is like a delicate balance that waits for a certain small shift of pressure to swing suddenly from one extreme to the other.

In the North Atlantic the Aleutian low has become faint and has spread over northern Canada, and at the same time the northwestern high of the winter has been erased.

Air masses and fronts

Polar front theory

The concept of *cyclones* and *anticyclones* as major spiral-shaped circulation patterns has been widely developed, and the forecasting of weather conditions is predicated on the paths that these circulations take. The cyclones, or low pressure centers, migrate in paths that are fairly predictable once the low has entered the continent or taken form over the continent. Their movement is generally eastward, and certain typical paths have been named such as are shown on Figure 15-7. The average rate of movement is 20 to 30 mph in the summer and 30 to 40 mph in the winter. The anticyclones over North America are apt to become stationary, and their progress sometimes resembles spreading rather than traveling. The relation of anticyclones to cyclones may be confusing at this point, and rightfully so. It was not until the Polar front theory was introduced that meteorologists began to make sense of the enigmatic details of highs and lows. So now we deal with masses of air that move southward and eastward from northern regions and become in-

Figure 15–6 Generalized view of the world's barometric pressure for July.

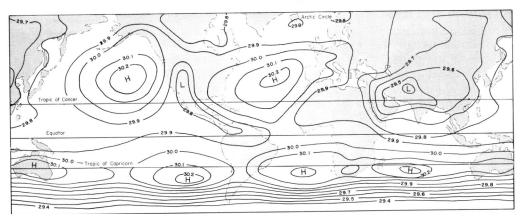

volved in Coriolis force deflections. The leading edges of these air masses are the foci of storms and demand detailed study.

The highs crossing the United States and Canada begin as surges or wedges moving southward from the polar regions and are regarded as moving masses of cold air in the lower troposphere. Their depth is ordinarily only 1 to 2 miles, and they result in cold air near the surface. The warm air which the cold air displaces is forced up and on top of the wedge of cold air. Most remarkable is the lack of much mixing of the two air masses, and a distinct interface of measurable proportions is maintained.

An *air mass* may be defined as a large sheet of air that may cover a part of a continent or ocean and that has a sameness of temperature and humidity at a given elevation. Air masses have distinct boundaries and have properties different from those of adjacent air masses. The boundary between two air masses is called a *front*, and most bad weather occurs along fronts rather than deep inside the air masses.

Sources of air masses

When an air mass remains stationary or moves very slowly, it gradually takes on the characteristics of the surface over which it passes, whether land or ocean. Air over a cold, snow-covered Arctic landmass will lose heat by radiation on the long winter nights and become extremely cold. Consequently it will hold little moisture. Such, then, is the nature of an air mass originating in northern Canada, its source region. In contrast, air that stagnates for a while over a warm tropical ocean will be warmed by radiation and will absorb much water vapor by evaporation. The water vapor will diffuse upward rapidly until the entire layer will be moist, and thus the characteristics of air from a tropical ocean region become well defined.

These air masses migrate because of the earth's prevailing circulation. The cold Arctic air moves southward into the belt of prevailing westerlies, and the warm Gulf of Mexico air moves northward; often, the cold dry air meets the warm moist air in complex frontal arrangements, generating severe storm conditions.

Classification of air masses

By American usage, air masses are classified first by the latitude of their origin and second by the nature of their place of origin, whether over land or ocean. In regard to latitude four types are recognized: arctic (A), polar (P), tropical (T), and equatorial (E). The differences between arctic and polar air are minor, as are those between tropical and equatorial. Arctic and equatorial serve as superlatives to emphasize coldness and warmness of the air masses and to point up extremes of places of origin.

In regard to the place of origin, whether over land or ocean, the air masses are subdivided into *continental* (c) and *maritime* (m). Meteorologists are also interested in whether an air mass is being heated or cooled by the surface over which it flows, because heating favors instability and cooling favors stability. Therefore, if the air mass is warmer than the surface it is denoted by (w) and if colder, it is denoted by (k). The classification by latitude, such as polar, would be prefixed by c, if continental, and the symbol would be written cP. If the air mass is colder than the surface over which it flows, the symbol would have the suffix k, and would appear as cPk. Table 15–1 shows the various classes of air masses with their symbols and source regions.

The source of the air masses is shown on the map of Figure 15–8, as well as the general path of travel of each. Let us first consider the polar continental type, which originates in the vast region of the 50th parallel and from Hudson Bay to Alaska. This type of air mass invades southern Canada and the United States in winter, passing down the Great Plains, generally east of the Rockies, and into the Mississippi Valley and across the Great Lakes region. They bring very cold air and freezing temperatures at times even to New Orleans and Florida, but in passing over the central United States they gradually warm so that the temperatures of the air by the time it reaches the southeastern states

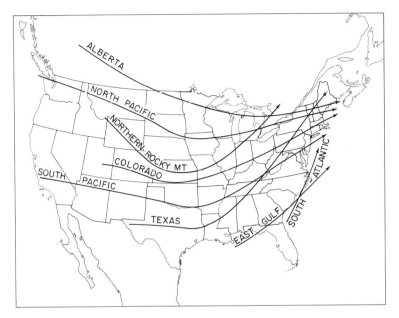

Figure 15–7 Typical paths of cyclones across the United States. [After Thomas A. Blair and Robert C. Fite, Weather elements, Englewood Cliffs, N.J., Prentice-Hall, 1965, p. 160.]

only occasionally falls below freezing, and then only by a few degrees. That part of the polar air mass that pours out across the Great Lakes picks moisture from the lakes and then drops it immediately to the southeast and over the Appalachian Mountains.

Equally important to the central and eastern United States are the tropical air masses that originate in the Gulf of Mexico, Caribbean Sea, and adjoining parts of the Atlantic Ocean. This air is warm, moist, and unstable as it moves northward into the Mississippi Valley, and supplies the water vapor for most of the precipitation there. It meets the continental cold, dry polar air with which it contrasts strongly. The warm tropical air masses also conflict with the

TABLE 15–1 Air Masses of North America

TYPE	SYMBOL	SOURCE REGION
Polar continental	cPk cPw	Canada, Alaska, Arctic
Polar maritime	mPk mPw	Northwestern Atlantic, North Pacific, particularly in the vicinity of the Aleutian low.
Tropical continental	cTk cTw	Southwestern U.S. and northern Mexico, only in summer.
Tropical marine	mTk mTw	Sargasso Sea, Caribbean Sea, and Gulf of Mexico.

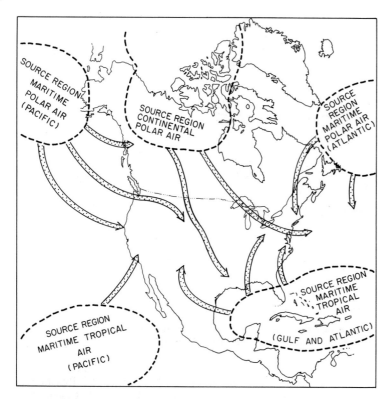

Figure 15–8 *Source regions of air masses affecting North America, and their general paths. [Modified after Strahler, 1963.]*

relatively mild dry air masses from the Pacific, which start moist but in crossing the Rockies lose much of their moisture.

The western coast of Canada and the United States and the Rocky Mountains owe much of their weather to the air masses that originate in the Aleutian low over the north Pacific. This moist maritime air moves southward and eastward into the continent as shown in Figure 15–8 almost constantly at moderate temperature both in winter and summer. In the summer the air aloft is dry and stable whereas the shallow lower air is somewhat unstable, and low stratus and summer fogs result along much of the Pacific Coast. As the air moves inland it is characterized as mPk air, but soon is heated as it passes over the arid lands, mountains, and plateaus, and the low clouds are dissolved. The air by now is dry and generally clear, still being cooler than the surface over which it moves, and when it reaches the Great Plains it is indistinguishable from continental polar air. It thus starts as mP and ends as cP. In the winter a series of cold moist fronts move in from the northern Pacific and bring most of the snow to the Rockies. In passing into the Mississippi Valley they often encounter the warm moist air from the Gulf of Mexico with increased storm activity and snow fall. The paths of the low pressure centers in these fronts have been diagrammed in Figure 15–7.

In Figure 15–8 you will note a source region in the North Atlantic. From this region maritime polar air masses occasionally invade the northeastern states. They bring drizzling rains in the summer and often heavy snow in the winter. The winds, known as the northeasters, are counter to the prevailing westerlies; only occasionally do they make their way to the continent, and then they invade only the New England states.

The source area in the Pacific off Central America spawns air masses that only infrequently reach the southwestern United States. Winter is the time when they do come, and the warm moist air brings heavy rains to southern California. These often mark the end of summer drought periods, and in view of the semi-aridity of the area, often result in damaging floods and landslides.

An elongate low pressure area develops as summer begins in the Sonoran Desert of northern Mexico, western Texas, southern Arizona, southern Nevada, and southeastern California, and persists during most of the hot season. Here on many days our highest temperatures are recorded. The region does not originate air masses of any extent, but is more the seat of intense desert conditions. It is equivalent in a small way to the much larger hot, low-pressure region of northern Africa and the Arabian deserts.

Fronts

Fronts of air masses are shown on weather maps to be of four kinds: *cold, warm, occluded,* and *stationary.* They are designated by conventional symbols such as shown in Figure 15–9. The advancing wedge of a cold air mass displacing warm air is a cold front. In simple cross section it is shown in Figure 15–10. In limited map view the same is shown above the cross section. The advancing wedge pushes the warm air upward and over it. Ground friction slows the cold front somewhat, but even so the cold air flows rather freely ahead as a thin wedge with a slope of 1 in 40 to 1 in 80. This means that at its very edge it is very thin and at a distance of 40 to 80 miles behind the edge it is only 1 mile thick. Thus diagrams are prone to exaggerate the slope of the front. Actually, the entire forward sloping wedge with the warm

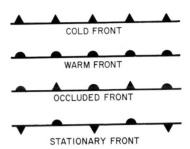

Figure 15–9 Symbols denoting types of fronts. Triangles and half-circles point in the direction of air mass movement.

air above is the *front.* Although the two air masses tend to mix at the front, it is a transition interface in which the temperature, humidity, and wind directions change within a short distance. If the warm air is maritime tropical, as is most often the case, it may break into spontaneous convection, producing dense cumu-

Figure 15–10 Advancing cold front. Cold front arrows fly with the wind.

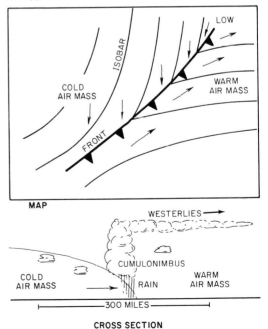

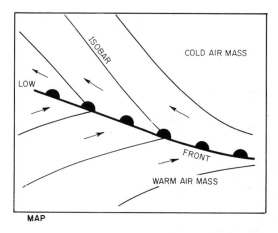

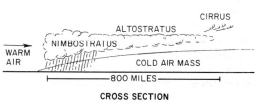

Figure 15–11 Advancing warm front. Warm front arrows fly with the wind.

lus and cumulonimbus clouds with attendant thunderstorms extending to extreme heights. A cold front often produces a line of thunderstorms 200 to 500 miles long. Such fronts pose a great problem to pilots because the storms extend too low to pass under and too high to fly over. If they must fly at all through the front, the pilots must try to discern the individual convection cells, which they do by radar, and steer a course between.

A warm front is produced when a mass of relatively warm air moves into a region of colder air. The cold air will remain close to the ground, while the warm air slides up over it as a broad, gently sloping front. The slope generally has a ratio of 1 to 300. If the warm air remains fairly stable, the slow, forced rise causes adiabatic cooling, and with condensation, stratiform clouds are the rule over the wide zone of the gentle slope. High cirrus and cirrostratus clouds mark the approach of the warm air. As this high fringe passes, altostratus clouds mark the advancing air mass, and finally nimbostratus clouds and a broad zone of light, steady precipitation (see Figure 15–11). If the warm air becomes unstable cumulus clouds and thunderstorms appear locally. If the cold air and the ground is below freezing, then the raindrops freeze while falling through the cold air to form ice-pellet sleet, or may freeze after striking the ground to form the well-known sleet or ice storms of the central and eastern parts of the continent. Warm fronts differ from the narrow and fast moving cold fronts. Warm fronts are 100 to 200 miles wide, move slowly, and bring cloudiness and precipitation over a long period of time. As the front passes the air temperature gradually rises and the winds shift to a southerly or westerly direction.

A third type of front is the *occluded front*. It may occur when a cold front overtakes a somewhat less-cold mass, and the two support a warm air mass above. The warm air has come from a different direction and has been separated from the ground by the other two. The situation is said to be a cold occlusion (see the drawing, Figure 15–12). The cold air masses both move to the right but the warm air comes from the distance (flows toward the reader). This is clearly a three-dimensional picture, and to understand the development of an occluded front we can next turn to the study of the remarkable satellite photographs of the cloud patterns, such as revealed in Figure 15–4. Another type of occlusion is the warm occlusion (see the lower cross section of Figure 15–12). Practically all fronts that enter the northwest coast of the United States are warm occlusions.

Weather from satellite photographs

The photographs and clouds

The first photographs on a regional scale of the cloud patterns were taken from rockets from vantage points many miles above the troposphere. Such a photograph taken from an Atlas rocket, in 1959, showed a cold front extending east of Cape Hatteras and opened a

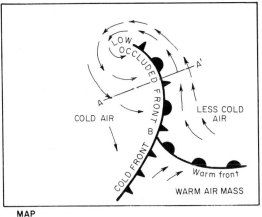

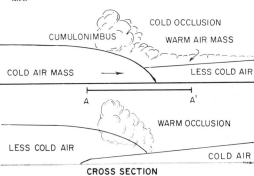

Figure 15–12 *In a cold occluded front, imagine that the cold air front south of point B is crowding and overtaking the northward advancing warm front. The result is lifting of the warm air mass north of B and separating it from the ground, thus producing a warm air layer above the two cold masses. This is an occluded front. In a warm occlusion, the overtaking air mass is warmer, and the relations are shown in the lower cross section.*

new era in the study of the weather. Less than eight months later the satellite Tiros I (Apr. 1, 1960) provided meteorologists for the first time with a large scale view of the weather picture. Since then other Tiros weather satellites have been placed in orbit. The first eight weather satellites orbit at altitudes of 350 to 600 miles, and have projectories inclined 48° to 58° to the equator. These orbits do not offer the best photographic coverage but were imposed by launching vehicle limitations. Tiros IX and X

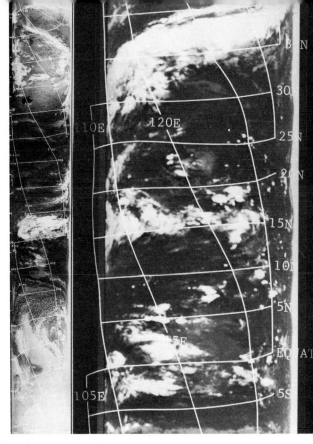

Figure 15–13 *Photographs of Nimbus I (HRIR) showing, by the superposed latitudes and longitudes, the orbital paths of the satellite. Left, a strip composed of several exposures; right, is an individual exposure. The Tiros IX photograph, Figure 7–27, is composed of vertical strips of photos, taken in the same manner as those of Nimbus I. These photographs were taken at night by high-resolution infrared radiometer.*

and the following Nimbus I and II have near-polar orbital paths being at nearly right angles to the equator. This provides for full global coverage and photographs at the same local time on each and every orbit. Nimbus I had an orbit that ranged from 423 km (perigee) to 933 km (apogee). It had an orbital inclination 98.7° and was northbound on the daylight side of the earth, crossing the equator at near local noon. It had performed successfully for 26 days when a malfunction of the solar paddles occurred. Nimbus II has taken the place of Nimbus I (study Figure 15–13 to understand the satellite orbits).

Latest is a new series of satellites of the National Space and Aeronautics Administra-

Figure 15–14 *Pacific Ocean cloud patterns taken on May 1, 1967, by the synchronous satellite ATS-I at an altitude of 22,000 mi. The subsatellite point is near the equator at long. 152° W. The photograph covers a region of the earth about 7000 mi in diameter. [Courtesy of National Aeronautics and Space Administration and Dr. Verner E. Suomi.]*

tion, called Applications Technology Satellites, (ATS), or "Wonderbirds." They are 700-lb radio-TV vehicles and are placed in orbit at about 22,000 miles above the surface of the earth. They are said to be in a synchronous orbit inasmuch as they are above the equator and rotate in the same direction and at the same angular velocity as the earth. They are thus continuously in one place. The camera of ATS-I is over the Pacific and covers an area having a 3500 mile radius, which, as you can see in Figure 15–14, includes a vast region from the equator to the high latitudes. The photographs appear as if the entire earth were included, but not quite. Not only are photographs televised from these lofty points, but infrared readings of cloud-top temperatures are taken, thus helping to identify the layers in the cyclone complexes.

The overall problem is to discern the complete weather picture from the satellite photographs, but since only the clouds show, it is

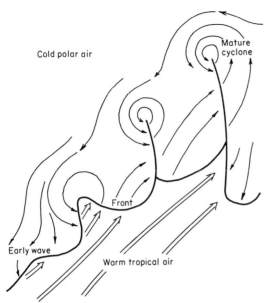

Figure 15-15 *A developing frontal wave may end with an occluded front.*

evident that the complete wind and air mass picture is not available. However, much can be discerned from the cloud patterns and much progress has been made in recognizing the wind directions, the air masses, the fronts, their movements and intensities.

One need only look at Figures 15-4 and 15-14 to realize that the cloud patterns are clear and impressive, and that they define, in part at least, great wind systems. The spiral nature of the cloud patterns, and hence the winds, are more forcefully expressed, perhaps, than many meteorologists had expected. But what of the large stretches of land and ocean not covered by clouds? What are the weather conditions there? It is obvious that the meteorologist must take the cloud patterns where they exist, try to interpret them and use this information to supplement all the other weather information available in making his forecasts.

Where clouds form

As we have learned, the primary cause of cloud formation is adiabatic cooling causing condensation, and hence clouds are formed where air rises. Conversely, cloud-free areas coincide with regions of downward air motion. Early studies had indicated that cyclonic spiral or vortex spins produce rising currents. Thus, the causes of vertical air currents as well as their relationships to cloud distribution should be well understood. The classical concept of cold fronts, warm fronts, and occluded fronts involve vertical components of air flow, and these concepts are now also involved in the interpretation of the cloud patterns. We should recall particularly that in the occluded front in which the cyclone circulation has developed to mature form, the warm air is forced further upward as one cold mass converges on another. The occluded warm air now becomes of great significance in the analysis.

Development of the cyclone

Frontal waves

We should recognize that the cyclonic circulations are most prominent in the middle and high latitudes. The beginning of their development is an indentation or wave in the front separating cold air from warm air. Such a disturbance is called a short *frontal wave*, and the cyclone that develops around each indentation is called a wave cyclone. A developing frontal wave that ends with the occluded front is illustrated in Figure 15-15.

A frontal wave is very unlikely to intensify and develop into a major storm unless the air in the middle troposphere above the surface front becomes agitated in its cyclonic circulation. These perturbations normally accompany the further evolution of the cyclone and front.

With the enlargement of the frontal wave comes increasing vorticity. This means that the paths of air flow curve in ever smaller circles toward a vortex. And with vorticity comes a force to create upward or downward flows. The positions of maximum and minimum vorticity and of downward and upward air flows in relation to the frontal wave is shown in Figure 15-16. Also shown in the figure is the wave trough and wave ridge in relation to the flow lines or wind directions. In words, the region of

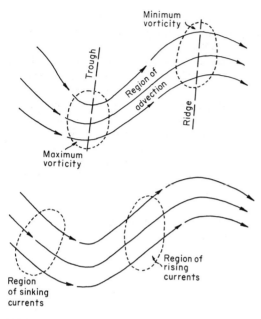

by hydrodynamic theory and mathematics. Conversely, the region to the west of the trough is one of descending currents where the clouds, if present, dissipate. Once the clouds develop, they are carried along considerable distances before getting involved in sinking currents in which they dissipate.[1] The process is illustrated in Figure 15–17. A good example of cloud development in association with flow lines through a trough as seen from a satellite is illustrated in Figure 15–18.

Levels of air flow

As the frontal wave develops (Figure 15–15), what is the nature of flow at different levels—at the ground, in the middle part of the troposphere, and at the tropopause level? The photograph of Figure 15–19 shows a cloud pattern which is considered to represent an early stage of a frontal wave. The schematic drawing of Figure 15–20 interprets the crescent-shaped cloud and gives the inferred front at the surface, the beginning of development of a low-pressure center and surface circulation around it, the bending of the lines of air flow in the middle troposphere starting a frontal wave, and the general high westerly circulation at the top of the troposphere flow just starting to be deflected into a weak trough. Note that the tropopause flow is hardly affected, if at all.

In the next stage, where a pronounced frontal wave has developed (Figures 15–21 and 15–22), strong vorticity has appeared in the cloud pattern. Relating this pattern with the fronts that existed below at the same time, we see that

Figure 15–16 *The frontal wave and areas of maximum and minimum vorticity and of sinking and rising currents.* [*After W. K. Widger, C. W. C. Rogers, and P. E. Shirr,* American Scientist, *September, 1966, p. 288.*]

maximum vorticity is near the trough in the westerly wind region of temperate latitudes, and the air flowing through this maximum leads to the ridge of the frontal wave, the region of minimum vorticity. The wind speed normally increases with altitude between the trough and ridge, and at the same time the wind is propelled upward, thus causing clouds. These flow-line directions can be demonstrated

Figure 15–17 *Formation, advection, and dissipation of clouds.* [*After Widger et al., 1966.*]

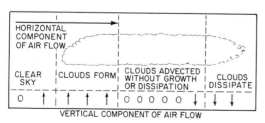

[1] The recent upper-air charts of the National Weather Service show lines of equal vorticity. Here, vorticity is computed from three variables, the horizontal wind shear (rate of change of wind speed in a direction normal to the wind direction), the amount of curvature of the pressure lines, and the Coriolis force at the particular latitude concerned. Using a computer, numerous points of vorticity are determined and then contoured to produce a vorticity analysis chart. By overlaying this vorticity chart on an upper-air chart the wind directions may be compared with the vorticity values. If the vorticity values increase upwind, it is implied that rising air is occurring. If they decrease upwind, then downward motion is implied.

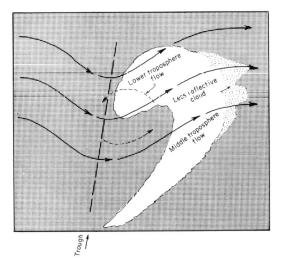

Figure 15–18 Schematic drawing of a crescent-shaped cloud pattern that develops east of the trough of a short wave, where rising currents cause condensation, and advection in the mid-troposphere carries the clouds into the beginning of a spiral. [After Widger et al., 1966.]

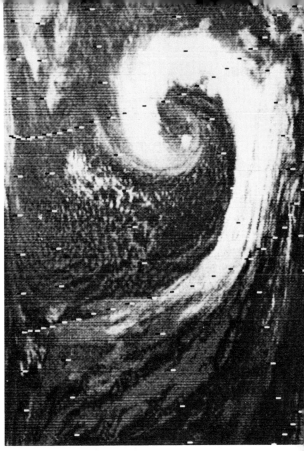

Figure 15–19 Early development of vortex. Photograph of Nimbus II with high-resolution infrared radiometer [Courtesy of National Aeronautics and Space Administration.]

Figure 15–20 Schematic view of an early cyclonic cloud pattern as photographed from a satellite, plus interpretation of fronts and winds. [After Widger et al., 1966.]

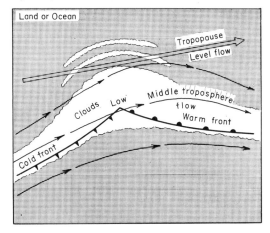

the clouds are mostly in the middle troposphere and reflect the air currents there. A good circulation is taking place around a low at the surface, and the tropopause flow has been bent into a distinct trough.

In the third stage, an occluded cyclone has formed. The relation of the clouds to the fronts is shown in Figures 15–23, 15–24, and 15–25. Vorticity is still strong in the middle troposphere flow, and a strong trough and ridge have formed in the tropopause flow. To be noted also is that some of the middle troposphere flow is leaving the spiral pattern. The tropopause level flow is marked by a strong trough and ridge. As the system occludes, the cold air being brought in from the north and west sinks due to its greater density and thus has very few clouds. Much of the cloudiness ahead of the warm occluded fronts is due to the warm moist air overrunning the cold air.

As the occluded cyclone reaches maturity and shows the beginning of deterioration, it appears

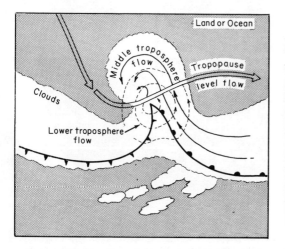

Figure 15–21 Schematic drawing of cloud pattern and interpretation of frontal wave approaching occlusion. [*After Widger et al., 1966.*]

Figure 15–22 Satellite photograph to match Figure 15–21. [*After Widger et al., 1966. Courtesy of NESC.*]

as in Figures 15–26 and 15–27. Note the scattering of the clouds and the diminishing vorticity of the middle troposphere flow.

A cyclone well along in the process of dissipation is shown in Figures 15–28 and 15–29. The clouds are dissipating rapidly, the occluded front has vanished and a new weak stationary front has been established at the southeast margin of the lingering spiral-cloud pattern. And thus the cyclone comes to an end. As it moves easterly, the air will become involved in a new cold front, a new short wave in the front will develop, and thus we are off on another cyclonic development.

Daily weather maps

Source and contents

The National Weather Service publishes a daily weather report in the form of a series of maps, all on one sheet about the size of a newspaper page. Anyone can subscribe to it. The most essential information used in the normal course of weather forecasting is contained on these maps. It is gathered by an elaborate system of stations in all parts of the continent and over the adjacent oceans, relayed quickly to Washington, analyzed, charted, published, and distributed within a few hours. Copies of weather maps should be available for perusal in your

Figure 15–23 Nimbus II photograph of two spirals, each approaching occlusion. [*Courtesy of NASA.*]

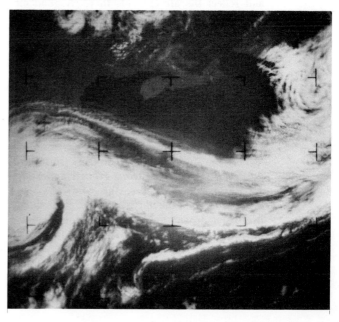

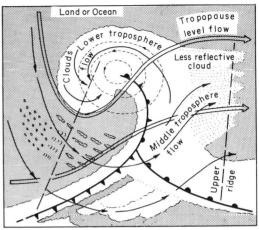

Figure 15–25 *Schematic drawing of cloud pattern and interpretation of an occluded cyclone. The entire system is moving eastward. T, upper trough; R, upper ridge. [After Widger et al., 1966.]*

in making day-to-day forecasts, and more important in making 5-day forecasts. The data for the upper-air map represent a time 6 hours before the lower-level map.

There are three additional small maps—one

Figure 15–26 *Schematic drawing of cloud pattern and interpretation of a mature cyclone in which the winds are beginning to lose velocity and the clouds to dissipate. T, upper trough; R, upper ridge. [After Widger et al., 1966.]*

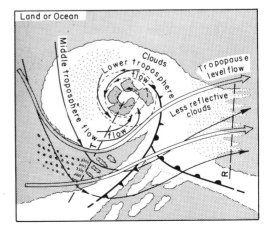

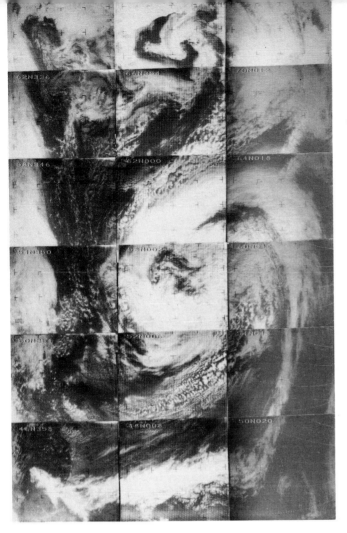

Figure 15–24 *Occluded cyclone. [Courtesy of NASA.]*

laboratory. The largest map on the sheet shows the surface weather, such as air masses, fronts, and cyclonic disturbances. An adjacent, somewhat smaller map contains the same kind of information, but for a time just 12 hours earlier. This is to show the developments during the past half day. Another map presents observations of the upper air, including a contouring of the 500 Mb surface. Since the pressure at sea level is about 1,000 Mb, the 500 Mb pressure surface indicates air at one-half that pressure; this occurs between 17,000 ft and 19,000 ft above sea level. Also included on this map are the temperatures (isotherms) of the 500-Mb surface. The data for upper air conditions are held to be as important as those for lower levels

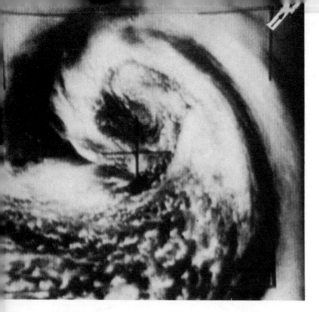

Figure 15–27 Satellite photograph of mature cyclone (compare with Figure 15–26). [After Widger et al., 1966.]

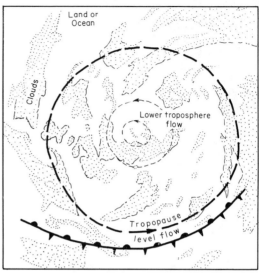

Figure 15–28 Schematic drawing of a dissipating cyclone, with interpretation of winds and stationary front. [After Widger et al., 1966.]

showing the maximum and minimum temperatures of all representative places in each state, another showing the clouds and the nature of precipitation for the day, and, finally, the Weather Service's forecast for the next 18 hours.

Meaning and use of the upper-air maps

Radiosonde rockets record and relay information on the air pressure, temperature, and humidity at desired altitudes. Meteorologists prepare maps from these data. Maps of several pressure surfaces are prepared, but the one used on the Daily Weather Maps of the Weather Service is at 500 Mb, which is the most significant when used alone, although other pressure surfaces are also used and related to the 500 Mb in certain analyses. These pressure-contour maps are also called constant-pressure maps. It must be understood that the contours define the *shape* of the surface at which a certain air pressure exists. They are thus like lines on a topographic map; they are not isobars, because all the lines on one surface are at the same pressure. The figures attached to the contours denote the elevation above sea level (see lower maps of the series in Figure 15–30).

The elevation contours of the pressure surface can be read directly as an indication of high pressure or low pressure, because of the following consideration: Referring to Figure 15–31, if you begin at A at the 17,600-ft altitude, which has a pressure of 500 Mb, and rise 1,000 ft vertically to 18,600 ft to point B, the air will be lighter at B and have a lower pressure than at C, also at the 18,600-ft altitude but having a pressure of 500 Mb. Thus the depressions in this contoured surface indicate regions of low pressure at approximately 18,000-ft altitude, and the elevations indicate regions of high pressure. These are the so-called lows and highs "aloft" of the TV weather reports. The air at such altitudes is the midtroposphere flow shown in the diagrams of developing cyclonic waves (Figures 15–18 to 15–29).

The air layer denoted by the properties recorded at 500 Mb surface, as well as at the 300 and 200 Mb surfaces, has come to be regarded as very significant in weather forecasting. Some features and relations to the lower troposphere air are:

1. The irregularities of the isobars at the ground, with their separate air masses and fronts give way to smooth, wavelike troughs and ridges at the 500 Mb surface.

2. The movement of these troughs, ridges, lows and highs aloft can be anticipated better than the irregular features at the surface.

3. The winds at 18,000 ft follow the isobars fairly closely, and are thus more regular.

4. Much is now known about the high altitude flow patterns and temperatures and the extent to which they control the surface air flow patterns; since the upper air variations are more regular, they become a significant guide in weather forecasting.

Example of daily weather forecasting

A set of maps representing weather changes over Canada and the United States for four days, October 13 through 16, 1964, is given in Figure 15–30. The data were taken from the Daily Weather Maps of the National Weather Service. In the figure, each day is represented by two maps, one of the surface weather conditions and one of the upper air conditions at the 500 Mb level. Starting with the surface weather on October 13, we note an air mass and front that has come from the low pressure region over the Gulf of Alaska and has moved into British Columbia. Let us call this front No. 1. At the south end it is a cold front, in the central part it is a warm front moving a little in the opposite direction, and to the north it becomes occluded. Some clouds and precipitation occur as the occluded front passes by. Across southern Canada is another front (No. 2) which is stationary along its western part, but along its eastern part is a cold front with cold air originating in the continental polar region of northern Canada and the Arctic. The remnants of a weak, rapidly vanishing warm front are over the Gulf of St. Lawrence. In southern Arizona and Texas, a weak cold front exists (No. 3) which by the following day (October 15) has moved eastward, merged with one off Cuba, and become stationary.

Figure 15–29 *Satellite photograph of dissipating cyclone (compare with Figure 15–28).*

On October 14, a new marine cold front from the Alaskan or Aleutian low (No. 4) has appeared and is likewise moving eastward. Front No. 1 has moved ahead to Alberta and across the northwestern United States, with the northern part continuing as an occluded front. The warm air aloft produces some precipitation. In northern Alberta, the southern Canada front (No. 2) appears now to branch from front No. 1—where the front is convex southward the cold air is moving southward, and where it is convex northward warm air is moving northward. The eastern part of the front has moved out into the Atlantic. Along the stationary front (No. 3) through the Gulf of Mexico an intense hurricane has evolved from the previous day's low. We will follow the progress of this subtropical storm during the next three days, but let us delay the enumeration of its characteristics for a moment.

On October 15, the cold marine front No. 4 has moved into western Canada and the northwestern United States, and the northern part is occluded. Front No. 1 has moved to the Dakotas and Wyoming with the northern segment remaining occluded and somewhat stormy. The Canadian front (No. 2) is becoming weak and is moving somewhat northeasterly. Hurricane

391

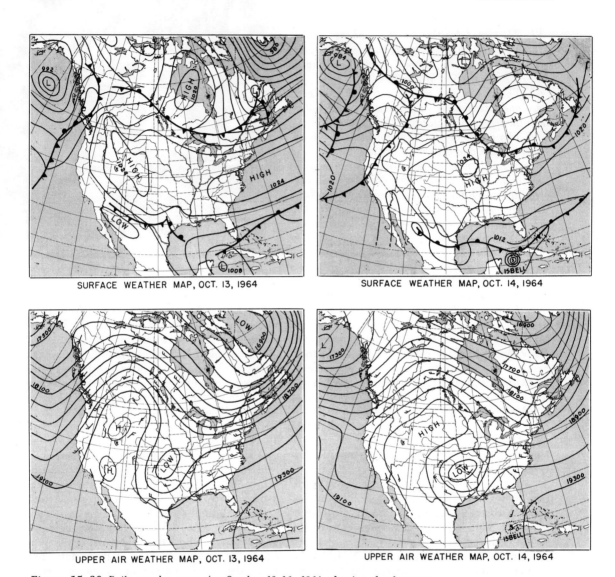

Figure 15–30 Daily weather maps for October 13–16, 1964, showing development of air masses, fronts, pressure changes, and Hurricane Isbell over a four-day period. Information taken from Daily Weather Maps of the National Weather Service. The solid lines of the surface weather maps (above) are isobars in millibars, and the

Isbell has moved possibly 100 miles to the northeast and now lies between Florida and Cuba.

On October 16, cold front No. 4 has shifted south and east to about the previous day's position of front No. 1, which is occluded, weak, and disappearing. Isbell and its associated weak front have moved further to the northeast, and her center now lies off the coast of South Carolina. On the next day, not shown on the maps, she loses much of her energy, and begins dissipating as a low over Virginia.

Practically all of the United States during these four days, except the northwest, has had

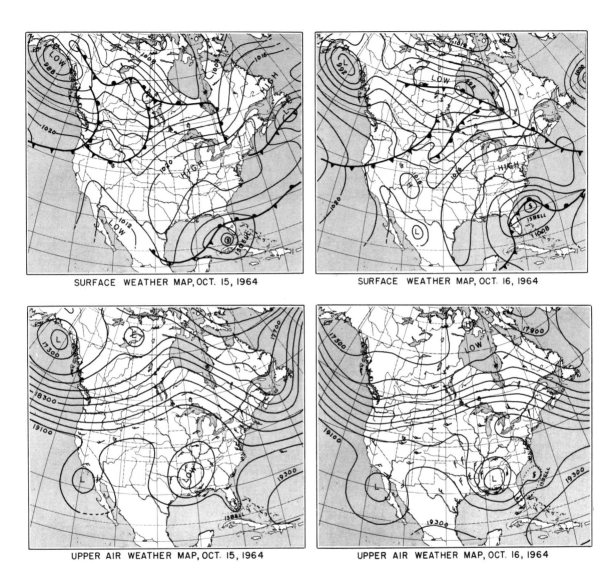

solid lines of the upper air weather maps (below) are elevation contours of the 500 Mb surface. Refer to Figure 15-9 for symbols of fronts used on surface weather maps. The symbol that resembles the letter F is used to show standard wind direction; the stem points in the direction of wind blow.

clear skies and sunshine because of widespread high barometric pressures. Note the eastward shift of the high from October 13 to 16, and the growth of another high in the west on October 16. Note also the prevailing Sonoran low over northwestern Mexico, southwestern Arizona and southern California. This is the region of very high summer temperatures and great aridity, but about now in October, the Sonoran low begins to weaken and disappear for the winter season.

It should also be noted that the polar front in western Canada has remained almost stationary throughout the period. This kind of air is shal-

Figure 15–31 *Reading the pressure contours in terms of high and low pressures. The air at B will have a lower pressure than at C. The air at A and the air at C each have the same 500 Mb pressures.*

low and, except for very strong and deep polar outbreaks, cannot cross the Rockies and move to the west. This is one of the reasons why the Pacific coast has such mild weather, even at higher elevations in the winter. Because of the westerly air flow from the Pacific much precipitation occurs in the mountains, and little moisture remains to be dropped over the Great Plains. The Pacific fronts approach the continent farther south in winter than in summer.

Referring now to the upper air maps of the four days, the chief display is a broad wave approximately across southern Canada. This separates the Arctic lows from a large high and low in the United States. The fronts from the polar regions of the lower troposphere (surface weather maps) are below the broad wave, and the large surface high over most of the United States lies below the gentle high and low of the upper air. As the cold fronts move across Canada and the northern United States, the broad wave aloft drifts along. With the upper-air map of October 13 established, it would seem reasonably sure that the broad wave would move easterly with the prevailing westerlies, but if a meteorologist were trying to diagnose the development of the surface fronts from the surface map of October 13, even if with much experience, he might be led astray. Particularly, it would be difficult to make a forecast for more than one day. Recognizing a connection between the upper air wave and the surface cold fronts, one feels more confident in a two- or three-day forecast, using the upper-air pattern.

An isotherm map of the 500 Mb surface (see Figure 15–32) reveals a broad wave about in the same position as the pressure-surface contour that follows the surface cold fronts. It, too, helps in predicting the courses of the surface fronts. Note that on the upper map of October 15 (Figure 15–30) showing the contours of the 500 Mb surface, hurricane Isbell is over southern Florida. The dots indicate a small circular isotherm (this is shown more clearly on the isotherm map of Figure 15–32). The surface low out of which Isbell evolved does not show on the upper-air map of October 13, but by the next day when the hurricane had gathered great vorticity and energy, it had warmed up a small circular area of the upper air directly above. This heating up of the upper air above the hurricane continued through October 16, after which the upper air was no longer affected. In other words, the hurricane occurred mostly below 20,000 ft.

As autumn turns into winter the cold continental air masses push farther southward across the continent, and the Aleutian maritime air surges into western Canada and the northwestern United States in a succession of fronts, although not regularly occurring either in time or in place. The experienced forecaster must take the seasonal variations of the weather into account, but by and large, he watches especially the new fronts that invade the continent. These are the cold dry continental Arctic air masses, the cold moist maritime northern Pacific air masses, and the warm, moist Gulf of Mexico and Caribbean air masses. With the aid of upper air data he predicts the course of the cyclonic storms, the development of the fronts, the movements of the highs and lows. All generally sweep across the country in an easterly direction. With the passing of air masses, he must predict the surface temperatures, the amount of cloud cover, and nature of precipitation. All told, the weatherman is a very important person in our economy, and his percentage score of correct predictions is, in this modern era of meteorological science, high.

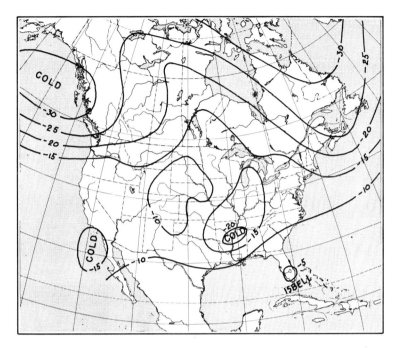

Figure 15-32 *Isotherms of the 500 Mb surface (about 17,000–19,000 ft above sea level) on October 15, 1964. Temperatures are in degrees Centigrade. [From Daily Weather Map, National Weather Service.]*

Hurricanes and tornadoes

Hurricanes are cyclonic storms of violent intensity and are bred over the oceans in the belt of tropical easterlies. They are nearly circular with very low pressure in the center and with high winds, dense clouds, and heavy precipitation (see Figures 15–33 and 15–34). The winds are light to moderate and gusty at the outer limits. They increase toward the center to squalls, then to furious gales, and immediately around the center in a fully developed hurricane to those of undescribable fury. On the oceans the high winds create tremendous seas and the tops of the waves are blown away in driving sheets and spray such that the sailor caught in the storm finds difficulty telling where sea ends and atmosphere begins.

Hurricane Betsy of August 12 to 18, 1956, alone caused property damage of about $1.5 billion in Florida and the Gulf States. In this storm the greatest property damage and loss of life arose from the sea surges over low-lying coastal areas in which all floating objects acted as gigantic battering rams against other objects.

The hurricanes that affect the United States originate in a wave or cyclonic low in the western Atlantic, and one of the courses they follow is westward across the Caribbean Sea and northward into the Gulf of Mexico, thence into the coast anywhere from Texas to Florida. The other course is from the Atlantic northwestward toward the coast of Florida, Georgia, and the Carolinas, either dissipating over land or turning out to the ocean again and fading to the north in the middle latitudes. We have mentioned in Chapter 10 the special oceanic temperatures in which hurricanes are spawned, and it

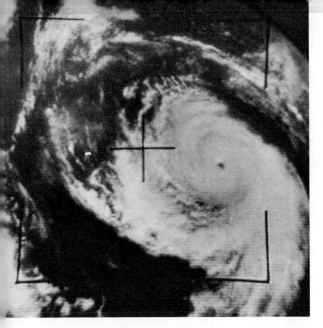

Figure 15-33 *Hurricane Ethel, photographed by Tiros VII on September 12, 1964. [Courtesy of NESC.]*

is fairly well demonstrated that hurricanes originate only over oceans with temperatures 79° F (26° C) or warmer and between latitudes 5° and 20° north or south.

The new Tiros satellite photographs have demonstrated that many hurricanes develop off the Pacific coast of southern Mexico and Baja California. These have generally gone unnoticed because they move northward over the relatively cold waters of the Japan Current and dissipate before invading southern California. Occasionally, however, a high-energy storm does move inland, perhaps once every 10 years. Strong west winds aloft are thought to shear off the tops of the hurricanes when they move out of the tropics, and thus help to dissipate them.

Not all tropical cyclones turn into hurricanes, but when the conditions are ripe the cyclone deepens rapidly, and a severe local storm is the result. We have seen on the weather maps of Figure 15-30 that a *low* turned into a full-fledged and frightening storm center in 24 hours. The closed cyclonic circulation had a diameter of 300 to 750 mi, and the violent winds ranged from 75 to 200 mph, directed counterclockwise in the Northern Hemisphere and clockwise in the Southern. A central calm or eye was found to be 12 mi in diameter in a Florida hurricane of 1945. It was measured by radar, and was marked by dense clouds which extended to a height of 18,000 ft with long rain-bearing tails spiraling around the center. This is also the view we obtain from satellites (see Figure 15-34). The weather maps of Figure 15-30 show that Isbell affected the 500 Mb surface by raising the temperature markedly, and thus we can assume that considerable turbulence extended upward to and through the 18,000-to-19,000-ft layer. The passing of the calm center, or eye, takes about 30 minutes but the violent portion of the hurricane lasts from 12 to 24 hours. The only good thing about hurricanes is that we now can observe them, watch their growth, follow their paths, and predict, at least a few hours in advance, what coastal areas and shipping lanes will be struck. Some thought has been given to the possibility of artificial dissipation of hurricanes, and one attempt has been made without striking success. The effects have not yet been evaluated quantitatively.

The tropical cyclones of the western Pacific, similar to our hurricanes, are called *typhoons*.

Tornadoes, like hurricanes, are cyclonic storms, but they differ from hurricanes in several respects. They develop over land and are especially frequent in the central part of the United States, although every state has reported them. The United States has about 150 of them a year, with Iowa having the greatest frequency: 2.8 tornadoes per year per each 10,000 sq mi. Twenty-nine confirmed tornadoes occurred along a cold front across Oklahoma during the single afternoon of May 1, 1954 (Blair and Fite, 1965). They are much smaller than hurricanes, with a dark funnel-shaped whirl extending downward from a low, heavy cumulonimbus. The cloud is attended with lightning and heavy rain. If and where the funnel reaches the earth, almost total destruction results in a path generally less than a quarter of a mile wide. The winds attain the highest velocities of any kind of storm, ranging from 200 to 500 miles per hour, and the updraft in the funnel is tremendous. The roar is deafening to those who are nearby. The average cross-country travel of the funnel is 35 to 45 mph, and thus it takes only about 30 seconds to pass by and totally ruin a barn or house. The length of the path of destruction is about 10 to 15 mi. The funnel may lift from the ground for a way, thus sparing a certain stretch from its effects.

The sudden drop of pressure toward the funnel creates an explosive effect within nearby buildings, because the inside pressure might become 70 to 400 lb per sq ft greater than the pressure at the funnel surface. Windows and walls fly toward the tornado. The lifting effect of the violent up-draft can pick up heavy objects, carry them hundreds of feet, and either slam them down with great destruction or set them down with miraculous ease and without injury. This is the vicious whimsy of a tornado.

Recent research indicates that the source of energy that causes tornadoes is electrical. It can be shown that purely hydrodynamic motion cannot adequately account for the very high wind-speeds, and thus an electrical energy mechanism is being considered. Specific measurements are needed but are difficult to come by, as you may well imagine.

Forecasting of tornadoes has proved a difficult task, because the violent convection lasts only a short time. The weather conditions that produce tornadoes are these: If warm, moist tropical air from the Gulf is invaded by an active cold front of polar air, a thunderstorm belt is produced, and if this is crossed by a strong high jet of cold air from the west, there may be some tornadoes. Thus, the conditions can be forecast, but the exact places at which the funnels may develop cannot be predicted. A radio and television alert can give warnings a few minutes ahead, in order that people can scurry to their cellars or basements or other underground shelters ahead of the storm.

Jet streams in the upper air

The *jet stream* is a narrow sinuous flow of swift westerly winds in the upper air. It has been observed at altitudes of 30,000 to 40,000 ft and is usually about 300 mi wide and 1 or 2 mi deep. Wind velocities become greatest to-

Figure 15–34 Satellite photograph of clouds in hurricane formation.

ward the center of the flow, where they may reach 300 mph. Planes taking advantage of this flow commonly attain speeds 100 mph greater than their normal ground speed. The position of the jet stream fluctuates widely, sometimes be-

Figure 15–35 Schematic diagrams of jet stream in the Northern Hemisphere, and the development of great waves in its course. [*After National Weather Service.*]

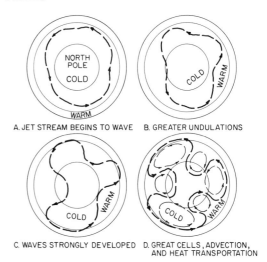

A. JET STREAM BEGINS TO WAVE   B. GREATER UNDULATIONS

C. WAVES STRONGLY DEVELOPED   D. GREAT CELLS, ADVECTION, AND HEAT TRANSPORTATION

397

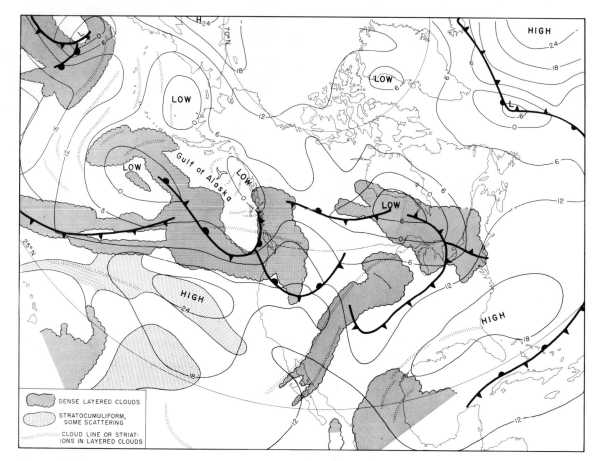

Figure 15–36 *Fronts, pressure contours, highs and lows at sea level, and the cloud patterns as photographed by the satellite ESSA 3 on January 12, 1967. The propellor-like symbol in the Gulf of Alaska is the center of a vortex.*

ing in the high latitudes of northern Canada, and sometimes as far south as the subtropical latitudes of the Gulf of Mexico. It often branches, producing two streams (Figure 15–37).

The jet stream seems to follow in a sinuous way along the leading edge of the cold polar air masses that move southward across the continent, and hence it is believed to be influential in steering the air masses, which in turn are responsible for the surface weather. It is strongest in the winter and on the east side of the continent, but for what reasons we are not sure. The flow develops from a broad wave course to a sinuous one, and finally into a meandering one, enclosing great pockets of warm air from the equator side and cold air from the polar side, as is shown in Figure 15–35. This is believed to be a vast mechanism of advection and mixing of cold and warm air in the upper air; it has the effect of keeping the temperatures of the earth in balance and within the familiar range.

Computerization of weather data

It is obvious to everyone who listens to weather reports on television that the forecasts

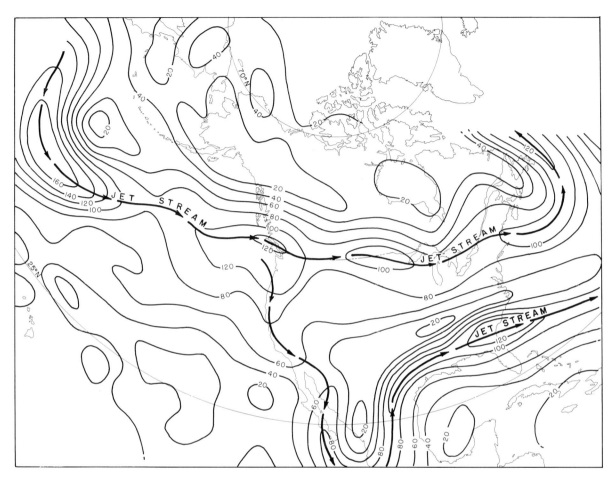

Figure 15–37 *Jet stream and air speed contours in knots for the 300 Mb pressure surface on January 12, 1967. [From National Weather Service.]*

are based in part on computer analysis. The positions of the troughs, ridges, lows, and highs aloft particularly are anticipated ahead by computer analysis of existing data. Since the winds and pressure distributions at the 500 Mb surface are more regular than those below, their patterns are more easily predictable, and general development patterns can be sought by the computer. Human judgment and experience, however, commonly enter into the official forecasts.

The National Weather Service has been releasing daily "today and tomorrow" forecasts, but beginning in February 1970, on the hundredth anniversary of the forecast services, 5-day forecasts were inaugurated. This has been made possible by recent advances in computer prediction.

Array of modern forecast maps

Although the basic forecast maps are produced in centralized Weather Service facilities, every major city has its own local Weather Service that utilizes the national maps, makes modifications at intervals as dictated by the developing weather, and serves the surrounding region. A weather laboratory is located at every

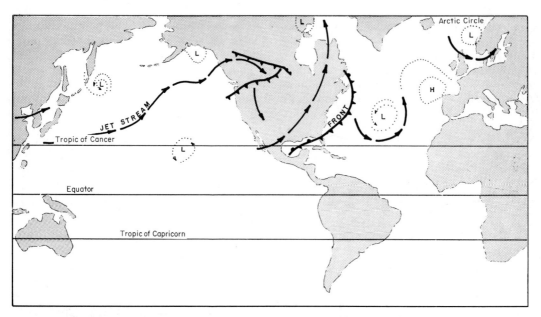

Figure 15–38 *The worldwide weather picture can be plotted on a map similar in projection to Figure 15–4.*

major airport to give pilots last-minute information on flight conditions ahead. To supply all the regional and local stations with the latest weather data, a system of communication has been perfected in which entire maps with the freshly recorded data and interpretations are sent over telegraph cables and reproduced by electronic and impulse-recording instruments. The maps printed out are called *facsimile copies*, and about a half-dozen instruments are constantly in use in each office, transmitting from 100 to 150 individual maps per day for the analysis of the local weather in all its aspects. Not only are these maps showing the changing weather transmitted at frequent intervals, but also the various kinds of maps that have been mentioned on previous pages, and a number of others still. The several surface maps, the several constant-pressure maps, maps showing vortices and vertical components of flow, maps showing thicknesses of air layers between the various pressure surfaces, and lastly, maps of the satellite cloud patterns superimposed on the common surface data. All these are multiplied 3 or 4 times because there are the Pacific, the Atlantic, the polar, the hemispheric, and the basic continental coverages.

An example of a Tiros IX satellite cloud-pattern map superimposed on the surface fronts and barometric pressure contours is given in Figure 15–36. An example of the jet streams taken from a 300 Mb map is shown in Figure 15–37. These are exhibits of the remarkable achievements in obtaining and compiling weather data at all desirable elevations.

Final look at the Tiros worldwide photograph

A composite of 450 individual photographs taken by the satellite Tiros IX was included on a previous page (Figure 15–4) to introduce the subject of the use of cloud patterns as seen from 400 to 22,300 miles above. Now that we have learned in a small way the nature of various cloud configurations, we should take another look at this remarkable composite and see what we can make of it. First, using the Daily Weather Maps and the facsimile copies, the jet streams of the day (January 13, 1965) were plotted on a map of similar projection as the photograph (Figure 15–38). Then, the fronts, lows, and highs were added. Compare these with the cloud patterns of the photograph and you will see much more than you had originally perceived, and thus you will realize the significance of satellite photography as an additional tool for use by the weatherman.

References

Blair, Thomas A., and Fite, Robert C. *Weather elements.* 5th ed. Englewood Cliffs, N.J.: Prentice-Hall, 1965.
Miller, Banner I. Characteristics of hurricanes. *Science*, Sept. 22, 1967.
Petterssen, S. *Introduction to meteorology.* New York: McGraw-Hill, 1958.
Strahler, A. N. *The earth sciences.* New York: Harper & Row, 1963.
Widger, W. K., Jr., Rogers, C. W. C., and Sherr, P. E. Looking down on the spirals in the sky. *American Scientist*, Sept. 1966.

Motion Pictures

Atmosphere: Envelope of life. Richard Cline, 1968, color, 16 mm, 22 min.
Formation of raindrops. American Meteorological Society, 1964, color, 16 mm, 26 min.
Planetary circulation. Massachusetts Institute of Technology, 1967, black and white, 16 mm, 27 min.
Sea surface meteorology. American Meteorological Society, 1967, black and white, 16 mm, 24 min.
Solar radiation: Sun and earth. American Meteorological Society, 1967, color, 16 mm, 18 min.

part four
environmental science

16

conservation, reclamation, and management of natural resources

Historical development

The environment can be defined as the surrounding conditions or forces that influence or modify. Biologically, the environment is the aggregate of all external conditions and influences affecting the life and development of an organism. Sociologically, it is the surroundings that affect human beings. A closely related term is *ecology*, the branch of biology which deals with the mutual relations of organisms and their environment. Environmental science, then, is defined as the study of all aspects of the environment.

Man has been aware for a century that he has been changing his environment, but only in the past few years has it become obvious that he is probably harming himself in so doing, and in some places has made his very existence precarious. We have not just conquered nature, but have achieved "over-kill." We see supine forests, exhausted and dirty rivers, and sullen air, and realize that to the victor belongs only the spoiled.

George Perkins Marsh, a Vermont attorney, wrote in 1864 in his book *Man and Nature*, "To disturb the balance of nature without calculating the consequences is to invite disaster. The web of life is made up of every single living organism, and the destruction of any part of the web might disrupt the whole biological community. Man is everywhere a disturbing agent. Wherever he plants his foot the harmonies of nature are turned to discord."

At the time of Marsh and to the beginning of the twentieth century there was no notion of a *land ethic*. Our grandfathers ruthlessly plowed, cut, mined, and killed wildlife. Progress was defined as expansion of the frontier, expansion of the railroads and urban communities, and increase in ships and population.

It was only during the administration of President Theodore Roosevelt that the National Forests were established and the conservation movement began to take shape. This was the first reaction against rugged Americanism. But during the time from Roosevelt to World War II conservation meant chiefly the maintenance of our forests and the protection of wildlife. Mineral production and water use also gradually came under surveillance, but initially conservation related only to some of our natural resources and their management.

In the last 25 years the conservation movement has begun to examine all aspects of our environment. A desperate urgency to protect the 75 percent of the people who live in the cities as well as to safeguard our national resources has seized the minds of many Americans; attention has spread to the urban centers from the rural areas. The administration has just announced its determination to make the nation's waters and air clean and healthy. It will also press for expanded and more beautiful recreational areas. "Quality environments" is now the watchword.

Social scientists emphasize today that we live in a world of two environments. One is the world of nature, the cradle of our species and the nourisher of our existence. The other is the world that man has fashioned—the world of technology. The task of environmental science today is to make the two worlds compatible. The problem is compounded by many conflicts of interest—industrial wastes pollute the waters and the air, yet our society depends on the output of industry. Automobile exhausts poison the air with carbon monoxide, yet without transportation our society could not exist. Progress is being made in controlling pollution, but in many places it increases at a greater rate than our capability of dealing with it.

The goal of environmental science has been thought of by some writers as the *management* of the environments for the maximum good of the people. But at the moment the goals revolve around the elimination of unsightly waste, pollution control of our waters and air, establishment of cleaner, more beautiful, and larger recreation areas, better management of urban growth, and abatement of soil erosion.

Scope of present treatment

Since environmental science embraces all studies that pertain to nature and man's well-being, it is apparent that the subject is so large that the scope of the present treatment must be circumscribed fairly carefully. Therefore only those aspects of environmental science for which a background has been laid in this book will be considered. In the simplest arrangement, these are: the waters of our rivers, lakes, and shorelines; the air we breathe; the soils that sustain us with food; and the natural beauties of our land. For these aspects a good foundation has been presented in preceding chapters. Subjects of biological nature, such as the vanishing species—grizzly bear, the red and gray wolves, the American bald eagle, the whooping crane, the California condor, the California redwoods, some of the great whales—must be passed by reluctantly. The controversial studies of pesticides and predator control can only be reviewed briefly. The biological aspects include the medical, in which the standards of control are determined, and in which also the measurement of human injury due to the pollutions is made. Then there are the philosophical, social, and economic aspects of environment management, about which many books and magazine articles have been written. These include the necessity for human population control along with environment management, a subject fraught with political and religious implications. Most of this material cannot be covered.

Where do the responsibilities rest—with the cities, counties, states, or the Federal government? Most environmental problems are national in scope, and thus the Federal government clearly has the responsibility for dealing with them. The administration and the Congress are now taking active leadership. The announced objective of the administration is to set standards and help finance the clean-up and beautification operations, but to turn over to the states the task of monitoring and control.

Another important aspect of conservation and environmental management is the economy of our natural resources. The assessment of our mineral wealth and how it should be produced and managed for the long-term good of all is surely part of the overall problem. Likewise the management of the national forests, wildlife, and ocean fisheries are intricately interwoven with environmental science. Again, the foundations for consideration of these subjects have been laid in previous chapters, but only modest attention can here be paid to them. The student must seek reading elsewhere, and the references at the end of the chapter will give him a good start.

U.S. government agencies involved in environmental management

The United States government has many bureaus, services, and offices that are concerned with the environment; so many, indeed, that they duplicate their effort in many specific fields. Five departments include most of the units that have to do with our natural resources and environments, namely: the Department of the Interior, the Department of Agriculture, the Department of Health, Education, and Welfare, and the Department of Housing and Urban Development, and the National Oceanic and Atmospheric Administration. A listing of the divisions of each that concerns the environment in one way or another will emphasize the great capability of the federal government as well as the urgent need of centralizing much of the effort.

Department of the Interior

Land resources

Bureau of Outdoor Recreation
National Park Service
Geological Survey
Bureau of Land Management

Fish and wildlife resources

Bureau of Sport Fisheries and Wildlife
Bureau of Commercial Fisheries

Water and electric power resources

Bureau of Reclamation
Federal Water Quality Control Administration
Office of Saline Water
Office of Water Resources Research
Bonneville Power Administration
Southwestern Power Administration
Southeastern Power Administration
Alaska Power Administration

Mineral and fuel resources

Bureau of Mines
Office of Coal Research
Office of Minerals and Solid Fuels
Office of Oil and Gas
Oil Import Administration

Human resources

Bureau of Indian Affairs
Office of Territories

Department of Agriculture

Soil resources

Soil Conservation Service

Forest resources

U.S. National Forests

Resource management, agricultural conservation, and aerial photographs

Agricultural Stabilization and Conservation Service

Land use

Research Division
Pesticides Regulation Office

Department of Health, Education, and Welfare

Authorized to set standards and enforce procedures on air and water quality.

Department of Housing and Urban Development

Authorized principally to work toward the elimination of slums, to improve housing for the masses, to improve the dreariness of core centers of large cities, and to plan esthetically for urban expansion.

National Oceanic and Atmospheric Administration

Environmental Science Services Administration

Formerly under the jurisdiction of the Department of Commerce.

National Weather Service (formerly U.S. Weather Bureau)
River Forecast Center

In addition to those listed there are undoubtedly other government bureaus and offices that engage in some kind of environmental work. Recently the President has established a high-level advisory council headed by the Undersecretary of the Department of the Interior and the administration's chief conservationist, to help in the campaign to clean up the environment. He has also asked Congress to give the various federal agencies a bigger voice in setting air- and water-quality standards.

The ordinary citizen does not realize how vast the federal agencies are that in one way or another have to do with the environment—their extent and budgetary requirements probably exceed his wildest imaginings. Even a congressman after a decade of tenure has only a partial comprehension of the responsibilities and activities of the various governmental branches. He can observe, however, simply by scanning the above list of bureaus, offices, divisions, and authorities, that much opportunity exists for duplication and inefficiency—and observers state that the duplication and inefficiency is enormous. Thus our attention must be focused not only on improving the environment but on cleaning up the governmental bureaucracy that deals with the environment.

Water management

Water conservation

General philosophy

Studies by the Department of the Interior's Geological Survey show that the U.S. water supply of 1,200 billion gallons per day is more than adequate for the nation's needs at present. These studies also indicate that there is no nationwide reduction of streamflow or lowering of the water table. These facts are reassuring and promise that our water problems can be solved by intelligent action; nevertheless, there is little reason for complacency. Almost every part of the United States has at least one major water problem, such as depletion of ground water, floods, droughts, and pollution of surface water. The cost of a clean and ample water supply is also increasing and it is becoming more difficult to meet the expanding and diversified demands for water.

Precipitation is the ultimate source of all water supply. An average of 30 in. of water falls as rain or snow in the conterminous United States each year. About 22 in. of water are evaporated or used by plants, and 8 in. run off into the oceans after traveling over or through the ground. The amount of water available for use is about 9 in., several times greater than the amount presently used in the United States (see Figure 16–1).

The distribution of this total U.S. water supply, however, is most uneven. The arid West, where the climate and soil are often highly favorable for agriculture, does not have enough water and perhaps never will. In the more humid areas of the country there is a surplus of water, but unfortunately, much of the surplus is of little use to man because it occurs in the form of floods which often do great damage. And even the humid areas are subject to occasional severe droughts, as in the 1930s, in 1949, and again in the 1960s.

Much available surface water is unusable for some purposes because it is badly polluted. Treatment of water to make it potable is expensive. The easily and cheaply available supplies of good quality water near the great centers of industry and agriculture are already developed, and development of additional supplies will cost more and more as time goes on.

In the past when more water was needed, the natural reaction was to increase quantity—to go to the tributaries for more water, to divert more water for irrigation, or to increase the size of reservoirs. Now we are learning that more water is not always the best answer to the problem. It is often better management of the water we already have.

Better management can "save" water, since less water can often be used to accomplish the same purpose and proper treatment of the

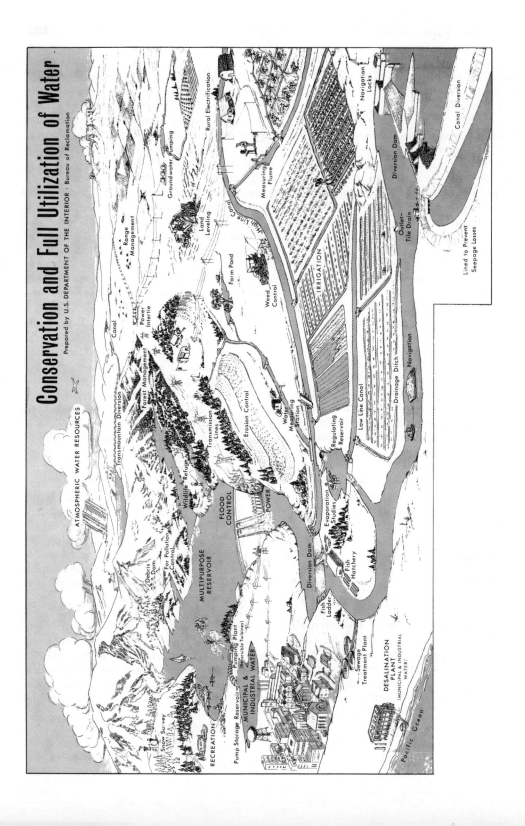

water actually used offers water of better quality for re-use downstream. In agricultural areas, water may be saved by more efficient irrigation methods, by controlling seepage, and by controlling evaporation from reservoirs. In industrial areas, manufacturers can save water by treating it and re-using it.

Balanced choices must be made between the best ways of using water. A personal budget includes necessities, savings, and a few luxuries. Similarly, the water budget must include water for domestic, industrial, and agricultural needs, for storage as insurance against drought, and finally, for recreation in protected wetlands, wilderness streams, and recreational areas. This is the true spirit of water conservation—protection and wise use.

Dam building

Man has learned to manage streams and rivers very well in one way, by the building of dams. Dams serve several purposes:

1. *They store water for irrigation and city use* (see Figure 16–2).
2. *They prevent floods.*
3. *They generate electrical power.*
4. *They facilitate shipping, not only by providing stretches of still water but also by extending the routes of shipping by means of locks.*
5. *They provide excellent recreational facilities.*
6. *They help in the propagation and conservation of wildlife.*

Figure 16–2 *Dam with irrigation canals, Gem County, Idaho. The irrigated land is below the canals.*

Often a single large river has a number of dams which are carefully integrated with the precipitation and with each other in serving human needs. Those that serve for power generation must be maintained above a certain minimum level, and thus are not allowed to fluctuate much. These are also best for recreational purposes. Those used for storage of irrigation water of necessity fill up and then perhaps drain completely. These are not very good for recreational purposes nor for game fish cultivation. The large dams serving for irrigation and power generation have been planned and constructed by the U.S. Bureau of Reclamation which receives its operating funds from leases and royalties of the federally controlled *public domain* lands.

Figure 16–1 *Schematics of the conservation and utilization of water. Development of the nation's water resources underlies every aspect of our daily lives. Dams built across rivers control and regulate streamflow to store water for irrigation, municipal, and industrial purposes; to generate hydroelectric power; protect land and people from floods; dilute and freshen polluted water; improve navigation, and for other uses. These reservoirs, along with surrounding land areas, provide outstanding recreational opportunities for people and create habitat and refuge for wildlife. Besides surface water, other sources being developed include underground aquifers, from which water is pumped for irrigation and other purposes; the sea, water from which is being desalted and converted to fresh water; and the atmosphere, from which, by means of weather modification, we seek to wring additional moisture to fall on our thirsty land. Thus, man exercises his ingenuity to conserve and utilize his most precious resource—water.* [Courtesy of the Bureau of Reclamation.]

Figure 16–3 Proposed dam site (black line) on the Provo River, Utah. Note the narrowed valley where the dam may be built, and the extensive storage area and volume above. At the right of the dam site is a cup or a seatlike form that may be a landslide. This is one of the aspects of the proposed site that must be investigated. [Courtesy of Bureau of Reclamation, Salt Lake City office.]

Figure 16–4 This tunnel in sandstone has been cut by a horizontal rotary drill. After the steel "slip form" is assembled inside the tunnel, concrete will be poured between the steel frame and the rock walls. The roof bolts and plates have been placed on the roof of the tunnel to prevent caving. Starvation Tunnel, Bonneville Unit, Central Utah Project. [Courtesy of Bureau of Reclamation, Salt Lake City office.]

In choosing a site for a dam the following requirements must be considered: (1) There should be a narrow place in the valley with a large storage area above (see Figure 16–3). (2) The rock foundations should be sufficiently strong to support the dam. It cannot be built on a landslide, either old or young, or small or large. Shale that may be present in the foundation rock must be tested for water-absorbent and expanding properties. If these occur, the dam must not be built at that site. Several dams have slid from their moorings, have breached, with destructive floods down the valleys, because of such shales. An example of a water-absorbent shale is given in the photos of Figures 16–4 and 16–5. Such shales are commonly bentonitic, which means that they were formed, partly at least, by volcanic ash whose feldspar content altered to a clay mineral called montmorillonite. This mineral is highly water-absorbent, resulting in a great deal of expansion. Also the bentonitic shales are slippery when wet, and thus a dam built on such rock is very insecure.

Limestones are commonly riddled with solution channels along joints and bedding surfaces, and thus are extremely porous. Some dams which have been built on limestone have collected little water back of the dam because the water has moved under the dam through the channeled limestone. The water then emerges in the stream bed farther down. Some dams built on limestone have filled, but with a conspicuous waste of water. Basalt flows, although very hard, are notorious for their porosity, and thus such lavas should be avoided in dam building or examined and tested very carefully. Even massive sandstone, like that of the great Glen Canyon Dam on the Colorado River, is a little porous and pervious, and now that the reservoir is filling up, much water is migrating laterally away in the sandstone walls of the reservoir. This can be determined by measuring the river inflow and noting the rise of the water level in the reservoir. If the reservoir does not fill up as fast as it should, then water is leaking out. This water is finding a place underground in the sandstone, as can be determined by checking the rising water table on either side of the reservoir. Probably all reservoirs lose water to some extent, but we should try to anticipate the extent before starting the construction. It should also be noted that in some of the plan-

Figure 16–5 Same tunnel as in Figure 16–4 but in a different place, where a bentonitic shale forms the lower two-thirds of the hole. Note the shale that has collapsed, the blocks of the sandstone above that have fallen off, and the ground water that has collected. These factors increase the cost of driving the tunnel. [Courtesy of Bureau of Reclamation.]

Figure 16–6 Water pollution and air pollution on a waterfront where an industrial complex has been built. Coal dumps and a steel mill in foreground.

ning not all the requisites can be met in finding a suitable dam site. But if the need for a dam at a certain site is great enough, then corrective measures must be planned that will make the dam a justified risk.

Obviously, the planning of a major dam with all its ramifications is an enterprise that requires the skills of an engineer, geologist, hydrologist, soils scientist, wildlife specialist, economist, an electric utility, and the government to finance the operation. What has been said is simply a brief introduction to those aspects of dam building that relate to the earth sciences.

Water reclamation

Degradation of water quality

Without water most industries could not function, and without industrial development present-day standards of living in the United States could not have been achieved. Many of the large industries are concentrated along a few great rivers (see Figure 16–6). At first the flows of these rivers diluted the wastes, the organic matter was oxidized, and the inorganic substances were dissolved or carried away; however, streams have a limited ability to purify organic wastes; if this natural or self-purifying ability of running water to assimilate and to stabilize wastes is overtaxed, the river becomes turbid. Sunlight then cannot penetrate into deep water, and green plants in the river bed die because they need sunlight for photosynthesis to remove carbon dioxide from the water and release oxygen to it.

What has brought pollution into public focus in recent years is the enormous increase in water use. Economic and population growth not only multiplies demand for water, it simultaneously swells the volume of waste materials reaching watercourses. Consider water demand: During April 1964 the United States was withdrawing water for use at a rate equivalent to 118 trillion (118,000,000,000,000) gallons a year. This tremendous amount is about one-tenth the volume of water in Lake Huron; yet, despite this large rate of use, national water needs are increasing at a rate of about 25,000 gallons per minute. Fortunately, much of the withdrawn water is not consumed but is returned after use to the river downstream or to the underground.

Degradation of water quality is, nevertheless, inherent in use. Water in the nation's river systems is withdrawn for many purposes, including transport of wastes. The transport of wastes from the home, factory, or farm to the sea is a beneficial use of water, but at times man offends nature. When such waste-disposal practice is abused, problems arise and controls become necessary for the benefit of the community as a whole. Realistically, the solution to the waste-water problem consists in learning how to handle wastes, not how to stop waste disposal. Obviously, streams must not be overloaded with wastes. Although treatment of waste is necessary, the degree of treatment required depends in part on the quantities of water available for dilution in the receiving waters. The most economical solution to the pollution-control problem may lie in partial treatment of waste in combination with an increase in the flow by means of reservoir releases. Where water is scarce, wastes may require complete treatment. Pondage of waste at one time for release at another time during periods of high flow is often practical.

Synthetic detergents

Synthetic detergents have contributed to waste-water problems. These surface-active agents, as they are called, were developed to overcome the disadvantages of ordinary soap in hard water. Soap in its cleansing action forms insoluble compounds, or scum, and often causes many washed fabrics to appear dingy. Synthetic detergents, on the other hand, maintain their efficiency in hard water without formation of scum, and 2.4 million tons are produced annually to satisfy the enormous demands of home and of industry. But though these detergents do the cleansing job well, the specific surface-active agent most widely used in synthetic detergents—a chemical called alkyl benzene sulfonate or ABS—causes foaming in streams, and in some localities a head of suds on water drawn from the tap. This foaming results from the fact that detergent wastes reaching sewage-treatment plants do not break down biologically; that is, ABS and compounds of similar structure are not decomposed by bacteria in the water. ABS, then, can travel long distances in streams and through the ground without losing its identity. In addition to foaming and frothing, detergent wastes retained in filtered municipal water supplies have caused taste and odor problems. The off-taste has been described as oily, fishy, or perfume-like.

Industry believes it has found the answer to foaming and related problems in chemicals that will replace ABS. One new compound, or group of compounds, is called linear alkylate sulfonate (LAS). The structure of LAS is such that detergent wastes containing this material decompose readily by bacterial action. Another new compound is nitrilotriacetic acid (NTA).

However, some of the new detergents still present a problem. Most of them contain 30 to 50 percent of various phosphate compounds, and LAS and NTA compounds do not alleviate the influence of phosphate in the water even though the new detergent materials are biodegradable. The phosphates serve as fertilizers to aquatic plants, principally algae, and nourish their growth, sometimes to the extreme. A reported substitute for LAS and NTA is a starch derivative which contains no phosphate, is cheaper, and is completely biodegradable. This subject will be further discussed in the sections dealing with *Sewage Disposal* and *Accelerated Eutrophication*. Results of actual use are not yet known.

Stream pollution from coal-mine waste

Where coal is mined, an associated mineral called pyrite, containing iron and sulfur, is exposed to water and air. As a result of weathering, the pyrite breaks down to form sulfuric acid and acid-producing compounds of iron. Through pumping or by natural drainage, the acid wastes flow into nearby streams (see

Figure 16–7 Strip mining for coal, Miller County, Arkansas. This also shows reforestation on the waste rock piles or rows. The old operations can be seen by the large trees and the most recent ones by the absence of vegetation. The latest dumps are mostly covered with water. If this water should seep away into a nearby creek, it would add considerable sulfuric acid to the stream. A reclamation measure now taken in some coal mining areas is to bulldoze off the tops of the waste strips before artificial planting.

Figure 16–8 Sewage treatment plant capable of removing 85 percent of the organic material and all of the mineral material for a population of about 200,000 people. Top view is an aerial photograph of the entire facility; lower left view shows the primary filters; right view, the secondary filters.

Figure 16–7). Unable to completely neutralize these acid wastes, many streams become unfit for use by man, and no longer can support fish life. Without extensive treatment the polluted water will not meet the requirements of most domestic and industrial uses. Other difficulties result from damage to concrete and metal in bridge piers, dams, turbines, and culverts by the corrosive acid water.

The Monongahela River, draining parts of West Virginia and Pennsylvania, is an example of an acid stream. On the average, this stream empties the equivalent of 200,000 tons of sulfuric acid each year into the Ohio River. This amount of acid is about twice that used annually in the manufacture of industrial explosives in the United States. Although assimilation of the acid by the Ohio River begins immediately, it is not until the Ohio has passed the mouth of the Muskingum River at Marietta, Ohio, 170 miles downstream from the Monongahela River, that the acid load has been neutralized completely.

The control of acid wastes from mines has no easy solution. The problem is the subject of extensive investigations by government, university centers, and industry. A joint federal-state study begun in 1964 at a demonstration site in the Roaring Creek–Grassy Run area near Elkins, West Virginia, may lead to a better understanding of the factors contributing to acid pollution. Several techniques for controlling acid wastes are known: Closing or sealing of abandoned mines, diversion of surface and ground water from mine workings, prompt removal of mine drainage from acid-forming materials, and regulating release of acid wastes to receiving streams.

Sewage disposal

The chief reclaimers of water are municipal sewage disposal plants throughout the country. These are remarkable facilities inasmuch as the inorganic and organic content of the sewage waters is separated out, and the water is brought to a quality good enough to be returned to lakes or rivers, or used for irrigation. The aerial photograph of Figure 16–8 shows the several structures in a new 9-million-dollar plant that serves 200,000 people. The inset views of Figure 16–8 show primary filter tanks and secondary sedimentation tanks. This facility, with its primary and secondary extraction units, is 85 percent effective. In other words, the equivalent of about 15 percent of the raw sewage passes out and into a sumplike bay.

The harmful components of sewage are the coliforms and the bacteriological oxygen-demand created by the organic matter. The coliforms are the bacteria and other organisms that originate in the colon of man and other animals and serve as a reliable index to such disease organisms as typhoid, diphtheria, dysentery, and polio. In the process of decomposing organic matter, the bacteria require oxygen, which they extract from the water. The higher the bacteriological oxygen-demand, the lower the available oxygen for fish. At a certain level the trout die and only the trash fish (such as carp and suckers) survive. At a still lower level, all fish perish, anaerobic bacteria take over the job of decomposition, and fetid odors are given off.

In sewage treatment plants, chlorine is used effectively to kill off the coliforms, leaving the effluent water from the secondary stage fairly clean. Bacteriological oxygen-demand, however, is a more difficult problem. The first and second stages of treatment reduce the organic content by about 85 percent. With 15 percent of organic matter remaining in the effluent from the second stage, this organic matter encapsulates the bacteria, which then are less vulnerable to the chlorine.

In the established water-quality classification, Class C is considered the minimum desirable for surface streams—suitable for domestic supplies after treatment, and for agriculture, wildlife, and industry, suitable without treatment. Class C water has a coliform count of 5000 per 100 ml and a bacteriological oxygen-demand of 5 mg per liter. Class D water is suitable only for irrigation of selected crops and has a base limit of 5000 coliforms and a bacteriological oxygen-demand of 25 mg per liter.

In order to obtain water classed as type C, an additional tertiary stage would have to be added to the plant of Figure 16–8. This would cost an estimated $18 million.

The President has asked for, and Congress has appropriated, $4 billion to assist communities in building sewage disposal plants. To obtain this aid the communities will be required to contribute $6 billion as matching funds, and the raising of this money will be augmented by a special federal group called the Environmental Financing Authority to help hard-pressed municipalities. It is estimated that the $10 billion will be sufficient to construct 1500 new treatment facilities, and to expand and upgrade 2500 plants.

Thermal pollution

It appears that government will force private industry to clean up its waters and at its own expense. Old plants not designed for water treatment units may find it rather expensive to install them but new industrial plants will undoubtedly plan to build in the necessary facilities. The engineering capability is already present. Added to chemical and particulate pollution by industrial waters is thermal pollution. Some of the water discharges are warm or hot, and the raised temperature when effluents are discharged into rivers or lakes is sufficient to cause the elimination of certain fish species. Thermal pollution constitutes a real problem because 94 percent of industrial water is used for cooling.

Steam-electric power plants, steel mills, petroleum refineries, and other industrial plants withdraw huge amounts of water daily from rivers and lakes for purposes of cooling. These withdrawals, slightly reduced in volume and warmer by as much as 30° F, are returned to the watercourse after use. The heat thus transferred reduces the ability of a stream to hold dissolved oxygen, because oxygen is more soluble in colder water. Increased temperature has a serious indirect effect. Oxygen combines more rapidly with organic wastes as the water temperature increases. Thus the dissolved oxygen may be consumed more rapidly than it can be replaced as the water flows downstream. The minimum oxygen content may be zero. Ideally, heated waters should be cooled before discharge so the receiving stream can maintain required oxygen levels. This cooling can be done through the use of still ponds, spray ponds, or cooling towers.

Besides sampling and analyses of the contaminated waters as a monitoring device, aerial photography is being used to advantage for detection of thermal pollution. Infrared photography is particularly good in detecting waters warmer than adjacent waters; if the warm waters are also polluted, then the polluted waters may be detected by this means. Another type of photography, called multispectral, utilizes four light bands of the spectrum: blue, green, red, and near-infrared. The four photographs are then composed in a special viewer in various degrees, and, in addition, stereoscopic sensing is added. Ground control of the hydrologic conditions of limited areas of rivers, lakes, and swamps may thus be extended over large areas.

A significant study of the effects of warming of the lower Connecticut River, initiated in 1965, will continue through 1972. This is focused on a 5-mile stretch of the river above and below Haddam Neck, where the Connecticut Yankee Atomic Power Company has built a nuclear power plant. Effluent from the plant is discharged at a temperature of 20° F above the influent river water. The term for artificial (nonchemical) warming of rivers and lakes is *calefaction*. The calefaction in the Haddam Neck section of the Connecticut River does not necessarily deserve to be called thermal pollution; for here no drastic biological effects have been noted. There have been changes in the river near the power plant, in the bottom fauna at the point of effluent discharge, and in the condition of the bottom habitat near the plant intake, but the economically most important fishes, the shad, which are migratory, have not been affected. The annual yield of most resident fishes showed an increase in 1969 in the plume of warmer water. Many factors are being studied, and the above conclusions are provisional.

The waters of Par Pond, near Aiken, South Carolina, are used for purposes of cooling in the Atomic Energy Commission's Savannah River plant, where nuclear materials such as plutonium 239 and tritium are manufactured. The lake contains about 180 billion gallons of water, and more than 1 billion gallons are used each day for cooling. Some of the water returning to the lake reaches 115° F. The fishes and turtles of this lake do not seem to mind the artificially warmed water. The largemouth bass have been noted in water up to 76° F, and they and the yellow-bellied turtle have been the special targets of study by researchers of the University of Georgia. For both, the researchers report extra-rapid growth, early maturing, and more eggs. It must be recognized that Par Pond is a clean, immature body of water, created just 10 years ago by damming. No sewage is dumped into it. Thus the good health of the fishes and turtles may be due to factors that are somewhat unique.

Another study concludes that if we must lose heat generated by nuclear energy, it should be spent in enhancing the rate of decay of sewage or in furthering the growth of agricultural ecosystems—and not dumped heedlessly outside civilization centers.

Farm-animal-waste disposal

The aerial photograph of Figure 16–9 shows the bay of a lake which is mostly swamp as a natural condition. Into it have been discharged the sewage of a town plus runoff irrigation water, and cattle corral and farm waters. They have combined to render the bay a health hazard, an inferior place for wildlife propagation, and a disgrace. The Bureau of Reclamation now plans to dike the bay off, divert the inflowing stream, pump the bay dry, and convert it into a truck-farm area. A sewage disposal plant will be built, and the cattle enclosures must be contained. By this is meant, watering tanks or troughs must replace the through-flowing ditches, and animal refuse must be contained in the individual lots and prevented from being carried to the bay. The

Figure 16–9 This aerial photograph shows a shallow bay of a fresh-water lake into which runs the sewage of a small town and the drainage from farms and cattle corrals. Under President Nixon's plan and under the management of the Bureau of Reclamation, a dike will be built across the mouth of the bay (left side of the photograph), the bay will be drained and converted into a truck farming community, a sewage disposal plant will be built, and the farm runoff will be contained.

practices that benefited the pioneers are not good enough for the present generation.

Disposal of liquid waste by underground injection

Great volumes of unwanted water or brine that are produced along with oil and gas have been injected into the underground. In the United States this amounts to 10,000 acre-feet yearly. In Texas alone there are 20,000 brine injection wells. Many other kinds of industrial wastes are being disposed of in ever increasing amounts by injection underground. This is a convenient disposal method but can we continue it with impunity?

Actually, injection is storage; it is not disposition. And what we are storing may come back at a later time to haunt us. In most areas there is little underground storage space available and it must be created by driving out existing water. We must review the various aspects of ground water, discussed in Chapter 8, in order to appreciate the conditions involved. We have to deal with the zone of saturation below the water table in which pore spaces are filled with water and also with homogeneous rock masses, on the one hand, and with inhomogeneous stratified sedimentary rocks on the other. In most cases of injection the pore water already present is displaced by the waste water and driven elsewhere. In some underground reservoirs, the resident water is stagnant, and in others it is circulating. It is evident that each underground reservoir must be well-studied and tested before injection procedures are permitted. After it has been decided that injection of certain volumes can take place, hydraulic engineers must establish safe injection pressures, and geochemists must anticipate the chemical and mineralogical changes that will take place when the waste liquids drive out the original water. If the ground waters are circulating, where will the waste liquid go, and who might be affected? This is a most pertinent question when radioactive waste liquids are pumped underground. These are problems for the professional geologists and hydraulic engineers, but responsible citizens must appreciate the problems of disposal and recognize the dangers involved in indiscriminately disturbing the existing ground-water environment.

An interesting, although unique, example, is that of forceful injection of waste liquids in the Rocky Mountain Arsenal Well, 10 miles northeast of Denver. The well was drilled to a depth of 12,045 ft and was provided with a casing to 11,975 ft. The fresh-water aquifers to 1,426 ft were cemented off. The well is bottomed in a metamorphic rock (gneiss), from which water flowed into the well and rose to a level of about 2,500 to 3,000 ft below the land surface. The liquid wastes were forcefully injected into the gneiss at a reported pressure of less than 2,000 lb per sq in. (psi) at the wellhead. Under this pressure the gneiss accepted more than 400 gallons per minute. From March 1962 to September 1963, and from September 1964 to February 1966, 165 million gallons of liquid waste were injected. In the seventh week of injection the seismograph at the University of Colorado recorded an earthquake of 1.5 magnitude. Then followed through August, 1967, a swarm of 1514 earthquakes in the same area with magnitudes ranging from 0.5 to 5.3. It has been reasonably well demonstrated that the earthquakes were caused by the injections. The injected liquids built up sufficient stress to cause rock movements relieving the expanding volume. An analysis of the seismograms suggests that there existed a natural regional stress field, whose release was triggered by the injections, and thus the swarm of earthquakes represented, at least in part, a strain relief of natural origin. But the important thing here is that the artificial injection triggered the earthquakes.

Accelerated eutrophication

Eutrophication is the natural biological enrichment of a body of water to the point where some forms of life flourish at the expense of food fish. Man's activities, which introduce excess nutrients, along with other pollutants, into many of our lakes, streams, and estuaries, are causing significant changes in aquatic environments. The excess nutrients greatly accelerate the process of eutrophication. Artificially accelerated eutrophication may lead to ecological disasters like those that occurred in the Great Lakes.

It is recognized that nutrients added to fish ponds result in increased production of fish, and thus it has been supposed that sewage introduced into estuaries and bays would be beneficial so far as fish yield is concerned. If eutrophication is accelerated, however, oxygen may become short in supply, and perhaps certain very desirable food fishes will be hurt while other species increase in abundance. In other words, fertilization may be overdone.

A notable and alarming example of artificially accelerated eutrophication is Lake Erie. Not only have the food fishes been seriously reduced in number by the lack of oxygen due to the high influx of organic materials, but the sewage and industrial wastes have excited the growth of algae (algal blooms) to the extent that even shipping is being impeded. Lakes Michigan and Huron are also experiencing decreased fish production due to the pollutants.

Federal scientists have reported that years of dumping 10 million tons per year of organic-rich sludge near the entrance of New York harbor, about 12 miles from land, has created a "dead sea" of ocean pollution, devoid of marine life. This barren sea now shows signs of invading the coasts of New Jersey and New York. To correct this situation a majority of the congressmen from New York and New Jersey have proposed legislation that would revoke the authority of the engineers to permit sludge dumping within 25 miles of Ambrose Light.

Nearby Jamaica Bay is a polluted, island dotted estuary with an Atlantic Ocean inlet. It lies between Brooklyn and Long Island's Rockaway Peninsula. From the standpoint of environmental management it may be the most important polluted estuary in the country, for two reasons: it is the object of great public and private competition; and it is in the midst of a heavily populated urban area where the quality of human life for most of the inhabitants is seriously deteriorating. Jamaica Bay has already been half filled, and of the filled area one third is occupied by Kennedy Airport. The airport needs more runway space but it is contended that the expansion would impede or make more costly an existing program to improve water quality. It would adversely affect an unusual wildlife sanctuary and impair its ecological integrity. Consequently, the Environmental Studies Board advised the Port Authority not to expand until further studies have been made, and the Authority agreed. The study group has contended that the problem is not birds versus planes or jobs versus pollution, and stressed that the environmental problem cannot be comprehended in such terms.

Galveston Bay is another example of an estuary in crisis. It is beset by a complex of problems which some view as beyond repair; others feel that the bay still has a chance to be rescued. The map of Figure 16–10 shows the main elements of the environmental problem concerned. The bay is a drowned river valley resulting from the rising sea level, which in turn is due to the dissipation of the last great ice caps. A barrier beach has been built by the waves, and the entrance to Galveston Bay is a tidal inlet. The bay covers 533 sq mi. It originally supported a rich growth of oysters, shrimp, crabs, and a number of valuable food fishes, such as sea trout and redfish. Half of the bay is now closed to oyster fishing, although the most productive reefs are still open to harvesting. This brackish bay is less salty toward the head where the freshwater streams enter and more salty near the mouth. Brackish water is essential for oysters and young forms of shrimp to survive. Young finfishes, croakers, anchovies, and menhaden, find a nursery in the bay but spend part of their adult life in the Gulf of Mexico. If the nursery is affected, then the fish in the gulf will be affected. The bay waters are enriched by nutrients brought in by the tributaries and flushed out of the shallows and marshes by tidal action. This nutrient has resulted in the richness of the bay life. What has happened to this classic ecological community?

The rapid growth of human population around the bay (Houston, Texas City, and Galveston) and the accompanying industrial growth have changed the bay's natural environment to the effect that life in the bay's waters has been greatly disturbed, dislocated, and reduced. The Houston Ship Channel was constructed up Buffalo Bayou in 1914. This made Houston a major port, surpassed only by New York and New Orleans. The ship channel has become lined with many industrial facilities, oil refineries, chemical and petrochemical plants, fertilizer factories, gypsum and cement plants, and a steel mill. As a result, the old Buffalo Bayou, 25 miles long, ranks as one of the filthiest stretches of water in the United States. It is reported that the industrial pollutants and huge volumes of poorly treated domestic sewage from Houston and its suburbs are imposing a daily waste that is equivalent to the raw sewage of 2 or 3 million people. Dissolved oxygen is totally lacking in parts of the channel at times. The channel would probably remain polluted for years even if all effluents were cut off immediately, because the bottom is covered with a 2-ft layer of putrid sludge.[1]

[1] See D. S. Greenberg, Test case of an estuary in crisis, *Science*, Feb. 20, 1970.

Galveston Bay is an example of disturbed ecologies that are extremely alarming in many wetlands elsewhere. Wetlands are those areas where no clear-cut border between land and water exists—they are the swamps, marshes, bogs, and tidal flats. The U.S. Fish and Wildlife Service estimates that there were about 125 million acres of wetlands in the United States when Columbus came, but now, due to reclamation practices of draining and filling there are only 60 million acres.

Conservationists are particularly worried about marshes—both salt and freshwater—because they are at the same time the biologically most productive and perhaps the least appreciated natural areas on earth. Most people, particularly those in urban areas, tend to regard marshes as "mosquito-infested swamps" that are best filled in. But ecologists point out that such areas teem with plant and animal life, and, together with adjacent tidal flats, creeks, streams and bays, any given marsh helps form a single biological system that is vital to the life cycles of many creatures that range far afield both on land and in water.

Salt marshes are the nurseries of the sea because they provide spawning and nesting grounds for birds, fish and mollusks that may live out most of their lives far away. It is calculated that two-thirds of the ocean's sport and commercial fish either begin their own lives in wetland areas or feed on other creatures spawned in wetlands.

Natural eutrophication systems maintain the productivity of the wetlands and thus we have to do nothing for them except leave them alone.

Land use and conservation

Soil, water, plant, and mineral management must reflect the concept of public ownership and must have as its major aim the preservation of these resources for future generations. The following paragraphs indicate who owns the land in the United States and what is being done to protect it. Conservation of the soils, management of the runoff, the preservation of our forests and scenic wonderlands, and the

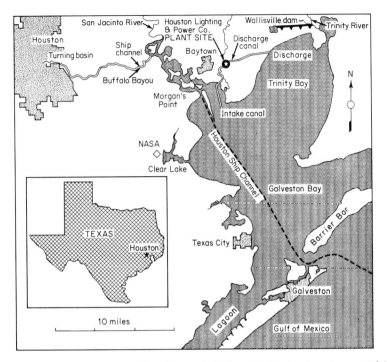

Figure 16–10 *Galveston Bay, Texas, including the barrier bar, lagoon, tidal inlet, ship channel and other features.* [From *Science, February 1970.*]

regulated production of our mineral wealth are the main subjects discussed.

Land ownership and use in the United States

Land surveys in colonial days followed land development and were unorganized. As the western frontiers opened and countless fertile acres beckoned, a need for a systematic land survey became urgent, and probably the most important service that the federal government rendered the new states, territories, and private owners was the development and execution of such an orderly division of the land. Much of the United States west of Pennsylvania at the time of the Revolutionary War was recognized as public lands, or the public domain, and so it still remains in the Rocky Mountain states. Most of this territory has been subdivided, or laid out, in rectangular tracts bounded by north, south, east, and west lines, each tract having a particular designation, such that it is impossible for patents or titles, as obtained from the Government, to conflict. The system was probably devised by General Rufus Putnam, an American officer in the Revolutionary War. It was first used in laying out the eastern portion of the state of Ohio in 1786–1787, then called the Northwest Territory. This was the first land owned and sold by the national government.

Much of the public domain west to the Rocky Mountain states has since passed into private ownership, principally, at first, by homesteading. Private lands fall under state jurisdiction. As can be seen from Figure 16–11, about half of the land in the Cordilleran region is still under federal control, with the United States owning 30 percent of the Montana acreage and 99.8 percent of Alaska (1967). Since Alaska achieved statehood, the federal government has

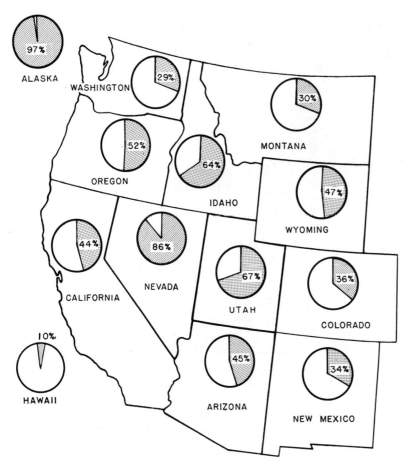

Figure 16–11 *Portion of land in the western states that was owned by the federal government in 1967. [Courtesy of Ward Roylance.]*

been turning over considerable land to the state.

The land *ownership* in a typical western state (Utah) is shown in greater detail in Table 16–1; land *use* is shown in Table 16–2.

Soil conservation

About one-third of the privately owned agricultural land in the United States is managed in order to prevent soil erosion and to maintain or enhance soil fertility, but two-thirds still needs conservation treatment. Erosion not only depletes the land from which soil is removed, it fouls the waters and damages flood plains and stream beds, or fills reservoirs and harbors with sediment. Measures taken to eliminate soil erosion and soil deterioration not only restore vitality and health to the environment but also create landscape patterns that are pleasing to the eye.

Congress in 1935 established the Soil Conservation Service in the Department of Agriculture to initiate and direct a national program of soil and water conservation. Under enabling acts subsequently passed by every state, rural citizens have organized soil and water conservation districts that now embrace 99 percent of

TABLE 16–1 *Ownership and Administration of Land in Utah, 1967*

	Total acreage	52,777,474 acres	
Nonfederal land	State land Counties, cities, misc.	3,577,090 acres 47,759 acres	7%
	Private land (almost all private land is completely assessed for taxes)	12,469,586 acres	23.5%
	Indian trust lands	2,044,876 acres	4%
Federal Ownership Administration	Dept. of Defense	1,817,407 acres	
	Department of Agriculture (All Forest Service)	7,689,846 acres	
	Department of the Interior National Park Service	549,599 acres	65.5%

	Bureau of Land Management
	  Coal withdrawals 1,629,100 acres
	  Oil shale withdrawals 1,015,600 acres
	  Water areas 123,790 acres
	  Reclamation, reservoir,
	    power site withdrawals 1,528,337 acres
		24,528,469 acres	

	Other federal agencies	53,109 acres	

rural properties but do not yet include all landowners in a cooperative program. The Agricultural Stabilization and Conservation Service administers the Agricultural Conservation Program, which shares with farmers the cost of carrying out needed conservation measures on their farms and creates state and local committees of farmers to administer farm programs. Two million of the nation's 3½ million farmers are formally cooperating with local soil conservation districts and Agricultural Stabilization and Conservation Committees to establish good conservation and land-use practices.

Soil erosion occurs where there is enough rain or snowmelt to cause runoff (Figure 16–12) or where the land is flooded by irrigation (Figure 16–13). To avoid erosion, the soil must be protected from moving water. Dense vegetation such as cover crops, grasses, trees, or mulches, will intercept rain and slow the runoff (refer to Chapter 7). Severe erosion where more than 75 percent of the original organically rich fertile soil has been lost is common in the Piedmont Province and upper coastal plain of the Atlantic slope. In the Midwest it is severe in many places on the loess soils of the Mississippi and Missouri rivers, and the prairie sections of Missouri, Illinois, Oklahoma, and Texas. In the West, the Palouse Hills of Washington and the coastal section of California have suffered severe sheet and gully erosion. About 11 percent of the soils used for crops in the United States are on sloping surfaces and are shallow over the rock mantle

TABLE 16–2 Land Use in Utah, 1967

Land used for farms		12,870,000 acres*
Cropland	2,070,000 acres	
(used for harvest or pasture)		
Woodland	820,000 acres	
(mostly used for pasture)		
Other farmland	9,970,000 acres	
(mostly pasture)		
(Irrigated farmland amounted to 1,100,270 acres in 1964)		
National forest lands (used for timber cutting, grazing, recreation, watershed conservation, etc.)		7,690,000 acres*
Water areas		170,000 acres
Federal and state parks		560,000 acres
Federal military installations		1,820,000 acres
Cities, towns, villages		260,000 acres
Federal mineral lands (withdrawals)		2,760,000 acres*
Federal lands withdrawn for reclamation, power sites, reservoirs		1,530,000 acres
Highways, airports, railroad lands		300,000 acres
Remaining land—federal, state, and private (most of this land is used primarily for grazing, if at all)		24,820,000 acres*
Total land area		52,780,000 acres

* Taken from Ward Roylance, *Utah's Geography and Counties,* private publication.

(Figure 16–14). When the topsoil is lost there is no chance for a new plow layer. In addition, some of the underlying shales or sandstones are too acid or too salty to sustain a healthy crop. Thus the toll of soil erosion in the United States has been great.

The common practice in the past has been to plow straight furrows parallel to the property lines or section lines, up and down the rolling hills. Thus the stage was set for rapid runoff down the furrows to the gully bottoms. But now contour plowing is a must. In this way the precipitation is caught in the horizontal furrows and is held or ponded (see Figure 16–15). Where the top soil is badly eroded, forest planting may be advisable. On mountain slopes terracing may be practiced, with the planting of grasses, shrubs, and trees of various kinds (see Figure 16–16). Gullies in prairie lands are often dammed with brush, logs, and rocks. If overgrazing is the cause of the erosion, then it is clear that the number of sheep, cattle, or deer, as the case may be, must be radically reduced or eliminated entirely, until the vegetation has been reestablished.

The federal government shares with farmers the cost of carrying out certain water, woodland, and wildlife conservation practices on farms and ranches throughout the nation. It provides this assistance to more than a million individual farms each year, at a cost of about $250 million. Farmers share two-thirds of the cost. Government employees render the following services:

1. *Establish protective cover, including forests (Figure 16–17).*
2. *Improve and sustain the cover.*
3. *Effect safe disposal of water.*
4. *Establish and sustain wildlife.*

Figure 16–12 This farm was plowed in the approved contour manner, but after a heavy rain that exceeded the holding power of the land, runoff and erosion resulted. The Soil Erosion Service seeks to estimate the maximum precipitation periods and to recommend terracing procedures that will check the runoff in times of high precipitation. [Courtesy of Soil Erosion Service and M. N. Boswell.]

Figure 16–14 Dry farm wheat fields in southern Idaho. This gently sloping piedmont had been dissected during the late Pleistocene, but sufficient soil formed on the flat-topped uplands to support wheat growing. It is apparent that some recent gullying has started, and measures should be taken to check this development.

Figure 16–13 A deep gully is the result of an irrigation ditch allowed to run unchecked straight down the slope. The flow was at the rate of 4.5 cu ft per sec. [Courtesy of Soil Erosion Service and M. N. Boswell.]

5. Protect soil and prevent soil erosion.
6. Construct small dams. About 2 million small earth dams have been constructed by the Agricultural Conservation Program in the past 30 years. These dams conserve water, reduce floods and siltation, add to the ground water, and save soil.
7. Construct terraces. About 30 million acres of land have been terraced to date.

Managing the runoff

Some streams are deepening their valleys, some widening, and some filling. Some are building levees. These processes concern man, especially in the populated regions. The many streams and the many different circumstances present a problem of exposition, however, beyond the scope of this book. It must simply be stated that the principles of stream activity, as they have been presented in Chapter 7, must be taken as a guide in determining what *has* hap-

Figure 16–15 Contour plowing and terracing in Nebraska.

Figure 16–16 Terracing in the high mountains where overgrazing had resulted in erosion and gullying. This is an expensive procedure but is the only recourse to save our watersheds, forests, and summer grazing lands. [Courtesy of Ed Shane and U.S. Forest Service. This photograph appears in Mountain Water, *a pamphlet of the Forest Service.*]

pened and what *will* happen. It must also be evident that in any engineering project along a river (dam, bridge, railroad, highway, levee, flood embankment) we must anticipate that the response of the river will be to pursue its customary course. A meandering river that has been artificially straightened will strive to meander again. Soil erosion caused by plowing might cause the stream to fill up its channel and flood the valley bottoms. A dam built across a river will create a reservoir which, in time, surely will be filled with sediment. A bridge abutment might be undercut by a meandering stream. A farmer on one side of a river may lose land as the river meanders, and a farmer on the other side may gain.

More basic than dam building to control runoff is the maintenance of good soil and vegetation cover. Many studies have shown that if we allow the vegetation (grass, brush, forests) to be depleted, then the soil will start to wash away. Thus we must constantly watch the soil and vegetation cover everywhere (see Figure 16–18).

National forests and wilderness areas

History of the U.S. national forest system

Forestry in the United States was still mostly in its dark ages at the beginning of this century. A few farsighted and public-spirited men had tried from time to time to arouse realization of the dangers that lay ahead if wasteful destruction of forests were not checked, but they were as voices crying in the wilderness. To most people it seemed the forests would last forever. Not until 1891, when the national forest system was started, did the conservation movement get under way on a nationwide scale. The establishment in 1905 of the Forest Service, U.S. Department of Agriculture, in its present form marked the real beginning of a national forest-conservation policy.

In March 1891, President Benjamin Harrison created the first forest reserve—the Yellowstone Timberland Reserve, an area of 1,239,040 acres in Wyoming. This reserve is now the Shoshone and Teton National Forests. Later on in the same year he signed a proclamation withdrawing 1,198,080 acres in Colorado, now the White River National Forest. Before President Harrison's term had expired he had set aside forest reservations totaling 13 million acres. In 1897 President Grover Cleveland proclaimed more than 20 million more of reserves. At the beginning of President Theodore Roosevelt's administration about 107 million acres had been set aside as forest reserves, and during his terms 148 million acres more were added. This is as much land as three large western states contain.

The first Chief Forester was Gifford Pinchot, and his Secretary of Agriculture wrote a letter to him saying:

In the administration of the forest reserves it must be clearly borne in mind that all land is to be devoted to the most productive use for the permanent good of the whole people, and not for the temporary benefit of individuals or companies. When conflicting interests must be reconciled, the question will always be decided from the standpoint of the greatest good of the greatest number in the long run.

An Act of Congress in 1960, called "The Multiple Use–Sustained Yield Act of 1960," reaffirms the policies stated in the letter to Gifford.

Conservation measures

Lumbering operations have until recently left the cut-over forest in a sad state. The primitive forest with its dead trees is littered with the slashings of the newly cut trees in great disarray. Many of the saplings are broken and crushed. The idea is to get the good logs out in the cheapest way and leave the rest.

In the national forests, the Forest Service marks the trees to be cut and enforces a certain amount of clean-up operation after the logging. Roads are watched and protected from runoff erosion. The newest and most attractive method of logging from the point of view of conservation is the clean-cut practice. Everything is cut and cleared away in a certain area, and new seedlings are planted (see Figure 16–19). In badly eroded forest areas, three watershed restoration practices are employed.

1. In areas that involve management programs designed to restore somewhat damaged watersheds, the following steps may be taken: (a) adjustment of livestock numbers and season of use to grazing capacity, (b) rest and rotation grazing, and (c) under certain conditions, complete elimination of use.

2. Seeding of grasses and shrubs, and an adequate management program. When deteri-

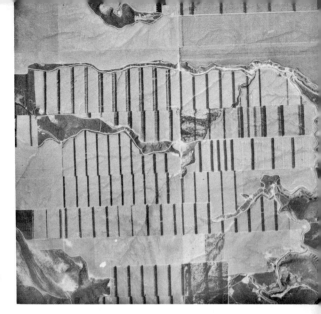

Figure 16–17 *Planting of trees and bushes in strips across the wind sweep helps to prevent silt and sand blowing and to reclaim the land for agriculture. [Malheur County, Oregon.]*

oration of the plant cover has reached an advanced stage natural seeding becomes necessary. This must be followed by complete protection, usually for 3 to 5 years, and thereafter use must be carefully managed to insure the maintenance of sufficient vegetation to prevent rainstorm runoff and soil erosion.

3. Structural measures such as contour trenches and seeding to stabilize the soil, followed by proper management. Where gullies are too large to be controlled by contour trenches, special gully-control structures may be necessary.

The shelterbelt program was started by President Franklin D. Roosevelt to restore the Great Plains prairie lands that had been wasted during the long drought of the early 1930s. The first tree in the shelterbelt program of the prairie region was planted near Mangum, Oklahoma. This was the start of the Prairie States Forestry Project, to lessen drought conditions, protect crops and livestock, reduce dust storms, and provide useful employment for drought-stricken people. Under this project the Forest Service cooperated with prairie farmers in planting strips of trees at right angles to the prevailing winds on farms in the Dakotas, Nebraska, Kansas, Oklahoma, and northern

Figure 16–18 *Good high-country pasture land, not overgrazed or eroded, is considered well managed. [Courtesy of Soil Erosion Service and Vicki Griffiths.]*

Texas. At first, funds came from appropriations provided for emergency conservation work, and then by the Emergency Relief Act. In 7 years, more than 217 million trees were planted; 30,000 farmers participated in the program. In 1942, the project was transferred to the Soil Conservation Service to be continued as an activity of the soil conservation districts. Many benefits have already been derived from the program.

Recreational use

The Forest Service has assumed the obligation of providing "developed" recreation sites, which have their purpose in improved observation, swimming, boating, camping, and picnic facilities. The developed recreation sites have been constructed where the public has been accustomed to concentrate. In the building of these sites good craftsmanship and rustic charm have been watchwords. The problems are litter and vandalism. The fire-patrol roads and trails have now been discovered as superb nature trails for hiking and packing. Educational literature is being published by the Forest Service.

Multiple use means skiing. The Forest Service has allowed private individuals under permit to develop skiing facilities in many of the National Forests. The skier sees in the mountain tops the grandeur of the winter forest, but the snowpack that he enjoys is the source of our drinking water. With 10,000 skiers in some ski resorts on a weekend it is evident that water contamination will ensue. Sewage disposal is the most critical problem and often, like garbage, it must be hauled by truck out of the watershed.

The sport of snowmobiling is new in the

realm of forest control. The big problem seems to be vandalism of cabins that were snowbound in the past but are now readily reached by snowmobile. Rescuing of the poorly equipped snowmobiles and coping with a new source of litter are the main chores of the rangers in wintertime.

Wilderness areas

Wilderness is an integral part of the American heritage. This nation was born in wilderness, and from it came the land and materials needed to build a country. Untamed lands spread from coast to coast. The freedom of wild lands, their great open spaces, and their grandeur are interwoven in America's history, art, and literature, and have strongly influenced the shaping of the national character. Many of these lands, majestic samples of primeval America, are parts of the national forests of the United States. Here, as wild and just as free as ever, over 14 million acres of wilderness are held in trust by the U.S. Department of Agriculture's Forest Service for the use, enjoyment, and spiritual enrichment of the American people.

Many of the early conservationists were in the Forest Service. Led by Aldo Leopold, who was one of the nation's most distinguished naturalists, their thinking influenced early recognition of wilderness values and benefits to the nation. So began the Forest Service concept of wilderness land management—designating as wilderness those lands predominantly valuable as wilderness so as to manage and maintain them indefinitely for their out-of-the-ordinary values.

Once an area is designated a wilderness by the Secretary of Agriculture on the recommendation of the Chief of the Forest Service, no permanent roads may be built into it, nor is mechanized equipment permitted. Timber cutting and other commercial uses, except already established grazing of livestock, are not allowed. Resorts, organization camps, summer homes, and similar uses are not permitted. Moreover, positive management in the form of

Figure 16–19 Lumbering in the coastal ranges of Oregon by removal of everything preparatory to new planting.

controlled grazing of pack stock, trail building and maintenance, fire protection, and supervision of use and clean-up of choice camping areas is necessary to prevent damage to the soil and vegetation. Such steps keep wilderness in its natural state for the enjoyment of all who wish to journey beyond civilization.

There are now 84 separate areas of land, about 8 percent of the more than 186 million acres of the national forest system, which are designated as wilderness. They are located on parts of 73 National Forests in 14 states and their combined area, over 14 million acres, exceeds the total land area of New Hampshire, Connecticut, and Massachusetts. Their management is included in the overall program of multiple-use management for the national forest system, which comprises 154 national forests and 19 national grasslands in 41 states and Puerto Rico.

National forest wilderness lands embrace desert country and brushland, virgin timber stands, great valleys and alpine meadows, and some of the nation's loftiest mountains. They contain countless streams and lakes. They abound in fish, small animals and birds, and big game animals such as deer, elk, moose, mountain goats, and bighorn sheep. Some are the province of the grizzly bear, which needs the wilderness to survive.

National parks, national monuments, riverways, national shorelines, and recreation areas

The national parks in the United States now number 35 and contain approximately 30 million acres of land. They embrace the major scenic spectacles of the country, and in addition preserve some remarkable natural ecologies and evidence of prehistoric man. To date there are also 82 national monuments in the United States, which, like the national parks, preserve significant and highly scenic and esthetic segments of our natural heritage. Canada has taken the same action as the United States and has set apart large tracts of land in the west and the east as national parks, to preserve the outstanding scenic values of the land for us and for future generations. The United States, in addition to the national parks and monuments, has established riverways and recreation areas. For instance, a strip of land about 5 miles wide along each side of the Glen Canyon reservoir on the Colorado River has been set aside, under government protection and administration, as a recreation area. Few roads will traverse these strips, but pleasure boats can reach hundreds of miles of shorelines. There are at present 12 recreation areas and 2 riverways in the United States.

In managing, developing, and protecting this estate, valuable beyond calculation, the National Park Service acts as trustee. A long series of laws, starting with the act of March 1, 1870, that established Yellowstone National Park as the first national park in the world, have set the pattern of management. When the National Park Service was established in 1916, Congress laid upon it the mandate to conserve and to provide for public enjoyment of "the scenery, the natural and historic objects and the wildlife" in the national parks, monuments, and reservations "in such manner and by such means as will leave them unimpaired for the enjoyment of future generations."

In these parks and monuments it has been necessary to make some modification of the natural scene. Certain provisions have to be made for the public that comes to obtain the enjoyment for which the park was established. The Government has had to provide certain administrative structures and facilities necessary for the performance of the various functions of the National Park Service; it has had to make provision for safe water supplies for its own employees and the public, for sanitation, for sewage disposal, for camp grounds, for roads and trails; through concessioners it has had to see that overnight accommodations, eating places, and other facilities essential to the public's satisfactory enjoyment of the park were provided. These are developments needed wherever people congregate, and all of them have involved some changes in the landscape. They have actually affected only a small fraction of this great area. The wilderness, scarcely touched or modified, lies but a short distance away. The birds, the game animals, the predators, all the wildlife of the park follow their normal existence with virtually no interference from man.

Highways and esthetics policy

With the coming of automobiles, roads connecting towns and cities became a necessity. First came one-lane, then two-lane, then four-lane highways; first hardly graded at all, then well-graded, and then paved highways. Since World War II the United States has embarked upon a massive program of high-speed, nonstop interstate highway construction, which in a few years will be completed, at least as first planned. The esthetic guidelines of this huge engineering program were pronounced in an Act of Congress called the Highway Beautification Act (Public Law 89–285), passed on October 22, 1965. It has three divisions:

1. *control of outdoor advertising*
2. *control of junkyards*
3. *landscaping and scenic enhancement*

The basic rule about commercial signs, displays, and devices is that they must be at a distance greater than 660 ft from the right-of-way of the highway. This applies to the Interstate System and the Primary System of highways, whose construction the Federal government largely subsidizes. Failure of the states to enforce this rule results in a 10 percent penalty of the amount which would otherwise be appropriated.

Junkyards are proliferating along the main highways, especially on the approaches to the cities. The 1968 regulation provides that such junkyards must now be located at least 1000 ft from the highway right-of-way, and that they be effectively maintained so that they do not become eyesores. Effective control means that the junk must be screened off so as not to be visible from the highway.

In order to effect landscaping and scenic enhancement, the Secretary of Commerce may approve as part of the construction of the federally aided highways the costs of "landscape and roadside development, including acquisition and development of publicly owned and controlled rest and recreational areas and sanitary and other facilities reasonably necessary to accommodate the traveling public."

The California Division of Highways contends that esthetics have value and must be a major consideration when locating a highway during the planning stage. It should also be kept in mind that much can be done in the way of esthetics at little or no additional cost.

During the studies to determine the location of a route, the planners should develop an intimate knowledge of the history of the area with which they are dealing. This may seem somewhat removed from esthetics; however, the reactions on one's emotions are just as real, just as strong, as the emotions that come into play when viewing a beautiful scene. In this instance, the emotions are experienced more by the community than the highway user.

In rural areas, every effort must be made to achieve compatibility with the area through which the highway travels. This is of prime importance to both the highway user and the highway viewer. In rolling or mountainous areas, a curving alignment can be used effectively to preserve existing features of the terrain. Adequate sight distance, however, cannot be sacrificed to the extent that the safety of the highway user is reduced.

The highway should fit the terrain. In the rural area this means that regardless of the height of cuts or fills, the natural ground should appear to begin at the edge of the shoulder, not at the right-of-way line. If the highway is built on flat land, the slopes should be flat. Ditches should be nonexistent or at least with very flat slopes. In rolling country the slopes should roll and warp. The cuts and fills should blend into the existing ground so they appear as part of the country, not as part of the roadway. In rugged, steep terrain, the cuts should be rugged and steep if the bedrock permits and the steep slopes will not cave.

The slopes in a forest area should normally be flattened and rounded when a highway is built through it. The slopes should be gentle enough that trees will grow anew. A slope, however, can be warped to save a particularly fine specimen or group of trees. The U.S. Forest Service has been very helpful in this respect. They are interested in the long-range planning program for the forest areas. Certainly the Forest Service should be brought into the picture, both in the early stages of a project and during construction.

Whenever possible on divided highways, separate alignment and grade should be used. This, in most cases, will make the highway appear to fit into the country better.

Many features of design of the highways, such as horizontal and vertical curves, cuts and fills, guard rails, slope rounding and end grading, drainage structures including side ditches, guide markers and other hardware, and overpasses and underpasses must be left to the engineer, but there are esthetic qualities to all of these. A review of existing ditches shows that more than half are unnecessary. Many of them provide surfaces of rapid erosion (Figure 16–20). Revegetation should be considered on every project regardless of size. It is not only important from the standpoint of erosion con-

Figure 16–20 *Erosion along the side of a new highway that has been unseeded and left unprotected.* [Courtesy of Soil Erosion Service and M. N. Boswell.]

trol but adds greatly to appearance. A grass cover will quickly make a raw slope blend into the terrain.

The steepness and texture of a cut slope is very important where natural revegetation is expected to take place. There is a critical point on slopes where the slight sloughing of the material will prevent natural revegetation. All slopes other than rock cuts should be laid back far enough to prevent this sloughing and thus allow the natural seeds to germinate. A flat slope will normally revegetate better than a steep slope.

Many of our highways follow the valley floors, because along these routes the roads and highways are cheaper to construct, but in taking these routes much valuable farmland has been used up. In planning new highways the nonagricultural lands should be utilized as much as possible.

We all are aware of certain stretches of the highways that are always in need of repairs. The damage is usually due to one of three causes:

1. swampy or wet foundation materials
2. foundations of shale or clay that absorb water and swell
3. landslides

Nearly all state highway engineering departments recognize the need of geologists in the route-planning stage. Geologists can recognize the above conditions and alert the engineers. If possible, the geologic conditions are avoided, but if not, precautionary or stabilizing measures are taken. In some deteriorating stretches of old highways, when the cause of sagging or heaving has been ascertained, it may be decided that the rerouting of the highway is almost impossible and that the cheapest and best procedure is simply to continue the repairs each year. This would not be the wise thing to do, obviously, if lives might be endangered.

Urban expansion

The urban sprawl

Most cities have grown without an overall plan. Not only has there been an absence of a plan, but effective zoning and building regulations have been wanting. What does uncontrolled growth result in? It leads to the chopping up of large green spaces into inadequate patches and to monotonous rows of small houses, which soon become bedroom suburbs. This leads in turn to decay of land values and to loss of pride and sense of well-being. Then come lowered ethical and moral values along with high rates of crime. Uncontrolled growth also leads to high-rise apartments where city dwellers live an elbow-to-elbow existence. The more fortunate breathe filtered air at home and at work, but the others wheeze and cough through smog-filled days and nights. Urban sprawl commonly eats up valuable agricultural land, never to be reclaimed.

Planning and control

A beautiful city can be produced with an area-wide master plan. This should provide for a pleasing and livable central-city area and self-contained satellite communities which have preserved and beautified the natural terrain. Rapid-transit facilities should be included in the plans. Abundant green space in and around the city must also be provided. This is done by creating city parks and playgrounds, by providing easements along river banks where the natural vegetation may be retained and cultivated, footpaths and bicycle trails provided, and additional recreational and picnic grounds built up. Planning and zoning may provide attractive and efficient clustering of industry to the benefit of both city and industry. The living patterns should be oriented not to technology but to people. The family unit is the most important consideration.

Other problems of urban sprawl

At the present, urban expansion is mostly a suburban affair. The central cores of some cities are losing population and are deteriorating. And as the suburbs grow good croplands, orchards, pastures, and forests are eaten up. Highways, airports, factories, power plants, and mills require much land—in 1970 it was estimated that the land required for such purposes was 215 percent that of 1960. About 3 million acres of productive agricultural land was removed from cultivation in 1968 by urban and highway expansion.

City and industrial growth result in increased runoff with increased erosion in some places and sedimentation in others. The areas eroded are those that have been cleared and stripped of all plant cover, but not protected. Thirty percent of the sediment of the Potomac River originates within the metropolitan area of Washington, D.C.

The purchasers of homes should be protected by law from the hazards of the area that go unmentioned by the builders and sellers. In California any new proposed real-estate development must have been inspected by a certified geological engineer, who will write a report on the land involved. The report must point out unsafe features, if any, such as potential landslides, faults, and earthquake potential, filled or marshy ground with foundation subsidence problems, and the potential of flood damage. The report is judged by the state government before building permits are issued. The geology profession has strongly recommended that all prospective property buyers seek the advice of a certified geologist before making a purchase.

Urban renewal

Some old core-sections of cities are so decayed that the only remedy is to tear down the buildings, usually slum apartments, and start to rebuild from scratch. This takes well-conceived and experience-based plans. Other core areas may be less run-down, and less difficult and expensive to restore. Streets can be closed off and made into tree-lined walkways. Covered shopping areas with central parking lots can be made very attractive. Benches and kiosks are recommended. Downtown San Francisco has solved a difficult space problem involving people and parking. In the hub of the shopping district is an attractive 2.6-acre park, planted with palms, Irish yews, boxwood and bright flowers. A person can withdraw to this pleasant park from the busy flow of cars and shoppers. Underneath the park is a four-story parking ramp which accommodates 1000 cars, with a turnover of 2700 cars a day. According to Aldo Leopold,

A thing is right only when it tends to preserve the integrity, stability, and beauty of the community, and the community includes soil, water, fauna, and flora, as well as people.

Conservation of minerals

The wise use of minerals grows in importance as our country becomes more industrialized and as its population increases. Our dependence on machines made from refined metals, mineral fuels to drive the machines, and an array of everyday utensils and building materials made from various minerals becomes more acute each day. With each new discovery or invention our mineral needs become more complex. At the end of World War I, only 50 of the 92 naturally occurring elements were used, and only 20 of these in significant amounts. Today nearly all are used. Seventy are produced in commercial amounts, and 34 are used in sizable amounts. In addition to the elemental materials about 50 kinds of rocks and minerals are used and marketed. Our economy in the past 100 years has been based on our mineral resources. In the ultimate analysis nearly every dollar of our gross national product is dependent on a person operating a machine, which in turn is made from a mineral product and is powered by a mineral fuel.

Some mineral commodities are very abundant. One such mineral is coal. North America, Eurasia, and Australia have especially vast reserves. It has recently been estimated that the world's supply will last 200 to 300 years, even at the growing rate of consumption. Coal will remain our chief source of energy. The discovery of new oil reserves in the United States plateaued out in 1957, and if discovery and consumption trends are projected we must conclude that our oil will be used up soon after 2000, if our search efforts are not drastically stepped up. Our natural gas will be gone in 1990.[2]

Of 41 principal minerals necessary to support the economy of the United States, 15 will last 50 years, 3 will last 33 to 40 years, 25 will last less than 20 years, 9 will last less than 10 years, and for 8 there are no reserves. The most important nonferrous metals, copper, lead, and zinc, have a life expectancy of about 20 years. These periods will be greatly extended undoubtedly because we will import those minerals and metals of which we are in short supply, and technology and reclamation will also help extend the life of the reserves. Sophisticated geophysical and geologic exploration will find new deposits, too, but the lesson is clear—our reserves are exhaustible and we already are forced to look more and more to foreign sources for our supply.

[2] Future petroleum provinces of the United States. A summary report by National Petroleum Council, 1970, p. 138.

Minerals and mineral fuels are nonreplenishable. Once mined they are gone and do not regenerate like plant crops growing from the soil. Used metals, to some extent, may be reclaimed, but not coal, oil, and gas, at least for this geologic age.

The heritage of the people of the United States and Canada is the free enterprise system, and certainly the ways of the system were at first the almost untrammeled exploitation of our natural resources until about the end of the nineteenth century. The great iron, copper, and coal mines, the railroads, and the lumbering industry were the basis of the United States' industrial growth to the position of foremost industrial power in the world. All this was generally considered commendable. Conservation-minded elements began to rise at the beginning of the twentieth century, and the national forests and parks were first established. Regulatory measures on oil and gas production were soon enacted. The federal government, especially since 1930, has become more conservation-minded and has taken steps to stem the wasteful production of minerals and oil and to insure that mining and drilling do not leave unsightly scars on the landscape. At present a most pressing problem is how to manage the production of our mines and oil fields, our water supplies, how far into the future to program the production of the nonreplenishable mineral reserves, and how to shape our foreign relations to provide for a supply of those minerals and fuels which we now need or may need later.

The problem of free exploitation versus conservation has become especially pointed in the federal government's attempts to establish new National Parks, to add acreage to the ones already in existence, or to create new wilderness areas. National Parks and wilderness areas have traditionally not been available to exploration and mineral production, so a struggle has resulted between the politically strong oil, mining, and lumbering interests and the conservationists, who also enjoy strong political support. As a result, the philosophy of multiple use of lands has been proposed and partly carried out. By multiple use is meant that a scenic recreational area may at the same time serve for mineral and oil production. In certain cases a limited time has been given private enterprise to find the minerals or oil before the land is to be withdrawn permanently from exploitation.

The problem of control of exploration and exploitation on federal and state lands is further illustrated by the oil shale situation in Colorado, Utah, and Wyoming. The organic matter in the shale, called *kerogen*, may be converted to readily usable petroleum products. Further, the shale crops out in canyons and hill slopes or may be reached by drilling at relatively shallow depths (see Figure 16–21). The deposits are of staggering size, even compared with the world's largest oil reserves, and contain oil worth at least hundreds of billions of dollars and possibly a few trillions of dollars. A single conventional lease (say 2500 acres in size) would contain almost as much oil as the largest live oil field yet discovered in the United States. But technology has not yet perfected a way to produce the oil competitively.

The potential worth of the oil-shale lands to the federal government is so great that some economists have dreamed of paying the national debt with the royalties; needless to say a number of prominent people have urged the government to close all mineral leasing on the acreage and proceed with great caution and after thorough study. About 77 percent of the oil shale lands are owned by the United States. The rest is owned by Colorado, Utah, and Wyoming and by private individuals and companies. The federal lands and most of the state lands were withdrawn from leasing in 1930, but with new technological advances in recent years strong pressure is being exerted on the Department of Interior to open the lands again to leasing.

The situation is really much more complicated than so far suggested. Free enterprise is backed up by business groups and chambers of commerce who want to see the oil shales developed as quickly as possible. The oil companies

Figure 16–21 Oil shale in northwestern Colorado, showing the nature of the terrain, and the outcrops. Insert is a sample of the oil shale, which would run about 30 gallons of oil per ton of rock.

claim they will not spend the millions of dollars required to develop the extraction processes unless they own the mineral right to the shale. Others have suggested that small leases be granted for research and development work. Still others argue that the technology will come whether or not the land is leased, and that after commercial feasibility is demonstrated, the government will be able to lease the land at much higher prices. Conservationists take the position that the large companies' desire to lease is based simply on their traditional desire to control the land and that the mountain slopes will be devastated with vast pits, cuts, and wastepiles.

Then there are the differences within the oil interests themselves. One difference resolves about cheap imports of oil from the Persian Gulf, North Africa, and South America. This group is likely to resist the inroads into the oil market from oil shale. The companies having large reserves and refineries in foreign lands, and whose import quotas are short, will be in no hurry to see the oil-shale program proceed. All this is recited to show you that the controversy that develops between the "rugged Americans" and the conservationists is not at all simple, and that the basic arguments may get sidetracked on many other issues. The Secretary of the Interior, as a newspaper editor has said, whoever he is, will need to understand multiple use, the sophistries of mineral finance, and the strategies of world resources. He must also be able to walk on water, or at least tread very fast, while juggling the conflicting interests and aspirations of those engaged in mining, petroleum, forestry, grazing, and recreation on the one hand, and on the other, the hard determination of the conservationists to preserve the natural ecologies.

Pesticides in the environment

Great concern has arisen in recent years over pesticides, or insecticides and herbicides, and their effects on the environment. In killing pests, including the nontarget pests, the ecological balance has been disturbed and especially several species of birds have been seriously affected. Concern has been expressed that pesticide residues may be absorbed in food crops and thus may affect human health. Some fish crops have unusual concentrations of the residues. Certain lots of canned tuna fish and swordfish, for instance, contain alarming amounts of mercury, and have been condemned by the Food and Drug Administration. It is explained that the mercury compounds first lodge in algae in very small amounts; then small fishes eat the algae and their bodies gain a somewhat concentrated dose of mercury. The big fishes eat the small fishes and the concentration of the poison becomes even greater. Tuna and swordfish are at the top of the food chain or ladder.

The excellence of analytical methods in detecting minute amounts of pesticide residues in waters, soils, plants, and animals has lead to a better understanding of this environmental problem than others. The pesticide levels retained in the human body can be measured, but medical science has not yet decided what observed levels are harmful.

Most pesticides are biodegradable and therefore do not persist in the environment. Their effects are limited to the specific areas of application and are of short duration. No undesirable long-term health effects are known. But the chlorinated hydrocarbon insecticides are persistent, and may be detected over wide areas. The most widespread of these compounds is DDT. Available data show that residues in the soils in most cases are not harmful either to the crop or to the microflora. In the United States most surface waters contain chlorinated hydrocarbon insecticides, but the concentrations are in parts per billion or less, and as far as is known, they are not harmful. Little is known about contamination of the air by insecticides, but in any case all monitoring programs of the soils, crops, and waters should continue, education on the use of pesticides should be thorough, and enforcement of existing laws and regulations should not be relaxed. We must be prepared to recognize harmful effects, should they begin to occur, immediately.

It is worthwhile to examine the effect of DDT on birds in a little more detail. Four species, the bald eagle, the osprey, the brown pelican of California, and the white-faced ibis of Mexico and Utah are known to be laying eggs with fragile shells that get crushed by the mother, or eggs with strange, soft, leathery shells. The ibis nests in marshes and feeds both in the marshes and in nearby agricultural areas. Two nesting colonies in the Bear River marshes of Utah have been studied where the fragile and leathery shelled eggs have been observed. Here DDT has been used repeatedly to combat mosquitoes in the marshes and against other insects in the nearby fields. The birds of these colonies fly to Mexico for the winter where DDT is also used to spray the crops. In 1968 the eggs of one colony of 100 pairs of birds were entirely lost because their shells were too thin, but later the birds renested and the eggs were successfully hatched. The other colony of 2500 pairs had many hatching failures due to thin shells. In 1969 25 percent of the females in both colonies laid both thin-shelled eggs and soft, leathery-shelled eggs. The latter were somewhat wrinkled and when pressed with the thumb dented with an audible pop. These soft-shelled eggs generally contained embryos and were incubated by the parent birds for the full period, but they usually failed to produce chicks. The 5200 birds in the two colonies in 1968 dropped to 1175 in 1969. The bald eagle, the osprey, and the brown pelican have also produced soft-shelled eggs, with associated hatching failures.

It is thought that DDT must have an impact on the birds' ability to produce normal amounts of calcium because shells of eggs collected from the colonies contained abnormal amounts of DDT and its derivatives, DDE and DDD. With repeated ingestion the DDT must result in the substitution of some other material to produce the soft shell.

DDT was replaced in Arizona in 1969 by other insecticides, and it is reported that the honeybees were nearly wiped out. Many crops need the bees for pollination. At the time the honeybees were being eliminated, to the detriment of many crops, Arizona's hay, grain, milk, milk fat, and beef—all heavily laced with DDT residues a few years ago—now meet both state and federal tolerances. But the insecticides attack the target pests as well as many other insects, and the ecology suffers. A widely written-about alternate to the use of insecticides is a biological control, namely, the use of insects to control insects. The remarkable praying mantis is a good insect in the garden because it feeds on several plant-injurious insects and helps maintain nature's balance. It is reported to feed on aphids, tomato worms, lace bugs, lice, mites, borers, maggots, and flies.

Another useful insect, the European wasp, deposits its eggs in the larva of the alfalfa weevil; when the eggs hatch, the wasp larvae feed on their host. Other useful wasps are the almost-microscopic one, *Trichogramma*, that controls the apple worm codling moth, and *Dendrosoter protuberans*, a parasite of the European elm bark beetle. The ladybug or properly the lady beetle, is particularly fond of aphids and greenbugs. The green lacewing (*Chrysopa*) larva eats the bollworm and budworm of cotton plants and the fruitworms of tomato plants.

Creating sterile strains is another method of insect control and has been used successfully. And another method is to lure undesirable species to feed on germ-packed baits, whereupon the insect dies in epidemic numbers. Japanese beetles are attacked by milky spore disease, a bacterium. The cabbage looper, army worm, and other truck-farm crop pests succumb to a certain bacillus.

Air pollution

The air we breathe concerns us all. We both pollute it and suffer from the pollution. Tomorrow our very survival may depend on the effectiveness of our combat against air pollution. The smog that plagues the cities of the world not only irritates our eyes, nose, and throat, but it causes disease and death, reduces crop production, discolors paints, deteriorates and corrodes metals and building stones, and impedes visibility. It costs us billions of dollars each year. And the costs will continue to rise until an aroused public acts to halt the dispersal of the materials into the atmosphere that cause smog.[3]

Nature of air pollutants

The materials that contaminate the atmosphere may be classed as particulates and invisible gases. By particulates is meant the finite mineral particles, the pollen grains, grains of organic matter from the soils, and the carbon particles of smoke. These particles range in size from over 40 microns, which settle out of the air fairly rapidly, to 0.1 micron, which are nearly gaseous and tend to remain in the atmosphere indefinitely. Dusts, unless present in large amounts, are caught by the cilia of the respiratory tract and do not reach the lungs, and hence generally are not very harmful. The fall of rain and snow cleans the air of the particulates fairly effectively. We can see the particulates and are most disturbed by them, but the invisible gases are the most poisonous to man and plants. Among the gases that man contributes to the atmosphere, which can be considered pollutants, are nitric oxide and nitrogen dioxide (NO and NO_2), hydrocarbons and peroxyacetyl nitrate, ozone (O_3), sulfur

[3] Michael Treshow, *Whatever happened to fresh air?* University of Utah Press, 1970.

dioxide (SO_2), carbon monoxide (CO), and fluoride (CaF_2, HF, and others). Mercury and cadmium have recently been detected in alarming amounts in certain foodstuffs. Toxic compounds of these metals are introduced into the air from smelters, and mercury is used in certain sprays. Radioactive fallout is both particulate and gaseous, and is classed by some authorities in a third category. Some of the gaseous pollutants condense into aerosols or particulates. In large cities odors represent various gaseous contaminants, and altogether about 100 man-made additions to the natural air have been identified. It is estimated that 300,000 manufacturing and smelting establishments contribute about 133 million tons of air pollutants annually to the air of the United States.

The word *smog* was used first in London to describe the combination of coal smoke and fog. The Los Angeles smog is known to come mostly from automobile exhaust, and is more accurately designated as photochemical pollution. London smog is worse in damp cold weather, but the smog of Los Angeles and several other large U.S. cities is worse when the sun is shining and when a temperature inversion occurs. The conditions for a temperature inversion have previously been described. More will be said on the subject of smog in the following paragraphs on the sources of the pollutants.

Sources of the pollutants

Carbon dioxide and related combustion products

The air contains a small amount of carbon dioxide (460 ppm by weight) but, as explained in Chapter 14, CO_2 is most important in absorbing the outgoing thermal radiation from the earth's surface, and thus this small amount is important. According to Plass[4] the latest calculations show that if the CO_2 content of the atmosphere were doubled, the surface temperature would rise 3.6° C, and if the amount were cut in half the temperature would fall 3.8° C. Some climatologists estimate a rise of 3.6° C would bring a tropical climate to most of the earth, and a lowering of 3.8° C would precipitate another ice age. We are not sure about these conclusions but we can become quite concerned when we learn that the combustion of coal, oil, gas, and wood added 6 billion tons of CO_2 to the atmosphere in 1954 and 12 billion in 1964. The rate has been doubling each ten years. Without this annual addition there would be a balance between natural gains and losses of CO_2 which amount to about 0.1 billion tons on each side of the ledger. Hot springs, fumaroles, and volcanoes add about 0.1 billion tons annually, and the deposition of limestone and dolomite and peat deposits subtract this amount. The human annual increment amounts to an addition of CO_2 in the atmosphere of 30 percent in a century, and thus the possibility that the climate has been warming up and will continue to do so is real. We have already noted a recession of the glaciers in almost all parts of the world in the last 60 years. The average annual temperature has risen 2° C in Norway since 1930, and the International Geophysical Year reports estimate that the Greenland ice sheet is suffering a net loss of 22 cu mi of ice per year.

In the immediate future we will undoubtedly be concerned with the products of combustion other than CO_2 because they are toxic and result from a burning process which is only about 20 percent efficient. The products of this partial combustion are wasted into the atmosphere in various gaseous and particulate forms. Carbon monoxide, sulfur dioxide, sulfur trioxide, and nitric oxide are the chief toxic compounds emitted as waste to the atmosphere from general factory, power plant, and garbage-disposal operations, and from home and office heating. It is estimated that on a national basis, industry contributes about 17 percent of the pollutant load, power generating plants 14 percent, and space heating and refuse disposal about 9 percent. The rest comes from automobiles. The methods of minimizing the emissions of indus-

---

[4] Gilbert N. Plass, Carbon dioxide and the climate, *American Scientist*, July 1956, pp. 302–316.

trial plants are largely beyond the scope of the present writing, but it can be said that the emissions *can* be effectively controlled. Electrostatic precipitators, scrubbers, and various filter systems do the job, but many plants, especially the old ones, are not well designed for the installation of these systems, and because of the high expense, they have not been installed.

It should not be concluded that all the sulfur in the atmosphere is a result of unnatural conditions. Much salt is picked up from the oceans and transported in the atmosphere, and some of this is sulfate (mostly $CaSO_4$). The decay in swamps also produces H_2S which is oxidized and soon becomes part of the sulfate content. Isotopes of sulfur are being used to distinguish between sulfur introduced by industry and natural sulfur, and to identify the individual industrial sources.

Automobile pollutants

It is estimated that 100 million automobiles in the United States contribute 60 percent of the products of air pollution. The internal-combustion engine threatens our environment in two ways; it consumes a lot of oxygen from the atmosphere and it releases several toxic gases. The toxic gases contain lead, barium, boron, nickel, copper, iron, molybdenum, and titanium. Of these lead is the most serious. Carbon monoxide and nitric oxide are also bad. Hydrocarbons include ethylene, aldehydes, acids, phenols, polycyclic and aromatic hydrocarbons. Probably worse is the reaction product of the hydrocarbons and nitric dioxide, peroxyacetyl nitrate, commonly referred to as PAN. On freeways when the traffic jams and stalls, the CO content of the air has been measured as 100 ppm, seven times higher than when traffic is moving. SO_2 and NO_2 deteriorate rapidly in nature, but CO has a human lifetime of duration.

Some of the obnoxious fumes can be eliminated by "positive crankcase ventilation." The principle is as follows: While any automotive engine is in operation, a certain amount of the fuel and exhaust fumes pass through the piston rings and into the crankcase. Called blow-by, these fumes must be removed to prevent severe contamination of the engine oil. The traditional way is through a road draft tube leading from the crankcase. Most cars also have a fresh air inlet on the oil fill pipe. Movement of the vehicle creates a slight vacuum at the draft tube outlet under the car and a slight pressure of air under the hood around the oil fill pipe. These pressure differences draw fresh air in to ventilate the crankcase and exhaust the contaminated blow-by. So that no blow-by air will be exhausted into the atmosphere, positive crankcase ventilation (PCV) systems have been developed which recycle engine blow-by fumes back into the engine's combustion chambers where these hydrocarbons can be burned. In California, where the problem of air pollution has reached national prominence, all new cars sold since 1961 have had to be equipped with a PCV system. Evidence of the effectiveness of PCV is that most cars built in this country since 1963 have been so equipped.

But the blow-by is only part of the exhaust. Officials of Ford and General Motors say that improvements are possible which could virtually eliminate the emission of pollutants, but since the cost of each new car would be increased about 10 percent, and fuel consumption would also be increased, they predict that smog-free air in the cities is quite far in the future.

To meet the existing exhaust-emission standards for hydrocarbons and carbon monoxide, the automobile manufacturers have used two basic methods. The first is to inject air into the exhaust manifold near the exhaust valves, where exhaust-gas temperature is highest, thus inducing further oxidation of unoxidized or partially oxidized substances. In this approach, carburetion and spark timing are also adjusted to reduce the amounts of pollutants emitted. The second basic method is to design the cylinders and adjust the fuel-air ratio, spark timing, and other variables to reduce the amounts of hydrocarbons and carbon monoxide in the exhaust to the point where air injection is not required. Most 1968 model cars in the United States use the second basic method, with air injection being used mainly on vehicles with manual transmissions. Beginning in the fall of 1969, all new cars sold in California were equipped by law with devices to eliminate 85 percent of the hydrocarbons of the exhaust.

These methods should be capable of meeting the 1970 federal standards for hydrocarbon and carbon monoxide emissions, but to achieve even lower emissions would require more complex systems. Two devices that have been studied extensively for use in such systems are the exhaust manifold thermal reactor and the catalytic converter. Both are still in the development stage, but have been shown to be able to achieve very low emissions of hydrocarbons and carbon monoxide.

Leaded (ethyl) gasoline, which enhances the performance of the car, has been rated a major contributor to air pollution. Since the high compression-ratio motors require leaded gasoline, and since high compression means more power, efficiency and gasoline economy, it is going to be difficult to eliminate the use of such fuel. Also, the presence of lead in the exhaust makes it difficult to lower the emissions of carbon monoxide and the hydrocarbons by means of catalyst and afterburner installations. Ford and Chrysler officials say they are prepared to redesign and produce motors with low compression ratios to use low-octane, lead-free gasoline. Standard Oil of Indiana has a premium brand of gasoline that has a 100-octane rating, but it costs a penny more per gallon. Atlantic Richfield says it could put suitable fuel on the market if there were sufficient demand.

Fluoride

Fluoride, a chemically complex, highly toxic material, is released to the atmosphere probably in both particulate and gaseous form. The fumes of smelters, blast and steel furnaces, phosphate fertilizer plants, ceramic furnaces, and aluminum plants are particularly laden with the extremely toxic and corrosive fluoride products, since the mineral *fluorite* (CaF) occurs in the ores that are smelted and is added in the treatment of iron ore. *Fluorine* is a contaminant of phosphate rock, and occurs in many clays used in the manufacture of ceramic products and aluminum. The treatment of the ores, phosphate rock, and clays occurs at high temperatures and some of the fluorine unites with the metals, is volatilized, and passes off into the atmosphere. Some may be released as HF, which is extremely dangerous. Since SO_2 is a common by-product of the fluorides, the areas around the above mentioned types of installations have felt the effects of air pollution badly, and much effort has been expanded in curtailing the emissions, both by private industry and by the government.

Ozone and crop damage

Ozone, it has been discovered, prevents normal plant growth. Its presence is noted first by flecking and bleaching of the leaf surface. Weather fleck of tobacco plants is a serious damage, and it extends from Ontario to the southeastern states. The disease can be definitely traced to ozone emanating from urban centers. Onion blight, grape stipple, and spinach bleach are the effects of ozone on other crops. Now it is reported that the ponderosa forest in the mountains east of Los Angeles is being killed by ozone in the smog. Prevailing winds carry smog from the megalopolis 60 miles eastward into the San Bernardino and San Jacinto Mountains where trees 100 feet tall and 700 years old are losing their needles, are attacked by insects, and die. Seventy-five percent of about 1.7 million trees are being destroyed. With the loss of the stately ponderosas the whole forest ecology suffers—grouse, quail, squirrels, chipmunks, and man, who builds houses with the ponderosa pine lumber.

Radioactive waste in the atmosphere

The radioactive materials that are released into the atmosphere, such as those formed when a nuclear bomb or weapon is detonated, are both particulate and gaseous, but the gases probably soon lose their radioactivity. The particles carry the radioactivity for some time. The particles may be inhaled and they may settle on the body or clothing and cause skin burns.

Under extreme conditions a passing cloud laden with the particles may give off injurious gamma radiation. Since the radionuclides are transported freely by the atmosphere it is estimated that about two-thirds of them eventually fall into the oceans and one-third on land. Much of those that settle on land fall on plants, and thus crops receive a good deal of the radioactive fallout. This becomes concentrated in the food the cattle eat and the milk and cereals that man consumes. The surface streams drain off the radionuclides and the water we drink may become radioactive. Measurements of radiation intensity are made in roentgens (as in X rays), and by 1960 it had been decided that the safe maximum exposure for individuals over 18 years old should not exceed 0.5 roentgens (r) per year over the whole body, and that the average dose to the general population over a 30-year period should not exceed 5 r to the gonads (sexual organs where germ cells are created). The average natural radioactivity to which we are exposed amounts to about 3 r in 30 years.[5] The potential hazard of a release of nuclear energy depends on the way the radioactive materials are diluted and transported in the atmosphere, on the time that has elapsed, and on the mechanisms by which the contaminants deposit on surfaces. Nearly all of the man-made effluents before the advent of the atomic bomb were into the lower layer of the troposphere, known as the friction layer by the micrometeorologists. This layer extends only 300 ft up from the ground and has received much study because of industrial contaminants and chemical warfare agents. But since 1944 atomic bombs have been discharged at heights of more than 100,000 ft. If atomic energy is used to power future satellites, space probes, and manned spaceships, we will be confronted with the release of radioactivity into the stratosphere and ionosphere. Most of our knowledge concerning the mixing of radioactive materials with the atmosphere applies to the troposphere.

When a gas vapor or aerosol is introduced into the atmosphere it dilutes either by molecular or turbulent diffusion. We can neglect molecular diffusion because it is so much slower than turbulent diffusion that it may be disregarded. Eddies in the atmosphere range in scale from almost microscopic to cyclonic, and thus mixing motions range from a few centimeters to hundreds of kilometers. The rate of diffusion depends on the size of the eddy. This may be stated in terms of lapse rates: the greater the lapse rate, the stronger the turbulence (see Chapter 14). Observations of diffusion have been made on smoke emissions from stacks, and the results are instructive (observe Figure 16–22). It is concluded that the maximum concentration downwind from a stack will be inversely proportional to the square of the stack height, directly proportional to the rate of emission, and inversely proportional to wind speed. It is noted that the concentrations downwind from the stack depend on *emission* rate of the contaminant rather than concentration of the contaminant in the effluent air. In other words, the concentration in the plume is a function of the size of the plume, and this is little affected by the volume of the exhaust air which is an insignificant addition to the size of the

[5] The above is a very incomplete statement of the standards of maximum exposure. For occupational workers the standards are higher. It is concluded that people younger than 18 should not be exposed more than that of the natural radioactivity. The problem is complex, and much has been written about it. For further reading, see Merril Eisenbud, *Environmental radioactivity*, McGraw-Hill, New York, 1963.

Figure 16–22 Dissipation of stack plumes. Note the effect of the buildings on the plume of the lower stack. The plume of the taller stack lacks the downwash of the lower one.

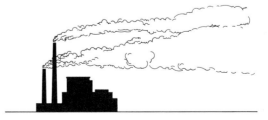

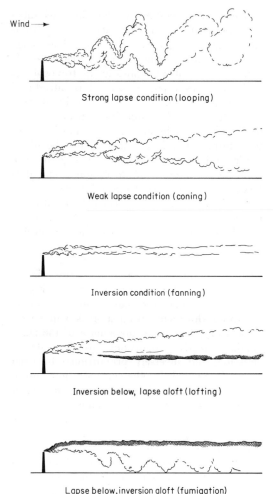

Figure 16–23 *Stack plume behavior with various temperature and lapse rates. See Chapter 14.*

plume (see Figure 16–23). It may be concluded that in the accidental discharge of radioactive waste into the atmosphere from plants that handle radioactive materials or reactors, the radioactive materials will become diffused to marginal intensities at the surface within tens of miles.

It has been noted by Robert C. Pendleton, Director of Radiological Health Laboratory at the University of Utah, that radioactivity in the Salt Lake Valley smog gradually builds up, and in the case of the mid-December smog of 1969 the activity was five times that of normal. Fortunately the smog only lasted seven days. In past smog build-ups ten times the normal activity was reached. As the general particulates increase so does the radioactivity. The radioactivity is determined by counting the number of nuclear disintegrations per minute by means of a scintillometer. At five times the normal background there were 4500 disintegrations per minute, which is comparable to that experienced in the uranium mines, and it is noted that uranium miners had a rate of lung cancer ten times that of the average U.S. population. According to Pendleton a small amount of the radioactivity in the smog came from the French and Chinese atmospheric nuclear bomb testing, which could be noted throughout the state, but the greatest part of the radioactivity of the smog comes from the burning of coal by industrial plants in Salt Lake City. Coal carries various amounts of uranium, radium, and thorium, and in certain places the coal had been considered a source of uranium during the uranium boom.

During a wintertime inversion that results in a smog build-up, which in American cities averages 60 percent automobile exhaust, coal smoke also builds up and contributes the chief amount of radioactivity.

There are few data to predict the diffusion of radioactive waste ejected into the stratosphere by an atomic bomb—it would probably be spread on a global scale. Based on observations prior to 1958 the residence time of debris ejected into the lower stratosphere in tropical latitudes is about 2 to 3 years. In polar latitudes the residence time appears to be 8 months. The chief radioactive material in atomic bomb detonations is strontium 90 and this, together with other fallout particles, is deposited where greatest rainfall occurs. When the radioactive dust from the stratosphere enters the troposphere it is carried chiefly by the west-to-east winds of the temperate latitudes where it is deposited.

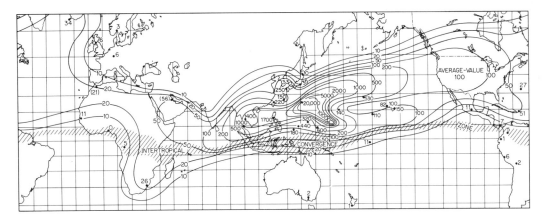

Figure 16–24 Radioactive fallout from a multimegaton thermonuclear explosion in the Marshall Islands in November 1952. The contours depict the intensity of fallout in terms of millicuries per 100 square miles during the time period from 2 to 35 days after the detonation.

The mean residence time of the debris in the troposphere in these latitudes is judged to be about 30 days, and the fallout seems to settle preferentially.

A study has been made of two explosions in 1958 above Johnson Island in the Pacific, one at an altitude of 125,000 ft and one at 250,000 ft. It was concluded that the average residence time at these altitudes was about 7 and 15 years, respectively. The fallout that reached the surface of the earth from the multimegaton thermonuclear explosion in the Marshall Islands in 1952 is shown in Figure 16–24.

The President's plan

On February 10, 1970, President Nixon proposed to Congress and the nation that the government lay a heavier hand on those who pollute our air and water. He asked that the Federal agencies be given a bigger voice in setting air and water quality standards, and proposed that court-imposed penalties be authorized of up to $10,000 per day for polluters. He asked for tougher auto-exhaust emission standards, beginning with the 1973 model vehicles, and also for Federal controls on motor-fuel composition. He has ordered Federal research into the ways that the internal-combustion-engine exhaust could be eliminated altogether. The President's proposals for water purification have been cited previously. This all shows the gravity of the environmental situation in the United States today.

Some bright spots in the pollution picture

Air pollution, according to one authority, is a feedback signal that man has overgrazed his ecological niche. The signal has been recognized by many and has stimulated a real awareness not only of the air but of the total environment. Our leaders in science and government are now planning to achieve a status of equilibrium with the total environment. Air pollution could thus be a blessing in disguise.

We note another encouraging sign. London's smoky, sulfurous air was foul enough 18 years ago to cause 4000 deaths in a single winter. Now the thick black soot from countless chimneys and the strangling peasoup fogs have gone, and 8 million Londoners are getting 50 percent more sunshine than they used to. Not only is there better air over London, but the Thames Estuary is clearing up. Awash in human and industrial slime no fish could live in it along a 40-mile stretch above and below London Bridge in 1958. Now 43 varieties of fish are noted in it, including the fastidious sea trout. One $3\frac{3}{4}$-pound individual, entirely healthy, was caught in late 1970, and the event marked such a significant change for the better that the specimen was placed in the Museum of Natural History. The Royal Commission on Environmental Pollution observes everything from sewage sludge, radioactive waste, oil slicks, derelict lands, and auto exhausts, to sound levels. Three large ministries fall under it, and each minister is forced to know what the others are up to. It is called "total strategy," and the system promises to be even more successful than that that brought about the London improvements.

The thesis has been advanced by some that already man exceeds the capacity of the planet to support him. What, then, is the future of the underdeveloped nations? Or if the present world's people and more can be supported, all at a comfortable level of living, what is the population level at which we must stabilize? How do we measure the planet's capacity to support man? This may be the guiding question for scholars in the next generation.

References

Chorley, Richard J., ed. *Water, earth and man.* London: Methuen, 1969.
Cleaning our environment, the chemical basis for action. American Chemical Society, report, 1969.
Disposal of liquid waste by injection. U.S. Geol. Survey, Circular 631, 1969.
Greenberg, D. S. Test case of an estuary in crisis. *Science,* Feb. 20, 1970.
Hoult, David P., ed. *Oil on the sea.* New York: Plenum Press, 1970.
Hubbert, M. King, NASA, and NRC. In *Energy resources.* San Francisco: W. H. Freeman and Co., 1969.
Merriman, Daniel. Calefaction of a river. *Scientific American,* May 1970.
Plass, Gilbert N. Carbon dioxide and the climate. *Amer. Scientist,* July 1956, pp. 302–316.
A primer on waste water treatment. U.S. Geol. Survey, 1969.
A primer on water quality. U.S. Geol. Survey, 1968.
Treshow, Michael. *Whatever happened to fresh air?* Salt Lake City: University of Utah Press, 1970.
Wilderness. U.S. Dept. of Agriculture, U.S. Forest Service, PA-459, 1969.

Filmstrip

Environmental pollution: our world in crisis. Rochester, N.Y.: Ward's Natural Science Establishment.

For further information about environments and conservation, *viz.,* federal and state agencies and foundations working on and sponsoring aid, literature and films, projects sponsored, etc., write to The Conservation Foundation, 1250 Connecticut Avenue, N.W., Washington, D.C. 20036.

index

index

Page numbers in *italics* indicate illustrations. Page numbers followed by the letter "t" indicate tabular information.

Aa, 55, *56*
Abrasion,
  by glaciers, 193, *195*
  by waves, 246
  by wind, 225
  See also Erosion
Abyssal cones, 285
Abyssal hills, 285
Abyssal plain(s), 285
  Atlantic, 304, *306*
  ocean floor, 282
  terriginous sediments of, 304–305, *306*
  Tufts, 303
  turbidity-current deposition on, 304, *307*
Acadian fold belts, 311, *313*
Adiabatic chart, 364, *365, 370*
Advection, 357
Aeromagnetic map, *131*
Aeronomy, 346
Aerosols, 341
Africa, northern, sand seas of, 228
African plate, stability of, *332*
Agassiz, Louis, *199*
Age, geologic, 137
Agencies, in evironmental management, 406–407
Agricultural Conservation Program, 421
Air, cleaned by rain and snow, 435
  liquefaction of, 339
  mobility of, 342
  origin of, 277–280
  quality standards for, 441
  saturated, 361
  solar radiation effects on, 355
  upper, jet streams in, 397–398
    maps of, 390–391, *393*
  warming over land, 357
  water vapor in, 361–362
  See also Atmosphere
Air currents and flow, levels of, 386–388
  vertical, 385
    vorticity in, 385, *386*
  See also Wind(s)
Airglow, 349
Air mass(es), Arctic, 394
  Caribbean, 394
  classification of, 378–381
  continental, 378
  definition of, 378
  fronts and, 377–382
  maritime, 378
  North American, 379t, *380*
  polar continental, 378, 379t
  polar maritime, 379t, 380
  sources of, 378, *380*
  stable, 365
  tropical continental, 379, 379t
  unstable, 365
  weather prediction and, 373–401
Air pollution, 435–442
  automobiles and, 437–438
  awareness of, 441–442
  carbon dioxide in, 436–437
  combustion products in, 436–437
  corrective measures in, 441–442
  fluoride in, 438
  nature of, 435–436
  ozone in, 438
  radioactive materials and, 436, 438–441
  smog and, 367–368
  sources of, 436–441
Air pressure, barometric, *see* Barometric pressure(s)
Airy, George B., 132
Alaskan Earthquake, 163, *169–172*
Alaskan low, 375–377, 380, 391
Albedo, 355
Aleutian Islands, 289
Aleutian lows, 375–377, 380, 391
Algal heads, 71, *72*
Alkyl benzene sulfonate (ABS), 412
Alluvial fan(s), 215–217
  block faulting and, 183, *186*
  definition of, 183
  delta versus, 183
  ground water in, *219*
Alluvium, unconsolidated, meandering and, 179
Alpha rays, 148
Altitude, atmospheric pressure and, 343, *343, 349*
  high, physical phenomena at, 348–350
Amundsen Glacier, 192
Anderson, Don L., 105
Andesite, 329
Andesite line, 329
Andros Island, 301

Angle of incidence, definition of, 354
  of sun's rays, 353–355, *353*, *354*
Angstrom, 22, 346
Anion, 25
Anorthosite, 19
Antarctic Bottom Water, 270, 271
Antarctic Circumpolar Current, 266, 270
Antarctic Ice Cap, 189, *190*
Antarctic Intermediate Water, 271
Anticline(s), *79*, 80–83, *83*
  coal in, *84*
  definition of, *79*, 80
  outcrop patterns of, 81, *82*
Anticyclone(s), definition of, 360
  mapping of, 390
  movement of, 377
  wind circulation around, 359, 360
  See also High(s)
Anticyclonic circulation in Northern Hemisphere, *360*
Apatite, *35*
Apennine Mountains, 17, *18*
Appalachian Mountains, erosion of, 317
  formation of, 311, 317
  skin-deep deformation of, 319
Aquiclude(s), 211, 220
Aquifer(s), definition of, 211
  salt water in, 220
  sandstone, 217–219, *220*
Arabian Desert, 228, *228*
Archean Era, 277, 278
Arches, wave-cut, 247
Archipelagos, volcanic, deep-sea trenches and, *292*
Arctic, air masses from, 394
Arctic high, 377
Arctic low, 394
Arctic Ocean Basin, 329
Arête, *196*, 197
Argillite, 67
Argon, 339, 339t
Arguelo deep-sea fan, *296*, 303

Arid regions, valleys in, 182, *185*
Armstrong, R. E., 280
Artesian water pressure, 211–212, 216
Artesian wells, *213*, *219*, 220
Asama, Mount, 55, *57*
Ash, Mazama, 303, *304*
Ash beds, volcanic, *140*
Ash falls, 55
Asian high, 376, 377
Aso-San, 48, 56
Asteroids, 2–8
  location of, 3, *5*
  orbits of, *5*
  sizes of, 3
Astoria Canyon, 303
Astoria deep-sea fan, 303
Atlantic City, 248, *251*
Atlantic Coastal Plain, 219, *220*, 301
  erosion of, 421
Atlantic Ocean, currents in, 265, *266*, *271*
  deep circulation in, 270–272
  fit of continents bordering, *329*
  geomorphic divisions of, 282, *283*
  longitudinal section of, 270, *272*
  oxygen in, 270, *272*
  profile of, *292*
  red clays in, 297
  salinity in, *260*, *261*, 270, *272*
  sea-floor spreading in, 299
  poles of, 332
  subsidence in, 285
  temperatures of, 263, *264*, 270, *272*
  water masses in, 270–272, *272*
Atlantic Ocean basin, 319–325
  formation of, 329
  magnetic anomalies in, 319–322, *327*
  sea-floor spreading in, 319–322
Atlantic shelf, 300, *300*, 302
Atlantis II, 308
Atmosphere, 337–401
  advection in, 357

carbon dioxide in, 340–341
characteristics of, 338–350
composition of, 339–342
  variations in, 346–347
conduction in, 357
convection in, 357, see also Wind(s)
divisions of, 345–346
dust in, 339, 341–342
early, 279
gas laws applied to, 343–345
of Jupiter, 5
meteor destruction by, 345
moisture condensation in, 361–372
molecular makeup of, 339, *341*
origin of, 277
permanent gases in, 339–340
regional circulation in, 358
regions of, 347
solar, 345
solar radiation absorbed by, 275, 348, 351, 355
  preferentiality in, 352
stratification of, 345–350, *345*
sulfur in, 436, 437
ultraviolet radiation absorption by, 345, 347, 352
upper, clouds in, 348
  composition of, 347
  definition of, 345
  layers of, *348*
  probing of, 346
  winds in, 348
upper boundary of, 345
Van Allen belts in, 349–350
water vapor in, 340, 355
  See also Air
Atom(s), in atmosphere, *341*
  combining properties of, 23–25
  definition of, 21
  mass number of, 23
  measurement of, 22
  oxygen, 25
  sizes of, *22*, 23
  structure of, *22*, 24t, *25*
  weight of, 23
Atomic bomb, atmospheric radioactivity from, 440, *441*

Atomic number, 22, 23
Atomic theory, 21–23
Auroras, 348–349, *349*
Automobiles, air pollution from, 437–438
Autumnal equinox, 354
Avalanche, rock, 161
    See also Landslide(s)
Avogadro, Amedeo, 339
Axis, earth's, inclination of, *353*
Azores high, 375

Bacon, Francis, 322
Bacteria, to kill insects, 435
  in streams, 414
Badlands, *186*
Bagnold, R. A., 223
Bahama Islands, 301
Baja California, 330
Baltic Sea, 261
Bar(s), *250*, 252
  along beach, 244, *247*
  barrier, 248, 249, *251*, *253*
  bay-mouth, 247
  definition of, 345
Barchan, 227, *227–230*, *229*, *232*
Barometers 343, *344*
Barometric pressure(s), altitude and, 343, *343*, *349*
  global, *376*, 377
    surface winds and, 375–377
  measurement of, 343, *344*
  over Northern Hemisphere, *375*
Barrier, 248
Basalt, dam sites and, 410
  dikes of, *42*
  in earth's crust and mantle, 107
  formation of, 47
  lunar, 19
  magma of, 41
  sill of, *43*
  transformation to granite, 47, *48*
Basalt glass, 35
Basaltic uplifts, 315, *325*
Basin, catchment, 190, *198*
  ocean, *see* Ocean Basin(s)
  wind-carved, 225

Batholith, definition of, 38
  geologic time and, 141
  Sierra Nevada, 45, 311
Batholith belt, 311
Bathometer, 281
Bathymetric chart, of ocean floor, *287*
Bauxite, 159
Bay of Fundy, 240, 273
Bayous, 111
Beach(es), arcuate, *254*
  bar(s) along, 244, *247*
  barrier, definition of, 248, *251*
    formed by alongshore currents, *252*
    of Galveston Bay, 417, *419*
    hurricane effects on, 249, *252*, *253*
    origin of, 249, *251*
  coast topography and, 240
  curved, 247, *250*, 252
  definition of, 239
  erosion along, 244, *247*
  factors affecting, 240
  hurricane damage of, *253*
  maintenance of, 252
  oil pollution of, 252
  profile of, 244–247
  rilled sand on, *243*
  storm, 231
    definition of, 244
  tidal effects on, 240, 244
  Waikiki, 239, *242*
  wave effects on, 240
Beardmore Glacier, 192
Becquerel, Antoine, 148
Bed(s), sandstone, 64, *65*
Bedding, false, 65
  graded, 67, *69*
Bedding planes, 65, *67*
Bed load, definition of, 180
  wind transport of, 222
Bell Canyon, *196*
Bench marks, uplifts indicated by, 187
Benches, along Atlantic shelf, 300, *300*
Benioff, Hugo, 107
Benioff zone, 329, *332*
Ben Lomond Mountain, 250
Bentonite, 68

Bergschrund, 196, *198*
Bering Sea, 305
Berm, 244
Bermuda Islands, 287
Bermuda Rise, 285, 287, 288
Beta rays, 148
Biela's comet, 9
Big Horn Mountains, 315
Biotite, 34
  in geologic dating, 151
  in granite, *35*
  in schist, 92
  weathering of, 157
    See also Mica
Birds, DDT effects on, 434
Black Hills, 217, *220*
Black Sea, 262, 305
Blake Plateau, 301, 307
Blowouts, 232, *234*
Bode's Law, 12
Bonneville salt crust, 235
Bossons Glacier, 191
Bouguer, Pierre, 131
Bouguer correction, 131–132
Bouguer gravity map, *134*
Boulder Dam, 177, *181*
Bowen's reaction series, 47, *47*
Boyle, Robert, 342
Boyle's Law, 342
Bradley, W. C., 29
Bragg, Laurence, 29
Bragg, W. H., 29
Breaker(s), 238–239
  damage caused by, 242, *245*
Breccia, fault, 86
  volcanic, 37, *39*, *40*
Breezes, land, *357*, 358
  mountain, 358, *359*
  sea, *357*, 358
  valley, 358, *359*
    See also Wind(s)
Bridal Veil Falls, *197*
Buffalo Bayou, 418
Bullard, E. C., 104, 121

Calcite, 69
Calcium carbonate, dams from, in caves, 214
  precipitation during Archean Era, 278
  precipitation in caves, 214

INDEX **447**

Calcium carbonate (*Continued*)
  in spring water, 215, *219*
  in streams, 171
Caldera, 56, *57*
Calefaction, 415
California, Baja, rafting of, 330
  earthquakes in, *112*
Cameras, for deep-sea photography, *288*, 291
Cameron Cave, *215*
Canadian high, 375
Canadian national parks, 428
Canadian Shield, 204, 205, 309, *310, 311*
Canals, irrigation, *409*
Canyon(s), formation of, 171, 182
  submarine, 282–285, *284*, 303
  See also specific canyon, for example, Grand Canyon
Cape Hatteras, 300, *300*
Cape Kellett, *250*
Cape May, 300, *300*
Capillary attraction, 208
Carbon-14, in geologic dating, 152
Carbonate(s), 30, 31t
  sediments of, 301
Carbonate rocks, water in, 212–214
Carbon dioxide, as air pollutant, 436–437
  atmospheric, 339t, 340–341
  life and, 340
  in oceans, 263, 279
Carbon monoxide, atmospheric, 436–438
Caribbean Sea, 305, 394, 395
Carlsbad Caverns, 213
Caspian Sea, 305
Cassiterite, 307
Catastrophism, in geologic dating, 138
Catchment basin, 190, *198*
Cathode rays, 27
Cathode tube, early, *30*
Cation, 25
Cave(s), cliff, 246
  columns in, 214, *216*
  deposits in, 214

draperies in, 214, *216*, *217*
formation of, 156, 207, 213, *215*
limestone, 213, 214, *215*
pools in, 214, *216*
stalactites in, 214, *216*
stalagmites in, 214
terraces in, 214, *216*
wave-cut, 247
well-known, 213
  See also specific cave, for example, Cameron Cave
Caving, erosion by, 171, 172, *177*
Cement, sandstone, 63–64, *64*
Censorinus Crater, *16*
Central Atlantic Water, 272
Centrifugal force, 360
Centripetal force, 360
Cephalopods, fossil, 145
Ceres, 3
Chalk cliffs, 252, *253*
Channels, levied deep-sea, 285
Charles, Jacques, 342
Charles's Law, 343
Charleston Earthquake, 111
Charlotte Harbor, *251*
Chatter marks, glacial, *195*
Chernozems, 159
Chert, 72–73, *247*
Chesapeake Bay, 252
Chilean Earthquake, 97
Chlorides, 30
Chlorinity, sea-water, 261
Chlorite, 92
Chrysopa, 435
Church, E., 193
Cinder cones, 37, *39*
  volcanic, 52–53, *52*
Cinders, 36
Cirque, *191*, 196, *196*, *199*
Cirque-basin lakes, 197
City planning, 430–431
Clavius, 14
Clay, deposits in deltas, 181, 185
  glacial, *78*, 191
  ground water in, 211
  red, *see* Red clay
  water penetration of, 208

well water from, 217
  See also Shale
Clay particles, river deposition of, 179
  river transport of, 169, 173, *178*
  wind transport of, 222
Cleavage, slaty, 90, *90*, 92
Cleveland, Grover, 424
Cliff(s), chalk, 252, *253*
  erosional processes attacking, 245
  in headlands, 247, *249*
  hydraulic pressure against, 246
  sea, *250*
  undercut, 241
  wave-cut, 245, *248*, *254*
  waves breaking on, 240–241, *241*
Cliff cave, 346
Climate(s), atmospheric carbon dioxide and, 341
  glacial, of Pleistocene epoch, 201–205
  humid, effects on valleys, 182
  past, soils as indicators of, 159
  soil and, 158
  temperature and, 341
Cloud(s), 368–371
  advection of, 386, *386*, *387*
  altocumulus, 370
  altostratus, 370, 382
  cirrocumulus, 369, 370
  cirrostratus, 369, 370, *371*, 382
  cirrus, 368, 370, 382
  cumulonimbus, 368, 370, 382
  tornadoes and, 396
  cumulus, *368–371*, 369, 370, 381, 382
  condensation in, 370–371
  convection currents in, 370–371
  cumulus fractus, 369
  dissipation of, 386, *386*
  formation of, 385, 386, *386*, *387*
  heat reflected by, 355, *355*
  mother-of-pearl, 348

Cloud(s) (*Continued*)
  night-glowing, 348
  nimbostratus, 370, 382
  relation of fronts to, 387, *388*, *389*
  storm, convection, *371*
  stratocumulus, 368, 369, *369*, 370
  stratus, 368
  types of, *367*
  unstable air and, 365
  in upper stratosphere, 348
Cloud, Preston E., 279, 280
Cloud patterns, cyclonic, *387*, *389*
  frontal wave and, *388*
  over Pacific Ocean, *384*
  satellite photographs of, 382–385, *386*, *387*, *388*, 398
  vorticity in, *386*, *387*
Coal, 71–72, *73*
  in anticline, *84*
  geologic formations and, 145
  location of, 310
  radioactive materials in, 440
  reserves of, 432
  strip mining of, land reclamation and, *413*
Coal mines, water pollution from, 412–413
Coast, definition of, 239
  emerging, *253*
  topography of, beach formation and, 240
  See also Shore(line)
Coastal plains, ground water in, 219–220
Coast Ranges, formation of, 313
Cobalt, 306
Col, *196*, 197
Cold occlusion, 382, *383*
Colluvium, 161, *162*
Colorado Plateau, sand dunes of, 229–231
  sedimentary strata in, 139, *141*
Colorado River, 181, *181*, 186
Columbia River, 303, *304*
Columns, in cave, 214, *216*
Comb ridge, 197

Combustion products, air pollution from, 436–437
Comet(s), 8–9
  orbits of, *5*, *8*
Common Water, 271
Compass, 115
  declination of, 115–116, *120*
  dip needle, 123, *125*
  dip of, 117
  inclination of, 116–117
Concordant contact, 38
Concretions, in red clay, 297
Condensation, in cumulus clouds, 370–371
  definition of, 361
Condensation level, lifting, 363
Condensation nuclei, 366
Conduction, 357
Cone(s), abyssal, 285
  cinder, 37, *39*
  volcanic, 52–53, *52*
Cone of depression, of water table, 211, *211*
Conglomerates, 62, *63*
  water in, 207
Connate water, 207
Conservation, 404–442
  by farmers, 422
  history of, 404–405
  land use and, 418–435
  of minerals, 431–434
  of soils, 420–423
  U.S. government agencies in, 406–407
  of wildlife, 422
  dams and, 409, *409*
Continent(s), bordering Atlantic Ocean, fit of, *329*
  formation of, 322
Continental areas, ocean salt transported to, 233–235
Continental crust, North American, 309–319
Continental Divide, *187*
Continental drift, 322–325
  of Europe, *334*, 335
  magnetic pole changes and, 120
  major plate of crust in, 331–335

Continental rise(s), definition of, 282
  in ocean floor, *284*, 285–288, 297
Continental shelf, 282, *283*, 284
  near Corpus Christi, *302*
  deltas on, 299, *300*
  due to reef build-up, 301, *303*
  origin of, 302–303
  submarine canyons along, 303
  terriginous sediments on, 299–302, *300*
  types of, 301
Continental slope, *284*, 302
  definition of, 282
  deposits on, 303–304
Contour plowing, 422, *423*, 424
Contour trenches, 425
Convection, in atmosphere, 357
  free, 366
  See also Wind(s)
Convection currents, crustal deformation and, 326
  in cumulus clouds, 370–371
  in mantle, 319, *326*
Copernicus Crater, *17*
Copper (ore), dip needle used to locate, 123
  formation of, 205
  subsea deposits of, 306–308
Coral reefs, 71
Cordillera, Western, 310–315
Core barrels, 281, 291, *297*
Coriolis, Gaspard de, 268
Coriolis force, *268*, *269*
  movement of highs and lows affected by, 378
  ocean currents and, 267, 268, 271
  winds and, 358–360, *359*, 373, 386
Corona, solar, 345
Corpus Christi, continental shelf near, *302*
Corrasion, wave, 245, 246
Correlation, guide fossils and, 144–145
Cotopaxi, 51
Cove, *254*
Crater(s), definition of, 50
  on earth, 5

Crater(s) (*Continued*)
  lunar, 5, 14–16, *15, 18*
    See also specific crater, for example, Copernicus Crater
Crater Lake, 56, *59*, 303
Craters of the Moon, *54, 56*
Creep, 161–162, *165*
Cretaceous period, tectonic map of, *317*
Crevasses, glacial, *192, 200*
Crookes, William, 27
Crops, radioactive fallout on, 439
Cross-bedding, *66, 68, 74, 76*
  definition of, 65
Crust, *108*
  across United States, *110*
  composition of, 45
  deformation of, convection currents and, *326*
  elements in, 32t
  equilibrium of, 132–133, *134*
  layers of, *110*
  major plates of, continental drift and, 331–335
  physical properties of, *106*
  transitional layer between mantle and, 315
  velocity layers of, 108t
  See also Mantle
Cryogenics, 339
Crystal(s), 25–26
  forms of, 26–27, *29*
  hygroscopic, 366
  quartz, *30*
  seed, 33
  structure of, 27–29
  types of, 26, *28*
Crystalline basement, 309
Crystallography, X-ray, 27–29
Curie, Marie, 148
Curie, Pierre, 148
Curie point, 121
Current(s), ocean, see Ocean(s), currents in
  rip, 243, *246*
  turbidity, 283, 304, *307*
    See also specific current, for example, Humboldt Current

Cusp, 249–250, *253, 254*
Cyclone(s), 360, *360*
  definition of, 360
  development of, 385–386, *387–391*
  dissipation of, 388, *389–391*
  mapping of, 390
  movement of, 377
  occluded, 387, *388*, 389
  paths across United States, 379
  satellite photograph of, *390*
  See also Low(s)
Cyclonic circulation, 359, *360*

Dakota Sandstone, 217, 219
Dalton, John, 21
Dam(s), beaver, 168
  breaks in, floods from, 175
  calcium carbonate, 214
  in flood control, 409, *409*
  from landslide, *167*
  purposes of, 409, *409*
  river base level changes and, 177
  sites for, 410, *410, 411*
  in water conservation, 409–411
Darwin, George, 13
Dating, see Geologic time and dating
Day length, solar radiation and, 353–355
DDD, 435
DDE, 435
DDT, 434
Debris, glacial, 191
Deception, Mount, *198*
Declination, 115–116, *120*
  changes in, *125*
  compass compensated for, *120*
  definition of, 116
  map of, *119*
Décollement surface, 82
Deepstar, 307
Deflation, 224–225
Delaware Bay, *251*, 252
Delgada deep-sea fan, *296*, 303
Delta(s), 181–182
  alluvial fan versus, 183
  birdsfoot, *251*

continental shelf formed from, *300*
on continental shelves, 299
definition of, 65
Mississippi River, 181, *185*, 248, 300
tidal, *251*
Delta River, *236*
Denudation, rates of, 184–187, *187*
  definition of, 184
  uplift versus, 187
Desert(s), erosion, 223
  global location of, 228
  See also specific desert, for example, Arabian Desert
Desert pavement, 224, *225*
Desert soil, 159
Detergents, water pollution and, 412
Devonian period, tectonic map of, *313*
Dew point, 362, 362t
Diapirs, 83
Diatom(s), 294, *295*
Diatom ooze, 294, *296, 298*
Dike(s), basalt, *42*
  definition of, 40
  geologic time and, 141
  intrusive, metamorphism and, 94
  pegmatite, 37
  radiate, from volcanoes, 56
  relation of sedimentary rock to, *41*
Dinosaur tracks, 77
Dip, to find magnetic ore, 117
  See also Inclination
Dip needle, 123, *125*
Discharge, water, 168
Discontinuity, 107–109, *108*
Discordant contact, 38
Distributaries, 181
Doldrums, 373
Dolomite, 69–70
  atmospheric carbon dioxide and, 340
  joints in, water flow through, 213
  weathering of, 155

Dome(s), circular, *83*
  definition of, 80
  indicated by rock outcrops, 81
  salt, 83
Draperies, cave, 214, *216*, *217*
Dredges, to obtain ocean floor samples, 291
D-Region, 347
Dreikanter, 225, *225*, *226*
Drift, continental, *see* Continental drift
  glacial, 198
  stratified, 198
Drilling, deep-sea, 325
Drizzle, 367
Drought prevention, 425
Drumlin, 199, *202*
Dune(s), gypsum, 229, *229*
  sand, *see* Sand Dunes
Dune grass, 231
Dust, air polluted by, 435
  atmospheric, 339, 341–342
  composition of, 341
  from meteors, 341
  sand separated from, by wind, 224
  solar radiation absorbed by, 355
  volcanic, 341
Dust bowl, 222
Dust storms, 222, *223*, 233, *238*
Dynamo theory, 121
Dyne, 345

Eagle, DDT effects on, 434
Earth, age of, 279
  angle of incidence of sun's rays on, 353–355, *353*, *354*
  axis of, inclination of, *353*
    polar versus equatorial, 129
  as bar magnet, 115
  collision of comet with, 9
  core of, 103, *103*, 104, *106*
  craters on, 5
  crust of, *see* Crust
  distance from sun, 353
  geographic poles of, 127, 132
  gravitational field of, 125–136, *see also* Gravity
  heat from, reflected by clouds, 355
  history of, 137–153
  hospitableness of, 8
  layers of, 107
  magnetic field of, *see* Magnetic field(s)
  magnetic poles of, 121
  mantle of, *see* Mantle
  orbit of, 352
  origin of, 277
  polar flattening in, 129
  polar shift in, 113
  rotation of, tides and, 12–13, 273
  rotational inertia of, 104
  seismologic study of, 102–109
  shape of, 129, 132, *132*, *134*
    gravitational variations and, 129
  solar energy received by, 274, *275*
  in solar system, 2–20
  solids of, 21–29
  surface of, cooling of, evaporation and, 361
    geoid, 131
    water content of, 259
    winds along, 360–361, *361*
Earth-moon system, evolution of, 12–14
Earthquake(s), 95–114
  along oceanic trenches, 105, 291
  building construction and, 114
  in California and Nevada, *112*
  classification of, 107
  deep-seated, 105–107, *107*
  definition of, 95
  distance between seismograph and, 100
  epicenter of, definition of, 96
    locating, 101
    relation to focus, 97
    in Rocky Mountains, *113*
  examples of, 96–98
  faulting in, *see* Fault(ing)
  focus of, definition of, 96
    depth of, 101–102
    wave propagation from, *100*
  forecasting, 95, 111–114
  intensity of, 102
  intermediate, 107, *107*
  lakes formed by, 111, 162
  locating, 100–102, *104*
  in North America, 110–111
  oceanic zones of, 332
  polar shift and, 113
  in populated regions, 97–98
  recording of, 98
  rock deformation preceding, 113
  rock fall from, *163*
  shadow zone of, 104, 105, *106*
  shallow, 107, *107*
  study of, *see* Seismology
  tidal waves caused by, 273
  from underground liquid waste injections, 416
  underwater landslide from, 304, *307*
  volcanic activity and, 57, 329
    See also specific earthquake, for example, Alaskan Earthquake
Earthquake belt, 110, *111*
Earthquake code specifications, 114
Earthquake waves, types of, 99
  velocities of, *102*, *106*, 107
Earth waves, effects of core on, 103, *103*
East Indian Ocean, 306
East Pacific Ridge, 298, 325
  microfossils in, *300*
  San Andreas Fault and, 330–331, *331*
Easterly, 269, 373
Eclogite, 107
Ecology, 404
Edwards Limestone, 220
Ekman spiral, 269, 270, *270*
El Capitan, *197*
Electricity generation, dams and, 409, *409*
Electromagnetic spectrum, 352t
Electron(s), definition of, 21, 28
  mass of, 22
  valence of, 24
Electron dot formula, 24t

Element(s), abundant, in earth's crust, 32t
  atmospheric, 339–342
  definition of, 22
  in sea water, 259t
  transmutation of, 149
Ellef Ringes Island, 83
Ellesmere Island glaciers, *190*
El Misti, 48
Elsasser, Walter M., 104, 121
Emery, K. O., 302, 303
Energy, solar, see Solar radiation
Eniwa Volcano, 58
Environment, definition of, 404
  landslide, 164–165
  pesticides in, 434–435
Environmental Financing Authority, 414
Environmental science, 403–442
  definition of, 404
  scope of, 405–406
  task of, 405
Epidote, *35*
Epoch, 137
Equinox, 354
Equitorial Current, 265, 267
Equitorial Easterlies, 269
Equitorial low, 373
Era, 137
E-Region, 347
Eros, 5
Erosion, of Appalachian Mountains, 317
  of Atlantic coastal plain, 421
  along beach, 244, *247*
  by caving, 171, 172, *177*
  of circular dome, *83*
  of cliffs, 245
  definition of, 154
  glacial, 193–197
  of Grand Canyon, 186
  of headlands, 247, *249*, *250*, *254*
  of joints, *80*
  of platform, 246
  river, 171–173
    fault scarp and, 182–184, *186*
    headward, 175–176, *179*, *180*
    river velocity and, 173–176, *178*
  of shoreline, 253
  soil, see Soil(s)
  by solution, 172
  thrust sheet after, *86*
  from turbidity currents, 283
  wind, 222–225
    desert formed from, 223
Erosion surfaces, dissected, 184, *187*
Erratics, 205
Escarpment, see Fault scarps
Esker, 200, *202*, *203*
Estuaries, 240, 252
Eugeosyncline(s), 311
Europe, continental drift of, *334*, 335
Eutrophication, 417–418
Evaporation, atmospheric water vapor and, 340
  definition of, 361
  solar radiation and, 356
  water supply and, 407
Everest, George, 132
Everest, Mount, 259
Exfoliation, 157, 159, *159*
Exploration, dip used in, 117
  geologic maps for, 145
  gravity anomalies in, 133–136
  magnetic field used in, 123–125
  for minerals, 145, 432
  seismology in, 109
  submarine, 305

Fallout, radioactive, 436, 439
Fan(s), alluvial, see Alluvial Fan(s)
  submarine, 282–285, *296*, 303
Fanglomerate, *219*
Faraday, Michael, 27, 339
Farmers, conservation by, 422
Farming, soil classification and, 145
Fault(ing), border, *324*
  definition of, 83
  folding and, 85
  horizontal displacement with, 84–85, *88*
  normal, 84, *87*, *88*
  ocean floor, *129*
  overthrust, *81*, 83
  springs along, 215
  transform, 288, *291*
  of volcanic ash beds, *88*
  See also specific fault, for example, San Andreas Fault
Fault block(s), alluvial fans and, 183, *186*
  of Great Basin, 317, 330, *325*
  tilted, *88*
Fault breccia, 86
Fault scarps, 85–86, *89*, *96*, *97*
  alluvial fans and, *186*
  dissected, *88*, *186*
  formation of, 96
  fresh, 110, *114*
  in Hebgen Lake Earthquake, 97, *97*
  river erosion and, 182–184, *186*
  submarine, 287
Feldspar(s), 33
  calcium-rich, 45, 47
  in granite, *35*
  types of, 34
  weathering of, 157
Felsenmer, *155*
Fennoscandia, 188
Fernandian Beach, *301*
Fetch, 237
Fish, eutrophication and, 417
  pesticide poisoning of, 434
  water calefaction and, 415
  water pollution and, 414, 418
Fissure, 40
Fjord, 197, *199*
Flint, 72–73, *247*
Flood control, 409, *409*
Flood plain(s), 181, 233
Flood stage, 175–176
  definition of, 181
Florida Current, 267
Florida escarpment, 301
Florida Keys, 248
Flume, see River(s), channel(s) of
Fluorides, atmospheric, 438
Fluorine, commercial, 339

Fluorite, 438
Fog, 366–367
    Pacific Coast, 380
Fold(ing), 80–83, *81*
    competent and incompetent, 81
    décollement, 82, 84, *86*
    diapiric, 83
    of Jura Mountains, *86*, 319
    mountain building and, 147
    overturned, *82*
    of sedimentary rock, 76, *85*
    types of, *85*
    unconformities in, 86–88
Fold belt(s), Acadian, 311, *313*
    Laramide, 313
    Nevadan, 311
    North American, 310–319
    Precambrian, 309
    Taconic, 311, *313*
        See also Tectonic unit(s)
Food chain, radioactive fallout and, 439
Foraminiferan ooze, 296
Foredune, 232, *234*
Forests, conservation measures for, 425–426
    national, 424–427
        history of, 424–425
        vandalism in, 426, 427
    recreational use of, 426–427
    sand dune encroachment on, *235*
    solar radiation absorption and, 356
Formation, definition of, 21
Fossil(s), cephalopod, 145
    in East Pacific Ridge, 298, *300*
    evidence of lunar orbits from, 13
    guide, correlation and, 144–145
    marine, 277
    in Paris Basin, 143–144
    systematic occurrence of, 142
    as time markers, 138, 142–145
Fracture zones, submarine, 288, *292*, *294*, *295*

Franciscan Group, deformation of, sea-floor spreading and, 331, *332*
Franklin, Benjamin, 266
F-Region, 347
Front(s), 381–382
    clouds and, 387, *388*, *389*
    cold, 381, *381*, 388, 394
    definition of, 378, 381
    designation of, 381, *381*
    kinds of, 381
    occluded, 382, *383*, 388, 391
    Pacific, 394
    polar, 377–378
    satellite photograph of, *398*
    warm, 382, *382*
Frontal waves, 385–386, *385*, *386*, *388*
Front Range, 315
Frost action, 154–155
    on cliffs, 245
    in soil formation, 159
    on volcanic rock, *156*
Fujiyama, 48, *49*, 51, 55
Fumaroles, 436

Gabbro, 34, 47
Gal, 129
Gales, 395
Galveston Bay, 417, *419*
Gamma rays, 148
Gas(es), atmospheric, commercial uses of, 339
    characteristics of, 338
    gravity anomalies and, 136
    invisible, air pollution from, 435
    natural, *see* Natural Gas
    nature of, 338–339
    noble, 339
    pressure-volume relationships of, *342*
Gas laws, 342–343
    applied to atmosphere, 343–345
Gay-Lussac, Joseph, 342
Geiranger Fjord, *199*
Gellman, H., 104
Geographic poles, different altitudes of, 132

gravitational acceleration at, 127
Geoid, 131, *134*
Geologic maps, *see* Map(s)
Geologic time and dating, 137–153
    catastrophism and, 138
    by classic methods, 138–142
    divisions of, 137
    fossils in, 138, 142–145
    of organic matter, 152
    radioisotopic, 148–153
        carbon-14 in, 152–153
        fission-track method of, 153
        lead methods of, 149–151
        potassium-argon method of, 151–152
        rubidium-strontium method of, 152
        zircon lead-alpha method of, 151
    of submarine lavas, 153
    superposition and, 138–142
    unconformities and, 138–142, *142*
    uniformitarianism and, 138
Geologic time scale, *139*
Geological Survey, 146
Geosynclines, composition of, 311
    definition of, 310
    in North America, 310–319
Gerstenkorn, Horst, 13
Geyser(s), 220–221
Geyserite, 221
Giant's Causeway, 42, *44*
Glacial milk, 191
Glacial stages, 202
Glacier(s), 189–206
    abrasion from, *195*
    braided streams from, 181, 200
    catchment basins from, 190, *198*
    continental, *190*, 192–193
    definition of, 189
    formation of, 202–205
    crevasses in, *192*, *200*
    debris from, 191, *192*
    deposits from, 198–201
    ground water in, 217

Glacier(s) (*Continued*)
 erosion due to, 193–197
 expanded-foot, 189, *191*
 flow of, 191, *192*, 193
 hanging valley formed by, *196, 197*
 lakes formed by, 198–200, 201, *201*
 lobate feet of, 189
 melt waters from, 181, 191–192, 197
 moraines formed by, *see* Moraine(s)
 outwash gravels from, *201*
 piedmont, 189, *201*
 Pleistocene climate for, 201–205
 plucking action by, *195*, 196, 198
 polar-type, *190*
 receding, *192*, 199
 rock, 160
 stagnant, *200*
 striations made by, *195*
 talus from, 197, *199*
 types of, 189
 U-shaped valleys formed by, 193, 195, 196, *196, 197*
 valley, definition of, 189
  features of, *196*
  nourishment of, 190–191
  section of, *198*
   See also specific glacier, for example, Hubbard Glacier
Glacier National Park, 204
Glaciofluvial deposits, 200–201
Glauconite, 152
Glen Canyon Dam, 410
Globigerina, 296
Globigerina ooze, 294, 296, 298
Glomar Challenger, 291, 292
Gneiss, 82, 92–93, *92–94*
Gold, subsea deposits of, 307, 308
Gondwanaland, 322
Gorda Ridge, 303
Gorge(s), 171, *186*
Governmental agencies, in environmental management, 406–407

Graben, *88*, 163, *169*
 submarine, 288, *291, 292*
Graben rule, 164, *172*
Grand Banks turbidity current, 304, *307*
Grand Canyon, 186
Granite, *94*
 basalt transformation to, 47, 48
 composition of, 32, 34, *35*, 62
 exfoliation of, 159
 joints in, *79*
 structure of, 33
 weathering of, 62, 157, 158, *158*
Gran Marais Embayment, *234*
Grass of sea, 296
Gravel, glacial, 197, 198, 200, *201*
 water in, 207, 217
 wind transport of, 224
Gravimeter, 127, *131, 132*
Gravitation, Newton's law of, 125, 131
Gravitational acceleration, 127–129
Gravitational field, 125–126
Gravity, anomalies of, exploration and, 133–136
 in Himalayan Mountains, 132, *134*
 in Paradox Valley, 135, *136*
 definition of, 125
 at different sea levels, 131
 maps of, 129
 measurement of, 126–127
 polar increase of, 127–129
 rock movement induced by, 159–165, *161*
 unit of, 129
 variations in, 129–131
  corrections used in determining, 131
Gravity map, Bouguer, 133, *133, 134*
Gravity tectonics, 319
Great Barrier Reef, 69, 71, 301
Great Basin, 315
 block faulting of, 317, *325*, 330
 formation of, 317

 salt collection in, 235
 seismic layering of, 315, *324*
Great Egg Bay, 248
Greater Appalachian Mountain System, 317
Great Lakes, eutrophication in, 417
 Pleistocene stages of, *204, 205*
 predecessors of, 206
 sand dunes along, 231
Great Plains, dust bowl in, 222
Great Salt Lake, *70–72, 72, 76, 77*, 237
 formation of, 235
Great Salt Lake Desert, *229*
Greenland ice cap, 189, *190*, 205
Green River shale formation, 71
Ground water, 206–221
 advantages over surface water, 212
 in alluvial fan, 215, *219*
 artesian, 211–212, *see also* Artesian wells
 in Atlantic Coastal Plain, *220*
 beds impervious to, 211
 capillary fringe of, 209
 confined, 211
 definition of, 206
 discharge of, 209
 examples of, 215–220
 geysers and, 221
 in glacial deposits, 217
 hydrostatic pressure of, 211
 from irrigation, 210
 legal aspects of, 211, 220
 recharge of, 209–211
 reservoirs of, 210, 220
 in rocks, *209*
  carbonate, 212–214
  sedimentary, 211
  stratified, 211–212
  without barriers to movement, 208–211, *210*
 in sediment, *209*
 in seeps, 214–215
 in springs, 214–215
 table of, 208–209
 zones of, 209
Gulf Coastal Plain, 219
Gulf of Bothnia, 261

Gulf of California, 330
Gulf of Mexico, 300, 301, 305
Gulf of St. Lawrence, 252
Gulf Stream, 266–267, 270
   fog due to, 367
   temperatures of, 264
Gullying, *423, 424,* 425
Gutenberg, Beno, 104
Guyots, 285, *288*
Gypsum, 70–71, *83*
Gypsum dunes, 229, *229*

Hadley's Rill, *18*
Half Dome, *197*
Halite, 26, 27, 29, 30
Halley's Comet, 9
Hanging valley, *196, 197, 199*
Harrison, Benjamin, 424
Hawaiian high, 375
Hawaiian Rise, 287, 288
Haze, moist, 367
Headlands, erosion of, 247, *249, 250, 254*
   truncated, *250,* 252
Heat, cloud reflection of, 355
Heat budget, 351–353
Heat of sublimation, latent, 361
Heat of vaporization, latent, 361
Hebgen Lake Earthquake, 96–97, *96,* 110
   fault from, 85, *89*
   isoseismals of, *105*
   lake formed by, 162
   landslide from, 162, *163, 166, 167*
Helium, atom of, 22, *22*
   commercial use of, 339
   in upper atmosphere, 347
Hidalgo, 5
High(s), 374–376
      See also Anticyclone(s)
Highway(s), building of, esthetic guidelines for, 428–430
   soil erosion prevention along, 429, *430*
Highway Beautification Act, 428
Hills, abyssal, 285
Himalayan Mountains, 132, *134,* 325
Holmes, Arthur, 151

Honeybees, insecticide effects on, 435
Hood, Mount, 51
Horn, 197, *199*
Hot springs, 340, 436
Hottorf, Johann, 27
Howard, Luke, 368
Hubbard Glacier, *191*
Hudson River, submarine canyon from, 282, *284*
Humboldt Current, 265
Humid climates, coastal dunes in, 231–232
Humidity, 362
Hurricane(s), 395–397
   barrier beaches affected by, 249, *252, 253*
   definition of, 395
   satellite photograph of, *396, 397*
   shore damage from, *252, 253*
   tides affected by, 273
   weather mapping of, 391, *392, 394*
      See also Typhoon(s)
Hurricane Betsy, 395
Hurricane Ethel, *396*
Hurricane Isbell, 392, *392–393*
Hydration, in soil formation, 159
   as weathering agent, 157–158
Hydroelectric power, 409, *409*
Hydrogen, atom of, 22, *22*
   commercial use of, 339
   isotopes of, 23
   in upper atmosphere, 347
Hydrologic cycle, 206–207
   of salt, 233
Hydrostatic pressure, in confined water, 211
Hygroscopic crystals, 366

Ibis, DDT effects on, 434
Icebergs, 189, 297
Ice cap, Greenland, 189, *190,* 205
   North American, 204–205
   Scandinavian, 202–204, *203*
Ice flow, 189, 190
Icelandic low, 375, 376

Igneous rocks, 32–47
   chemical composition of, 35
   classification of, 34–37, 36t
   concordant intrusion of, 40
   in crust, 107
   dating of, 152
   definition of, 32
   extrusive, 38, 41–42
   flows of, structures of, 42–44
   intrusive, 38–41
      metamorphism and, 94
   kinds of bodies of, 37–44
   mineral composition of, 33–34
   origin of, 44–47
   record of magnetic field locked in, 117
   representative, *35*
   silicates in, 33
   texture of, 33
      See also Volcanic rock(s)
Imbrium Mare, 17
Inclination, 116–117, *120*
   changes in, *125*
   map of, *123, 124*
      See also Dip
Indian Ocean, 296
   currents in, 265
   sea-floor spreading in, 321, 332
Indian Ocean Basin, 329
Indonesian Archipelago, 305
Industry, water in, 411, *411*
Infrared radiation, carbon dioxide absorption of, 340
Inlets, tides and, 240
Insecticide, biological control versus, 435
Insolation, angle of incidence and, 354, *354*
   definition of, 274, 352
   length of day and, 353–355
Interplanetary particles, in red clay, 297
Intrusion, concordant, 40, *40*
Inversion, 364
Ion(s), 25
Ionosphere, 347
Iron (ore), dip needle used to locate, 123
   formation of, 205, 279, 280
   in laterite, 159

Iron (ore) (*Continued*)
 in sandstone, 63
 subsea deposits of, 306, 308
Irrigation, canals for, *409*
 dams and, 409, *409*
 ground water from, 210
 soil erosion from, 421, *423*
Island arcs, 289–291
Isobars, 357, *357*, *358*
Isobaths, *284*
Isogonics, 116
Isoseismal lines, 102, *105*
Isostasy, 132–133, *134*
Isotherms, 389
Isotopes, 23
 radioactive, 149, 150t

Jamaica Bay, 417
Japan Current, 396
Jasper, 72
Jet stream(s), 397–398, *397*, *399*
Joint(s), erosion of, *80*
 in granite, *79*
 in limestone and dolomite, 213, *215*
 in sedimentary rock, 77–79, *78–80*
 in volcanic rock, *81*
 water in, 207, 213
Juno, 3
Jupiter, 3, 5, 6, 258
Jura Mountains, *86*, 319
Juvenile water, 207

Kaolinite, 67
Karst topography, 213–214
Katmai, Mount, 53
Kermadek Earthquake, 100, *101*, *103*
Kerogen, 433
Kettles, 200
Kharga Oasis, 227
Kilauea, 51, 54, *57*, 58, 59
Kilimanjaro, 51
Klippe, 84, *87*
Knick point, 182, *186*
Kodiak Island, 254
Krakatoa, 273, 341
Krypton, 339

Labrador Current, 267
Laccoliths, 40, *43*, 141
Lacewing, green, 435
Ladybug, 435
Lagoon, *251*, *253*
Lagoon Nebula of Sagittarius, *11*, 277
Lake(s), cirque-basin, 197
 formed from earthquakes, 111, 162
 glacial, 198–200, 201, *201*, 206
 Great, *see* Great Lakes
 morainal dam, 199
 oxbow, 180, *182*, 184
 rock basin, *196*, 197
 wind effects on, 237
Lake Erie, 237, 417
Lake Lucero, 229
Lake Mead, 177, 181, *181*
Lake Michigan, *206*
Lake Superior, 279
Lake Winnepeg, 206
Laminations, shale, 68
Lamont Geological Laboratory, *282*, 298, 321
Land, ownership of, 419–420, 421t
 pasture, high-country, *426*
 private, 419
 public, 419
 solar radiation storage by, 352, 356
 surface of, water table and, *209*
Land breezes, *357*, 358
Landslide(s), in Alaskan Earthquake, 163, *169–172*
 creep phenomenon in, 160
 dam made by, *167*
 environment for, 164–165
 glide block, 160
 graben rule for, 164, *172*
 in Hebgen Lake Earthquake, 97, 162, *166*, *167*
 moisture effects on, 160
 permafrost conditions and, 161
 prevalence of, 160
 prevention of, 165
 rapid, 162–164, *168*

rock particle size and, 160
 skirting lava plateau, 163, *168*
 slump block, 160
 translation gliding, 163, *172*
 underwater, 285, 304, *307*
Land survey, 419
Land use, conservation and, 418–435
 in United States, 419–420
 in Utah, 422t
Lapilli, 36
Lapse rate(s), 364–366, *365*
Laramide fold belt, 313
LAS, in water management, 412
Laser, in earthquake forecast, 113
Latent heat of sublimation, 361
Latent heat of vaporization, 361
Laterite, 159
Latitudes, *132*
Lava, definition of, 32
 submarine, dating of, 153
 vesicular pillow, *46*
Lava dome(s), 51–53, *51*, *52*
 definition of, 51
 lunar, 17, *17*
Lava flows, 55
Lava plateau, 163, *168*
Lava rock, 32, 36
Lawsuits, concerning ground water, 211, 220
Layering, seismic, 315–317, *324*
Leaching, *see* Solution
Lead, 308
Lehman Cave, 213, *215–217*
Leonardo da Vinci, 179
Leopold, Aldo, 427, 431
Levees, 181
Leverett Glacier, 192
Lewis overthrust, 84
Libby, W. F., 152
Libyan Desert, 227
Life, atmospheric carbon dioxide and, 340
 in Milky Way, 12
Lifting condensation level, 363
Lignite, 72
Limestone, 69–70
 atmospheric carbon dioxide and, 340
 caves in, 213, 214, *215*

Limestone (*Continued*)
  dam sites and, 410
  dating of, 152
  joints in, 213, *215*
  marble formed from, 89
  plastic deformation of, 89
  reefs of, 71, *73*
  sandstone interlayered with, 73
  shale interlayered with, 74
  weathering of, 155
    See also specific limestone formation, for example, Madison Limestone
Linear alkylate sulfonate (LAS), 412
Lines of force, 115
Liquid, vapor pressure of, 262, *262*
Lisbon Earthquake, 97
Lithium, 22, *22*
Little Cottonwood Canyon, *196*
Liv Glacier, 192
Llano Uplift, *310*
Lodestone, *116*, 155
Loess, 232–233, *237*, *238*
  See also Silt
London smog, 436, 441
Long Beach, wave damage at, 242, *245*
Long Island, 248
Los Angeles smog, 436, 438
Low(s), 374–376
  wind circulation around, 359, *360*
  See also Cyclone(s)

Madison Limestone, 217
Magma(s), basalt, 41
  definition of, 32
  geologic time and, 141
  lunar, 19
  metamorphism and, 94
  primary, 45
  in Sierra Nevadas, 313, *322*
  variations of, 44–45
Magmatic differentiation, 45–47
Magnetic anomalies, 123, 319–322, *326*, *327*

Magnetic field(s), 115–125, *116*
  ancient, 120
  changes in, 117–120, *126*
  dynamo theory of, 121
  electric currents and, *127*
  in exploration, 123–125
  intensity off California, *295*
  local surveys of, 117
  record of, in igneous rock, 117
  reversal of, 121, *321*, *327*
  source of, 121–123
  Venusian, 121
Magnetic ore, 117
Magnetic pole(s), changes in location, 117, 120, *126*
  differences in, 121
Magnetite, in granite, 35
  magnetic field record locked in, 117
  magnetic properties of, 115, *116*
  subsea deposits of, 307
Magnetometer(s), 123, *129*, 281
Mammoth Cave, 213
Manganese deposits, 305–307, 308
Mantle, boundary between crust and, 107–109, *108*, 315
  composition of, 109
  convection currents in, 319, *326*
  definition of, 104
  low-velocity zone in, 104–105, *106*, *108*
  physical properties of, *106*
  rock, 158
  See also Crust
Map(s), aeromagnetic, *131*
  of declination, *119*
  geologic, 145–148
    of angular unconformity, 147, *147*
    interpretation of, 146–148
    mineral discovery and, 145
    preparation of, 146
  gravity, Bouguer, 133, *133*, *134*
  of inclination, *123*, *124*
  isotherm, 394, *395*
  of lows, 390
  tectonic, see Tectonic map

  upper air, 390–391, *393*
  weather, see Weather maps
Marble, formation of, 89
  weathering of, 155, *158*
Marcus Baker, Mount, *192*
Mare, lunar, 14, 16–17
Mariana Trench, 259
Marine environment, mineral resources of, 305–308
Marius Crater, *18*
Mars, 7, 10, 258
Marsh(es), 418
Marsh, George Perkins, 404
Mass movement, gravity-induced, 159–165, *162*
Mass transport, of surface water, 237, *240*
Matagorda Peninsula, 249, *252*
Matterhorn, 197, *199*
Mauna Loa, *49*
Mazama, Mount, 303
Mazama ash, 303, *304*
Meander(s), 179–180, *182*
  soil erosion and, *183*, 424
  stream deposition and, *183*
  stream entrenchment and, 180, 182, *184*
Mediterranean Sea, 261, 305
Mendocino fracture zone, 288
Mercalli scale, 102, *105*
Mercury (planet), 3, 6, 10, 258
Mercury compounds, fish poisoning by, 434
Meridian, local magnetic, 117
Mesa Central, 315
Mesosphere, 345
Mesozoic time, 329
Metamorphic rock(s), 88–94
  changes in, evidence of, 88–91
  characteristics of, 91t
  in crust, 94, 107
  definition of, 88
  kinds of, 91–93, 91t
Metamorphism, 93–94, 151
Meteor(s), 341, 345
Meteoric water, 207
Meteorite(s), composition of, 104
  definition of, 5
  types of, 10

Meteorological Rocket Network, 346
Mica, 34
Mica schist, 92
Microfossils, 298, *300*
Microscope, petrographic, 27
Mid-Atlantic Ridge, 288, *291*, 299, 319–322, *327*
    magnetic anomaly over, *326*
    sediments on, *299*
Middle America Trench, *296*
Milky Way, life in, 12
Milligal, 129
Mineral(s), 21–60
    abundance of, 30–32
    classification of, 30–31
    conservation of, 431–434
    definition of, 21, 26
    ferromagnesian, 33, 157
    glacial transport of, 197–198
    groups of, 30–32
    locations of, 310
    optical properties of, 27
    phosphorescence of, 148
    from recreational areas, 433
    reserves of, 432
    rock-forming, 31t
    in shale, 67
    in siltstone, 67
    stability ranges of, 93
    submarine, 305–308
    weathering of, 157
    X-ray identification of, 29
Mines, coal, water pollution from, 412–413
Mining, strip, *413*
Miogeosyncline, 311
Mississippi River, 267
    changing channels of, 176
    delta of, 181, *185*, 248, *300*, *302*, *303*
    gradient of, 167
    sediment carried by, 169
Moenkopi Formation, 67
Moenkopi Plateau, *232*
Moho discontinuity, 107–109, *108*
Mohole Project, 108
Mohorovičić, A., 107
Moisture, atmospheric condensation of, 361–372
    landslides and, 160, 161
    solar radiation absorption by, 355
    See also Water
Molecules, 23
    in atmosphere, *341*
Monsoon system, 377
Monterey deep-sea fan, *296*, *303*
Monterey Formation, 246
Monterey submarine canyon, *282*, *287*
Monte Somma, *54*, 56
Montmorillonite, 410
Monument(s), National, 428
Monument Valley, *227*
Moon(s), acceleration of, *14*
    age of, 19
    composition of, 10, 17–20
    craters of, 5, 14–16, *15*, *18*
    evolution of, 12
    of Jupiter, 5
    lava domes on, 17, *17*
    maria of, 14, 16–17
    mountain chains on, 17
    orbit of, 13
    physiographic divisions of, 14
    of Saturn, 6
    Sea of Tranquility on, 17, 19, *19*
    surface of, 14–17
    temperatures of, 8
    tides affected by, 272, *273*
        rotation of earth and, 12–14
    of Uranus, 6
Moraine(s), age of, 203
    definition of, 198
    drainage system of, 203
    glacial lakes and, 198–200
    ground, 199
    lateral, 197, *200*
    medial, 197, *200*
    recessional, 199, *206*
    streams in, 199
    terminal, *192*, *196*, 198, 199, *201*
        glacial periods indicated by, 202
Mountain(s), skin-deep deformation of, 319
    terriginous sediments in, 298
    volcanic, see Volcano(es)
        See also specific mountain, for example, Fujiyama
Mountain breezes, 358, *359*
Mountain building, 110, 187
    folding and, 147
    study of, 95
Mud cracks, 75, 77
Murray Fracture Zone, 288, 330
Murray ocean floor escarpment, *129*
Murzuch, giant dunes of, *228*
Muscovite, 34, 92
    See also Mica
Myojinsho, 57, *59*

Nansen bottles, 261
Nantucket Island, *253*
National forests, wilderness areas in, 427
National parks, 428, 432
Natural gas, location of, 310
    in reefs, 71
    reserves of, 432
    in sedimentary rocks, 72
    seismologic discovery of, 109
    from shale, 68
    See also Petroleum
Natural resources, conservation of, see Conservation
Nebula, Lagoon, *11*, 277
Neptune, 6, 258
Neutron(s), 21, 22
Nevada, earthquakes in, *112*
Nevadan fold belt, 311
New Madrid Earthquake, 111
Newton, Isaac, 125, 127
Nickel, 306
Niigata Earthquake, 113
Nimbus satellites, 383, *383*
    See also Satellite photograph(y)
Nimitz Glacier, 193, *195*
Nitric oxide, atmospheric, 436, 437
Nitrilotriacetic acid (NTA), 412
Nitrogen, 339, 339t
Nixon, Richard M., 441

North America, air masses affecting, 379t, *380*
  continental drifting of, *334, 335*
  crust of, 309–319
  earthquakes in, 110–111
  fold belts of, 310–319
  geosynclines of, 310–319
  Precambrian basement of, *311*
  tectonic units of, 309–310
North American Ice Cap, 204–205
North Atlantic Deep and Bottom Water, 271
Northeasters, 381
Northern Hemisphere, cyclonic and anticyclonic circulations in, *360*
  global air circulation for, *374*
  jet streams in, *397*
  movement of highs and lows across, 377
  oceans of, heat budget for, 276, *276*
  quadrants of, 375, *375*
  surface pressures and winds over, *375*
Northern lights, 348–349, *349*
Norwegian Sea, 270
Nuclear explosions, seismograms of, 105
Nuée ardentes, 53

Oases, 225, 227
O'Brien, M. P., 242
Obsidian, 35
Ocean(s), 257–335
  carbon dioxide in, 279
  composition of, 277
    weathering and, 278
  conditions needed for, 258
  currents in, 265–270
    causes of, 267, *268*
    Coriolis force and, 267, 268, *271*
    cross-bedded sedimentary rock and, 74
    Ekman spiral and, 269–270, *270*
  earthquake zones in, 105, 332
  effect on temperature, 274
  equatorial, warming of, 267, *268*
  evaporation from, 276
  floor of, *see* Sea floor
  gases in, 262–263
  heat budget of, 274–276, *276*
  heat loss from, 276
  level changes in, rivers affected by, 176
  origin of, 276–280
    hypotheses of, 277
  petroleum beneath, 305
  salinity of, 261, *261*, 277
  salt from, *see* Salt(s)
  sand deposition and, 65
  solar radiation storage by, 274, 275t, 352, 356
  temperatures of, 263–265, *265*
  trenches in, *see* Trench(es)
  uniqueness of, 258–259
  waves in, *see* Wave(s)
  See also Sea(s)
Ocean basin(s), constitution of, 319–331
  See also specific ocean basin, for example, Atlantic Ocean Basin
Ocean city, 248
Oceanic crust, 281
Oceanographic research, 281, *282*
Oceanographic ships, 291
Oceanus Procellarum, 20
Oil, *see* Petroleum
Old Faithful, *221*
Olivine, 45, 47
Oolite, 69, *70*, *71*, 77
Ooze(s), 294, 296, 298, *298*
Orbit(s), of asteroids, 5
  of comets, *5*, *8*
  lunar, 13
  of planets, *5*, 6
Ore(s), from juvenile waters, 207
  See also specific ore, for example, Iron ore
Organic compounds, in sea water, 263
Orthoclase, 34
Orthoquartzite, 91

Osprey, DDT effects on, 434
Outcrop patterns, of anticlines, 81, *82*
  of elliptical uplift, *83*
  of synclines, *82*
Outwash plains, 200
Overgrazing, 422, *424*, 425
Overthrust, 84, *87*
Owens Valley, 217
Oxidation, in soil formation, 159
Oxides, 30, 31t
Oxygen, atmospheric, 339, 339t, 347
  atom of, *25*
  commercial use of, 339
  isotopes of, 23
  origin of, 279
  in sea water, 262, 262t, 270, 272
Ozark Dome, *310*
Ozone, crop damage from, 438
  molecular makeup of, 339
Ozonosphere, 346

Pacific Coast, 302, 380, 394
Pacific Ocean, cloud patterns over, *384*
  currents in, 265
  deep-sea drilling in, 325
  fracture zones of, *294*, 330
  midoceanic ridges in, 325
  northwestern, old crust of, 325–329
  radiolarian ooze in, 296
  red clays in, 297
  salinity of, 261
  sea-floor spreading in, 321, 332
  subsidence in, 285
  temperatures of, 264
  trenches bordering, *321*, 329, *330*
  tsunamis in, 274
  volcanic arcs in, 329, *330*
  volcanoes bordering, *294*, *321*, 329
Pacific Ocean Basin, 325–331
  age of, 327
Pacific shelf, 302, 303
Pahoehoe, 55, *56*
Pallas, 3

Palouse Hills, 421
Pangaea, 325, 329, 335
Paradox Valley, gravity anomalies in, 135, *136*
Paricutín Volcano, *39*
Paris Basin, fossils in, 143–144
Parks, national, 428, 432
Particulates, 435
Pasture land, high-country, *426*
Pavement, desert, 224, *225*
Pegmatite, 33, *37*
Pegmatite dike, *37*
Pelagic sediments, 292–299, *298*
Pelée, Mount, 42, 53
Pelican, DDT effects on, 434
Pendleton, Robert C., 440
Pendulum, gravity and, 127
Peninsula, Matagorda, 249, *252*
Pennsylvania period, tectonic map of, *315*
Peoria Loess, *238*
Peridotite, 35, 107
Period, geologic, 137
Permafrost, 161
Permeability, of rock, 208
Peroxyacetyl nitrate, atmospheric, 437
Pesticides, 434–435
Petroleum, beach pollution by, 252
  gravity anomalies and, 136
  location of, 310
  offshore drilling for, 254
  in reefs, 71
  reserves of, 432
  in sedimentary rocks, 72
  seismologic discovery of, 109
  from shale, 68, 71, 433, *433*
  subsea, 305
Péwé, T. L., 193
Phoebe, 6
Phosphates, subsea deposits of, 307
  in water pollution, 412
Phosphorescence, 148
Photogrammetry, 146, *146*
Photography, deep-sea, camera for, *288*, 291
  to detect thermal pollution, 415

satellite, *see* Satellite photograph(y)
Photosynthesis, atmospheric carbon dioxide and, 340
  in sea water, 263
Phyllite, 92
Pinchot, Gifford, 425
Pipes, volcanic, 56
Plagioclase, 34
Plain(s), abyssal, 282, 285
  coastal, water in, 219–220
  flood, 181
  outwash, 200
Planet(s), 2–8
  gaseous, 5
  oceans on, 258
  orbit(s) of, *5*
  orbital speeds of, 3
  physical data on, 5t
  sidereal periods of, 3
  sizes of, *3*
Plankton, 272, 296
Plass, Gilbert N., 436
Plates, crustal, continental drifting and, 331–335
  in sea-floor spreading, 331, *332*
Platform(s), definition of, 309
  erosion of, 246
  in headlands, 247, *249*
  wave-cut, 244, 246, *247*, *249*, *253*
Platinum, 307
Pleistocene epoch, glacial climates of, 201–205
  North American Ice Cap formation in, 204–205
Plowing, contour, 422, *423*, *424*
Plucking, glacial, *195*, *196*, 198
Plug, gypsum, *83*
Pluto, 3, *5*, 6
Podzol(s), 159, *160*
Polar front, 393
  definition of, 375
  theory of, 377–378
Polar high, 373
Pollution, *see* Air pollution; Water pollution
Pools, in caves, 214, *216*
  terraced, *216*
Porcelainites, 246

Porosity, 207, *208*
Porphyry, 33
Pothole, *177*
Power generation, dams in, 409 *409*
Prairie States Forestry Project, 425
Praying mantis, 435
Precambrian basement of North America, *311*
Precambrian rocks, 309, *310*
Precipitation, from different cloud types, 370
  measurement of, 371–372
  water supply and, 407
    See also Rain; specific types
Pressure, *see* specific type, for example, Vapor pressure
Pressure ridge, 163
Proton(s), 21, 22
Psychrometer, 362
Pteropod ooze, 294, 296
Pteropods, 296
Public domain lands, 419, *420*
Puerto Rican Trench, 305
Pumice, 36, *39*
Pumice falls, *58*
Putnam, Rufus, 419
Pyroclastics, 56
Pyroxene, 47

Quartz, *30*, 31, 34, *35*, 62
Quartzite, 91
Quebec Earthquake, 111

Radiation, infrared, solar, 340
Radioactive decomposition, 149
Radioactive fallout, on crops, 436, 439
  food chain affected by, 439
  from thermonuclear explosion, 441, *441*
Radioactive materials, air pollution from, 438–441
  in coal, 440
Radioactivity, standards of maximum exposure to, 439
Radioisotopes, 149, 150t
Radiolaria, 296

Radiolarian ooze, 294, 296, *298*
Radium, 150t
Rain, air cleaned by, 435
   unstable air and, 365
   See also Precipitation
Raindrop impressions, *67*
Rainfall, 371
Rain gauge, 371
Rainier, Mount, 303
Recreation areas, 428
   dams and, 409, *409*
   forests for, 426–427
   minerals from, 433
Recrystallization, 89
Red clay, 292, 296, 297, *298*
   accumulation rate of, 298
   composition of, 297
Red Fish Pass, *251*
Red Sea, 261, 308
Reef(s), coral, 71
   Great Barrier, 69, 71, 301
   limestone, 71, *73*
Reef building, continental shelf-forming and, 301, *303*
Reforestation, to prevent soil erosion, 422
   on strip-mine lands, *413*
Refraction, of waves, 241–243
Refraction coefficient, of waves, 242
Refraction diagram, of waves, 242, *245*
Regalith, 17
Relative humidity, 362, 363t
Research, oceanographic, 281, *282*
   See also Exploration
Reservoirs, 410
Resonance theory, 13
Revegetation, 425, 429, 430
Rhyolite, 35
Richer, Jean, 127
Richter, C. F., 110
Richter scale, 102
Ridge(s), arête, *196*
   comb, 197
   continental, *see* Continental rise(s)
   midoceanic, 285–288, *291*, *292*
      earthquake zones and, 332

Rift valley, *88*, 163, *169*
   submarine, 288, *291*, *292*
Rip current, 243, *246*
Ripple marks, in rock, 74–75, *76*, *77*
Ripples, 65, *67*
Rise(s), *see* Ridge(s)
River(s), abrasion by, 171–172, *177*
   acids in, 413
   activity of, analysis of, 176–184
   alluviation of, 182
   banks undercut by, 171, 172
   base levels of, 174, 176–179, *180*
      dams and, 177
   bed load of, 180
   braided, 180–181, *184*
      dust storms and, 233, *238*
      glacial, *200*, *201*
   buoyancy of, 173
   calcium carbonate in, 171
   channel(s) of, changing, 176–177
      roughness of, 168
      shape of, flow velocity and, *175*
      waterfall abrasion of, 172
   competence of, 176
   deltas of, *see* Delta(s)
   deposition by, 173–176, 179
      meander and, *183*
   dissected erosion surfaces due to, 184, *187*
   distributaries of, 181
   downcutting of, 171
   drainage patterns of, 166, *173*, *174*
   economic importance of, 165
   entrenchment of, 180, 182, *184*
   erosion by, *see* Erosion, river
   in flood stage, 175–176
   flow velocity of, 166–168
      channel shape and, *175*
      deposition and, 173–176
      erosion and, 173–176, *178*
      factors in, 167
      stream load and, 173–176
      water impact and, 173, *177*

   formation of, 206
   frictional drag of, 173
   graded profile of, 174
   gradient of, 167–168
   longitudinal profile of, 176, *178*, 179
   man and, 165–188
   management of, with dams, 409–411
   meandering, *see* Meander(s)
   melt-water, 191
   in morainal complex, 199
   ocean levels and, 176
   rejuvenated, 182
   steeply graded, *175*
   terraces formed by, 182, *185*
   transportation by, *see* Stream load(s)
   transportation zone of, 174
   turbidity currents from, 285
   underground, 213
   valleys widened by, 182
   water discharge from, 168
   water impact of, 173, *177*
   See also Stream(s)
Riverways, 428
Roaring forties, 374
Robert Scott Glacier, 192
Rôches moutonnées, *195*
Rock(s), carbonate, ground water in, 212–214
   creep by, 161
   crystalline, 33
   decomposition of, 154
   definition of, 21
   deformation of, before earthquake, 113
   disintegration of, 154
   folding of, *see* Fold(ing)
   glacial striations on, *195*
   glacial transportation of, *192*, 197–198
   gravity-induced movement of, 159–165
   ground water in, 207–214, *209*
   in icebergs, 297
   igneous, *see* Igneous rocks
   intrusive, *40*
   joints in, *see* Joint(s)
   lava, 32, 36

Rock(s) (*Continued*)
  mass movement of, 160–161
  metamorphic, *see* Metamorphic rocks
  outcrops of, 81
  particles of, landslides and, 160, *161*
  permeability of, 208
  plastic deformation of, *see* Rock flow(age)
  porosity of, 207–208
  Precambrian, 309, *310*
  sedimentary, *see* Sedimentary rocks
  slides of, *see* Landslide(s)
  soil from, 158
  stratified, 211–212
  texture of, 33
Rock avalanche, 161
  See also Landslide(s)
Rock basin lake, *196*, 197
Rockets, to explore atmosphere, 346, 347
Rock fall, 161, *163*
Rock flour, 191
Rock flow(age), 81
  excessive plasticity in, 86
Rock glaciers, 160, *164*
Rock mantle, 158
Rock sea, *155*
Rocky Mountains, earthquakes in, *113*
  formation of, 311, 313, 315
Roentgen, William, 28
Roosevelt, Franklin D., 425
Roosevelt, Theodore, 404, 424
Rotational inertia, of earth, 104
Rubey, William W., 278, 279
Runoff, city growth and, 431
  definition of, 206
  management of, 423–424
Rutherford, Ernest, 148

Sahara Desert, 228, *228*
Saint Elias, Mount, *192, 200*
Saint Francis Dam, 176
Salt(s), in Atlantic Ocean, *260* 261, 270, *272*
  common, 24, *25*
  formations of, 70–71
  in Great Basin, 235
  hydrologic cycle and, 233
  in Mediterranean Sea, 261
  in Pacific Ocean, 261
  in sea water, 259–262, *261*
  transported to continents, 233–235
Saltation, sand movement by, 224
Salt domes, 83
San Andreas Fault, 84, *89*, 313
  earthquakes along, 98, 110
  Eastern Pacific Ridge and, 330–331, *331*
  horizontal displacement along, *292*
Sand, abrasion by, 225
  in deltas, 181, *185*
  deposition of, in layers, 65
  dust separated from, by wind, 224
  formation of, *157*
    by waves, 231
    from glaciers, 197, 198, 200
  grains of, 62, *64*
  movement of, by saltation, 224
  rate of fall through air, 223 *223*
  river transport of, 169, *176, 178, 178*
  water in, 207, 208, 211, 217
  wave transport of, 243, *246*
  wind transport of, 222, 224
Sand dunes, 65, *369*
  barchan, 227, *227–230*, 229, *232*
  coastal, in humid climates, 231-232
  of Colorado Plateau, 229–231
  dome-shaped, 229
  from flood plains, 233
  forest endangered by, *235*
  form of, 226–227, *226*
  along Great Lakes, 231
  longitudinal, 230, *232*
  in Monument Valley, *227*
  of Murzuch, *228*
  parabolic, 229, *229*, 230, *230, 232, 234*
  repose slope of, angle of, 226
  sandstone formed from, 229, *231*
  shape and size of, 226
  along shores, 231–232
  slip face of, 226
  transverse, 229, *229*, 230, *230, 232*
  types of, 229, *229*, 230
  vegetative stabilization of 231, 232, *232*
  wind formation of, 225–226
Sand seas, 228
Sandstone, 62–64, 369
  aeolian, *159*
  aquifer from, 211, 217–219, *220*
  cross-bedded, *66, 68*
  Dakota, 217, 219
  dam sites and, 410, *410*, 411
  exfoliation of, *159*
  formed from sand dune, 229, *231*
  graded bedding of, *69*
  joints in, 78, *79, 80*, 82
  limestone interlayered with, 73
  quartzite, 64
  shale interlayered with, 73, 75
  strata of, 64
  water in, 207, 208, 211
  weathering of, 155, 157, *157*
San Francisco Earthquake, 98
San Nicolas Island, 241, *244, 245*
Santa Barbara oil spill, 254
Santa Catalina Island, *302*
Santa Cruz Island, 246, *249*, 250, *254*
Sargasso Sea, 261, 269, 272
Satellite(s), Applications Technology, 384
Satellite photograph(y), of clouds, 382–385, *388, 398*
  of front, *398*
  of hurricanes, 396, *396, 397*
  of mature cyclone, *390*
  of weather, 376, *376*
  worldwide, 400–401, *400*
Saturn, *3, 6*, 258

Scandinavian Ice Cap, 202–204, *203*
Scarp, fault, *see* Fault scarps
Schist, 92
Scoria, 36, *37*
Sea(s), grass of, 296
  sand, of North Africa, 228
  See also Ocean(s)
Sea breezes, *357*, 358
Sea floor, abyssal plain of, 282
  bathymetric chart of, *287*
  geomorphic divisions of, 282–291
  minerals from, 305–308
  samples from, 291–292, *297*
  sediments of, 281–308
    classification of, 292
    rates of accumulation of, 298
    terriginous, 299–305
  topography of, 281–308
Sea-floor spreading, 299
  in Atlantic Ocean basin, 319–322
  continental drift and, 322, *see also* Continental drift
  dating of, 152
  Franciscan Group deformation and, 331, *332*
  magnetic variations and, *129*
  plates involved in, 331, *332*
  poles of, 332
Seamounts, 285, *294*
Sea of Japan, 305
Sea of Okhotsk, 305
Sea of Tranquility, 17, 19, *19*
Sea water, chemical nature of, 259–263
  chlorinity of, 261
  elements in, 259t
  ionic constituents of, 259
  nonconservative constituents of, 263
  organic compounds in, 263
  photosynthesis in, 263
  salinity of, 259–262
  samples of, 261
Seaweed, 272
Sediment(ation), carbonate, 301
  on continental shelves, 299–302, *300*
  deformation of, 76–77, *78*
  ground water in, *209*
  on Mid-Atlantic Ridge, *299*
  pelagic, 292–299, *298*
  river base level fluctuation and, 176, *180*
  river transport of, 169, *176*
  sea-floor, *see* Sea floor, sediments of
  terriginous, 298, *298*, 299–305
    of abyssal plains and deep trenches, 304–305, *306*
    accumulation rates of, 298
    in mountains, 298
  types of, 299
Sedimentary rocks, 61–77
  chemical, 69–71
    definition of, 61
  classification of, 61
  clastic, 62–69
    definition of, 61
  in Colorado Plateau, 139, *141*
  along continental shelves, 300
  cross-bedding in, 74, *76*
  definition of, 61
  in earth's crust, 107
  folding of, *see* Fold(ing)
  in geosynclines, 311
  ground water in, 211
  interlayering of, 73–74, *75*
  intrusive rock and, *40*
  joints in, 77–79, *78–80*
  markings in, 75
  oil and gas in, 72
  organic, 71–73
    definition of, 61
  plastic deformation of, 89
  relation of dikes and sills to, *41*
  ripple marks in, 74–75, *76, 77*
  shearing of, *85*
  structures of, 74–88
    primary, 74
    secondary, 74, 79–88
Seeps, 214–215
Seich, 97
Seismic layering, of Great Basin, 315, *324*
  of western United States, 315–317, *324*
Seismic profiles, 301, *301*, 302
  of Atlantic abyssal plain, 304, *306*
Seismogram(s), definition of, 96
  of nuclear explosions, 105
  reading of, 98–100
Seismograph, 98, *99, 100*
  definition of, 96
Seismologist, 96
Seismology, 95–114
  definition of, 96
  earthquake studies with, 95, 100
  in exploration, 109
Sentinal Range glaciers, *195*
Serenitatis Mare, 17
Seven Sisters, *253*
Sewage treatment, 413–414, *413*
Shackleton Glacier, 192
Shadow zone, of earthquake, 105, *106*
Shale, 67–69
  bentonitic, 410, 411, *411*
  dam sites and, 410, *410*, 411, *411*
  definition of, 67
  in diapirs, 83
  formation of, 90, *90*
  limestone interlayered with, 74
  oil from, 68, 71, 433, *433*
  permeability of, 211
  sandstone interlayered with, 73, *75*
  water in, 208, 211, 410
  weathering of, 155
  See also Clay
Shearing, of sedimentary rock, *85*
Sheep backs, *195*
Shelterbelt program, 425
Shenandoah Caverns, 213
Shiprock, 56
Shishaldin, Mount, *50*, 51, 55
Shock pressure of waves, 240
Shore(line), definition of, 239
  deposits along, 247–250

Shore(line) (*Continued*)
  emerging, 250–251, *253*
  environment of, 252–255
  erosion of, *253*
  hurricane damage of, *252*
  national, 428
  oil drilling along, 254
  submerging, 252
  waves and, 235–255, *see also* Wave(s)
  See also Coast
Shoshone National Forest, 424
Showa-Shinzan, *51*, 52, *52*
Siberian high, 375
Sierra Madre Occidental, 315
Sierra Nevada, 311
  crust in, magmatic invasion of, 313, *322*
  formation of, 317, *325*
Silica, in geyser water, 221
  organic extraction of, 73
  in sandstone, 63
Silica tetrahedron, 31–33, *33*, 34
Silicates, 30, 31–33, 31t
Sill, basaltic, *43*
  definition of, 40
  geologic time and, 141
  metamorphism and, 94
  relation of sedimentary rock to, *41*
Silt, in deltas, 181, *185*
  dust storms and, 233, *238*
  glacial deposits of, 191, 198, 200
  in glacial lakes, 201
  rate of fall through air, 223, *223*
  river transport of, 169, *178*, 179
  water movement through, 208
  wind transport of, 222, 224, *236*
  in Yukon Valley, 233
  See also Loess
Siltstone, 66–67, 208
Simpson's window, 275
Sink holes, 214, *216*
Skiing, 426
Sky, mackerel, 370

Slate, 91–92
  shale formed from, 90, *90*
Slaty cleavage, 90, *90*, 92
Sleet, 382
Slickenside, 86, *89*
Slides, *see* Landslide(s)
Slip face, of sand dune, 226
Slope profile, *187*
Smith, James P., 145
Smith, William ("Strata"), 142–143, 144, 145, 147
Smog, 367–368, *367*, 435, 436, 438, 440, 441
  definition of, 367
  See also Air pollution
Smoke stack emissions, 439, *439*, *440*
Snow, air cleaned by, 435
  solar radiation absorption and, 356
  See also Precipitation
Snowfall, factors in, 190
  measurement of, 371
Snowmobiling, 426
Soddy, Frederick, 148
Sodium chloride, 24, *25*
Soil(s), bauxite, 159
  chernozem, 159
  classification of, 145
  climate and, 158
  conservation of, 420–423
  creep by, 161, *165*
  definition of, 158
  desert, 159
  formation of, 158–159
  glacial transport of, 197
  as indicators of past climates, 159
  laterite, 159
  loess, 232–233, *237*, *238*, *see also* Silt
  podzol, 159, *160*
  solar radiation absorbed by, 356
  types of, 158–159
  volcanic, 55–56
Soil erosion, 158
  city growth and, 431
  due to irrigation, 421, *423*
  due to runoff, *423*
  in forest areas, 425

  gullying and, *424*
  along highways, 429, *430*
  meandering and, 424
  overgrazing and, 422, *424*
  prevention of, 420, 429
    contour plowing in, 422, *423*, *424*
    reforestation in, 422
    revegetation in, 425, 429, 430
    terracing in, *423*, *424*
Soil moisture, zone of, 209
Soil profile, 158, *160*
Solar atmosphere, 345
Solar constant, 351
Solar corona, 345
Solar radiation (energy), 351
  absorption of, 353–355
    by atmosphere, 275, 348, 351, 352, 355
    by desert air, 355
    evaporation and, 356
    by land surface, 356
    by snow, 356
    by water surfaces, 356–357
  air temperature and, 355
  atmospheric intensity of, 351 357
  atmospheric intensity of, 351
  earth's receipt of, 274, *275*
  electromagnetic spectrum of, 352t
  land storage of, 352
  ocean storage of, 274, 352
  reflection of, 353–355
    by clouds, 355, *355*
  wave-length distribution of, 274, *274*
Solar storms, 349
Solar system, 2–9
  earth in, 2–20
  origins of, 9–12, 277
    clues to, 9–10
    early concepts of, 10–12
Sole, 83
Solifluction, 161
Solstices, 354
Solution, erosion by, 172
  in soil formation, 159
  as weathering agent, 155–156, 158

Somma, Mount, 56
Sonoran low, 393
South America, drifting of, *334*, 335
South China Sea, 305
Southern Hemisphere, reconstruction of, for Mesozoic time, *329*
South Pacific Ridge, 325
Spatter cone, volcanic, 53, *54*
Spectrum, electromagnetic, 352t
Speliology, 213
Spelunker, 213
Spits, 247, *250*, 252, *252*
Springs, 214–215, *218*, *219*
 hot, 340, 436
Squalls, 395
Stacks, 247, *250*, *254*
Stalactites, 214, *216*
Stalagmites, 214
Steno, Nicolaus, 27
Stock, 40
 geologic time and, 141
Storm(s), air masses and, 378–380, 382
 along cold front, 381
 dust, 222, *223*, 233
 frontal waves and, 385
 ice, 382
 solar, 349
 along warm front, 382
 weather map of, 391, *392*
 wind, *see* specific type, for example, Hurricane(s)
Storm beaches, 231, 244
Stratified rocks, 211–212
Stratopause, 347
Stratosphere, 345, 347
Strato-volcanoes, 51
Stream(s), effluent, 210
 influent, 210
  See also River(s)
Stream load(s), 168–171
 bed, 169–171
 clay particles in, 173
 dissolved, 171
 distribution of, *176*
 flow velocity and, 173–176
 particle size and, 169, *176*, 179
 source of, 172–173

 suspended, 169
 transport of, 168–171
Stream profile, 174, *178*
Strip mining, *413*
Stromboli, 53
Strontium-90, atmospheric, 440
Subatomic particles, 21
Sublimation, 361
Submarine canyons, 282–285, *284*
Subpolar low, 374
Subsidence, 285
Subtropical high, 373, 374
Suess, Eduard, 322
Sulfates, 30, 31t
Sulfides, 30, 31t
Sulfur, atmospheric, 436, 437, 438
Sun, earth's distance from, 353
 earth's orbit around, *352*
 rays of, angle of incidence of, 353–355, *353*, *354*
 tides affected by, 272, *273*
 See also Solar entries
Sunset Crater, 53
Sunsets, volcanic dust and, 341
Superposition, geologic dating and, 138–142
 law of, 138
Surface-active agents, 412
Surface water, *see* Water
Survey, for map making, 146
Swamps, *216*
 formed from earthquake, 111
 hydrogen sulfide from, 437
Swash, 238, *242*, *243*
Swells, 236
Synclines, 80–83
 definition of, 80
 outcrop pattern of, *82*

Taconic fold belts, 311, *313*
Talus, definition of, 155
 formation of, 155, *156*
 glacial, *164*, 197, *199*
 sandstone, *157*
Talus cones, 155, *157*
Talus fans, 155
Tamayo fracture zone, 330
Tarumai Volcano, *58*

Tectonic map, of Cretaceous period, *317*
 of Devonian period, *313*
 of Pennsylvanian period, *315*
 of Tertiary period, *317*, *319*
Tectonic unit(s), 309
 See also Fold belt(s)
Tectonics, gravity, 319
Tehran Earthquake, 97
Temperature(s), atmospheric stratificaton according to, *345*
 changes in, adiabatic, 362–364, *364*
 lapse rates in, 364–366, *365*
 climate and, 341
 effect of oceans on, 274
 of Gulf Stream, 264
 of oceans, 263–265, *264*, 270, *272*
 of planets, 258
 of stratosphere, 347
 of thermosphere, 347
 of troposphere, 347
Terrace(s), along beaches, 244, *246*
 in caves, 214, *216*
 along emerging shore, 250
 around pool, *216*
 to prevent soil erosion, 422, *423*, *424*
 stream, 182, *185*
Terriginous sediments, *see* Sediment(ation)
Tertiary period, tectonic map of, *317*, *319*
Tetlin River, *238*
Teton Glacier, *201*
Teton National Forest, 424
Texture, of rock, 33
Thermal, 371
Thermosphere, 345, 347
Thomson, Joseph, 28
Thorium, 149
Three Sisters, 55, *55*
Thrust(ing), 83–84
Thrust sheet, after erosion, *86*
Tidal waves, 273
Tide(s), 272–274, *273*
 barrier bars and, 249, *251*
 in Bay of Fundy, 240

Tide(s) (*Continued*)
  beaches affected by, 240, 244
  delta formed by, *251*
  earth's rotation and, *273*
  effects of sun and moon on, 272, *273*
  estuaries and, 240
  high, 240
  hurricane effects on, 273
  inlets and, 240
  lunar, 272
    rotation of earth and, 12–14
  neap, 272
  spring, 272
Till, 198
Tiltmeter, 58
Time, geologic, *see* Geologic time and dating
Tin, 307
Tiros satellites, 383, *383*
    See also Satellite photograph(y)
Titan, 6
Titanite, 307
Topography, Karst, 213–214
Tornadoes, 395–397
Tracks, in sedimentary rocks, 75, 77
Trade winds, 269, 272
  definition of, 373
Transmutation, of elements, 149
Transpiration, 207
Travertine, 215, *219*
Trench(es), bordering Pacific Ocean, *321, 329, 330*
  contour, to control gullies, 425
  ocean, 289–291
    earthquakes along, 105, 291
    terriginous sediments of, 304–305
    volcanic archipelagos and, *292*
    See also specific trench, for example, Puerto Rican Trench
Trichogramma wasps, 435
Trilobites, 144, 145
Trojan, 3
Tropic of Cancer, 353

Tropic of Capricorn, 354
Tropopause, 347
Troposphere, 345, 347
Trough, wave, 235
Tsunamis, 273
Tuff, 36, 42
Tufts abyssal plain, 303
Turbidity current, 283, 304, *307*
Turtles, 415
Tycho Crater, 15, *16*
Typhoon(s), 396
    See also Hurricane(s)

Uinta Mountains, 315, *324*
Ultraviolet rays, atmospheric absorption of, 345, 347, 352
Uncompahgre Plateau, 124
Unconformities, 86–88, *90*
  angular, 147, *147*
  definition of, 86
  geologic dating and, 138–142, *142*
Undertow, 243
Uniformitarianism, 138
United States, Bouguer gravity map of, 133, *133*
  cyclone paths across, *379*
  western, seismic layering of, 315–317, *324*
Uplift(s), basaltic, 315, *325*
  denudation versus, 187
  elliptical, *83*
  indicated by bench marks, 187
  rates of, 187–188
Uranium, 148–149, *149*, 150t
Uranus, 6, 258
Urban expansion, 430–431
Urey, Harold C., 10, 277
Utah, land use in, 421t, 422t

Valence, 24
Valence electrons, 24
Valley(s), alluviated, 200
  in arid regions, 182, *185*
  entrenched, *180*
  formation of, 156
  hanging, 195, *196, 197*
  in humid climates, 182
  knick point of, 182, *186*

rift, *88*, 163, *169*
  submarine, 288, *291, 292*
  river, drowned, 252
  stream widening of, 182, *185*
  U-shaped, *193*, 195, 196, *196, 197*
  V-shaped, 182, *184*
Valley breezes, 358, *359*
Valley walls, weathering of, 182
Van Allen, J. A., 349
Van Allen radiation belts, 349–350
Vandalism, in national forests, 426, 427
Vapor, in equilibrium with its liquid, 262, *262*
  water, *see* Water vapor
Vaporization, heat of, latent, 361
Vapor pressure, 361
Variation, 116
Varves, 201, 204
Venus, composition of, 10
  magnetic field of, 121
  physical characteristics of, 6
  water on, 258
Vernal equinox, 354
Vesta, 3
Vesuvius, 48, 53, *54*, 56
Visibility, in fog, 366–367
Volcanic bombs, 37, *39*
Volcanic breccia, 37, *39, 40*
Volcanic cone, definition of, 50
  truncated, submarine, 285, *288*
Volcanic rock(s), frost action on, *156*
  in geosynclines, 310, 311
  joints in, *81*
  sedimentary rocks interlayered with, 74, *75*
    See also Igneous rocks
Volcano(es), 47–59
  ash from, *38*, 55, *140*
  beds of, *88*
  atmospheric carbon dioxide from, 340, 436
  belt of, 110
  bordering Pacific Ocean, *294, 321, 329, 330*
  cauldron subsidence in, 56–57, *59*

Volcano(es) (*Continued*)
  classification of, 50
  cinder cone of, 52–53, *52*
  composite, 51
  dust from, 297, 341
  earthquakes and, 57, 329
  eruption of, 54, 57–59
    tidal waves from, 273
    vent, 53
  lava flows from, 55
  necks of, 56
  new, *38*
  along oceanic trenches, 291, *292*, *296*
  pipes of, 56
  radiate dikes from, 56
  shield, *49*, 50–51
  soils formed from, 55–56
  spatter cone from, 53, *54*
  submarine, 57, *59*
  types of, 50–53
    See also specific volcano, for example, Vesuvius
Von Jolly, Philipp, 104
Von Laue, Max, 29
Vulcano, 53

Waikiki Beach, 239, *242*
Warm occlusion, 382, *383*
Wasatch Mountains, 315, 317, *325*
Wasps, 435
Waste(s), acid, control of, 413, *413*
  coal mine, water pollution from, 412–413
  disposal of, water in, 412
  farm-animal, disposal of, 415–416
  liquid, underground injection of, 416–417
  radioactive, air pollution from, 438–441
Water, amount needed, 411
  bacteria in, 414
  chlorine treatment of, 414
  connate, 207
  conservation of, 407–411, *409*
  dams in, 409–411

  general philosophy of, 407–409
  ground, see Ground water
  heavy, 23
  industries and, 411, *411*
  juvenile, 207
  management of, 406–418
    linear alkylate sulfonate (ALS) in, 412
    nitrilotriacetic acid (NTA) in, 412
  on Mars, 258
  melt, from glaciers, 197
  in metamorphism, *93*
  meteoric, 207
  origin of, 277–280
  quality of, classification of, 414
    degradation of, 411–412
    standards for, 441
  reclamation of, 411–416
  in rock joints, 207
  rock penetration by, 208
  in rock pores, 207–208
  salt, in aquifer, 220
  shallow, waves in, 238–239, *241*–*243*, see also Wave(s)
  specific yield of, in sandstone, 208
  surface, 259
    ground water advantages over, 212
    mass transfer of, 237, *240*
    solar radiation absorption by, 356–357
  surfaces of, wind drag on, 237, *240*
  utilization of, *409*
  on Venus, 258
  in waste disposal, 412
Water-air contact, 361
Water discharge, 168
Waterfall(s), from hanging valley, *197*, *199*
  river channel abrasion by, 172
Waterfowl, effects of oil on, 254
Water masses, of Atlantic Ocean, 270–272, *272*
Water pollution, 411, *411*
  in Buffalo Bayou, 418

  from coal-mine wastes, 412–413
  detergents and, 412
  eutrophication and, 417–418
  fish affected by, 414, 418
  in Galveston Bay, 417
  in Great Lakes, 417
  in Jamaica Bay, 417
  phosphates in, 412
  of surface water, 407
  thermal, 414–415
Water pressure, artesian, 211–212, 216
Watershed restoration, 425
Water supply, 407
Water table, cone of depression of, 211, *211*
  definition of, 209
  effect of well on, 211
  recharging of, 212
  relation of land surface to, *209*
Water vapor, atmospheric, 340, 361
  effect on solar radiation absorption, 355
  condensation nuclei and, 366
  See also Moisture
Wave(s), abrasion by, 246
  alongshore transport by, 243–244, *246*
  arches formed by, 247
  base of, 247
  beaches affected by, 240, 242, *245*
  breaker, 238–239
    damage by, 242, *245*
  breaking on walls and cliffs, 240–241, *241*
  caused by wind, 236
  caves formed by, 247
  cliffs cut by, 245, *248*, *254*
  corrasion by, 245, 246
  crest of, 235
  cross-bedded sedimentary rock and, 74
  cusps formed by, 249–250
  fetch and, 237
  as geologic agent, 222–235, 239–243
  height of, 236, 238, 239, *241*

INDEX

Wave(s) (*Continued*)
  length of, 235, 238, *241*
  in open water, 235–238
  parts of, *239*
  period of, 235, 238
  platform cut by, 244, 246, *247, 249, 253*
  refraction coefficient of, 242
  refraction diagram of, 242, *245*
  refraction of, 241–243, *243–246*
    definition of, 241
  sand produced by, 231
  in sea, 236
  in shallow water, 238–239, *241–243*
  shock pressure of, 240
  shoreline activity of, 235–255
  size of, 237–238
  superimposed, 236, *243*
  surf, 239
  terraces along beaches from, 244, *246*
  tidal, 273
  trough of, 235
  velocity of, 236, 238
  water particle movement in, 237, *239, 240*
Wave form, 235–236
  ideal, 236, *239*
  water particle movement and, 237, *239, 240*
Weather, atmospheric dust and, 341
  Pacific Coast, 394
  satellite photography of, 376, *376, 382–388*
Weather forecasting, 373–401
  computers in, 398–399
  daily, 391–394, *392–393*
  maps in, *see* Weather maps
Weathering, 154–159
  of biotite, 157
  definition of, 154
  erosion and, *see* Denudation
  of feldspar, 157
  of ferromagnesian minerals, 157
  frost action in, 154–155
  of granite, 157, 158, *158*
  of limestone, 155
  of marble, 155, *158*
  ocean constituents and, 278
  of sandstone, 155, 157, *157*
  of shale, 155
  soil formation by, 158–159
  solution in, 155–156, *158*
  of valley walls, 182
Weather maps, daily, 388–401, *392, 393*
  contents of, 388–390
  source of, 388–390
  upper air maps and, 390–391, *393*
  isotherm, 394, *395*
  jet stream information on, *399*
  of storm, 391, *392*
  types of, 399–400
Weddell Sea, 265, 270, 271
Wegener, Alfred, 322–325
Well(s), artesian, 213, *219*, 220
  effect of water table on, 211
  flowing, 212, *219*, 220, *220*
  nonflowing, 212, *213*
Westerlies, prevailing, 374
Western Cordillera, 310–315
Wet adiabatic rate, 363
Wetlands, 418
Whales, 296
White Mountains, 317
White River National Forest, 424
White Sands National Monument, 228–229, *229*
Wichita Mountains, *310*
Wilderness areas, 427
Wildlife conservation, 409, *409*, 422
Wind(s), 357–361
  abrasion by, 225
  basins carved by, 225
  circulation around highs and lows, 359, *360*
  circulation patterns of, 373–377
  clay transported by, 222
  convective system in, 357–358, *357*
  Coriolis force and, *268, 269*, 358–360, *359*, 373, 386
  deflation by, 224–225
  deposition by, 225–231
  drag of, on water surface, 237, *240*
  dreikanter formed by, 225, *225, 226*
  dust carried by, 297, 341
  erosion by, 222–225
  frictional drag of ground on, 360, *361*
  global, basic model of, 373–375, *374*
  gravel transported by, 224
  in jet stream, 397
  lake levels affected by, 237
  loess formation and, 233
  oases formed by, 225
  ocean salt transported by, 233–235
  regional, influences on, 358
  sand dunes formed by, *see* Sand dunes
  sand transported by, 222, 224, *226*, 227
  silt transported by, 222, 224, *236*
  surface, 360–361, *361*
    barometric pressures and, 375–377
    over Northern Hemisphere, *375*
  trade, 269, 373
  transport by, 222–225, *236*
  in upper atmosphere, 348
  velocity of, sand movement and, 224, *226*, 227
  waves caused by, 236
  See also Air currents and flow
Windbreaks, 425
Wind River Mountains, *324*
Wind storms, *see* specific type, for example, Hurricane(s)
Wizard Island, 56
Wollard, George P., 133
Wonderbirds, 384

Xenon, 339
X rays, 28

Yellowstone National Park, 221, *221*, 428
Yellowstone Timberland Reserve, 424
Yosemite Valley, *197*

Yucatán escarpment, 301
Yukon Valley, 233

Zinc, 307, 308
Zion National Park, *231*, *369*

Zircon, 151, 307
Zone of aeration, 209
Zone of saturation, 209
Zone of soil moisture, 209